State of the Art and Future Perspectives in Smart and Sustainable Urban Development

State of the Art and Future Perspectives in Smart and Sustainable Urban Development

Editor

Tan Yigitcanlar

MDPI • Basel • Beijing • Wuhan • Barcelona • Belgrade • Manchester • Tokyo • Cluj • Tianjin

Editor
Tan Yigitcanlar
School of Architecture and
Built Environment
Queensland University of
Technology
Brisbane
Australia

Editorial Office
MDPI
St. Alban-Anlage 66
4052 Basel, Switzerland

This is a reprint of articles from the Special Issue published online in the open access journal *Sustainability* (ISSN 2071-1050) (available at: www.mdpi.com/journal/sustainability/special_issues/Smart_Sustainable_Urban_Development).

For citation purposes, cite each article independently as indicated on the article page online and as indicated below:

LastName, A.A.; LastName, B.B.; LastName, C.C. Article Title. *Journal Name* **Year**, *Volume Number*, Page Range.

ISBN 978-3-0365-3840-2 (Hbk)
ISBN 978-3-0365-3839-6 (PDF)

Cover image courtesy of Xiang Ji

Contents

About the Editor

Tan Yigitcanlar

Tan Yigitcanlar is an eminent Australian researcher with international recognition and influence in the field of urban studies and planning. He is a Professor of Urban Studies and Planning at the School of Architecture and Built Environment, Queensland University of Technology, Brisbane, Australia. Along with this post, he holds an Honorary Professor role at the School of Technology, Federal University of Santa Catarina, Florianopolis, Brazil, and the Founding Director position of the Australia–Brazil Smart City Research and Practice Network. His research is clustered around the following three interdisciplinary themes: smart technologies, communities, cities, and urbanism; sustainable and resilient cities, communities, and urban ecosystems; and knowledge-based development of cities and innovation districts. His research findings are disseminated in over 250 articles published in high-impact journals, and 19 key reference books published by esteemed international publishing houses. Amongst the urban and regional planning scholars, 2020 Science-wide Author Databases of Standardised Citation Indicators ranked him as #1 most highly cited researcher in Australia, and #7 worldwide. For this achievement, he was recognised as an "Australian Research Superstar" in the Social Sciences Category at The Australian's 2020 Research Special Report.

Preface to "State of the Art and Future Perspectives in Smart and Sustainable Urban Development"

In the age of climate, disaster, and pandemic catastrophes, it is of utmost important to transform our cities into sustainable, resilient, robust, and liveable metropolises. Ensuring, smart and sustainable urban development is critical to meet our growing needs for natural resources, industrial products, energy, food, transportation, shelter, and effective waste management, while conserving and protecting environmental quality and the natural resource base, which is essential for future life and development.

This book contributes to the conceptual and practical knowledge pools in order to improve the research and practice on smart and sustainable urban development by presenting an informed understanding of the subject to scholars, policymakers, and practitioners. This book presents contributions—in the form of research articles, literature reviews, case reports, and short communications—offering insights into the smart and sustainable urban development by conducting in-depth conceptual debates, detailed case study descriptions, thorough empirical investigations, systematic literature reviews, or forecasting analyses. This way, the book forms a repository of relevant information, material, and knowledge to support research, policymaking, practice, and the transferability of experiences to address urbanization and other planetary challenges.

The scope of this book—which compiles the cutting-edge work of researchers that have investigated the state of the art and future perspectives in smart and sustainable urban development—includes the following broad areas and other relevant topics:

- Theoretical and conceptual underpinnings and analytical and policy frameworks of smart and sustainable urban development;
- Methodological and technical approaches for the evaluation and forecasting of smart and sustainable urban development;
- Technological progress, developments, and trials concerning the quadruple bottom-line development of smart and sustainable cities;
- Global best and good practice on smart and sustainable urban development case investigations, demonstrations, and reports;
- Smart and sustainable urban development planning, design, applications, and governance models to deliver desired urban outcomes;
- Premises, pitfalls, implications, and impacts concerning the future of urbanization and smart and sustainable urban development.

Tan Yigitcanlar
Editor

sustainability

Editorial

Towards Smart and Sustainable Urban Electromobility: An Editorial Commentary

Tan Yigitcanlar

School of Architecture and Built Environment, Queensland University of Technology, 2 George Street, Brisbane, QLD 4000, Australia; tan.yigitcanlar@qut.edu.au; Tel.: +61-7-3138-2418

In the age of anthropogenic climate change, developing smart and sustainable transport systems is among the most popular urban policy debates. Given that the future of urban mobility must be based on renewable energy resources, planning on how to move towards smart and sustainable urban electromobility is critical, and forms the rationale of this paper. In this editorial commentary: (a) The important role of smart and sustainable transport systems in addressing urban and environmental challenges is discussed; (b) The need for and the drivers of smart and sustainable urban electromobility are presented, and; (c) The key issues for establishing a new research agenda for smart and sustainable urban electromobility are listed. The editorial commentary also highlights the overall contributions of the papers of the Special Issue on the state-of-the-art and future perspectives in smart and sustainable urban development.

Today, we are living in a rapidly urbanizing world that accommodates human populations beyond the carrying capacity of our planet [1]. In many developed countries, at present, urbanization levels have reached as high as 80–90%. While generally urban densification is the desired form of growth, only a fraction of urban development is happening in the form of infill development [2]. Most of the growth is expansion towards greenfield—i.e., urban fringes and peri-urban areas—that is ending up forming a large metropolitan conurbation [3]. The result is an unsustainable sprawl and a serious threat to highly sensitive urban ecosystem services [4].

The perils of globally common expansionist urban development patterns are not limited to pollution and urban ecosystem loss only [5,6]. Anthropogenetic climate change, deforestation, biodiversity loss, socioeconomic inequalities, increased natural disasters and health hazards are among some of the consequential negative externalities of such development types [7]. There is, hence, an urgent need for rethinking the sustainability of urban systems and finding ways to reshape, if not rectify, our urban spaces to minimize the undesired consequences of unsustainable anthropogenic activities [8].

For instance, the combination of private motor vehicle dependency, sprawling urban form, and vehicle combustion engine technology makes urban transport one of the major contributors to global greenhouse gas (GHG) emissions [9,10]. Addressing urban transport externalities, whether it is environmental or societal, is a highly challenging task [11,12] It requires sound government policy, commitment, and determination, along with automotive industry transformation and community behavioral change and support [13].

Advanced technological innovations also offer invaluable opportunities to help urban administrators tackle the challenges in establishing smart and sustainable urban transport systems [14]. For example, developments in transport technologies have brought intelligent transport systems a long way from automated traffic signalization to autonomous and connected vehicles [15]. Today, these smart mobility systems are in service in a growing number of cities across the globe—especially in the form of autonomous public shuttle buses [16].

Nonetheless, technological innovations alone are not adequate to turn our transport systems into truly smart and sustainable ones. A critical component of smart and sustain-

Citation: Yigitcanlar, T. Towards Smart and Sustainable Urban Electromobility: An Editorial Commentary. *Sustainability* **2022**, *14*, 2264. https://doi.org/10.3390/su14042264

Received: 11 February 2022
Accepted: 12 February 2022
Published: 16 February 2022

able urban transport is from which source energy of the vehicle is obtained. Electrification in this sense is a good example, as electricity produced from fossil fuels is not sustainable at all [17]. Thus, an electric vehicle is only as sustainable as the source of its power—whether it comes from fossil fuels or renewables, such as solar, wind, tidal and so on [18].

The mode of transport will have to invariably be powered by renewable energies due to serious environmental and societal pressures we are facing today [19]. Electric vehicles, also including hydrogen fuel cell electric vehicles, are an important step towards achieving the smart and sustainable urban transport goal. However, the current global market share of electric vehicles is only about 1%. Besides in many countries, even including developing nations, this figure is very low, such as 0.4% for Australia [20]. Similarly, renewable energy is only responsible for about 29% of global electricity generation, where in many countries this figure is much lower than 20% [21].

Nevertheless, increasing global pressures, especially during the 26th UN Climate Change Conference of the Parties (COP26), have pushed many climate action passive countries, including Australia, for a pro-climate policy change. For example, most recently, the Australian Government adopted a 'technology-led approach' to emissions reduction as one of the most prominent national climate change strategies—others include emissions reduction incentives, regulating emissions, and reporting on emissions [22]. Subsequently, government subsidies for technocentric solutions to sustainability issues, including electric mobility or electromobility uptake, will soon start to become commonplace [23].

In recent years, exponential technological advancements, under the pressure of climate crises, have given birth to many innovative solutions not only in the automotive and energy sectors, but also in urban infrastructure and construction sectors [24,25]. Examples include autonomous and connected vehicles, smart roads and highways, Mobility-as-a-Service (MaaS), photovoltaic solar energy conversion, battery storage of photovoltaic systems, smart grids, smart buildings, and smart homes, and so on [26–29].

Bringing sustainability to urban transport systems via electrification that is sourced from renewables is widely seen as a novel approach [30]. Nonetheless, vehicle electrification is only the tip of the iceberg. In other words, an electric vehicle is not the only required component of electromobility. Achieving electromobility, in the context of sustainable urban transport, also requires system thinking. This includes, but is not limited to, road infrastructure, power transfer systems, renewable energy generators, urban form, user behavior and so on [31–34]. Figure 1 illustrates the main drivers of smart and sustainable urban electromobility.

So far, numerous studies highlighted the challenges of electric vehicle uptake ranging from the cost of the vehicle, distance travelled with a single charge, availability of superfast charging stations, government incentives, battery life, second-hand market, source of electric energy, and so on [35–37]. Besides the vehicle cost issue, the second obstacle concerns vehicle charging matters [38].

The most recent attempts to address the limitations of smart and sustainable urban electromobility see it as an ecosystem that goes behind the vehicle alone, incorporate state-of-the-art technological solutions to the charging problems [39]. To be precise, the installation of a flexible dynamic wireless power transfer system on the main roads facilitates efficient, safe, on-demand, reliable and bi-directional power transfer to electric vehicles [40,41]. In such an ecosystem, another requirement is to develop incentive-based demand management for electric vehicles, including V2G and G2V transfer of power while maintaining stability of the electricity grid, V2G and G2V [42,43].

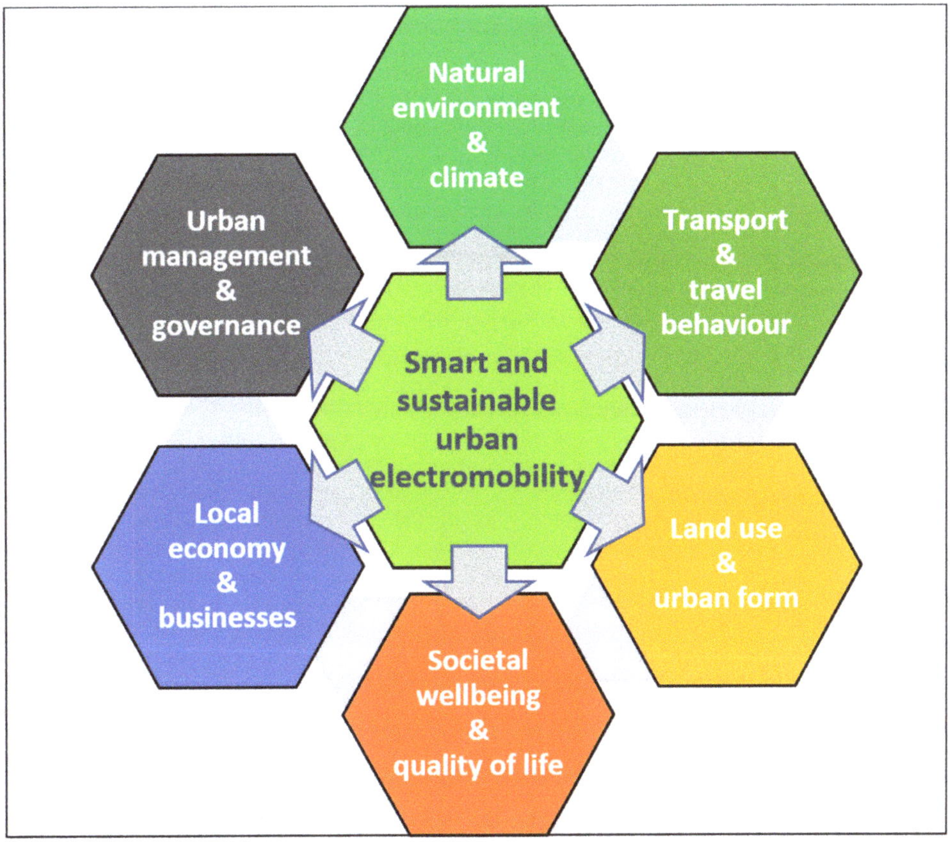

Figure 1. Drivers of smart and sustainable urban electromobility.

Along with the innovative vehicle and power technologies, sustainable electromobility also requires the development of an efficient transportation network with multi objective optimization framework incorporating electrical demand incentives, electric vehicle communication and charging and/or discharging, and urban infrastructure constraints [44–47].

Additionally, redevelopment or restructuring the urban form and main land use destinations, as well as understanding the underlining factors and changing traveler behaviors, accordingly, are important elements of a smart and sustainable urban electromobility ecosystem [48–50].

Given the urgency of establishing smart and sustainable transport systems in cities across the globe, there is a need for thoroughly exploring efficient methodologies towards a smart and sustainable urban electromobility transformation without burdening transportation and electrical grids. Besides, the need for developing technologies, infrastructures, and systems to support the mobile energy needs of urban mobility networks is becoming evident [51]—e.g., Mobile-Energy-as-a-Service (MEaaS).

In this regard, the following questions will be helpful in forming a much clearer view of how to plan for smart and sustainable electromobility in our cities. These questions also contribute to establishing a new research agenda for moving towards smart and sustainable urban electromobility.

- Why is smart and sustainable urban electromobility needed, and how can it help in addressing some colossal urban and environmental problems?

- Which technologies, infrastructures and systems are needed for developing smart and sustainable urban electromobility networks?
- When can smart and sustainable urban electromobility become relatively affordable for wider adoption in cities?
- Where are the most suitable locations to trial and operationalize smart and sustainable urban electromobility systems?
- Who are the key stakeholders of smart and sustainable urban electromobility to support the development and uptake?
- What do experts, authorities, service providers, electric vehicle users, and community at large think on smart and sustainable urban electromobility?
- How can the development and adoption of smart and sustainable urban electromobility systems be incentivized and regulated?

Against the above editorial commentary, the Special Issue on the state-of-the-art and future perspectives in smart and sustainable urban development supports the efforts in improving research and practice in smart and sustainable urbanism. The Special Issue contributes to the conceptual and practical knowledge pools to improve the research and practice on smart and sustainable urban development by bringing an informed understanding of the subject to scholars, policymakers, and practitioners. It offers insights into smart and sustainable urban development by conducting in-depth conceptual debates, detailed case study descriptions, thorough empirical investigations, systematic literature reviews, or forecasting analyses. The Special Issue, hence, forms a repository of relevant information, material, and knowledge to support research, policymaking, practice, and transferability of experiences to address the urbanization and other planetary challenges.

The Special Issue includes the following 20 commentaries, viewpoints, case reports, reviews, and research papers with the input of 76 urban scholars from across the globe:

1. Yigitcanlar, T. Towards Smart and Sustainable Urban Electromobility: An Editorial Commentary.
2. Ullah, I.; Shah, M.; Khan, A.; Maple, C.; Waheed, A.; Jeon, G. A Distributed Mix-Context-Based Method for Location Privacy in Road Networks.
3. Gulati, B.; Weiler, S. Risk, Recessions, and Resilience: Towards Sustainable Local Labor Markets through Employment Portfolio Analysis.
4. Cai, C.; Guo, Z.; Zhang, B.; Wang, X.; Li, B.; Tang, P. Urban Morphological Feature Extraction and Multi-Dimensional Similarity Analysis Based on Deep Learning Approaches.
5. Niemann, L.; Hoppe, T. How to Sustain Sustainability Monitoring in Cities: Lessons from 49 Community Indicator Initiatives across 10 Latin American Countries.
6. Taylor, J.; Jokela, S.; Laine, M.; Rajaniemi, J.; Jokinen, P.; Häikiö, L.; Lönnqvist, A. Learning and Teaching Interdisciplinary Skills in Sustainable Urban Development—The Case of Tampere University, Finland.
7. Guo, N.; Chan, E.; Yung, E. Alternative Governance Model for Historical Building Conservation in China: From Property Rights Perspective.
8. Boguniewicz-Zabłocka, J.; Capodaglio, A. Analysis of Alternatives for Sustainable Stormwater Management in Small Developments of Polish Urban Catchments.
9. Gurieff, N.; Green, D.; Koskinen, I.; Lipson, M.; Baldry, M.; Maddocks, A.; Menictas, C.; Noack, J.; Moghtaderi, B.; Doroodchi, E. Healthy Power: Reimagining Hospitals as Sustainable Energy Hubs.
10. Slater, K.; Robinson, J. Social Learning and Transdisciplinary Co-Production: A Social Practice Approach.
11. Salvia, R.; Egidi, G.; Salvati, L.; Rodrigo-Comino, J.; Quaranta, G. In-Between 'Smart' Urban Growth and 'Sluggish' Rural Development? Reframing Population Dynamics in Greece, 1940–2019.
12. Serrano, I.; Calvet-Mir, L.; Ribera-Fumaz, R.; Díaz, I.; March, H. A Social Network Analysis of the Spanish Network of Smart Cities.

13. Song, Y.; Stead, D.; de Jong, M. New Town Development and Sustainable Transition under Urban Entrepreneurialism in China.
14. Fieuw, W.; Foth, M.; Caldwell, G. Towards a More-than-Human Approach to Smart and Sustainable Urban Development: Designing for Multispecies Justice.
15. Malek, J.; Lim, S.; Yigitcanlar, T. Social Inclusion Indicators for Building Citizen-Centric Smart Cities: A Systematic Literature Review.
16. Wijesiri, B.; Bandala, E.; Liu, A.; Goonetilleke, A. A Framework for Stormwater Quality Modelling under the Effects of Climate Change to Enhance Reuse.
17. E, J.; Xia, B.; Buys, L.; Yigitcanlar, T. Sustainable Urban Development for Older Australians: Understanding the Formation of Naturally Occurring Retirement Communities in the Greater Brisbane Region.
18. Lim, S.; Malek, J.; Yussoff, M.; Yigitcanlar, T. Understanding and Acceptance of Smart City Policies: Practitioners' Perspectives on the Malaysian Smart City Framework.
19. Sabatini-Marques, J.; Yigitcanlar, T.; Schreiner, T.; Wittmann, T.; Sotto, D.; Inkinen, T. Strategizing Smart, Sustainable, and Knowledge-Based Development of Cities: Insights from Florianópolis, Brazil.
20. Yigitcanlar, T.; Cugurullo, F. The Sustainability of Artificial Intelligence: An Urbanistic Viewpoint from the Lens of Smart and Sustainable Cities.

This collection of these 20 papers focused on answering the following overall questions of this Special Issue: (a) What are the critical theoretical and conceptual underpinnings and analytical and policy frameworks of smart and sustainable urban development? (b) What are the critical methodological and technical approaches for the evaluation and forecasting of smart and sustainable urban development? (c) What are the critical technological progresses, developments, and trials concerning the quadruple bottom-line development of smart and sustainable cities? (d) What are the critical global best and good practice smart and sustainable urban development case investigations, demonstrations, and reports? (e) What are the critical smart and sustainable urban development planning, design, applications, and governance models to deliver desired urban outcomes? (f) What are the critical premises, pitfalls, implications, and impacts concerning the future of urbanization and smart and sustainable urban development?

Funding: This research received no external funding.

Conflicts of Interest: The author declares no conflict of interest.

References

1. Gao, Q.; Fang, C.; Cui, X. Carrying capacity for SDGs: A review of connotation evolution and practice. *Environ. Impact Assess. Rev.* **2021**, *91*, 106676. [CrossRef]
2. Yigitcanlar, T. *Rethinking Sustainable Development: Urban Management, Engineering, and Design*, 1st ed.; IGI Global: Hersey, PA, USA, 2010.
3. Mortoja, M.; Yigitcanlar, T. How does peri-urbanization trigger climate change vulnerabilities? An investigation of the Dhaka megacity in Bangladesh. *Remote Sens.* **2020**, *12*, 3938. [CrossRef]
4. Bueno-Suárez, C.; Coq-Huelva, D. Sustaining what is unsustainable: A review of urban sprawl and urban socio-environmental policies in North America and Western Europe. *Sustainability* **2020**, *12*, 4445. [CrossRef]
5. Mahbub, P.; Goonetilleke, A.; Ayoko, G.; Egodawatta, P.; Yigitcanlar, T. Analysis of build-up of heavy metals and volatile organics on urban roads in Gold Coast, Australia. *Water Sci. Technol.* **2011**, *63*, 2077–2085. [CrossRef] [PubMed]
6. Csete, M.; Horváth, L. Sustainability and green development in urban policies and strategies. *Appl. Ecol. Environ. Res.* **2012**, *10*, 185–194. [CrossRef]
7. Sotto, D.; Philippi, A.; Yigitcanlar, T.; Kamruzzaman, M. Aligning urban policy with climate action in the global south: Are Brazilian cities considering climate emergency in local planning practice? *Energies* **2019**, *12*, 3418. [CrossRef]
8. Mortoja, M.; Yigitcanlar, T. Local drivers of anthropogenic climate change: Quantifying the impact through a remote sensing approach in Brisbane. *Remote Sens.* **2020**, *12*, 2270. [CrossRef]
9. Yigitcanlar, T.; Dodson, J.; Gleeson, B.; Sipe, N. Travel self-containment in master planned estates: Analysis of recent Australian trends. *Urban Policy Res.* **2007**, *25*, 129–149. [CrossRef]
10. Perveen, S.; Yigitcanlar, T.; Kamruzzaman, M.; Hayes, J. Evaluating transport externalities of urban growth: A critical review of scenario-based planning methods. *Int. J. Environ. Sci. Technol.* **2017**, *14*, 663–678. [CrossRef]

11. Wangsness, P.; Proost, S.; Rødseth, K. Vehicle choices and urban transport externalities: Are Norwegian policy makers getting it right? *Transp. Res. Part D Transp. Environ.* **2020**, *86*, 102384. [CrossRef]
12. Butler, L.; Yigitcanlar, T.; Paz, A. How can smart mobility innovations alleviate transportation disadvantage? Assembling a conceptual framework through a systematic review. *Appl. Sci.* **2020**, *10*, 6306. [CrossRef]
13. Dur, F.; Yigitcanlar, T. Assessing land-use and transport integration via a spatial composite indexing model. *Int. J. Environ. Sci. Technol.* **2015**, *12*, 803–816. [CrossRef]
14. Novizayanti, D.; Prasetio, E. Orchestrating innovation network: Case of sustainable transportation technology. *Int. J. Sustain. Transp.* **2021**, *4*, 35–41. [CrossRef]
15. Golbabaei, F.; Yigitcanlar, T.; Bunker, J. The role of shared autonomous vehicle systems in delivering smart urban mobility: A systematic review of the literature. *Int. J. Sustain. Transp.* **2021**, *15*, 731–748. [CrossRef]
16. Dennis, S.; Paz, A.; Yigitcanlar, T. Perceptions and attitudes towards the deployment of autonomous and connected vehicles: Insights from Las Vegas, Nevada. *J. Urban Technol.* **2021**, *28*, 75–95. [CrossRef]
17. Monteiro, V.; Afonso, J.; Ferreira, J.; Afonso, J. Vehicle electrification: New challenges and opportunities for smart grids. *Energies* **2019**, *12*, 118. [CrossRef]
18. Domínguez-Navarro, J.; Dufo-López, R.; Yusta-Loyo, J.; Artal-Sevil, J.; Bernal-Agustín, J. Design of an electric vehicle fast-charging station with integration of renewable energy and storage systems. *Int. J. Electr. Power Energy Syst.* **2019**, *105*, 46–58. [CrossRef]
19. Letnik, T.; Marksel, M.; Luppino, G.; Bardi, A.; Božičnik, S. Review of policies and measures for sustainable and energy efficient urban transport. *Energy* **2018**, *163*, 245–257. [CrossRef]
20. Foley, B.; Degirmenci, K.; Yigitcanlar, T. Factors affecting electric vehicle uptake: Insights from a descriptive analysis in Australia. *Urban Sci.* **2020**, *4*, 57. [CrossRef]
21. Li, H.; Edwards, D.; Hosseini, M.; Costin, G. A review on renewable energy transition in Australia: An updated depiction. *J. Clean. Prod.* **2020**, *242*, 118475. [CrossRef]
22. Australia's Climate Change Strategies. Available online: https://www.industry.gov.au/policies-and-initiatives/australias-climate-change-strategies (accessed on 5 February 2022).
23. Wang, N.; Tang, L.; Pan, H. A global comparison and assessment of incentive policy on electric vehicle promotion. *Sustain. Cities Soc.* **2019**, *44*, 597–603. [CrossRef]
24. Leydesdorff, L.; Alkemade, F.; Heimeriks, G.; Hoekstra, R. Patents as instruments for exploring innovation dynamics: Geographic and technological perspectives on "photovoltaic cells". *Scientometrics* **2015**, *102*, 629–651. [CrossRef]
25. Faisal, A.; Yigitcanlar, T.; Kamruzzaman, M.; Paz, A. Mapping two decades of autonomous vehicle research: A systematic scientometric analysis. *J. Urban Technol.* **2021**, *28*, 45–74. [CrossRef]
26. Fahrenbruch, A.; Bube, R. *Fundamentals of Solar Cells: Photovoltaic Solar Energy Conversion*; Elsevier: New York, NY, USA, 2012.
27. Mahmud, K.; Hossain, M.; Town, G. Peak-load reduction by coordinated response of photovoltaics, battery storage, and electric vehicles. *IEEE Access* **2018**, *6*, 29353–29365. [CrossRef]
28. Butler, L.; Yigitcanlar, T.; Paz, A. Smart urban mobility innovations: A comprehensive review and evaluation. *IEEE Access* **2020**, *8*, 196034–196049. [CrossRef]
29. Li, W.; Yigitcanlar, T.; Erol, I.; Liu, A. Motivations, barriers and risks of smart home adoption: From systematic literature review to conceptual framework. *Energy Res. Soc. Sci.* **2021**, *80*, 102211. [CrossRef]
30. Cao, J.; Chen, X.; Qiu, R.; Hou, S. Electric vehicle industry sustainable development with a stakeholder engagement system. *Technol. Soc.* **2021**, *67*, 101771. [CrossRef]
31. Schipper, L. Sustainable urban transport in the 21st century: A new agenda. *Transp. Res. Rec.* **2002**, *1792*, 12–19. [CrossRef]
32. Goldman, T.; Gorham, R. Sustainable urban transport: Four innovative directions. *Technol. Soc.* **2006**, *28*, 261–273. [CrossRef]
33. Pojani, D.; Stead, D. Policy design for sustainable urban transport in the global south. *Policy Des. Pract.* **2018**, *1*, 90–102. [CrossRef]
34. Mukherjee, S.; Ryan, L. Factors influencing early battery electric vehicle adoption in Ireland. *Renew. Sustain. Energy Rev.* **2020**, *118*, 109504. [CrossRef]
35. Abotalebi, E.; Scott, D.; Ferguson, M. Why is electric vehicle uptake low in Atlantic Canada? A comparison to leading adoption provinces. *J. Transp. Geogr.* **2019**, *74*, 289–298. [CrossRef]
36. O'Neill, E.; Moore, D.; Kelleher, L.; Brereton, F. Barriers to electric vehicle uptake in Ireland: Perspectives of car-dealers and policy-makers. *Case Stud. Transp. Policy* **2019**, *7*, 118–127. [CrossRef]
37. Yao, J.; Xiong, S.; Ma, X. Comparative analysis of national policies for electric vehicle uptake using econometric models. *Energies* **2020**, *13*, 3604. [CrossRef]
38. Gong, D.; Tang, M.; Buchmeister, B.; Zhang, H. Solving location problem for electric vehicle charging stations: A sharing charging model. *IEEE Access* **2019**, *7*, 138391–138402. [CrossRef]
39. Panchal, C.; Stegen, S.; Lu, J. Review of static and dynamic wireless electric vehicle charging system. *Eng. Sci. Technol.* **2018**, *21*, 922–937. [CrossRef]
40. Mou, X.; Gladwin, D.; Zhao, R.; Sun, H. Survey on magnetic resonant coupling wireless power transfer technology for electric vehicle charging. *IET Power Electron.* **2019**, *12*, 3005–3020. [CrossRef]
41. Zhang, B.; Carlson, R.; Smart, J.; Dufek, E.; Liaw, B. Challenges of future high power wireless power transfer for light-duty electric vehicles: Technology and risk management. *eTransportation* **2019**, *2*, 100012. [CrossRef]

42. Mozafar, M.; Amini, M.; Moradi, M. Innovative appraisement of smart grid operation considering large-scale integration of electric vehicles enabling V2G and G2V systems. *Electr. Power Syst. Res.* **2018**, *154*, 245–256. [CrossRef]
43. Tang, Y.; Chen, Y.; Madawala, U.; Thrimawithana, D.; Ma, H. A new controller for bidirectional wireless power transfer systems. *IEEE Trans. Power Electron.* **2017**, *33*, 9076–9087. [CrossRef]
44. Siskova, M.; Van den Bergh, J. Optimal urban form for global and local emissions under electric vehicle and renewable energy scenarios. *Urban Clim.* **2019**, *29*, 100472. [CrossRef]
45. Das, R.; Wang, Y.; Putrus, G.; Kotter, R.; Marzband, M.; Herteleer, B.; Warmerdam, J. Multi-objective techno-economic-environmental optimisation of electric vehicle for energy services. *Appl. Energy* **2020**, *257*, 113965. [CrossRef]
46. Perera, P.; Hewage, K.; Sadiq, R. Electric vehicle recharging infrastructure planning and management in urban communities. *J. Clean. Prod.* **2020**, *250*, 119559. [CrossRef]
47. Solanke, T.; Ramachandaramurthy, V.; Yong, J.; Pasupuleti, J.; Kasinathan, P.; Rajagopalan, A. A review of strategic charging–discharging control of grid-connected electric vehicles. *J. Energy Storage* **2020**, *28*, 101193. [CrossRef]
48. Donada, C. Leadership in the electromobility ecosystem: Integrators and coordinators. *Int. J. Automot. Technol. Manag.* **2018**, *18*, 229–246. [CrossRef]
49. Narassimhan, E.; Johnson, C. The role of demand-side incentives and charging infrastructure on plug-in electric vehicle adoption: Analysis of US States. *Environ. Res. Lett.* **2018**, *13*, 074032. [CrossRef]
50. Curiel-Ramirez, L.; Ramirez-Mendoza, R.; Bustamante-Bello, M.; Morales-Menendez, R.; Galvan, J.; Lozoya-Santos, J. Smart electromobility: Interactive ecosystem of research, innovation, engineering, and entrepreneurship. *Int. J. Interact. Des. Manuf.* **2020**, *14*, 1443–1459. [CrossRef]
51. Wu, M.; Bao, Y.; Chen, G.; Zhang, J.; Wang, B.; Qian, W. Hierarchical distributed control strategy for electric vehicle mobile energy storage clusters. *Energies* **2019**, *12*, 1195. [CrossRef]

sustainability MDPI

Review

Towards a More-than-Human Approach to Smart and Sustainable Urban Development: Designing for Multispecies Justice

Walter Fieuw [1,†], Marcus Foth [1,*] and Glenda Amayo Caldwell [2]

1 QUT Design Lab, School of Design, Queensland University of Technology, Brisbane 4000, Australia
2 QUT Design Lab, School of Architecture and Built Environment, Queensland University of Technology, Brisbane 4000, Australia; g.caldwell@qut.edu.au
* Correspondence: m.foth@qut.edu.au
† Deceased 23 January 2021.

Abstract: The term 'sustainability' has become an overused umbrella term that encompasses a range of climate actions and environmental infrastructure investments; however, there is still an urgent need for transformative reform work. Scholars of urban studies have made compelling cases for a more-than-human conceptualisation of urban and environmental planning and also share a common interest in translating theory into practical approaches and implications that recognise (i) our ecological entanglements with planetary systems and (ii) the urgent need for multispecies justice in the reconceptualisation of genuinely sustainable cities. More-than-human sensibility draws on a range of disciplines and encompasses conventional and non-conventional research methods and design approaches. In this article, we offer a horizon scan type of review of key posthuman and more-than-human literature sources at the intersection of urban studies and environmental humanities. The aim of this review is to (i) contribute to the emerging discourse that is starting to operationalise a more-than-human approach to smart and sustainable urban development, and; (ii) to articulate a nascent framework for more-than-human spatial planning policy and practice.

Keywords: urban planning; more-than-human; post-Anthropocene; environmental humanities; multispecies justice; sustainable cities; smart cities

Citation: Fieuw, W.; Foth, M.; Caldwell, G.A. Towards a More-than-Human Approach to Smart and Sustainable Urban Development: Designing for Multispecies Justice. *Sustainability* **2022**, *14*, 948. https://doi.org/10.3390/su14020948

Academic Editor: Nikos A. Salingaros

Received: 6 December 2021
Accepted: 10 January 2022
Published: 14 January 2022

1. Introduction

Urban development processes are increasingly under the spotlight due to their potential for contributing to lowering the carbon footprints of cities, to achieving greater levels of sustainability, and to restoring sensitive ecologies and biodiversity hotspots. However, urban studies, in a similar way to many other social sciences, suffer from human exceptionalism in the ontological framing of these issues. So far, human concerns such as comfort and convenience usually take priority over ecological imperatives and the urgent need to avoid a planetary ecocide [1,2].

Scholars in environmental humanities as well as in urban design and planning are calling for a more-than-human approach to smart and sustainable urban development in order to grapple with how nonhuman agencies shape geographies and urban places. Other future directions for research include the role of planning in learning from Indigenous knowledge systems and cultures such as 'Caring for Country' [3–5]. These issues have been acknowledged as high priorities by the Planning Institute of Australia (PIA), and PIA's Indigenous Planning Working Group continues to identify new planning approaches [6]. In this article, we respond to industrial, governmental, societal, and environmental needs by addressing an identified gap in knowledge within the urban development domain.

Here, we present a horizon scan [7–10] of recent investigations and studies offering a more-than-human perspective on urban planning. We elaborate upon these concepts in the main literature review section below. By way of an introduction, 'more-than-human' is considered to be an umbrella term that encompasses a diverse set of theories and practices with origins in and interrelationships across diverse fields of study including critical geography, urban studies, ecofeminism, new materialism, science and technology studies (STS), and decolonial studies. Just as the term 'nature-based solutions' has become an umbrella term that encompasses a range of environmental infrastructure interventions in response to anthropogenic climate change, the term 'more-than-human' is considered to be an approach to account for nonhuman agencies [11,12], recognise humanity's entanglements with ecosystems and the planet [13], work towards multispecies justice [14], and design for cohabitation [14–16]. Such new ways of conceptualising sustainable development—perhaps the most pressing global agenda in the wake of what Earth scientists call the 'Holocene (or sixth mass) extinction'—are a pressing and urgent endeavour [17].

The focal lens we apply to this scholarship domain is spatial planning. Spatial planning is a critical component in sustainability strategies, because at the command of the spatial planner is a set of diverse procedural, regulatory, design, and implementation tools that are usually premised on and enacted by government policy frameworks [18–20]. The purpose of spatial planning is to balance human development and economic activities within ecological boundaries [21]. Spatial plans should have foresight based on credible evidence that includes population growth modelling [22], urban development pressures, ecologically vulnerable areas, and other landscape considerations such as regenerative design approaches to deal with the environmental legacy of exhausted quarry and abandoned mine sites [23]. These plans should be backed by normatively 'good' design principles contained within design and planning traditions of new urbanism, smart growth, sustainability, and ecological urbanism. Since spatial planning implicates the allocation of scarce resources, it is inherently a political activity by nature [19,20,24–26], and it also has the role of setting agendas for harmonious human–environment interactions. Plans and design guidance inform a range of urban and environmental planning instruments such as strategic environmental assessments, population growth and distribution, housing, ecological corridors, land use planning, and transport planning [22].

The significance of developing a more-than-human approach to smart and sustainable urban development in this article is derived from two domains, that is, more-than-human scholarship and spatial planning. While some human geography and urban theorists, for example, [13,27–29], have engaged with the possibilities of more-than-human cities, so far, there has been no concerted effort to strategically review the literature at this intersection with the aim of informing smart and sustainable urban development and spatial planning practices drawing on the domains of knowledge identified in more-than-human scholarship. By offering this horizon scan literature review, we endeavour to stimulate and extend debate and discourse about an emerging, yet still nascent, more-than-human spatial planning framework [30–32]. This article contributes to creating more nuanced understandings of sustainable and smart cities, which are stated objectives in the Sustainable Development Goals and the New Urban Agenda of the United Nations.

Our motivation in presenting this horizon scan is to start to articulate some of the possible implications of enacting and subscribing to a more-than-human sensibility in the practice of spatial planning. We draw attention to some of the key works emerging in the field as possibly trendsetting and game changing, and we start to organise their proposed ideas towards a nascent framework that can be utilised to design for multispecies justice [13–15,30]. Such a framework cannot be accommodated within the limited scope of this one review. As such, we can only offer a nascent research agenda, and accordingly, our aim is not and cannot be to attempt to present a fully formed framework. This would likely require a monograph-length treatment as well as many contributions from colleagues across different fields. Nonetheless, in this paper, we acknowledge that this movement is now underway, and our hope is that the findings of this review will recruit further supporters and contributors to this urgent task.

A more-than-human framework to smart and sustainable urban development can shape and inspire the practices of planners, geographers, and designers, who are working on a variety of spatial scales. These practitioners might be working in government agencies, design firms, environmental peak organisations, community organisations, or could be interested members of the community. While there has been a recent spike in interest in design competitions, conference proceedings, and special research teams, we believe there is still a need for this academic scholarship to be made more accessible to practitioners.

Scholars have identified the ontological and epistemological roots of a more-than-human approach in the environmental humanities, social sciences, and design. Such studies tend to be grounded in constructivism and are influenced by theoretical orientations in critical geography and humanities such as decolonial theory, deep ecology, posthumanism, urban studies, and human geography. The guiding objective of this review is to start orientating spatial planning practice towards such more-than-human sensibilities. This strategic horizon scan of the literature identifies the ontological foundation, on which a more-than-human approach to spatial planning practice can be built. We suggest that this ontological foundation can offer fertile ground for timely and urgent research questions to be posed by the urban studies research community at large. These can include questions such as:

- How can a more-than-human perspective transform neighbourhoods into more ecological habitats?
- What are the essential components of a more-than-human approach to spatial planning?
- How does such an approach to spatial planning alter urban development?
- What are the outcomes and impact of a more-than-human spatial planning praxis?

While the limited scope of this review does not allow us to offer a book-length treatment of these questions that can offer a satisfactory scope for answering them, we also suggest that what is important is not so much the scope and scale of any one output but a strong commitment to engaging, on an ongoing basis, with these questions with a more-than-human sensibility. As such, we want to be modest in offering this review as one step on the way, whilst, at the same time, acknowledging the many other steps other colleagues have taken and their contributions. By jointly working towards informing this ontological framework, our objective is to contribute to proliferating more-than-human perspectives on spatial planning across disciplines. The transformational change agenda implicit in this work is to provide evidence of the potential, the merit, and the urgency for reforming smart and sustainable urban development processes in spatial planning praxis to create and design post-anthropocentric urban futures [1,33].

2. A Horizon Scan of More-than-Human Approaches to Smart and Sustainable Urban Development

This review is not a conventional systematic literature review that uses a replicable and rigid search algorithm to perform a meticulous appraisal of all primary sources on a given research question over a longer time period. The specific literature relating to more-than-human approaches to smart and sustainable urban development that we want to capture for this review is far too recent, fast paced, and interdisciplinary to be adequately captured by the rigid and limiting methodology of a systematic literature review. Instead, in this review, we conduct a specific type of literature review called a horizon scan [7–10] in order to identify research work and studies of relevance, with the aim of contributing to emerging debates in the fields of urban planning and development about trends, gaps, opportunities, and implications. For this review, horizon scanning is a useful method, because it allows us to capture a diversity of recent research publications at an early stage of their development, and the resultant list of references offers a rich collection of publications for the reader to draw upon for follow-up reading and to form interdisciplinary connections, which, in itself, is a contribution of this study.

Our search method employs a broad horizon scan of relevant and recent papers. We used databases such as Web of Science, Scopus, ACM Digital Library, and Google Scholar. In addition to the urban studies literature, we also purposefully included papers from other disciplines such as STS, sociology, cultural studies, human geography, environmental humanities, and policy. This enabled us to establish a transdisciplinary perspective.

We position this horizon scan at the intersection of spatial planning and more-than-human theory. After providing contextual background to cities in the age of the Capitalocene (Section 2.1) and the responding emergence of sustainable smart cities (Section 2.2), the origins of the more-than-human sensibility is reviewed as considered in the social sciences and environmental humanities (Section 2.3). The purpose, here, is to provide definitional clarity and differentiation from other perspectives on human–environment interactions. In the final part of the review (Section 2.4), the convergence of domains of studies is presented in a structured form in order to sketch the beginnings of a nascent and emerging framework for more-than-human spatial planning and design that invites other readers to expand and contribute.

2.1. Cities in the Age of the Capitalocene

The impact of human activity on the environment is fundamentally changing ecological processes, and at the heart of this so-called 'Capitalocene' is the urbanisation of humanity [34–37]. Cities occupy between 2–3% of the global land coverage, which is set to triple by 2030 based on the current trajectory of low-density urban sprawl [38]. It is estimated that 1.8–2.4% of croplands will be converted to urban areas by 2030, resulting in widespread wildlife habitat loss, reduced biodiversity, and further food insecurity [39,40]. Cities are also the economic drivers of the world economy, accounting for 80% of the global GDP output with a commensurate environmental cost; cities consume over two-thirds of the world's energy and account for more than 70% of the global CO_2 emissions [41].

Opportunities for improved living conditions are found in urban environments, and thereby, become strong 'pull factors' driving the 'second wave of urbanisation' in Asia and Africa at a scale unprecedented since the industrial revolution [42]. The so-called secondary cities of less than 500,000 inhabitants are the fastest growing, and they account for 75% of urban dwellers. The UN [43] has estimated that, by 2050, approximately two-third of the global population (i.e., 6.68 of 9.77 billion people) will live in cities. Ninety per cent of urban land cover is near coastlines, raising concerns about resilience to flooding caused by rising sea levels and other impacts of climate change [44].

Environmental activism in the wake of the threat of Holocene extinction (or the sixth mass extinction) has placed the rights of nature on the global agenda [45–49]. Some countries have already enacted rights of nature, including Ecuador's 2008 constitutional reforms and Bolivia's 2010 'Rights of Mother Earth' act. In India, the Ganga and Yamuna Rivers have been recognised with legal personhood [50], and in New Zealand, ecosystems such as the Whanganui River and the Urewera Forest have been granted legal rights [46]. Environmental advocacy and community activism groups such as the Australian Earth Laws Alliance (AELA) are lobbying national governments to follow suit [47]. The rights of nature agenda implicitly invoke the wisdom and cultural heritage of Indigenous populations past, present, and future, which provides cues to rethinking of spatial planning as ecological stewardship [48,51,52] grounded in an ethos of 'Caring for Country' [4,5,28,53–55].

Fighting for quality natural environments in cities ought not to be just an agenda of environmentalists. Studies have proven that healthy ecologies make people happier, healthier, more inquisitive, and more productive [56–60]. Local communities tend to experience a greater sense of local custodianship when urban green and blue belts are protected, healthy, and intact. They are also more likely to participate economically and co-produce outcomes such as ecotourism opportunities [59,61]. Conservationists and planners also have a responsibility to retain social access to these environments, since the phenomena of 'green gentrification' and elitist enclaves further entrench existing inequalities and access restrictions [59,62–66].

2.2. Sustainable Smart Cities

The imperative to grow cities of the future more sustainably is a global policy agenda, and the dedicated UN Sustainable Development Goal (SDG) 11 seeks to "make cities and human settlements more inclusive, safe, resilient, and sustainable." The United Nations' member states enacted the New Urban Agenda (NUA) in 2016, which requires national governments to work with regional and local governments to formulate urban plans in accordance with the International Guidelines on Urban and Territorial Planning [67]. These guidelines seek to standardise city growth processes and establish normatively 'good' practices. Urban and environmental planning instruments are, therefore, in the spotlight for their potential to contribute to greater levels of biodiversity conservation, protection of waterways, and reduction in climate change risks within and beyond the urban footprint.

The investment case for sustainable cities and its associated infrastructure and human settlements have also recently been assessed. A new report by the Coalition for Urban Transitions, titled 'Climate Emergency, Urban Opportunity' [68] is one of the most comprehensive assessments on the investment case for sustainable cities. The findings of the report, representing the views of 50 leading organisations (such as the C40 Cities Climate Leadership Group and the World Resources Institute), show that 90% of carbon emissions from cities can be cut by using existing technologies. This would require an investment of 2% of the global GDP (or 1.8 trillion USD) per annum. An expected return of 23.9 trillion USD by 2050 could be achieved and support the creation of 87 million jobs by 2030 in sectors such as waste, energy, and transport.

The transition towards sustainable cities is being enabled by the rapid evolution of 'smart city' technologies such as the Internet of Things (IoT), ecological infrastructure, and alternative energy [69–72]. There is a temptation to think that technology is a panacea for solving wicked problems, and scholars have critiqued the ontological roots of the smart city movement from critical science and technology studies (STS) and broader humanities, arts and social sciences (HASS) disciplines [1,73–78].

Green building rating tools (GBRT) are influential in shifting the urban development industry towards more sustainable outcomes [79,80], and could potentially incorporate calls for net-positive outcomes [81] and more-than-human design [31].

2.3. Encountering the More-than-Human City

Despite the global recognition of a need for sustainable cities, the perpetuation of the unsustainable status quo of the built environment necessitates bold and brave new responses to urban and environmental planning [82]. Drawing on empirical urban ecology research (e.g., street cats of Singapore and fruit bat colonies in Sydney), Franklin [27] argued for a more-than-human perspective to urban studies underpinned by theoretical constructs of 'becoming/worlding cities' and 'urban ecology' that went beyond previous approaches to design and planning such as biophilic design [83–85] and carbon-positive design [86]. Metzger [87] shared an optimism for a new focus in urban studies, but critically asked, "Are there any signals indicating that planning methodology is moving in such a direction? Not many would be my short answer" (p. 1005).

Studies in more-than-human, post-humanist, and post-anthropocentric perspectives have emerged from critical human geography and deep ecology, and have provided compelling examples of how a 'more-than-human turn' in social science and design research has been starting to grapple with the agencies of nonhuman actors [11,30,88–91]. When nonhuman agencies are better understood, complex problems such as conflict resolution in the law-space nexus of land management can be better managed, as illustrated, for example, in the research by Brown et al. [92] on wildlife corridors. Yigitcanlar et al. [1] proposed that designing post-anthropocentric cities of the future required a deeper engagement with the possibilities of ecological human settlement. This, in turn, requires increased attention to the temporalities, the wisdom of alternative knowledge systems (e.g., decolonising design and learning from Indigenous knowledge), and more sensitive design pedagogy [3,4,40,93–95].

The need to incorporate urban ecology concerns with the sustainable cities and smart cities movement is imperative to avoid a planetary ecocide [1] and transcend human exceptionalism [93]. Furthermore, cities are the next frontier in biodiversity conservation [96]. In Australia, cities support 30% of the threatened plant and animal species, which is the highest proportion of land use on a unit-area basis [97]. Scholars are calling for novel approaches to the scientific premise of a more-than-human city [27]. Biodiversity sensitive urban design (BDUD) considers intricate ways in which human and nonhuman lives are entangled in urban spaces [82,98]. In landscape theory, provocations for animal-aided design (AAD) seek ways in which conservation can be incorporated into the master plans of new urban development [99]. Incorporating more-than-human design and planning through new practices such as BDUD and ADD could contribute to the rewilding and realisation of more-than-human cities [100], but obstacles remain such as citizen's willingness to coexist with animals and wildlife conflicts [11,101–104].

Urban studies have suffered from human exceptionalism for too long [100,105]. The challenge for urban theorists is to formulate responses to this perceived 'turn' in social sciences emanating from human geography, ecofeminism, and critical STS, highly influenced by the more-than-human geography of Bruno Latour, Sarah Whatmore, and Donna Haraway [100,106–109].

2.4. *Towards a More-than-Human Approach to Smart and Sustainable Urban Development*

Houston et al. [13] argued that the recasting of urban development processes from the more-than-human perspective could result in innovations such as more responsive and improved climate-adaptive planning tools and narratives for diverse forms of future city growth. The authors concluded by arguing that, "planning theory requires a thicker, relational and more responsive form of post-humanism to imagine and enact just and sustainable cities in a time of global environmental uncertainty and change" (p. 203). This was further corroborated by Roös [110] who argued for the merits of employing a design pattern language approach to embed biophilia in planning practice. This design pattern language, see also [111–113], provides a foundation to better recognise human–nature interactions and inform a new theory for the sustainable development and planning of human settlements and cities, which is also supported by Liaros [40,94,95].

Our horizon scan has identified a need for scholars of urban studies to investigate the possibilities of a spatial planning regime that considers more-than-human perspectives. The approaching planetary ecocide urges us to recast spatial planning across the built environment and environmental professional competencies from a more-than-human perspective. Spatial planning is a vital task in guiding sustainable development outcomes with its associated technologies and design instruments [28,32,114]. Urban planning intends to create a rational and transparent 'development envelope' at the local scale to guide the activities of public and private interests [115,116].

Adams and Tiesdell [117] understood the purpose of local plans and associated development processes under four broad categories:

- Market shaping involves interventions to create a vision, sets development parameters, and integrates disparate investments to avoid the fragmentation of the urban fabric.
- Regulatory instruments are legislative checks and balances to compel, manage, and prevent urban development outcomes.
- Stimulus packages are utilised to kick-start desired outcomes.
- Finally, building capacity requires a shift from regulation and planning towards proactive coordination and leadership of development that contributes to improved placemaking.

Cities require more compelling visions of environmental harmony in which human activities can be seen as contributing to the restoration of sensitive ecologies, the net reduction of carbon emissions, and an overall push towards what Birkeland called net-positive design and development [81]. These city visions are required to be translated into refined strategies and methods at the neighbourhood and/or district level. The neighbourhood unit is deemed to be an appropriate scale where urban sustainability can be proactively planned for in terms of public services, networked infrastructures, and greater community participation [80,118,119]. A new generation of Neighbourhood Sustainability Assessment (NSA) tools have the potential to better measure the sustainability of urban systems [120], and these assessments of built environment sustainability performance could benefit greatly from a more-than-human perspective on integrative planning and development [31]. By combining these recent thrusts in urban studies, a framework can be developed to articulate a more-than-human spatial planning praxis. Such a framework requires a taxonomy of concepts and emerging practices, similar to the way that the *Feral Atlas* [121], for example, has documented the diversity of shared encounters and entanglements between humans and nonhumans in the form of a digital encyclopaedia grounded in 79 ethnographic field reports "on diverse topics as 'radioactive blueberries' and the spread of coffee rust, all accompanied by drawings, music, and films." [122]. Applying and expanding the framework, translating it into practice, evaluating its impact, and on that basis, refining its methods may start to address the knowledge gap of how to overcome human exceptionalism in urban studies [123].

3. Conclusions

In this horizon scan, we have demonstrated how subscribing to a more-than-human sensibility recognises the intricate ways in which human and nonhuman existences are entangled in urban space. This sensibility is premised on the notions of 'becoming'/ 'worlding' [13,103,124], which are established concepts within process philosophy. Furthermore, the process of becoming lends itself to novel approaches to design and planning practices that can bring about the transformational reform work needed to bring about genuine urban sustainability [125].

While the scope of this review is sadly not sufficient to present a fully formed framework, Figure 1 shows a nascent sketch of the emerging agenda towards a more-than-human approach to smart and sustainable urban development based on the review conducted above. The core component of this framework that now requires further collaborative work is the spatial planning body of knowledge and practices and the associated toolkit. We suggest that this joint effort could be unpacked and broken down further into three types of inquiries, each of which has already received some attention from both researchers and practitioners as follows: (1) more-than-human engagement [75,93], (2) more-than-human spatial planning practice [28,32], and (3) more-than-human governance and policy [126,127].

This succinct yet cogent horizon scan across the urban studies domain has reviewed emerging scholarships from diverse disciplines including critical feminism, geography, and STS, which question human exceptionalism. Scholars are calling for research agendas to consider novel approaches to the discipline of spatial planning from a more-than-human perspective [32,128].

How should spatial planners and urban designers respond to provocations that cities are to become more-than-human? In addition, how are such provocations different, or similar, to pressing agendas such as calls to rewild cities [100,106,129,130] and invest in nature-based solutions to counteract anthropogenic climate change?

To address this gap in knowledge identified by this review, further work is urgently needed to develop taxonomies of theoretical constructs and case studies of emerging practices to indicate what such a more-than-human spatial planning framework to guide smart and sustainable urban development and support multispecies justice could entail and how it could be adopted by industry and government. Enacting a more-than-human spatial planning framework would expedite and even leapfrog our transition towards not only more nuanced understandings but actual implementations and deployments of sustainable and smart cities, which are stated objectives in the Sustainable Development Goals and the New Urban Agenda of the United Nations.

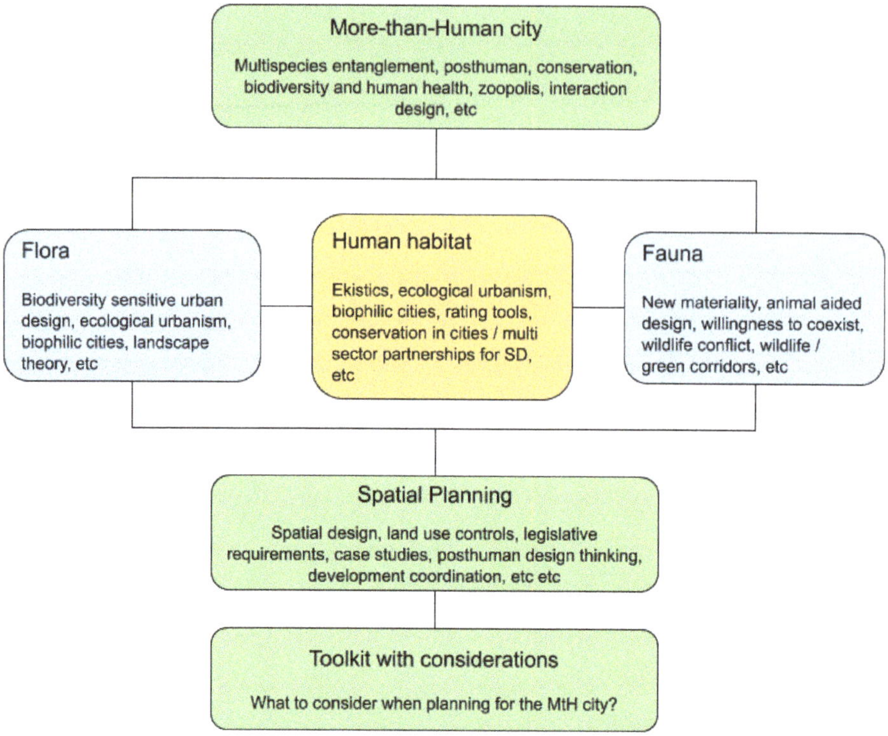

Figure 1. A nascent sketch of the emerging framework towards a more-than-human approach to smart and sustainable urban development.

Author Contributions: Conceptualization, W.F., M.F. and G.A.C.; formal analysis, W.F., M.F. and G.A.C.; funding acquisition, M.F.; investigation, W.F.; methodology, W.F.; project administration, W.F.; resources, M.F.; supervision, M.F. and G.A.C.; writing—original draft, W.F.; writing—review and editing, M.F. and G.A.C. All authors have read and agreed to the published version of the manuscript.

Funding: This research received no external funding.

Institutional Review Board Statement: Not applicable.

Informed Consent Statement: Not applicable.

Acknowledgments: We dedicate this article to our colleague, collaborator, and friend Walter Fieuw who sadly passed away in January 2021. Walter was a brilliant mind and a passionate advocate for genuinely sustainable cities and designing for multispecies justice. We are grateful for Walter's thought leadership, initiative, and research work. This article represents an edited version of Walter's writings for his PhD studies at QUT just before his passing.

Conflicts of Interest: The authors declare no conflict of interest.

References

1. Yigitcanlar, T.; Foth, M. Kamruzzaman Towards Post-Anthropocentric Cities: Reconceptualizing Smart Cities to Evade Urban Ecocide. *J. Urban Technol.* **2019**, *26*, 147–152. [CrossRef]
2. Light, A.; Powell, A.; Shklovski, I. Design for Existential Crisis in the Anthropocene Age. In Proceedings of the 8th International Conference on Communities and Technologies, Troyes, France, 26–30 June 2017; Association for Computing Machinery: New York, NY, USA, 2017; pp. 270–279.
3. Page, A.; Memmott, P. *Design: Building on Country*; Thames & Hudson: London, UK, 2021; ISBN 9781760761400.

4. Yunkaporta, T. *Sand Talk: How Indigenous Thinking Can Save the World*; Text Publishing: Melbourne, Australia, 2019; ISBN 9781925773996.
5. Latulippe, N.; Klenk, N. Making Room and Moving over: Knowledge Co-Production, Indigenous Knowledge Sovereignty and the Politics of Global Environmental Change Decision-Making. *Curr. Opin. Environ. Sustain.* **2020**, *42*, 7–14. [CrossRef]
6. Planning Institute of Australia. Improving Planners' Understanding of Aboriginal and Torres Strait Islander Australians and Recommendations for Reforming Planning Education Curricula for PIA Accreditation: A Discussion Paper. Indigenous Planning Working Group, Planning Institute of Australia: Kingston, ACT, Australia, 2010.
7. Hines, P.; Yu, L.H.; Guy, R.H.; Brand, A.; Papaluca-Amati, M. Scanning the Horizon: A Systematic Literature Review of Methodologies. *BMJ Open* **2019**, *9*, e026764. [CrossRef] [PubMed]
8. Foth, M.; Caldwell, G.A.; Fredericks, J. A COVID-19 Horizon Scan Looking for Post-Pandemic Implications for Design. In Proceedings of the Cumulus Roma 2021—Design Culture(s), Rome, Italy, 8–11 June 2021; Cumulus, International Association of Universities and Colleges of Art, Design and Media: Espoo, Finland. 2021. Available online: https://eprints.qut.edu.au/209728/ (accessed on 5 December 2021).
9. Schultz, W.L. The Cultural Contradictions of Managing Change: Using Horizon Scanning in an Evidence-based Policy Context. *Foresight* **2006**, *8*, 3–12. [CrossRef]
10. Könnölä, T.; Salo, A.; Cagnin, C.; Carabias, V.; Vilkkumaa, E. Facing the Future: Scanning, Synthesizing and Sense-Making in Horizon Scanning. *Sci. Public Policy* **2012**, *39*, 222–231. [CrossRef]
11. Tomitsch, M.; Fredericks, J.; Vo, D.; Frawley, J.; Foth, M. Non-Human Personas: Including Nature in the Participatory Design of Smart Cities. *Interact. Des. Archit.* **2021**, *50*, 102–130. Available online: https://eprints.qut.edu.au/227146/ (accessed on 5 December 2021).
12. DiSalvo, C.; Lukens, J. Nonanthropocentrism and the nonhuman in design: Possibilities for designing new forms of engagement with and through technology. In *From Social Butterfly to Engaged Citizen*; Foth, M., Forlano, L., Satchell, C., Gibbs, M., Eds.; The MIT Press: Cambridge, MA, USA, 2011; pp. 421–436. ISBN 9780262298445. [CrossRef]
13. Houston, D.; Hillier, J.; MacCallum, D.; Steele, W.; Byrne, J. Make Kin, Not Cities! Multispecies Entanglements and "becoming-World" in Planning Theory. *Plan. Theory* **2018**, *17*, 190–212. [CrossRef]
14. Celermajer, D.; Schlosberg, D.; Rickards, L.; Stewart-Harawira, M.; Thaler, M.; Tschakert, P.; Verlie, B.; Winter, C. Multispecies Justice: Theories, Challenges, and a Research Agenda for Environmental Politics. *Environ. Politics* **2021**, *30*, 119–140. [CrossRef]
15. Liu, S.-Y. Designing for Multispecies Collaboration and Cohabitation. In *Conference Companion Publication of the 2019 on Computer Supported Cooperative Work and Social Computing—CSCW '19, Austin, TX, USA, 9–13 November 2019*; ACM: New York, NY, USA, 2019; pp. 72–75.
16. Foth, M. The next Urban Paradigm: Cohabitation in the Smart City. *IT–Inf. Technol.* **2017**, *59*, 259–262. [CrossRef]
17. Ripple, W.J.; Wolf, C.; Newsome, T.M.; Galetti, M.; Alamgir, M.; Crist, E.; Mahmoud, M.I.; Laurance, W.F. World Scientists' Warning to Humanity: A Second Notice. *Bioscience* **2017**, *67*, 1026–1028. [CrossRef]
18. Yigitcanlar, T. Smart City Beyond Efficiency: Technology–Policy–Community at Play for Sustainable Urban Futures. *Hous. Policy Debate* **2021**, *31*, 88–92. [CrossRef]
19. Monno, V.; Khakee, A. Tokenism or Political Activism? Some Reflections on Participatory Planning. *Int. Plan. Stud.* **2012**, *17*, 85–101. [CrossRef]
20. Kamols, N.; Foth, M.; Guaralda, M. Beyond Engagement Theatre: Challenging Institutional Constraints of Participatory Planning Practice. *Aust. Plan.* **2021**, *57*, 23–35. [CrossRef]
21. Raworth, K. *Doughnut Economics: Seven Ways to Think Like a 21st-Century Economist*; Chelsea Green Publishing: Chelsea, VT, USA, 2017; ISBN 9781603586740.
22. Guaralda, M.; Hearn, G.; Foth, M.; Yigitcanlar, T.; Mayere, S.; Law, L. Towards Australian Regional Turnaround: Insights into Sustainably Accommodating Post-Pandemic Urban Growth in Regional Towns and Cities. *Sustain. Sci. Pract. Policy* **2020**, *12*, 10492. [CrossRef]
23. Roös, P.B. (Ed.) *Heal the Scar: Regenerative Futures of Damaged Landscapes*; Deakin University: Melbourne, Australia, 2020; ISBN 9780730001607.
24. Burton, P. Is Urban Planning in Australia Hindered by Poor Metropolitan Governance? *Urban Sci.* **2017**, *1*, 34. [CrossRef]
25. Foth, M.; Tomitsch, M.; Satchell, C.; Haeusler, M.H. From Users to Citizens: Some Thoughts on Designing for Polity and Civics. In Proceedings of the Annual Meeting of the Australian Special Interest Group for Computer Human Interaction, Brisbane, Australia, 28 November–1 December 2017; ACM: New York, NY, USA, 2015; pp. 623–633. [CrossRef]
26. Murray, C.K.; Frijters, P. *Game of Mates: How Favours Bleed the Nation*; Publicious: Gold Coast, QLD, Australia, 2017; ISBN 9780648061106.
27. Franklin, A. The More-than-Human City. *Sociol. Rev.* **2017**, *65*, 202–217. [CrossRef]
28. Metzger, J. A more-than-human approach to environmental planning. In *The Routledge Companion to Environmental Planning*; Taylor and Francis: Abingdon, UK, 2019; pp. 190–199. ISBN 9781351717403.
29. Metzger, J. Cultivating Torment: The Cosmopolitics of More-than-Human Urban Planning. *Cityscape* **2016**, *20*, 581–601. [CrossRef]
30. Shingne, M.C. The More-than-Human Right to the City: A Multispecies Reevaluation. *J. Urban Aff.* **2020**, 1–19. [CrossRef]
31. Loh, S.; Foth, M.; Caldwell, G.A.; Garcia-Hansen, V.; Thomson, M. A More-than-Human Perspective on Understanding the Performance of the Built Environment. *Archit. Sci. Rev.* **2020**, *63*, 372–383. [CrossRef]

32. Metzger, J.; Lindblad, J. *Dilemmas of Sustainable Urban Development: A View from Practice*; Metzger, J., Lindblad, J., Eds.; Taylor & Francis Group: Abingdon, UK, 2020; ISBN 9780429294457.
33. Wakkary, R. *Things We Could Design for More than Human Centred Worlds*; MIT Press: Cambridge, MA, USA, 2021.
34. Swilling, M. *The Age of Sustainability: Just Transitions in a Complex World*; Routledge: London, UK, 2019; ISBN 9780429603723.
35. Moore, J.W. The Capitalocene, Part I: On the Nature and Origins of Our Ecological Crisis. *J. Peasant Stud.* **2017**, *44*, 594–630. [CrossRef]
36. Bedford, L.; Mann, M.; Walters, R.; Foth, M. A Post-Capitalocentric Critique of Digital Technology and Environmental Harm: New Directions at the Intersection of Digital and Green Criminology. *Int. J. Crime Justice Soc. Democr.* **2021**, in press. Available online: https://eprints.qut.edu.au/213773/ (accessed on 5 December 2021).
37. Foth, M.; Mann, M.; Bedford, L.; Fieuw, W.; Walters, R. A capitalocentric review of technology for sustainable development: The case for more-than-human design. In *Global Information Society Watch 2020: Technology, the Environment and a Sustainable World—Responses from the Global SOUTH*; Global Information Society Watch (GISWatch), Finlay, A., Eds.; Association for Progressive Communications (APC): Melville, South Africa, 2021; pp. 78–82. Available online: https://eprints.qut.edu.au/203186/ (accessed on 5 December 2021)ISBN 9789295113404.
38. Seto, K.C.; Fragkias, M.; Güneralp, B.; Reilly, M.K. A Meta-Analysis of Global Urban Land Expansion. *PLoS ONE* **2011**, *6*, e23777. [CrossRef] [PubMed]
39. d'Amour, C.B.; Reitsma, F.; Baiocchi, G.; Barthel, S.; Güneralp, B.; Erb, K.-H.; Haberl, H.; Creutzig, F.; Seto, K.C. Future Urban Land Expansion and Implications for Global Croplands. *Proc. Natl. Acad. Sci. USA* **2017**, *114*, 8939–8944. [CrossRef] [PubMed]
40. Liaros, S. Circular Food Futures: What Will They Look Like? *Circ. Econ. Sustain.* **2021**, *1*, 1193–1206. [CrossRef]
41. Schindler, S.; Mitlin, D.; Marvin, S. National Urban Policy Making and Its Potential for Sustainable Urbanism. *Curr. Opin. Environ. Sustain.* **2018**, *34*, 48–53. [CrossRef]
42. UN Habitat. Planning Sustainable Cities: Global Report on Human Settlements 2009. Available online: https://unhabitat.org/planning-sustainable-cities-global-report-on-human-settlements-2009 (accessed on 5 December 2021).
43. UN. *World Urbanization Prospects: The 2018 Revision*; United Nations, Department of Economic and Social Affairs, Population Division: New York, NY, USA, 2019.
44. Revi, A.; Satterthwaite, D.E.; Aragón-Durand, F.; Kiunsi, R.B.R.; Pelling, M.; Roberts, D.C.; Solecki, W.; Balbus, J.M.; Cardona, O.-D. Urban areas. In *Climate Change 2014: Impacts, Adaptation, and Vulnerability. Part A: Global and Sectoral Aspects. Contribution of Working Group II to the Fifth Assessment Report of the IPCC*; IPCC: Geneva, Switzerland, 2014; pp. 535–612.
45. Boyd, D.R. *The Rights of Nature: A Legal Revolution That Could Save the World*; ECW Press: Toronto, ON, Canada, 2017; ISBN 9781770412392.
46. La Follette, C.; Maser, C. *Sustainability and the Rights of Nature in Practice*; CRC Press: Boca Raton, FL, USA, 2019; ISBN 9780429000393.
47. Graham, M.; Maloney, M. Caring for Country and Rights of Nature in Australia: A Conversation between Earth Jurisprudence and Aboriginal Law and Ethics. In *Sustainability and the Rights of Nature in Practice*; La Follette, C., Maser, C., Eds.; Routledge: London, UK, 2019.
48. Harden-Davies, H.; Humphries, F.; Maloney, M.; Wright, G.; Gjerde, K.; Vierros, M. Rights of Nature: Perspectives for Global Ocean Stewardship. *Mar. Policy* **2020**, *122*, 104059. [CrossRef]
49. Rühs, N.; Jones, A. The Implementation of Earth Jurisprudence through Substantive Constitutional Rights of Nature. *Sustain. Sci. Pract. Policy* **2016**, *8*, 174. [CrossRef]
50. O'Donnell, E.; Talbot-Jones, J. *Three Rivers Are Now Legally People—But That's Just the Start of Looking after Them*; The Conversation: Melbourne, Australia, 2017.
51. Buijs, A.; Fischer, A.; Muhar, A. From Urban Gardening to Planetary Stewardship: Human–nature Relationships and Their Implications for Environmental Management. *J. Environ. Plan. Manag.* **2018**, *61*, 747–755. [CrossRef]
52. Enqvist, J.P.; West, S.; Masterson, V.A.; Haider, L.J.; Svedin, U.; Tengö, M. Stewardship as a Boundary Object for Sustainability Research: Linking Care, Knowledge and Agency. *Landsc. Urban Plan.* **2018**, *179*, 17–37. [CrossRef]
53. Head, L. *Hope and Grief in the Anthropocene: Re-Conceptualising Human-Nature Relations*; Taylor and Francis Inc.: Abingdon, UK, 2016; ISBN 9781317576433.
54. Larsen, S.C.; Johnson, J.T. The Agency of Place: Toward a More-Than-Human Geographical Self. *GeoHumanities* **2016**, *2*, 149–166. [CrossRef]
55. Vásquez-Fernández, A.M.; Ahenakew, C. Resurgence of Relationality: Reflections on Decolonizing and Indigenizing "Sustainable Development". *Curr. Opin. Environ. Sustain.* **2020**, *43*, 65–70. [CrossRef]
56. WHO. *Urban Green Spaces and Health: A Review of Evidence*; Egorov, A.I., Mudu, P., Braubach, M., Martuzzi, M., Eds.; WHO Regional Office for Europe: Geneva, Switzerland, 2016.
57. Elmqvist, T.; Setälä, H.; Handel, S.N.; van der Ploeg, S.; Aronson, J.; Blignaut, J.N.; Gómez-Baggethun, E.; Nowak, D.J.; Kronenberg, J.; de Groot, R. Benefits of Restoring Ecosystem Services in Urban Areas. *Curr. Opin. Environ. Sustain.* **2015**, *14*, 101–108. [CrossRef]
58. Maller, C. *Healthy Urban Environments: More-Than-Human Theories*; Routledge: London, UK, 2018; ISBN 9781317217237.
59. Wolch, J.R.; Byrne, J.; Newell, J.P. Urban Green Space, Public Health, and Environmental Justice: The Challenge of Making Cities "Just Green Enough". *Landsc. Urban Plan.* **2014**, *125*, 234–244. [CrossRef]

60. Corburn, J. *Toward the Healthy City: People, Places, and the Politics of Urban Planning (Urban and Industrial Environments)*; MIT Press: Cambridge, MA, USA, 2009; ISBN 9780262013314.
61. Roseland, M.; Spiliotopoulou, M. Converging Urban Agendas: Toward Healthy and Sustainable Communities. *Soc. Sci.* **2016**, *5*, 28. [CrossRef]
62. Gould, K.A.; Lewis, T.L. *Green Gentrification: Urban Sustainability and the Struggle for Environmental Justice*; Routledge: London, UK, 2016; ISBN 9781315687322.
63. Hodson, M.; Marvin, S. Urbanism in the Anthropocene: Ecological Urbanism or Premium Ecological Enclaves? *Cityscape* **2010**, *14*, 298–313. [CrossRef]
64. Campbell-Arvai, V.; Lindquist, M. From the Ground up: Using Structured Community Engagement to Identify Objectives for Urban Green Infrastructure Planning. *Urban For. Urban Green.* **2021**, *59*, 127013. [CrossRef]
65. Anguelovski, I.; Connolly, J.J.T.; Garcia-Lamarca, M.; Cole, H.; Pearsall, H. New Scholarly Pathways on Green Gentrification: What Does the Urban "Green Turn" Mean and Where Is It Going? *Prog. Hum. Geogr.* **2019**, *43*, 1064–1086. [CrossRef]
66. Immergluck, D.; Balan, T. Sustainable for Whom? Green Urban Development, Environmental Gentrification, and the Atlanta Beltline. *Urban Geogr.* **2018**, *39*, 546–562. [CrossRef]
67. UN Habitat. International Guidelines on Urban and Territorial Planning. 2015. Available online: https://unhabitat.org/international-guidelines-on-urban-and-territorial-planning (accessed on 5 December 2021).
68. Colenbrander, S.; Lazer, L.; Haddaoui, C.; Godfrey, N.; Lobo, A.; Clarkson, H.; Huxley, R.; Parnell, S.; Smith, B.; Smith, S.; et al. *Climate Emergency, Urban Opportunity*; Coalition for Urban Transitions: Washington, DC, USA, 2019; Volume 160.
69. Bouzguenda, I.; Alalouch, C.; Fava, N. Towards Smart Sustainable Cities: A Review of the Role Digital Citizen Participation Could Play in Advancing Social Sustainability. *Sustain. Cities Soc.* **2019**, *50*, 101627. [CrossRef]
70. Heitlinger, S.; Bryan-Kinns, N.; Comber, R. *The Right to the Sustainable Smart City*; Association for Computing Machinery: New York, NY, USA, 2019; ISBN 9781450359702.
71. Yigitcanlar, T.K.; Foth, M.; Sabatini-Marques, J.; da Costa, E.; Ioppolo, G. Can Cities Become Smart without Being Sustainable? A Systematic Review of the Literature. *Sustain. Cities Soc.* **2019**, *45*, 348–365. [CrossRef]
72. Paulos, E.; Foth, M.; Satchell, C.; Kim, Y.; Dourish, P.; Choi, J.H.-J. *Ubiquitous Sustainability: Citizen Science and Activism*; Association for Computing Machinery Press: New York, NY, USA, 2008; Available online: https://eprints.qut.edu.au/14130/ (accessed on 5 December 2021).
73. Kitchin, R. Making Sense of Smart Cities: Addressing Present Shortcomings. *Camb. J. Reg. Econ. Soc.* **2014**, *8*, 131–136. [CrossRef]
74. Foth, M. Participatory Urban Informatics: Towards Citizen-Ability. *Smart Sustain. Built Environ.* **2018**, *7*, 4–19. [CrossRef]
75. Caldwell, G.A.; Fredericks, J.; Hespanhol, L.; Chamorro-Koc, M.; Barajas, M.J.S.V.; André, M.J.C. Chapter 13—Putting the people back into the "smart": Developing a middle-out framework for engaging citizens. In *Shaping Smart for Better Cities*; Aurigi, A., Odendaal, N., Eds.; Academic Press: Cambridge, MA, USA, 2021; pp. 239–266. ISBN 9780128186367.
76. Foth, M.; Hudson-Smith, A.; Gifford, D. Smart cities, social capital, and citizens at play: A critique and a way forward. In *Research Handbook on Digital Transformations (Research Handbooks in Business and Management)*; Zhegu, M., Olleros, F.X., Eds.; Edward Elgar Publishing: Cheltenham, UK, 2016; pp. 203–221. [CrossRef]
77. Roggema, R. (Ed.) *Designing Sustainable Cities*; Springer: Berlin/Heidelberg, Germany, 2020; ISBN 9783030546854.
78. Schraven, D.; Joss, S.; de Jong, M. Past, Present, Future: Engagement with Sustainable Urban Development through 35 City Labels in the Scientific Literature 1990–2019. *J. Clean. Prod.* **2021**, *292*, 125924. [CrossRef]
79. Blok, A. Worlding Cities through Their Climate Projects? Eco-Housing Assemblages, Cosmopolitics and Comparisons. *Cityscape* **2014**, *18*, 269–286.
80. Holden, M.; Li, C.; Molina, A. The Emergence and Spread of Ecourban Neighbourhoods around the World. *Sustain. Sci. Pract. Policy* **2015**, *7*, 11418–11437. [CrossRef]
81. Birkeland, J. *Net-Positive Design and Sustainable Urban Development*; Routledge: London, UK, 2020; ISBN 9780429290213.
82. Parris, K.M.; Amati, M.; Bekessy, S.A.; Dagenais, D.; Fryd, O.; Hahs, A.K.; Hes, D.; Imberger, S.J.; Livesley, S.J.; Marshall, A.J.; et al. The Seven Lamps of Planning for Biodiversity in the City. *Cities* **2018**, *83*, 44–53. [CrossRef]
83. Downton, P.; Jones, D.; Zeunert, J.; Roös, P. Biophilic Design Applications: Putting Theory and Patterns into Built Environment Practice. In *DesTech 2016: Proceedings of the International Conference on Design and Technology, Geelong, Australia, 5–8 December 2016*; Knowledge E: Dubai, United Arab Emirates, 2017; pp. 59–65. [CrossRef]
84. Gil, P.; Rossi, C.; Coral, W. Biophilic Evolutionary Buildings That Restore the Experience of Animality in the City. In Proceedings of the Biomimetic and Biohybrid Systems, Barcelona, Spain, 28–31 July 2015; Springer International Publishing: Berlin/Heidelberg, Germany, 2015; pp. 465–472.
85. Douglas, I.; Anderson, P.M.L.; Goode, D.; Maddox, D.; Houck, M.C.; Nagendra, H.; Tan, P.Y. *The Routledge Handbook of Urban Ecology*; Routledge: London, UK, 2020; ISBN 9781138581357.
86. Foth, M.; Kamols, N.; Turner, T.; Hearn, G.; Kovachevich, A. Brisbane 2032: The promise of the first carbon-positive Olympics for regenerative cities. In *Regenerative Design for Cities and Landscapes*; Roggema, R., Ed.; Springer: Berlin/Heidelberg, Germany, 2022; Available online: https://eprints.qut.edu.au/226785/ (accessed on 5 December 2021).
87. Metzger, J. Spatial Planning And/as Caring for More-Than-Human Place. *Environ. Plan. A* **2014**, *46*, 1001–1011. [CrossRef]
88. Robertson, S.A. Rethinking Relational Ideas of Place in More-than-Human Cities. *Geogr. Compass* **2018**, *12*, e12367. [CrossRef]
89. Forlano, L. Posthumanism and Design. *J. Des. Econ. Innov.* **2017**, *3*, 16–29. [CrossRef]

90. Foth, M.; Caldwell, G. More-than-human media architecture. In Proceedings of the 4th Media Architecture Biennale Conference; Zhigang, C., Ed.; Association for Computing Machinery: New York, NY, USA, 2018; pp. 66–75. [CrossRef]
91. Luusua, A.; Ylipulli, J.; Rönkkö, E. Nonanthropocentric Design and Smart Cities in the Anthropocene. *IT-Inf. Technol.* **2017**, *59*, 299. [CrossRef]
92. Brown, K.M.; Flemsæter, F.; Rønningen, K. More-than-Human Geographies of Property: Moving towards Spatial Justice with Response-Ability. *Geoforum* **2019**, *99*, 54–62. [CrossRef]
93. Clarke, R.; Heitlinger, S.; Light, A.; Forlano, L.; Foth, M.; DiSalvo, C. More-than-Human Participation: Design for Sustainable Smart City Futures. *Interactions* **2019**, *26*, 60–63. [CrossRef]
94. Liaros, S. Implementing a New Human Settlement Theory: Strategic Planning for a Network of Regenerative Villages. *Smart Sustain. Built Environ.* **2019**, *9*, 258–271. [CrossRef]
95. Liaros, S. A Network of Circular Economy Villages: Design Guidelines for 21st Century Garden Cities. *Built Environ. Proj. Asset Manag.* **2021**. [CrossRef]
96. Soanes, K.; Lentini, P.E. When Cities Are the Last Chance for Saving Species. *Front. Ecol. Environ.* **2019**, *17*, 225–231. [CrossRef]
97. Ives, C.D.; Lentini, P.E.; Threlfall, C.G.; Ikin, K.; Shanahan, D.F.; Garrard, G.E.; Bekessy, S.A.; Fuller, R.A.; Mumaw, L.; Rayner, L.; et al. Cities Are Hotspots for Threatened Species: The Importance of Cities for Threatened Species. *Glob. Ecol. Biogeogr.* **2016**, *25*, 117–126. [CrossRef]
98. Garrard, G.E.; Williams, N.S.G.; Mata, L.; Thomas, J.; Bekessy, S.A. Biodiversity Sensitive Urban Design. *Conserv. Lett.* **2018**, *11*, e12411. [CrossRef]
99. Weisser, W.W.; Hauck, T.E. Animal-Aided Design: Using a Species' Life-Cycle to Improve Open Space Planning and Conservation in Cities and Elsewhere. *biorxiv* **2017**, 150359. [CrossRef]
100. Steele, W.; Wiesel, I.; Maller, C. More-than-Human Cities: Where the Wild Things Are. *Geoforum* **2019**, *106*, 411–415. [CrossRef]
101. Rupprecht, C.D.D. Ready for More-than-Human? Measuring Urban Residents' Willingness to Coexist with Animals. *Fenn.-Int. J. Geogr.* **2017**, *195*, 142–160. [CrossRef]
102. Power, E.R. Border-Processes and Homemaking: Encounters with Possums in Suburban Australian Homes. *Cult. Geogr.* **2009**, *16*, 29–54. [CrossRef]
103. Westerlaken, M.; Gualeni, S. Becoming with: Towards the Inclusion of Animals as Participants in Design Processes. In Proceedings of the Third International Conference on Animal-Computer Interaction, Milton Keynes, UK, 15–17 November 2016; Association for Computing Machinery: New York, NY, USA, 2016; pp. 1–10.
104. Frawley, J.K.; Dyson, L.E. Animal Personas: Acknowledging Non-Human Stakeholders in Designing for Sustainable Food Systems. In Proceedings of the 26th Australian Computer-Human Interaction Conference on Designing Futures: The Future of Design, Sydney, Australia, 2–5 December 2014; Association for Computing Machinery: New York, NY, USA, 2014; pp. 21–30.
105. Nieto-Romero, M.; Valente, S.; Figueiredo, E.; Parra, C. Historical Commons as Sites of Transformation. A Critical Research Agenda to Study Human and More-than-Human Communities. *Geoforum* **2019**, *107*, 113–123. [CrossRef]
106. Hinchliffe, S.; Kearnes, M.B.; Degen, M.; Whatmore, S. Urban Wild Things: A Cosmopolitical Experiment. *Environ. Plan. D* **2005**, *23*, 643–658. [CrossRef]
107. Haraway, D.J. *Staying with the Trouble: Making Kin in the Chthulucene*; Duke University Press: Durham, NC, USA, 2016.
108. Haraway, D. Anthropocene, Capitalocene, Plantationocene, Chthulucene: Making Kin. *Environ. Humanit.* **2015**, *6*, 159–165. [CrossRef]
109. Latour, B. *Reassembling the Social: An Introduction to Actor-Network-Theory*; Clarendon Lectures in Management Studies; Oxford University Press: Oxford, UK, 2005; ISBN 9780199256044.
110. Roös, P.B. *Regenerative-Adaptive Design for Sustainable Development: A Pattern Language Approach*; Springer: Berlin/Heidelberg, Germany, 2020; ISBN 9783030532338.
111. Lyle, P.; Foth, M.; Choi, J. Design patterns for urban gardening. In *Citizen's Right to the Digital City: Urban Interfaces, Activism, and Placemaking*; Foth, M., Ojala, T., Brynskov, M., Eds.; Springer: Singapore, 2015; pp. 79–98. ISBN 9789812879172. [CrossRef]
112. Lyle, P.; Choi, J.; Foth, M. Designing to the Pattern: A Storytelling Prototype for Food Growers. *Multimodal Technol. Interact.* **2018**, *2*, 73. [CrossRef]
113. Schuler, D. *Liberating Voices: A Pattern Language for Communication Revolution*; MIT Press: Cambridge, MA, USA, 2008; ISBN 9780262195799.
114. Yigitcanlar, T.; Teriman, S. Rethinking Sustainable Urban Development: Towards an Integrated Planning and Development Process. *Int. J. Environ. Sci. Technol.* **2015**, *12*, 341–352. [CrossRef]
115. Hoch, C. *Pragmatic Spatial Planning: Practial Theory for Professionals*; Routledge: London, UK, 2019; ISBN 9780429021275.
116. Hopkins, L.D. *Urban Development: The Logic of Making Plans*; Island Press: Washington, DC, USA, 2001; ISBN 9781559638531.
117. Adams, D.; Tiesdell, S. *Shaping Places: Urban Planning, Design and Development*; Routledge: London, UK, 2012; ISBN 9781136254642.
118. Banfield, J. *Neighbourhood Planning: Place, Space and Politics*; Routledge: London, UK, 2019; ISBN 9780429534775.
119. Graham, S.; Marvin, S. *Splintering Urbanism: Networked Infrastructures, Technological Mobilities and the Urban Condition*; Psychology Press: Hove, UK, 2001; ISBN 9780415189651.
120. Jackson, S. A Summary of Urban Assessment Tools for Application in Australia. *Environ. Design Guide* **2016**, *84*, 1–18.
121. Tsing, A.L.; Deger, J.; Saxena, A.K.; Zhou, F. *Feral Atlas: The More-than-Human Anthropocene*; Stanford University Press: Palo Alto, CA, USA, 2020; ISBN 9781503615045.

122. Bonnett, A. Feral Atlas: The More-than-Human Anthropocene. *Soc. Cult. Geogr.* **2021**, *22*, 1326–1327. [CrossRef]
123. Forlano, L. Decentering the Human in the Design of Collaborative Cities. *Des. Issues* **2016**, *32*, 42–54. [CrossRef]
124. Abram, D. *Becoming Animal: An Earthly Cosmology*; Vintage Books: New York, NY, USA, 2011; ISBN 9780375713699.
125. Roggema, R.; Vermeend, T.; Van den Dobbelsteen, A. Incremental Change, Transition or Transformation? Optimising Change Pathways for Climate Adaptation in Spatial Planning. *Sustain. Sci. Pract. Policy* **2012**, *4*, 2525–2549. [CrossRef]
126. Sheikh, H.; Gonsalves, K.; Foth, M. Plant(e)tecture: Towards a Multispecies Media Architecture Framework for amplifying Plant Agencies. In *Media Architecture Biennale 20*; Association for Computing Machinery: New York, NY, USA, 2021; pp. 87–99. ISBN 9781450390484. [CrossRef]
127. Sheikh, H.; Foth, M.; Mitchell, P. (Re)imagining the ibis: Multispecies future(s), smart urban governance, and the digital environmental humanities. In *Routledge Digital Environmental Humanities Handbook*; Routledge: London, UK, 2022; Available online: https://eprints.qut.edu.au/226108/ (accessed on 5 December 2021).
128. Wiesel, I.; Steele, W.; Houston, D. Cities of Care: Introduction to a Special Issue. *Cities* **2020**, *105*, 102844. [CrossRef]
129. Mills, J.G.; Bissett, A.; Gellie, N.J.C.; Lowe, A.J.; Selway, C.A.; Thomas, T.; Weinstein, P.; Weyrich, L.S.; Breed, M.F. Revegetation of Urban Green Space Rewilds Soil *Microbiotas* with Implications for Human Health and Urban Design. *Restor. Ecol.* **2020**, *31*, 3997. [CrossRef]
130. Deary, H.; Warren, C.R. Trajectories of Rewilding: A Taxonomy of Wildland Management. *J. Environ. Plan. Manag.* **2019**, *62*, 466–491. [CrossRef]

sustainability

MDPI

Article

A Distributed Mix-Context-Based Method for Location Privacy in Road Networks

Ikram Ullah [1], Munam Ali Shah [1], Abid Khan [2], Carsten Maple [3], Abdul Waheed [1] and Gwnaggil Jeon [4,*]

1 Department of Computer Science, COMSATS University Islamabad, Islamabad 45550, Pakistan; ikram.comsats.cs@gmail.com (I.U.); mshah@comsats.edu.pk (M.A.S.); gallian92@gmail.com (A.W.)
2 Department of Computer Science, School of Computing, Engineering and Digital Technologies, Teesside University, Middlesbrough TS1 3BX, UK; abk15@aber.ac.uk or its.abidkhan@gmail.com
3 Secure Cyber Systems Research Group, WMG, University of Warwick, Coventry CV4 7AL, UK; cm@warwick.ac.uk
4 Department of Embedded Systems Engineering, Incheon National University, 119 Academy-ro, Yeonsugu, Incheon 22012, Korea
* Correspondence: gjeon@inu.ac.kr

Abstract: Preserving location privacy is increasingly an essential concern in Vehicular Adhoc Networks (VANETs). Vehicles broadcast beacon messages in an open form that contains information including vehicle identity, speed, location, and other headings. An adversary may track the various locations visited by a vehicle using sensitive information transmitted in beacons such as vehicle identity and location. By matching the vehicle identity used in beacon messages at various locations, an adversary learns the location history of a vehicle. This compromises the privacy of the vehicle driver. In existing research work, pseudonyms are used in place of the actual vehicle identity in the beacons. Pseudonyms should be changed regularly to safeguard the location privacy of vehicles. However, applying simple change in pseudonyms does not always provide location privacy. Existing schemes based on mix zones operate efficiently in higher traffic environments but fail to provide privacy in lower vehicle traffic densities. In this paper, we take the problem of location privacy in diverse vehicle traffic densities. We propose a new Crowd-based Mix Context (CMC) privacy scheme that provides location privacy as well as identity protection in various vehicle traffic densities. The pseudonym changing process utilizes context information of road such as speed, direction and the number of neighbors in transmission range for the anonymisation of vehicles, adaptively updating pseudonyms based on the number of a vehicle neighbors in the vicinity. We conduct formal modeling and specification of the proposed scheme using High-Level Petri Nets (HPLN). Simulation results validate the effectiveness of CMC in terms of location anonymisation, the probability of vehicle traceability, computation time (cost) and effect on vehicular applications.

Keywords: anonymity; formal modeling; location privacy; mix context; pseudonyms; traceability; VANETs

Citation: Ullah, I.; Shah M.A.; Khan, A.; Maples, C.; Waheed, A.; Jeon, G. A Distributed Mix-Context-Based Method for Location Privacy in Road Networks. *Sustainability* **2021**, *13*, 12513. https://doi.org/10.3390/su132212513

Academic Editor: Tan Yigitcanlar

Received: 1 October 2021
Accepted: 5 November 2021
Published: 12 November 2021

Publisher's Note: MDPI stays neutral with regard to jurisdictional claims in published maps and institutional affiliations.

1. Introduction

With the growth of wireless technology and intelligent transportation systems, vehicular ad hoc networks (VANETs) are attracting significant attention. Current goals are to make the ad-hoc network more efficient, secure and provide comfort to passenger on the road [1]. The main concern is to provide information regarding traffic congestion, collision notification, emergency, location services, weather conditions, and so on. VANETs can improve road safety and relief of vehicle drivers on the road. Traffic related information is analyzed and shared by vehicles in the network. VANETs is a subclass of mobile ad-hoc networks, which provide communication facilities to nearby vehicles in the road environment, which makes it different from others due to characteristics such as dynamic road topology, communication, sensing capabilities and transmission power for vehicles' function [2].

The basic architecture of vehicular networks consists of Road Side Units (RSU), On-Board Units (OBU), Event Data Recorders (EDR), various sensors, and navigation systems (such as GPS) [3]. RSUs are a road infrastructure that increase the communication connectivity to vehicles. The OBU is fixed in the vehicle with a temper proof device that protects the cryptographic credentials of vehicles. This is used for wireless communication among vehicles and with infrastructure [4]. An EDR archives various events of vehicles communication during a trip on the road. GPS can be used to provide geographical location, vehicle speed, movement direction, and acceleration for a specific time interval [5]. Obstacles on the road are detected with the aid of radar and sensors. In-vehicle, an omnidirectional antenna is fixed for wireless channel access in the network.

The deployment of onboard units permits communication among nearby vehicles and fixed road infrastructure, which make possible three communication models, i.e., Vehicle to Vehicle (V2V), Vehicle to Infrastructure (V2I), and hybrid communication model [3,6]. In the V2V model, there is no support of infrastructure and vehicles are communicated directly. For data and information gathering, vehicles communicate with road side infrastructure through the V2I model. In the hybrid model, vehicles do not communicate with infrastructure directly but communicate in a single or multihop manner, depending on the transmission range of vehicles. This enables long-distance communication between vehicles in the network. Various wireless technologies are suggested for communication in vehicular networks, such Dedicated Short Range Communication (DSRC), Cellular network, WiMax, WiFi, and VeMAC protocol [6,7]. Among the existing technologies, DSRC has the potential for use in wide range variety of applications, including collision avoidance which can save thousands of lives and billions of dollars annually [8].

The mobile node (vehicle) in a network can broadcast Basic Safety Messages (BSMs), Cooperative Aware Messages (CAM) or beacons to disseminate road environment information. The beacon message's contents consist of vehicle identity, velocity, position, and other information [9]. The vehicle broadcasts beacon messages in plaintext format, and so other entities in the network are able to learn the actual identity and location of vehicles by analyzing these beacon messages. Similarly, an adversary can obtain the personal information of a vehicle driver by collecting beacon messages and tracking the various locations visited, thus coming to know the behavior and activities of the vehicle driver. This has the potential to pose several types of threat to the vehicle driver, such as damage to social reputation, physical harassment, blackmailing, and property loss [10]. To protect the privacy of the vehicle, pseudonyms can be used in place of the real identity in the message, and this is a commonly accepted solution. The pseudonym is an alias or randomized identity of a vehicle inserted in the beacon message. However, the use of fixed pseudo-identity is not suitable for protecting the privacy of a vehicle, and it must be changed over time to guard against the linking the pseudonym of a vehicle over time.

For the protection of vehicle location privacy, various pseudonym-changing strategies have been proposed in the literature. Some techniques use the concept of a mix zone [11–15] to hide the vehicle identities in a zone created by the vehicle to change pseudonyms cooperatively. However, the performance of the mix zone is degraded in conditions of lower traffic density [16]. Techniques based on group signatures are introduced [17–20] to protect location privacy of vehicles in which the broadcast beacons are signed with a key assigned to a group to protect the identity of a vehicle in a group. However, the management of signatures in the group administratively burdensome [16]; large groups have difficulty with managing signatures while small group size impacts privacy protection. Schemes based on a silent period [21–23] can hide the identity of vehicles. However, these schemes have a detrimental effect for road safety applications. To overcome the onward limitations of existing schemes, we propose a novel scheme, the Crowd-based Mix Context (CMC) method that efficiently provides location privacy protection under diverse vehicle traffic conditions.

The existing pseudonym changing scheme, Ref. [24] addresses the problem of location privacy in mix zones under lower traffic density. It uses a concept of creating fake pseudonyms to anonymize the vehicle in the concerned region. However, generating fake pseudonyms in large quantities impact VANET applications and create computational overheads for the vehicle OBU. If a single-vehicle creates fake pseudonyms for anonymity purposes, the attacker may be able to find similarities in beacon messages, to ascertain that only one vehicle is on the road. Furthermore, fake pseudonyms create a liability issue in the network. The work undertaken in [25] is based on both mix zones and silent periods, where a large number of vehicles gather and anonymize identities in silent mode. However, while this works well in an urban scenario that consists of a higher number of vehicles, it is not consistent in the case of lower congested areas such as highways in which vehicles may rarely change pseudonyms [26].

Therefore, to preserve the location privacy of a vehicle, there is a need for an efficient approach that works under various traffic density conditions and also provides privacy outside mix zone areas. In this paper, we propose a distributed scheme CMC that offers privacy protection to vehicles in VANETs. In this paper, the terms network, road network, vehicular network, vehicular communication network and VANETs denote interchangeably and they all refer to vehicular ad-hoc networks. Our contributions in this paper are given below.

1. We introduce a virtual pseudonym exchange suitable method for a low number of vehicles in transmission range. This will mix vehicle identities to provide anonymisation to a target vehicle in that region.
2. Efficiently utilize the diverse traffic density according to the road context information to hide location privacy of a vehicle.
3. We utilize the road network context for the protection of location privacy, which reduces its impacts on road network applications.

The rest of the paper is organized as follows: Section 2 contains details of the existing literature on location privacy schemes. System models and goals are discussed in Section 3. The proposed solution is explained in Section 4. Formal modeling and specification of the proposed model are given in Section 5. Section 6 provides details of the experimental setup and evaluation criteria for location privacy. In Sections 7 and 8, we discuss the performance analysis and comparison of the proposed scheme, respectively. Finally, the paper is concluded in Section 10.

2. Related Work

As mentioned, a pseudonym can be used in beacon messages, rather than the real identity of a vehicle, as a means to provide privacy. However, some limited knowledge of the whereabouts of a vehicle means that the pseudonym of a vehicle can be identified, and all journeys of that vehicle can be recovered. For this reason, pseudonyms are changed periodically. However, simple pseudonym-exchange schemes may suffer from pseudonym linkability. That is, an adversary can discover the relationship between pseudonyms and hence recover vehicle journeys. Several schemes of pseudonym changing are proposed and tested in the literature; here, we review some of these. A taxonomy of location privacy schemes is shown in Figure 1. Table 1 contains details of a comparative analysis of existing schemes in a vehicular network.

A group navigation, in combination with a random silence period, is proposed in [21]. The vehicle remains silent, not broadcasting beacons in the network for a random period to avoid linkability. The vehicles are restricted to forming groups on the road, and the group leader will broadcast messages while other members of the group remain silent. Similarly, an advanced version is presented in [27] again using the combination of silent periods with the group formation concept. The vehicles remain silent if a certain low-speed threshold is met (below 30 km/h) and should change pseudonyms during this period as given in [28]. This means that a vehicle will not broadcast heartbeat messages at slow speeds, with the justification that the possibility of an accident during lower speeds is low. In [29], a safe-distance metric is used to find an obfuscation radius in which the value of velocity, position, and direction is perturbed to enhance the privacy of vehicles.

In addition, if a vehicle did not find any other vehicle within a safe distance, it remains silent to preclude tracking. An autonomous pseudonym update mechanism is presented in [23], which takes the speed and direction as parameters. If a certain traffic weight threshold is met, the vehicle will update its pseudonym in a silent mode, otherwise the vehicle waits for one more silent period. In the context-based scheme, the vehicle adaptively enters and exits from a silent period, changing the pseudonym based on the number of silent neighboring vehicles [30]. A similar scheme is introduced in [31] in which vehicle changes its pseudonym based on the context of silent neighbors. It also assesses the presence of misbehaving vehicles in the network and checks the success of the pseudonym changing process. Another scheme is based on a silent period that uses the concept of permutation to exchange pseudonyms between vehicles to create confusion for an adversary attempting location identification [32]. Silent period-based schemes have certain limitations, i.e., it impacts road network applications, the management of silent period duration for a vehicle trip is difficult to utilize for location protection: the use of a short silent period can provide one way to measure the effectiveness of a pseudonym linkability attack, while, for a long silence period, the knowledge of spatial and temporal relationship makes it possible for an adversary to track the vehicle [33].

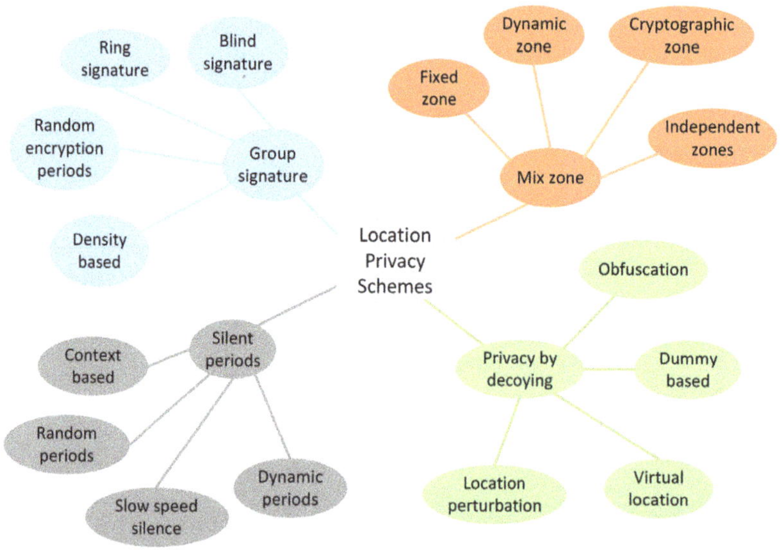

Figure 1. Categories of location privacy schemes.

A group communication method with a random encryption period is introduced in [20,34] for location privacy in VANETs. The scheme increases the confusion of the external attacker by creating an encryption zone around a vehicle's OBU. Another synchronized pseudonym changing protocol is proposed in [35] to provide unlinkability of the vehicle location tracks. The main aim is to break the spatial and temporal relationship of the vehicle pseudonyms. A data forwarding protocol is used in [36] for location privacy based on the social behavior of a vehicle driver. The social behavior of vehicles is collected from visiting social spots, i.e., shopping malls, intersections, and museums. If a vehicle visits a social spot, it can retrieve messages from RSU anonymously. The protocol achieves two things in parallel, i.e., preserves the location privacy and provides reliable transmission in the network. In [18], the concept of a cryptographic mix zone is used to hide the location information of vehicles that are based on one-time identity-based authentication. It has no dependency on trusted party, and the keys are updated within the zone. Any vehicle in the group may be a group key distributor in the cryptographic zone. A revocable group signature scheme is proposed in [37] for location privacy based on the Chinese remainder theorem and digital signature algorithm. It protects the anonymity of the vehicle as well as providing traceability to TA in case of a dispute of signature. However, the group signature method has certain issues regarding the management of signatures in the group. The signature of a large group is difficult to manage, while smaller groups affect the privacy protection level [33]. In [38], a dynamic grouping and virtual pseudonym exchange scheme is proposed in which diverse traffic conditions are utilized to update pseudonyms of vehicles for location protection.

Some techniques employ dummy data or inaccurate data to generate confusion for an adversary to identify the real location of a vehicle. In such a case, a perturbation algorithm is proposed in [39] that utilizes the reported position of two users at proximity and slightly modifies their position to create confusion for an adversary. The inaccurate beacon message is sent in between the periodic actual beacon messages to break the link between various locations of vehicles [40]. Dummy locations are generated in [41] for privacy protection of vehicles. In [42], the location of the neighboring vehicle is taken as a virtual shadow and sends two requests to LBS with two different locations. It will hide the actual location of a target vehicle from the location attacker. Similarly, virtual position points are generated in [43] that bridge between user and LBS. The sensitivity-based pseudonym changing scheme is introduced in [44] that takes regularities in vehicle movements to achieve personalised vehicle location protection. A new concept of multilevel obfuscation scheme is introduced in [45] to protect the location privacy of vehicles while communicating with LBS. The vehicle generates duplicate messages in connection with the surrounding vehicle to increase vehicle identity anonymisation in the vicinity. The concept of differential privacy and pseudonym permutation is used in [46] to hide the location trajectory of the vehicle. The trajectory of the user is divided into coarse-grained and fine-grained under the personalized user privacy requirements. Similarly, a new technique is introduced in [47] that protects the user's semantic location trajectory. It uses the concept of reinforcement learning based on differential privacy that randomizes the locations of the vehicle's trajectory. The optimized obfuscation policy is used in terms of privacy improvement and the loss of quality of the services. The obfuscation and hiding in the crowd concept are combined in [48] to increase confusion for an adversary trying to link vehicles' pseudonyms. The dummy-based location privacy scheme improves the level of privacy to some extent. However, there are certain problems related to these schemes, including management of dummy data being an issue, their impact on the quality of services, and generating overhead in the network.

Table 1. Comparative analysis of location privacy schemes in VANETs.

Ref.	Method	Adversary Model	Mode of Execution	Accountability	Preserve VANETs Applications	Privacy Metrics	Cost (Time, Computation, Communication)
[49]	Cooperative	General adversary	Infrastructure less	No	No	Protection rate	Not given
[13]	Mix zone	GPA	Infrastructure based	No	No	ASS, Location privacy gain	Reduced
[14]	Random selection	Passive adversary	Infrastructure based	No	No	ASS	Increased
[50]	Silent mode	Global passive adversary	Infrastructure less	No	No	ASS, entropy, tracking probability	Reduced
[32]	Scheme permutation	Global passive adversary	Infrastructure less	Yes	Yes	ASS, traceability, confusion, entropy	Not computed
[24]	Dummy data	External global attacker	Infrastructure less	No	No	Anonymity	Reduced
[12]	Cheating detection	Global passive adversary	Infrastructure based	No	No	ASS, entropy, attacker probability	Not mentioned
[30]	Silent mode	Global passive adversary	Infrastructure less	No	Yes	Anonymity, Traceability	Not mentioned
[31]	Silent mode	Global passive adversary	Infrastructure less	No	Yes	ASS, entropy, traceability	Not computed
[9]	Triggered based	External passive adversary	Infrastructure less	Yes	Yes	Anonymity, entropy, Tracking percentage,	Not computed
[51]	Dummy data	Global passive adversary	Infrastructure less	No	Yes	ASS, entropy, tracking probability	Not computed
[42]	Route confusion	General attacker	Infrastructure based	No	No	ASS, entropy, traceability	Not mentioned

A mix zone scheme is proposed in [52] which considers context information to change pseudonyms. The context information may be the number of neighbors, direction, and speed. The vehicles will find suitable opportunities to blend and be an anonymisation set with vehicles having similar properties. Julien et al. proposed to create mix zones at suitable areas to protect the location information of vehicles [53]. The privacy of vehicles in the mix zone is improved in [54] with the help of using a cryptographic concept. Here, the vehicle shares the status information only with neighboring vehicles. In [13], the pseudonym changing strategy is applied at a social spot which may be a road intersection or shopping malls, where several vehicles gather. The social spot becomes a mix zone to hide vehicle identities. The vehicles form a mix zone dynamically in [55] to guard against the linking of an old pseudonym to the new one. The messages of vehicles are encrypted in the zone. A similar scheme, introduced in [56], creates a mix zone dynamically and changes pseudonym based on the vehicle candidate location list. Abdelwahab et al. [57] introduced the concept of a silent mix zone, in which vehicles remain silent at the roadside intersection. An improvement is made in [15] to build an urban pseudonym changing strategy in silent zones; the vehicles exchange their pseudonyms in the silent zone.

Reputation-based schemes are proposed in [58,59] and these encourage the "selfish" vehicle behavior for pseudonym changing in the mix zone to protect location privacy. Pseudonym management and changing techniques are introduced in [60], where vehicles create a privacy zone at roadside infrastructure. The level of privacy protection is subject to a number of vehicles in the zone. In [61], a secure mix zone is created based on spatial and temporal factors. It has been shown that a temporal factor shift improves the privacy of vehicles. The virtual mix zone is created dynamically based on the expiry of pseudonyms [62]. A reputation model is also presented to encourage selfish vehicles to join the zone. The dynamic pseudonym changing technique proposed in [16] constructs multiple mix zones in the network. The privacy of the mobile object is protected with the help of the cryptographic methods in the communication. In [11], mix zones are planted at specific regions where vehicles change pseudonyms to hide their identities for the protection of vehicle privacy. A de-correlation privacy scheme is proposed in [63] that creates multiple mix zones in parking lots and traffic places. It achieves a high level of privacy protection of vehicle trajectory. Despite the useful features of the mix zone-based location privacy techniques, there are certain limitations. Firstly, in the mix zone, the level of privacy is degraded when operating in lower traffic density environments [17]. Secondly, privacy is provided to vehicles within the zone, and there is no privacy protection outside it. Thirdly, if the zones are deployed at fixed regions, only these areas provide the privacy protection and deployment costs increase the need to build a large number of zones with infrastructure support in the road network area.

Based on problems and limitations in the existing schemes for location privacy in a vehicular network, we propose a novel scheme using a crowd-based mix context that utilizes the diverse nature of vehicle traffic densities. The pseudonyms changing process depends on the number of neighbors in the transmission range and road context information. This improves the anonymisation of a target vehicle (a vehicle that an adversary wants to locate) in a crowd of similar-status vehicles in a concerned region.

3. Models and Goals

In this section, we discuss the system model and the adversary model. After that, the goals of this research work are explained. The first subsection provides a details of the entities used in the system model. The second subsection discusses assumptions about the strength of an adversary, and in the last subsection, we discuss the goals of the research work.

3.1. System Model

The system model is comprised of three things, namely the Trusted Authority (TA), RSU, and Vehicles. The depiction of the system model is shown in Figure 2. The Trusted

Authority (or certification authority or Government Authority) is assumed to be honest and will not take part in compromising vehicle location [62,63]. The vehicle must be registered with the TA before joining the ad hoc network. The TA provides a pseudonym pool to vehicles to be used for a number of days. The pseudonym is used for anonymous broadcast of beacon messages. On the expiry of its pseudonym pool, the vehicle can request another pseudonym pool from the TA. We assume that the vehicle has been registered with TA and is assigned a pseudonym pool. An RSU is a roadside infrastructure fixed on the road to increase the communication range of vehicles. It is a semi-honest entity in the system model, and it may or may not compromise the privacy of a vehicle. RSUs also play a role in the dissemination of data to other entities of the system model. We also assume that the authentication process is performed by each vehicle. The vehicle contains an OBU that is used for communication with other OBUs and infrastructure in the network. OBU records communication events of vehicles. The vehicle has a tamper-proof device that stores the key materials securely, such as anonymous identities and records of all communication events [64]. The vehicle is also equipped with GPS for precise location updates.

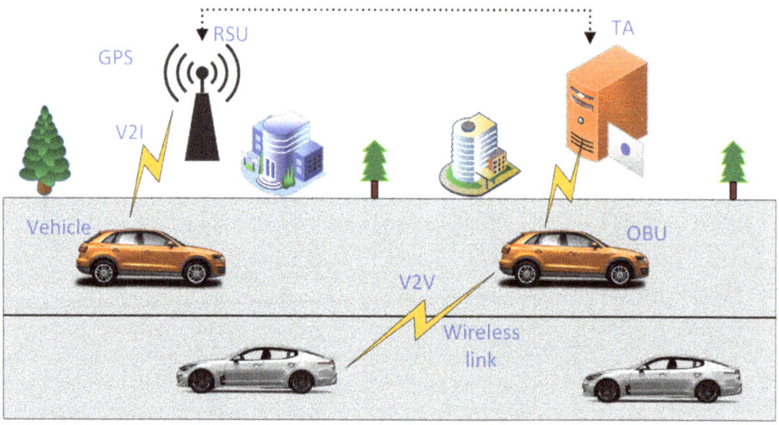

Figure 2. Basic system model.

3.2. *Adversary Model*

We assume a Global Passive Adversary (GPA) in this work. The detail of the adversary model is shown in Figure 3. The aim of the GPA is to track various locations to identify the journeys of a target vehicle in the network. In this model, we make various assumptions about the GPA. The GPA can deploy a low-cost radio transceiver to intercept broadcast beacon messages in the region of interest. The contents of the beacon message are pseudo-identity, speed, location, direction, and other headings. The adversary can capture a large portion of the network to catch the messages exchanged between vehicles. It can track the various locations of a vehicle with the help of eavesdropping vehicle communication [22]. It also has the ability to capture the pseudonyms of vehicles and can link the various pseudonyms of a vehicle used during a trip. The adversary captures beacon messages to try to correlate the old pseudonym with the newly changed pseudonym. By matching the different pseudonyms of a vehicle at different locations, the adversary gets knowledge of the target vehicle's behavior and could predict a vehicle's future locations.

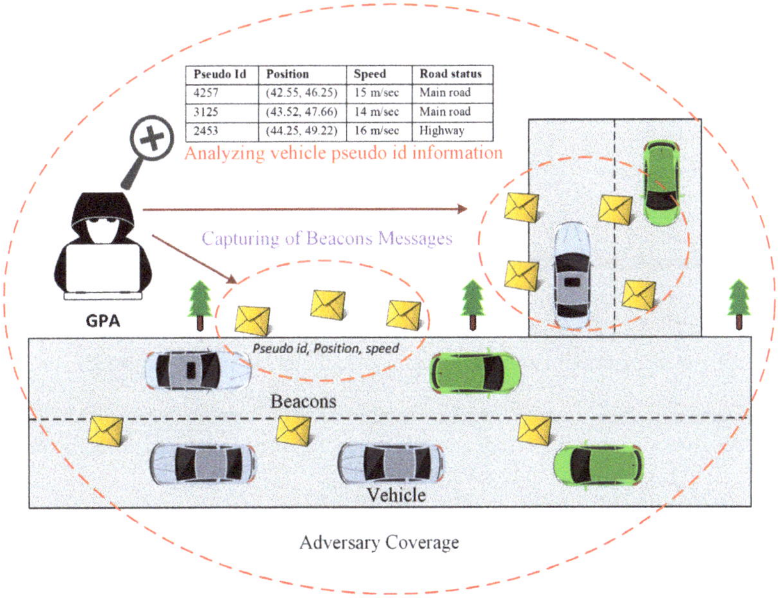

Figure 3. Adversary model.

3.3. Goals

Our primary concern in this research work is to achieve a high level of location privacy in a vehicular network. For this purpose, various factors and parameters are required. The factors include vehicle speed, traffic density, moving direction, and other road context information. The following goals are set to be achieved in this research work:

1. Location anonymisation at diverse vehicle traffic conditions.
2. Virtual pseudonym exchange process to protect location under low traffic conditions.
3. Dynamic mix context zone creation based on different road parameters such as vehicle speed, direction, and traffic density.
4. The pseudonym change and exchange process are based on the road context information.
5. Preventing an adversary from linking various pseudonyms of a vehicle at different location spots.
6. Reducing the impact of privacy protection on the efficacy of road network applications.

4. Proposed Solution

We propose a crowd-based mix context scheme which offers location privacy to vehicles in a vehicular network. The scheme comprises two cases based on the vehicle speed, direction, and traffic density. The first case in our proposed scheme is the road intersection or the situation in which vehicles have low speed and traffic congestion on the road. In the second case, vehicles have low speed and fewer neighbors within transmission range. We use a mix context method to hide the actual identity of the vehicle in a crowd. A crowd of vehicles of similar status neighboring vehicles is established where they mix their context and identities to blur the sensitive information of vehicles. Here, context means a vehicles' direction of movement, speed range, and the number of transmission range neighboring vehicles. The proposed scheme block diagram is shown in Figure 4. The vehicle senses the environment to find the number of neighboring vehicles in its range. Based on the road context information, vehicles change pseudonyms to mix it in the crowd of vehicles. Otherwise, the virtual crowd method is used to mix the identities of vehicles. In the first case, there is a higher number of neighboring vehicles in the vicinity, and the simple pseudonym changing process is used, while in the second case, there is a lower number of neighboring vehicles in the transmission range, so the target vehicle selects neighboring vehicles randomly to exchange pseudonyms with. In both of these cases, there is a need for the protection of location privacy of vehicles. The whole process of the crowd-based mix context procedure is explained with the help of Algorithm 1. *DenThreshold* is the vehicle traffic threshold and *NeigThreshold* is the number of transmission range vehicles in the vicinity of a target vehicle. The high-level flowchart of the mix context scheme is shown in Figure 5.

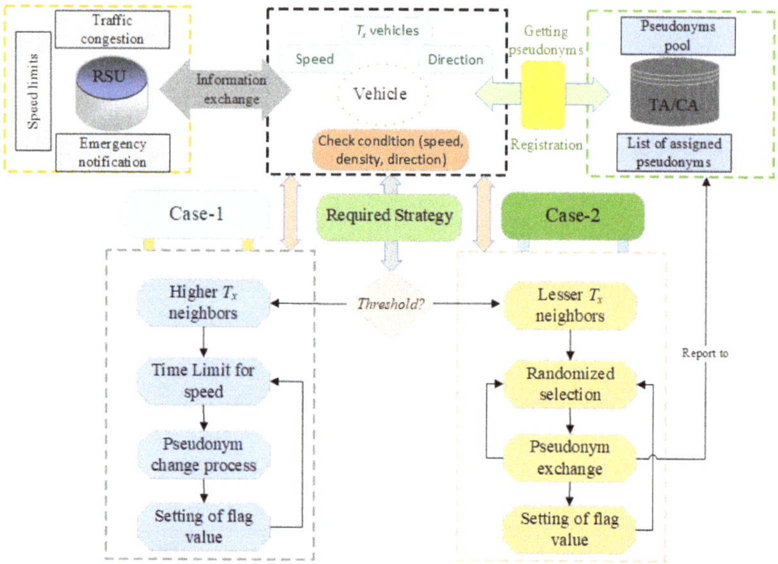

Figure 4. The block diagram of vehicles context mixing.

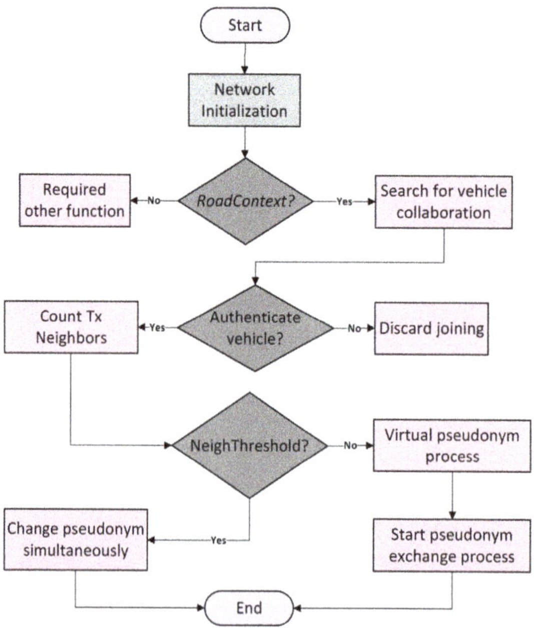

Figure 5. The high-level flow of the mix context method.

Algorithm 1 Crowd-based Mix Context.

Initialization: V_i: Any vehicle i, T_x: Transmission range, $DenThreshold$: Density threshold, $NeigThreshold$: Neighbor threshold, VD: Vehicle density, SP_R: Speed range, $LatencyBroad$: Beacon broadcast latency, D: Direction of vehicle
Input: $SP_R, D, DenThreshold, T_x$
Output: Successful pseudonym update process

```
for V_i = 1 → n do
    NeighborFunction(SP_R, D, T_x)
    if VD ≥ DenThreshold then
        Wait (TimeLimit)
        if Speed (lower) then
            Increase LatencyBroad(BSMs)
        else
            Normal broadcast
        end if
    end if
    if Neigbhors(V_i) ≥ NeigThreshold then
        Change pseudonym cooperatively
        Set PUpdate(V_i) to 0
    else
        Randomize selection of T_x neighboring vehicle
        Select V_j as virtualizer of V_i
        Exchange Msgs(V_i, V_j)
        Report Pseudonyms exchanged to TA
    end if
    Set timer for pseudonym change Pseudo(t)
        Expiry of Pseudo(t)
        Set PUpdate(V_i) to 1
        Start beacon transmission and wait for at least k context neighbors
end for
```

The neighbor function is used to search and count the number of neighboring vehicles in the transmission range. The procedure of neighbor function is given in Algorithm 2. The speed range, transmission range, and distance are given to the algorithm as input.

Algorithm 2 Neighbor Function.

Initialization: V_i: any vehicle i, T_x: Transmission range, SP_R: Speed Range, D: Direction of a vehicle, $CountV_{ID}$: Counting of number of vehicles
Input: SP_R, T_x, D
Output: Number of transmission range vehicles ($CountV_{ID}$)

```
for V_i = 1 → n do
    MessageReceived(M_i)
    Check(V_ID, Distance, SP_R)
    Calculate Distance (V_i, V_j)
    if (V_ID ≠ V_ID(i)  and  Distance ≤ 300 m) then
        CountV_ID ++
    else
        Check again(Limit)
    end if
end for
Return (CountV_ID)
```

4.1. Vehicle High Traffic Density at Low Speed

The vehicles sense the road environment and search for transmission range neighboring vehicles. In this case, the vehicle neighbor threshold is checked and verified, and based on the neighbor threshold, the pseudonyms of all vehicles are changed in the crowd. It mixes the context and pseudonyms of vehicles and confuses an adversary attempting to identify the vehicle in such a fluxed environment. This concept is shown in Figure 6. A crowd of vehicles is established when the vehicle's speed is reduced due to a roadside intersection or due to some traffic congestion situation occurring on the road. Every vehicle will broadcast a beacon message to ensure its presence in the congested vehicles' area and inform each neighboring vehicle about the pseudonym change. Each vehicle's neighborhood is verified based on the beacon message's information, i.e., transmission range, same direction, and same speed range. The vehicle will wait and continuously search for neighbors until a certain vehicle threshold is reached, when all vehicles start to change pseudonyms instantaneously. This means that each vehicle changes pseudonyms in the crowd of vehicles to anonymize itself. The contents of a beacon message include $BM(P_{ID}, T_x, V, NeighCount, D, DThresh, PUpdate)$, where P_{ID} is the pseudonym assigned to the vehicle, T_x is transmission range, V is the speed of the vehicle, $NeighCount$ counts the number of vehicles in the transmission range, D is the direction of the vehicle, $PUpdate$ is the updated pseudonym, and $DThresh$ is the vehicle density threshold.

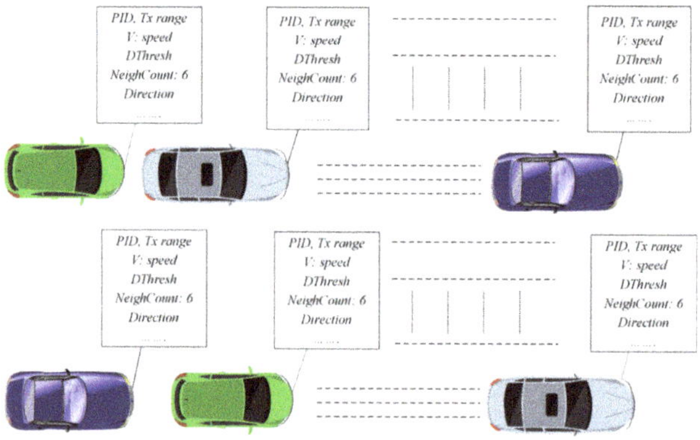

Figure 6. Mixing of context information in the crowd of vehicles.

The main procedure of case 1 is shown in Algorithm 3. The vehicle will monitor its speed and check neighboring vehicles in its transmission range. If the speed is less than a certain threshold, the vehicles will set the $Pupdate$ value to 1, which means that vehicles are ready to change pseudonyms. The vehicles will wait a certain amount of time if their speed remains low; the broadcast delay of beacon messages is increased. The delay in the broadcast of beacon messages reduces neighboring vehicles' burden for counting the number of vehicles in the vicinity. For pseudonym changing, all vehicles in the transmission range will change simultaneously and set $Pupdate$ to 0, which means that the pseudonym has been changed successfully. The pseudonym has a specific expiry time, and after this time, the flag is set to 1 and waits for another pseudonym change in the best context situation.

Algorithm 3 Mix context at higher density.

Initialization: V_i: Any vehicle i, T_x: Transmission range, $NeighThreshold$: Neighbor threshold, SP_R: Speed range, $TimeLimit$: waiting time limit, D: vehicle direction
Input: SP_R, direction, $NeighThreshold$, T_x
Output: Successful pseudonym change process

for $V_i = 1 \rightarrow n$ **do**
  $NeighborFunction(SP_R, D, T_x)$
  Set $PUpdate(V_i) = 1$
    Wait $(TimeLimit)$
  if SpeedCheck (lower) **then**
    Increase delay in Beacon broadcast
  else
    Otherwise normal broadcast
  end if
  if $Neighbors(V_i) \geq NeighThreshold$ **then**
    Change pseudonyms simultaneously
      $PUpdate(V_i)$ set to 0
      Set timer for pseudonym change $Pseudo(t)$
        After expiry of $Pseudo(t)$
        $PUpdate(V_i)$ set to 1
  else
    Case 2 function
  end if
end for

4.2. Vehicle Low Traffic Density at Low Speed

The second case of the proposed scheme involves low vehicle speed under low-traffic conditions. In this case, the virtual mix crowd method is used. The virtual crowd method scans the road environment for neighborhood vehicles in the surroundings, and based on the context of the vehicle neighbor's, the virtual pseudonym exchange process is executed. The target vehicle randomly selects one of the neighboring vehicles to exchange real and virtual pseudonyms with, as shown in Algorithm 4. Recall the beacon message include $BM(P_{ID}, T_x, V, NeighCount, D, DTresh, PUpdate)$ and is used to sense the vehicle traffic environment. In the virtual pseudonym exchange scheme, the vehicle randomly selects one of the transmission range neighbors to exchange pseudonyms and update its $PUpdate$ attribute. We discuss the virtual pseudonym exchange process with the help of an example. If vehicle V_1 wants to change the pseudonym and has V_2, V_3, V_4, and V_5 in its neighbor list. V_1 will randomly select one of the vehicles and start an exchange of pseudonyms with V_4. After completing the exchange process, both vehicles update their status of $PUpdate$ and publish it to the vehicle's crowd. Each vehicle that exchanges pseudonyms must report to the CA the pseudonyms exchanged. This process will reduce the liability issue in the network. The vehicle generates an image of its real pseudonym and exchanges it with its neighboring vehicle. The receiving vehicle verifies both of these pseudonyms, accepts the real one, and rejects the fake pseudonym. The vehicle generates such an image of a pseudonym that it is difficult for an adversary to identify it. Similarly, other vehicles such as V_2, V_3, and V_5 also apply the virtual pseudonym exchange process in the vicinity. As many vehicles take part in the virtual pseudonym exchange process, this produces a virtual crowd of vehicles in the network, which mixes each vehicle identity and location in the crowd. The pseudonyms exchange process is shown in Figure 7.

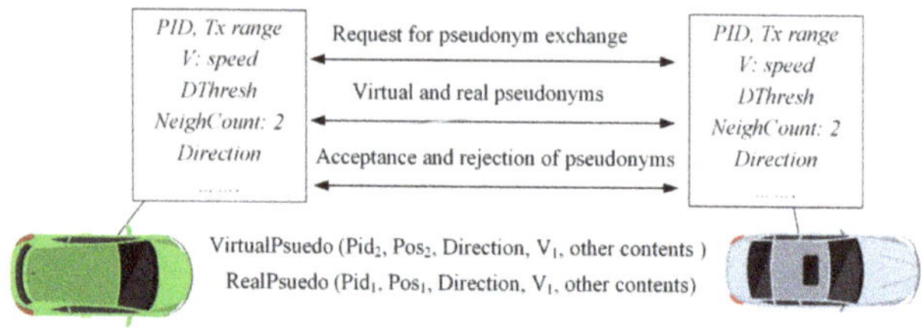

Figure 7. Virtual pseudonym exchange process between two vehicles.

Algorithm 4 Virtual Mix Context.

Initialization: V_i: Any vehicle i, T_x: Transmission range, $NeighThreshold$: Neighborhood threshold, SP_R: Speed range, D: Direction of vehicle, $LatencyBroad$: Beacon broadcast latency, $PUpdate$: Pseudonym update value
Input: $SP_R, D, NeighThreshold, T_x$
Output: Pseudonym Exchange process

for $V_i = 1 \rightarrow n$ **do**
    $NeighborFunction(SP_R, D, T_x)$
    if $Neighbor(V_i) \geq NeighThreshold$ **then**
        $CallAlgo2()$
    else
        Set $PUpdate(V_i) = 1$
        Increase $LatencyBroad$(BSMs)
    end if
    Randomize selection of a T_x neighboring vehicle
    select V_j as virtualizer of V_i
    Exchange $Msgs(V_i, V_j)$
        $Msg_1(P_{ID1}, POS_1, othercontents)$
        $Msg_2(P_{ID2}, POS_2, othercontents)$
    Set $PUpdate(V_i) = 0$
    Report Pseudonyms exchanged to TA
    Set timer for pseudonym change $Pseudo(t)$
        After expiry of $Pseudo(t)$
        Set $PUpdate(V_i)$ to 1
end for

5. Formal Modeling and Specification

High-Level Petri Nets (HLPN) is used for two reasons [65]: to simulate the proposed model and provide mathematical representation for analyzing the proposed model's behavior and structure properties. Formal modeling benefits are the interconnection of system components and processes, information flow among the processes, and information processing. We used HLPN for formal modeling and specification of the proposed scheme. HLPN is a set of seven tuples $(P, T, F, \varphi, R, L, M_0)$ as defined in [66].

In this section, we formally model and specify the proposed scheme CMC. We present the CMC scheme in HLPN in terms of mathematical properties (rules). For the representation of the system in HLPN, we define places and their associated data types; then, we specify a set of rules used in HLPN. Tables 2 and 3 contain details of the symbols and places used in the Petri nets. We design HLPN for Algorithms in the proposed model. Figure 8 shows the HLPN for the mix context scheme. The vertical bar shows transitions, and the circle shows places used in HLPN. The arrowheads in the diagram show the data flow in HLPN.

Table 2. Symbol description used in the mix-context method.

Symbol	Description
Reg-Request	Request for vehicle registration to TA
VRD	Vehicle registration data
AssignPool	Assign pseudonym pool to vehicles
PL	Pseudonym list
SM	Speed monitor
DC	Distance calculation
NS	Neighbor selection
NC	Neighbor count
NT	Neighbor threshold
SameID	Same vehicle (only one vehicle)
RC	Vehicle ready to change pseudonym
CPC	The cooperative pseudonym change process
ExpiryT	Pseudonym expiry time
PEP	Pseudonym exchange process
TxN	Transmission range neighbors
RSN	Random selection of neighbors
PE	Pseudonym exchange
BC	Broadcast beacon messages
LPN	Vehicle license plate number
V_{ID}	Vehicle identity
P_{ID}	Pseudo IDs of a vehicle
WT	Waiting time

Table 3. Places used in HLPN for mix-context method.

Types	Description
φ (Reg-Request)	$P(V_{ID} \times LPN)$
φ (VRD)	$P(V_{ID} \times PK_v \times P)$
φ (Pseudo-Request)	$P(V_{ID} \times LPN)$
φ (PL)	$P(P_{ID} \times PK_v \times P)$
φ (Road-Condition)	$P(SP_R \times D \times T_x)$
φ (SM)	$P(P_{ID} \times SP \times SP_R \times D)$
φ (DC)	$P(P_{ID} \times POSV_i \times POSV_j \times Dist)$
φ (NS)	$P(P_{ID} \times T_x \times Neigh)$
φ (NC)	$P(P_{ID} \times Dist \times NeighC)$
φ (NT)	$P(P_{ID} \times NeighC \times Dist \times Flag)$
φ (SameID)	$P(P_{ID} \times Dist)$
φ (RC)	$P(P_{ID} \times N_{ID} \times Thresh \times Indicator)$
φ (CPC)	$P(P_{ID} \times N_{ID} \times IndicatorS)$
φ (PUpdate)	$P(P_{ID} \times WT \times Flag)$
φ (ExpiryT)	$P(P_{ID} \times ThreshT \times WT \times Flag)$
φ (PEP)	$P(P_{ID} \times N_{ID} \times Thresh \times Indicator)$
φ (T_xN)	$P(P_{ID} \times N_{ID} \times IndicatorS \times RSN)$
φ (PE)	$P(P_{ID} \times N_{ID} \times Msg_i \times Msg_j)$
φ (BC)	$P(P_{ID} \times MsgP_i \times MsgP_j)$

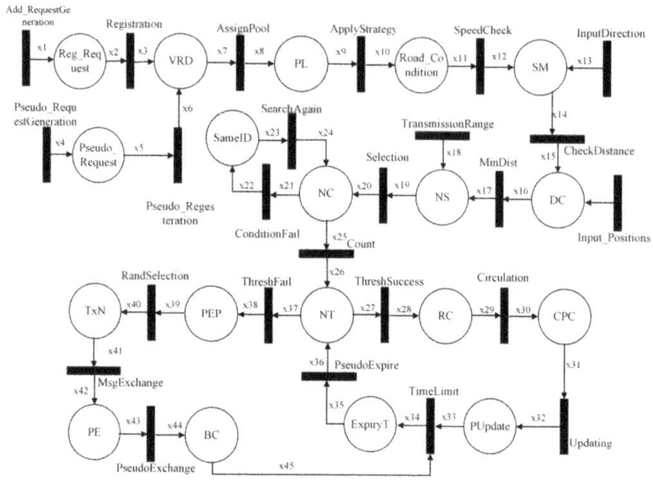

Figure 8. HLPN for mix context method.

The vehicle must register with TA before joining the network. For this purpose, the vehicle requests TA for registration. The TA processes the demand and provides Public Key (PK_v) and pseudonyms (P) in Equation (1). In Equation (2), a vehicle requests a pseudonym pool if it has expired. To assign the pseudonym pool to a vehicle, the vehicle's identity is verified, and the pseudonym pool is assigned to a vehicle as given in Equation (3). The identity of the vehicle is matched with V_{ID} and LPN in the vehicle registration list. The data in VRD is updated with a pseudonym pool assignment:

$$\begin{aligned} \boldsymbol{R(Registration)} &= \forall\, i2 \in x2 \wedge i3 \in x3 \mid i2[1] \neq i3[1] \\ &\wedge x3' := x3 \cup \{i2[1], i3[2], i3[3]\}. \end{aligned} \tag{1}$$

$$\begin{aligned} \boldsymbol{R(Pseudo-Registeration)} &= \forall\, i5 \in x5 \wedge i6 \in x6 \mid \\ &Add-request(x6' := x6 \cup \{i5[1], i6[1]\}). \end{aligned} \tag{2}$$

$$\begin{aligned} \boldsymbol{R(AssignPool)} &= \exists\, i7 \in x7 \wedge i8 \in x8 \mid compare \\ &(i7[1], i8[1]) = True \rightarrow x8' := x8 \wedge i8[1]. \end{aligned} \tag{3}$$

After the vehicle registration process, the concerned strategy of mix context is started as given in Equation (4), where vehicles sense the road conditions. First, vehicles will monitor the speed using speed check-in Equation (5). If the vehicle speed is in a particular range, then the distance (Equation (6)) is calculated with its neighboring vehicles. The distance calculation process takes position coordinates of transmission range vehicles. The vehicle will select another neighboring vehicle with a minimum distance for mixing road context information. The minimum distance is checked in (Equation (7)):

$$\begin{aligned} \boldsymbol{R(ApplyStrategy)} &= \forall\, i9 \in x9 \wedge i10 \in x10 \\ &\mid RoadCondition(i10[1], i10[2], i10[3]) \wedge \\ &x10' := x10 \cup \{i9[1], i10[1]\} \end{aligned} \tag{4}$$

$$\begin{aligned} \boldsymbol{R(SpeedCheck)} &= \forall\, i11 \in x11 \wedge i12 \in x12 \mid \\ i11[1] \in i12[3] &\wedge x12' := x12 \cup \{i12[1], i12[2]\}. \end{aligned} \tag{5}$$

$$\begin{aligned} \boldsymbol{R(CheckDistance)} &= \forall\, i14 \in x14 \wedge i15 \in x15 \mid \\ i14[1] \in i15[1] &\wedge Distance(i15[2], i15[3]) \rightarrow i15[4]. \end{aligned} \tag{6}$$

$$\begin{aligned} \boldsymbol{R(MinDist)} &= \forall\, i16 \in x16 \wedge i17 \in x17 \mid (i16[4] \\ &\leq i17[2]) = True \rightarrow x17 := x17 \cup \{i17[3]\}. \end{aligned} \tag{7}$$

After measuring the minimum distance between neighboring vehicles, the neighbor with minimum distance is selected as given in Equation (8). Next, the number of neighbors within the transmission range is counted in Equation (10). Certain conditions are required and, if the condition fails, as given in Equation (9), then finding of neighbors in transmission range starts again:

$$\begin{aligned} \boldsymbol{R(Selection)} &= \forall\, i19 \in x19 \wedge i20 \in x20 \mid (i19[1] \neq i20[1] \\ &\wedge i20[2] \leq 300) = True \rightarrow x20' := x20 \cup \{i20[3]\}. \end{aligned} \tag{8}$$

$$\begin{aligned} \boldsymbol{R(ConditionF)} &= \forall\, i21 \in x21 \wedge i22 \in x22 \mid \\ &(i21[1] = i22[1]) \wedge \exists\, (i22[1] = x22[1]). \end{aligned} \tag{9}$$

$$\begin{aligned} \boldsymbol{R(Count)} &= \forall\, i25 \in x26 \wedge i26 \in x26 \mid i25[1] \neq i26[1] \wedge \\ &(i25[3] \leq 300) = True \rightarrow x26' := x26 \cup \{i26[2]++\}. \end{aligned} \tag{10}$$

When the counting of neighbors is complete, the neighbor threshold is checked to determine the concerned case of mix context. In Equation (11), the neighbor threshold is met, and the process of cooperative pseudonym changing is circulated in the neighborhood as given in Equation (12):

$$\begin{aligned} \boldsymbol{R(ThreshSuccess)} &= \forall\, i27 \in x27 \wedge i28 \in x28 \mid (i27[2] \\ &\geq i28[3]) = True \rightarrow x28' := x28 \cup \{i28[2], i28[4]\}. \end{aligned} \tag{11}$$

$$\begin{aligned} \boldsymbol{R(Circulation)} &= \forall\, i29 \in x29 \wedge i30 \in x30 \mid \\ &(i30[3] = i29[4]) \wedge x30' := x30 \cup \{i30[3]\}. \end{aligned} \tag{12}$$

Next, Equation (13) shows the start of cooperative pseudonym updating process, and the vehicles in the transmission range change their pseudonyms; set a flag to 0 means that all neighboring vehicles change pseudonyms successfully. A time limit (Equation (20)) is set for the newly changed pseudonym, and after the pseudonym time expiry, vehicles set a flag to 1 (Equation (15)), which means that vehicles are ready for another pseudonym changing process:

$$\begin{aligned} \boldsymbol{R(Updating)} &= \forall\, i31 \in x31 \wedge i32 \in x32 \mid i31[1] = \\ &i32[1] \rightarrow x32' := x32 \cup \{i32[3] == 0 \wedge i32[2]\}. \end{aligned} \tag{13}$$

$$\begin{aligned} \boldsymbol{R(TimeLimit)} &= \forall\, i33 \in x33 \wedge i34 \in x34 \mid (i33[2] \\ &\geq i34[2]) = True \rightarrow x34' := x34 \cup \{i34[3]\}. \end{aligned} \tag{14}$$

$$\begin{aligned} \boldsymbol{R(PseudoExpire)} &= \forall\, i35 \in x35 \wedge i36 \in x36 \mid \\ &(i35[1] = i36[1] \wedge i35[3] \geq i35[2] = True \\ &\rightarrow x36' = x36 \cup \{i36[4] == 1\}. \end{aligned} \tag{15}$$

If the neighbor threshold is not met (Equation (16)), then the second case of mix context is chosen, in which a target vehicle in a vicinity randomly selects a neighboring vehicle (as given in Equation (17)) for a virtual pseudonym exchange process. The messages are exchanged between selected neighboring vehicles (Equation (18)) for a pseudonym update process. Next, vehicles exchange pseudonyms as given in Equation (19). In Equation (20), again a time limit is set for the expiry of the newly changed pseudonym:

$$\begin{aligned} \boldsymbol{R(ThreshF)} &= \forall\, i37 \in x37 \wedge i38 \in x38 \mid (i37[2] < \\ &i38[3]) = True \rightarrow x38' := x38 \cup \{i38[2], i38[4]\}. \end{aligned} \tag{16}$$

$$R(\boldsymbol{RandSelection}) = \forall\, i39 \in x39 \wedge i40 \in x40 \mid i39[4] = i40[3] \wedge x40' := x40 \cup \{i40[4]\}. \quad (17)$$

$$R(\boldsymbol{MsgExchange}) = \forall\, i41 \in x41 \wedge i42 \in x42 \mid if(i42[2] \in i41[4]) = True \rightarrow x42' := x42 \cup \{Exchange(i42[3], i42[4]\}. \quad (18)$$

$$R(\boldsymbol{PseudoExchange}) = \forall\, i43 \in x43 \wedge i44 \in x44 \mid Exchange(i43[3], i43[4]) = True \rightarrow x44' := x44 \cup \{Exchange(i44[2], i44[3])\}. \quad (19)$$

$$R(\boldsymbol{TimeLimit}) = \forall\, i45 \in x45 \wedge i34 \in x34 \mid (Exchange\ (i44[2], i44[3])) = True \rightarrow x34' := x34 \cup \{i34[3]\}. \quad (20)$$

6. Experimental Setup

We conducted various simulations of our proposed scheme to analyze its performance in vehicular networks. This section contains two parts: in the first part, we explain the simulation environment setup and parameters used in the simulation. In the second part, we discuss the evaluation metrics used for location privacy.

6.1. Simulation Setup

For the simulation of the proposed CMC scheme, we used Network Simulator 2 (NS2). The simulation parameters are described in Table 4. We run the simulations for 400 s in an urban environment. We deploy 200 vehicles on the real world map created with SUMO. The simulation area is approximately 5249 × 5053 square meters. The speed range is up to 10 m per second in the road network. We use SUMO for realistic mobility generation of vehicular networks. The OpenStreet map is used to provide a real world road scenario, as shown in Figure 9. The map is converted into SUMO to generate vehicle traffic on a real-world map. The vehicle mobility model file is generated with the help of SUMO. The simulation is run with diverse vehicle traffic densities.

Table 4. Simulation parameters for CMC.

Parameters	Value
Simulator	NS2, SUMO
MAP	OpenStreetMap
Routing protocol	AODV
Bit rate	6 MBPS
Simulation time	400 s
Number of Vehicles	200
Road area	5249 × 5053 m
Speed range	0–10 m/s
MAC protocol	IEEE 802.11p
Transmission range	500 m
Beacon interval	300 ms

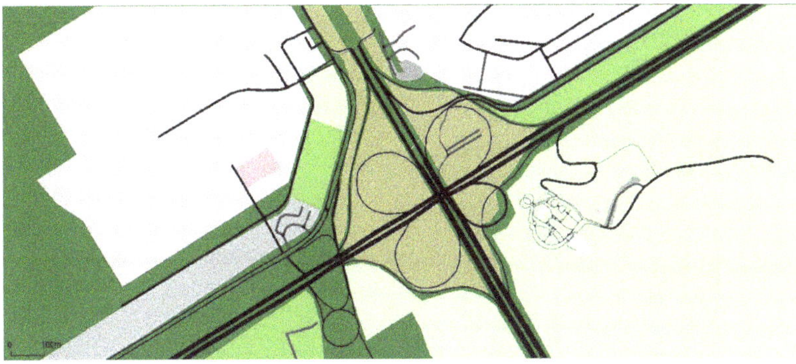

Figure 9. Real-world road network scenario using SUMO and OpenStreetMaps.

6.2. Evaluation Criteria

The evaluation criteria used in the majority of the literature are anonymity set size, entropy, and location traceability. These parameters are used to calculate the privacy protection level in VANETs. The privacy metrics are given below.

6.2.1. Anonymity Set Size

Anonymity is one of the significant metrics for evaluation of location privacy. It is the process of the grouping of users of similar status in a vicinity that hides the identity of a user in the group. Anonymity Set Size (ASS) is defined is the set of users/vehicles, including target users that are indistinguishable among the group of users. ASS achieves the anonymisation of the vehicle in a group of vehicles of similar status. In the road network, ASS is preferred, which hides the actual identities of a target vehicle during communication. Its value affects the location privacy of vehicles. The higher the ASS, the higher will be the level of vehicle privacy. We consider the arrival of vehicles at a particular point at the road is the Poisson process with rate λ and X is a random variable that denotes the number of vehicles gathered at the congested area for an interval of time T; then, the probability is calculated as follows [12]:

$$P(X = x) = \frac{(\lambda T)^x}{x!} e^{-(\lambda T)}, \tag{21}$$

where λT is the expected number of vehicles that change pseudonyms cooperatively. This can improve the anonymity of vehicles in the vicinity by updating pseudonyms at the same time. Hence, the expected number of vehicles during the time T is denoted as:

$$E(X = x) = \sum_{x=1}^{\infty} x \frac{(\lambda T)^x}{x!} e^{-(\lambda T)} = \lambda T. \tag{22}$$

After the computation of probability and the expected number of vehicles at a certain time, the anonymity set size is calculated as follows:

$$|ASS| = \sum_{i,j=1}^{n} V_i PC_j = \sum_{i=1}^{n} E(X_i = x) = \sum_{j=1}^{n} \lambda_j T. \tag{23}$$

where $V_i PC_j$ is the number of vehicle update pseudonyms that anonymize the target vehicle in the region.

6.2.2. Entropy

The entropy is used to evaluate the level of privacy of vehicles in the network. It calculates the degree of uncertainty in information from an adversary's perspective by linking various pseudonyms of a vehicle during communication and the pseudonym changing process. This measures the level of privacy achieved or anonymisation of vehicles. Let V_x be a set of vehicles that may be taking part in the pseudonym changing procedure, and V_y is a set of vehicles that successfully changed pseudonyms. Let $P(_{V_x \rightarrow V_y})$ be the probability of mapping the number of vehicles to the number of pseudonyms changed. The uniform probability of distribution evaluates the higher level of entropy and higher confusion for an adversary to identify the target vehicle. Consider the number of vehicles that have changed pseudonyms at time t; the entropy can be calculated as follows [12]:

$$H_t = \sum_{V_x, V_y \in V} P_{V_x} \rightarrow_{V_y} log_2 P_{V_x} \rightarrow_{V_y}. \tag{24}$$

where H_t is the entropy of vehicles anonymisation in the concerned area, and V is the total number of vehicles taking part in the pseudonym changing process. The average entropy can be calculated as follows:

$$H_{avg} = \frac{1}{V} \sum_{x,y \in V} H_t(x,y). \tag{25}$$

6.2.3. Vehicle Traceability

Traceability is another privacy metric that measures the tracking percentage of vehicles during a trip. It is inverse to the level of location privacy. It is the probability of an adversary determining the location spots of the vehicles. Let T_v be the tracing probability of an attacker itself, a measure of anonymity set. Traceability can be defined as given below [42]:

$$T_v = [1 - Pr(|ASS|)]. \tag{26}$$

Pr is the probability of vehicle anonymisation during the pseudonym change process. The value of T_v equal to 1 means that the adversary successfully tracked the target vehicle by linking its various pseudonym during various location spots. With increasing anonymisation, the traceability strength of an adversary reduces over time.

7. Performance Analysis

In this section, we analyze the proposed scheme's performance with the help of various experimental results. First, we discuss the performance of CMC under various traffic densities with the pseudonym update process. The proposed scheme CMC consists of both pseudonym change and exchange processes under diverse traffic conditions. We analyzed the anonymisation with the pseudonym change and exchange process as shown in Figure 10. During the pseudonym changing process, vehicles with high traffic density, i.e., have a higher number of neighbors in transmission range, the anonymisation of vehicles is improved. The pseudonym exchange process has lower anonymity due to fewer neighbors in transmission range. Depending on the process of pseudonyms update under various vehicle traffic conditions, the vehicle identity anonymisation is certainly different. A higher number of vehicles in a region that can change pseudonyms cooperatively will improve identity protection. Figure 11 shows the number of pseudonyms updated for various vehicle densities. We take two traffic densities. The lower traffic density case contains fewer vehicles in a road region, while higher density means a higher number of transmission range vehicles, as given in [52]. A higher number of vehicles taking part in the pseudonym update process increases identity protection. The figure clearly shows that a higher number of pseudonyms are updated during higher traffic conditions, which improves the anonymity of a vehicle in the crowd.

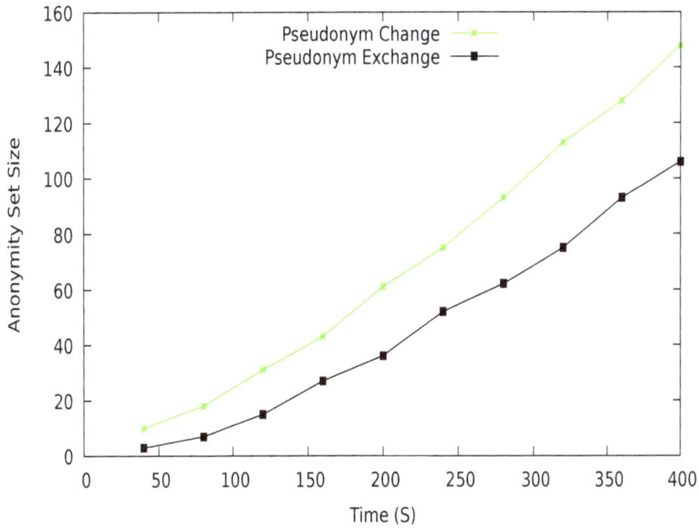

Figure 10. Anonymity of pseudonym change and exchange process.

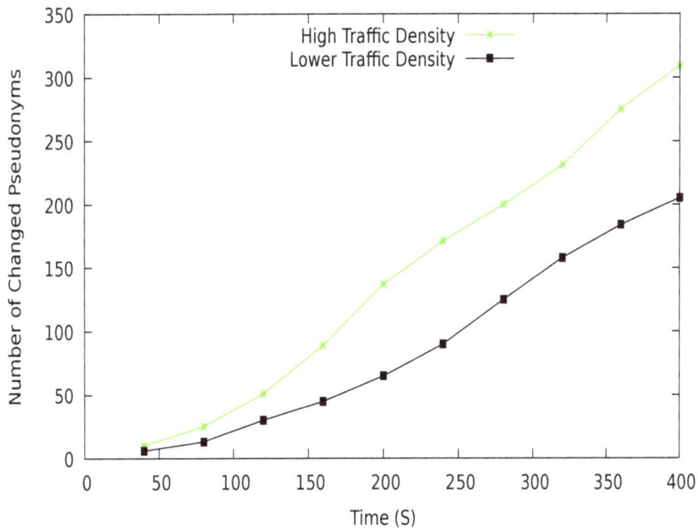

Figure 11. The number of pseudonyms updates at various vehicle densities.

8. Performance Comparison of CMC with Existing Schemes

The results of the proposed scheme CMC with existing schemes IndMZ [24] and TAPCS [25] in terms of ASS, entropy, and location traceability. The reason for choosing these two techniques is the similarity with our proposed scheme working with vehicle traffic conditions and used for privacy protection of vehicles in a road network environment. The values collected during simulations for anonymity set and entropy are given in Table 5. If a higher number of pseudonyms are changed by vehicles in a region, the target vehicle anonymisation is increased which ultimately increases entropy and confusion for an adversary to exploit the actual identity and location of a target vehicle. The location privacy scheme intends to create uncertainty in location information to generate confusion for an adversary. Here, confusion means adding uncertainty in the vehicle location information for an adversary that makes it difficult to identify a vehicle by linking old pseudonyms with newly changed pseudonyms. The uncertainty can be produced by the proper pseudonyms changing process. The proposed scheme CMC improves entropy and increases confusion for an adversary when compared with existing schemes IndMZ [24] and TAPCS [25] as shown in the table. The average ASS is evaluated based on vehicle density and simulation time. Figures 12 and 13 show the proposed scheme's results in comparison to existing schemes for vehicle anonymisation. CMC improves vehicle anonymity over IndMZ and TAPCS. The reason behind this is the efficient management of vehicles that took part in the pseudonym update process. Initially, the anonymity is low due to a smaller number of transmission range vehicles. After some time, the anonymity of vehicles increases due to the successful change of pseudonyms. In Figure 13, the proposed scheme's behavior is slightly undulating due to the lack of vehicle interest in cooperation with neighbors in the region. However, the overall process moves towards improvement in the anonymisation of vehicles.

Table 5. Values collected during simulations.

Number of Vehicles	Number of Pseudonyms Changed	Entropy of ASS	Adversary Confusion
	CMC		
100	71	8	40%
200	169	14.7	70%
	TAPCS [25]		
100	51	6.5	32%
200	135	12.6	56%
	IndMZ [24]		
100	36	4.8	14%
200	111	11.2	37%

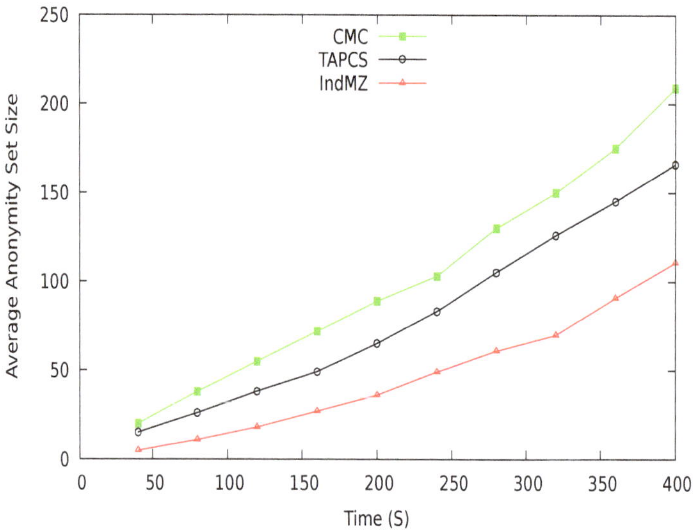

Figure 12. Average anonymity versus time.

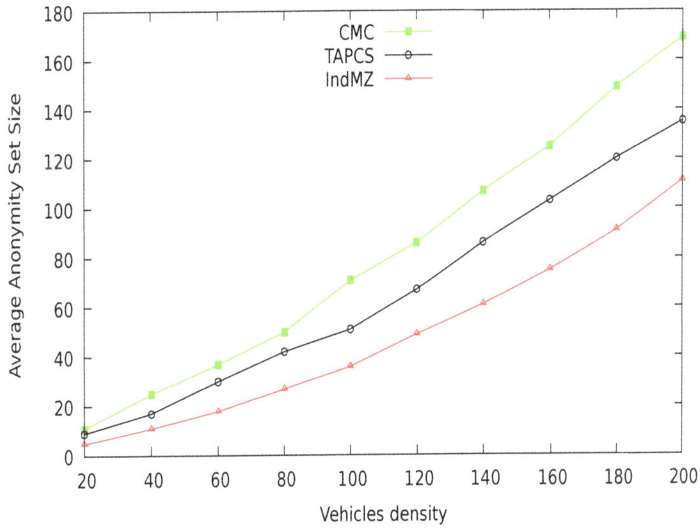

Figure 13. Anonymisation of vehicle identity at different vehicles' density.

Figure 14 shows the entropy of average anonymity set size at a certain time. The proposed scheme CMC beats the existing schemes TAPCS and IndMZ in terms of entropy. This shows the confusion generated concerning the target vehicle in the pseudonym updating process. CMC produces greater confusion than existing schemes for an adversary to find a target vehicle's actual identity in the communicating region. Ultimately, it improves the privacy protection level of vehicles. Similarly, in Figure 15, entropy is evaluated based on the number of vehicles took part in the pseudonym changing process. The entropy shows irregular behavior at different traffic densities; this is due to the lack of cooperation of vehicles in that region for the anonymisation process. CMC achieves higher confusion than TAPCS and IndMZ in various vehicle traffic. The achievement of the proposed CMC scheme regarding entropy is because of the efficient utilization of road context and pseudonym updating process. The IndMZ [24] generates fake pseudonyms in an individual manner and so the target vehicle can be easily identified by an adversary, which reduces the confusion level. The reduced performance of TAPCS compared with CMC regarding entropy is due to inefficient management of the pseudonym changing process in the silent period.

The vehicle tracking percentage during simulation time is shown in Figure 16. Initially, the proposed scheme CMC and TAPCS have similar tracking ratios at a certain amount of time. Over time, CMC reduces vehicle traceability for an adversary. The proposed scheme gets better results regarding reducing vehicle tracking probability compared with existing methods TAPCS [25] and IndMZ [24]. Similarly, tracking probability concerning the number of vehicles is shown in Figure 17. While there is a low number of vehicles at the start of the network, the tracking probability is also high. Whenever the number of vehicles is increasing, tracking probability is reduced to a certain level. CMC has a lower tracking percentage compared to existing schemes. The proposed scheme efficiently manages the road environment and makes use of any neighbor's cooperation in the pseudonym changing process, increasing the confusion for an adversary to track the pseudo-identities of vehicles.

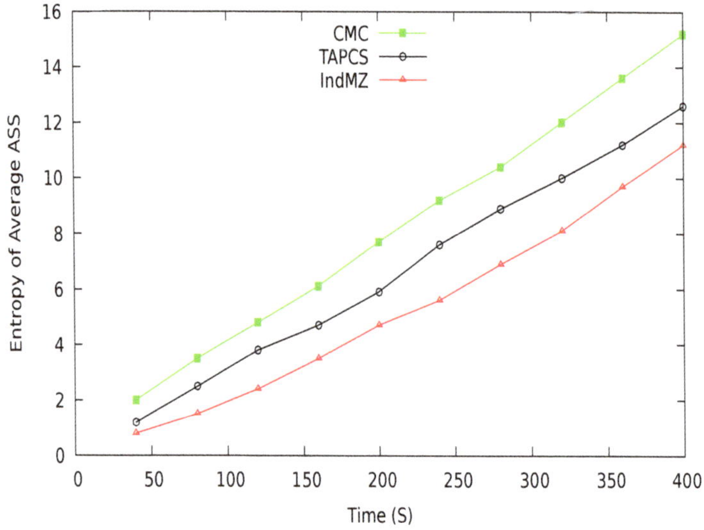

Figure 14. Entropy with different periods.

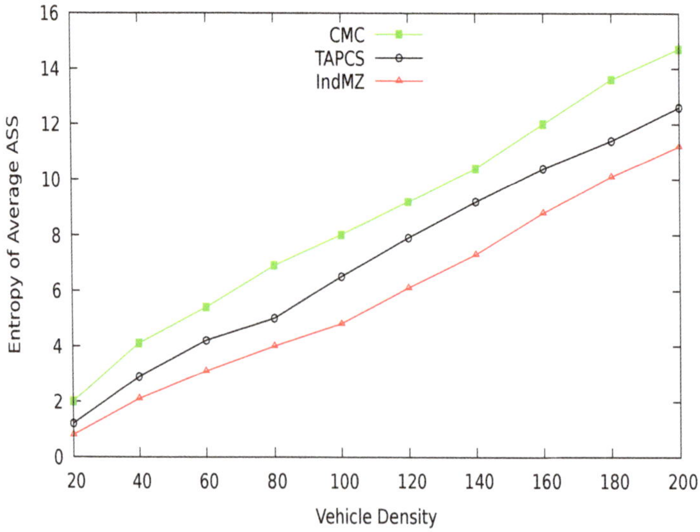

Figure 15. Entropy with different vehicle traffic density.

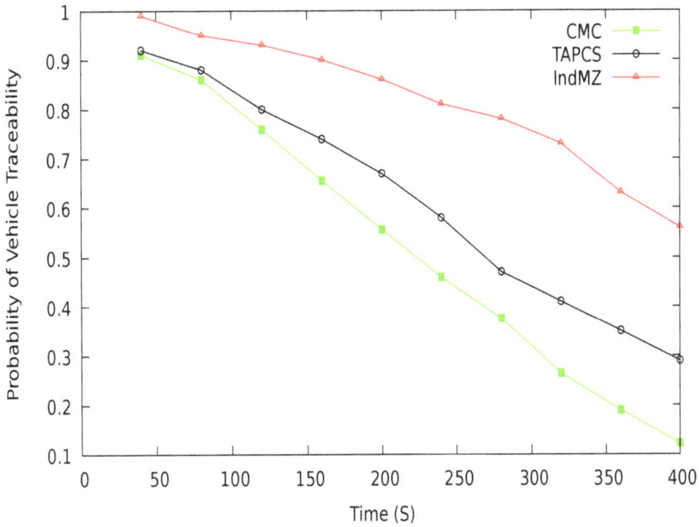

Figure 16. Vehicle tracing probability.

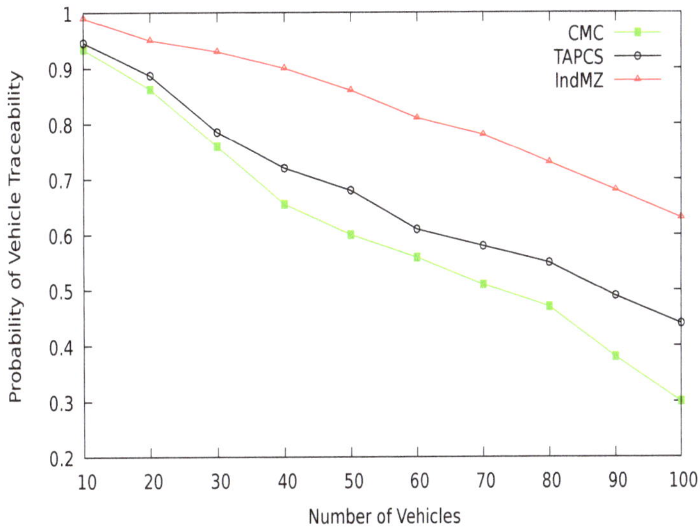

Figure 17. Tracking probability of the number of vehicles.

9. Analysis and Discussion

This section discusses the CMC's general security against GPA, the proposed scheme's impact, and its computation overhead. The details are given below.

9.1. Protection against GPA

The GPA can cover a large part of the network to capture beacon messages broadcast by vehicles. GPA analyzes the beacon messages for pseudo-identities of vehicles and matches them with old pseudonyms. If the adversary successfully matches these pseudonyms, then it can identify vehicles at various visited locations. Here, we examine the strength of GPA to extract the identity of a target vehicle. We investigate the strength of GPA with and without additional knowledge about a target vehicle. The additional information about the target vehicle may be collected at some road intersections or frequently visited locations. The information may be vehicle frequently visited roads, old pseudonyms, and location of interest. Based on this information, the adversary tries to match the pseudonyms of a target vehicle at the earlier locations with pseudonyms changed at the new visited locations. This knowledge improves an adversary's strength to identify a target, which may be used for matching vehicle pseudonyms. Figure 18 shows the average confusion per trace of GPA with and without additional knowledge. The GPA with additional knowledge has lower confusion in identifying a vehicle, and without additional knowledge, confusion is increasing at a higher rate. Our proposed scheme CMC increases an adversary's confusion with additional information because it efficiently mixes the vehicle context under diverse traffic conditions. Similarly, Figure 19 shows the confusion for both GPA with and without additional knowledge under different vehicle traffic densities. The increasing number of vehicles improves the average confusion rate for the GPA. Eventually, it increases the protection level of the location privacy of a target vehicle.

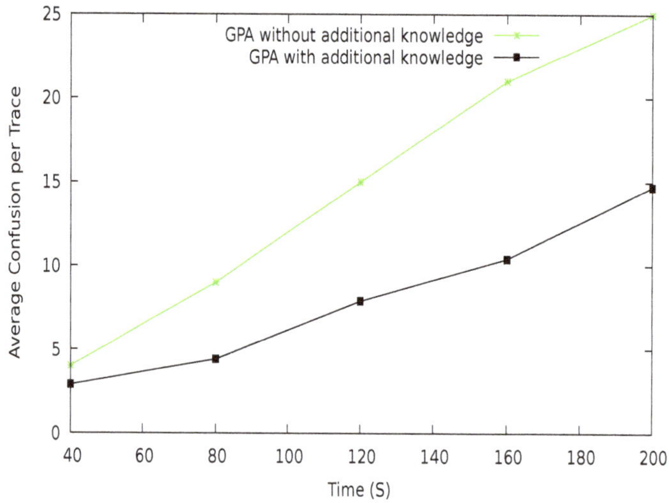

Figure 18. Adversary confusion for vehicle traces.

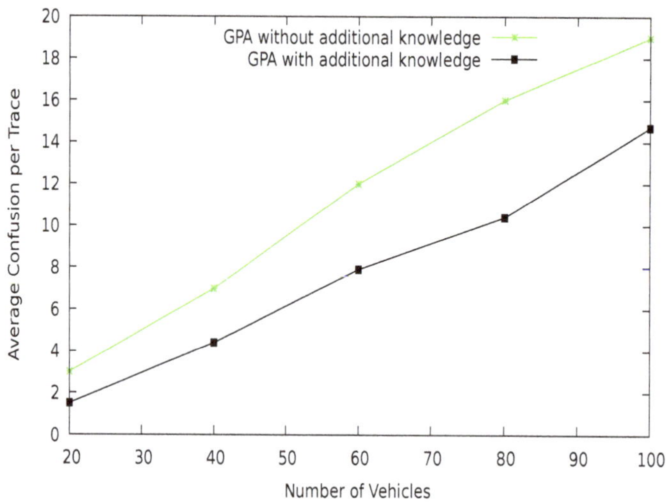

Figure 19. Adversary confusion with a different number of vehicles.

9.2. Impact on VANETs Applications

We analyze the impact of location privacy scheme on the vehicular network applications that may be safety-related and comfort applications. Achieving a higher level of privacy certainly will need to reduce its impact on road network applications. However, dummy data or wrong information is disseminated in the network for anonymizing a vehicle, which significantly reduces the application services of the vehicular network. Supposing inaccurate information is circulated, how would the vehicles efficiently utilize the applications of VANETs? The existing scheme IndMZ [24] takes the help of fake pseudonyms to anonymize the vehicles. This scheme falls into two problems. The first one is the computation burden on the vehicles. Secondly, it affects safety applications. The generation of fake pseudonyms for anonymisation reduces the application service quality. A large number of fake pseudonyms may also suffer communication among vehicles. The TAPCS [25] scheme is based on the radio silence period to preserve vehicles' location privacy. The use of silence periods during communication networks such as in VANETs is dangerous for vehicles' safety. During silent modes, mobile nodes do not broadcast road network information such as emergency, accident, the danger of lane changing, etc., and how the other vehicles will know about the road status information at a certain period. Safety application is critical, and delays in information dissemination compromise driver safety on the road. Thus, the TAPCS has a higher impact on road safety applications than the proposed scheme CMC. There is no concept of a silent period and generation of fake pseudonyms in our scheme CMC that impact road safety and comfort application. CMC disseminates road status information on a timely basis and utilizes road context information for vehicle anonymisation. CMC reduces privacy impact on safety applications compared with IndMZ and TAPCS.

9.3. Computation Overhead

The vehicle computation time for the pseudonyms changing process is analyzed in this section. Figure 20 shows the average computation time of the proposed scheme CMC using number of neighbors in the transmission range. A higher number of transmission range neighbors will take more time to compute the pseudonyms changing process as compared to lower transmission range neighbors. At the start of the simulation, there is a lower number of transmission range neighbors and a lower computation burden on the vehicles for the pseudonym changing process. The proposed scheme's average computation cost is compared with TAPCS [25] and IndMZ [24], as in Figure 21. The CMC creates a lower computation overhead than existing schemes. This is because of the utilization of road context information and pseudonym changing process for vehicle anonymisation. IndMZ scheme produces a much higher computation cost due to the generation of fake pseudonyms in larger quantities. The TAPCS generates a lower cost of computation than IndMZ by utilizing the vehicle traffic conditions in the road environment, although the proposed scheme CMC has a lower computation overhead as compared with the existing scheme. However, this overhead can be reduced further to optimize the computation and communication time of vehicles by applying a privacy scheme.

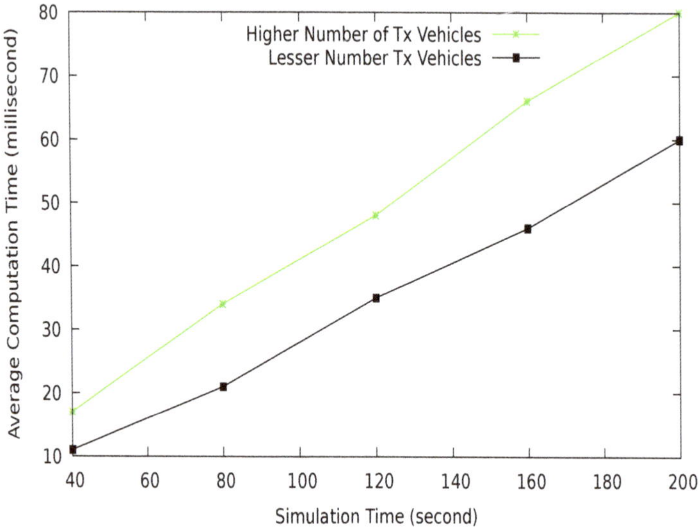

Figure 20. CMC computation overhead for the pseudonyms' changing process.

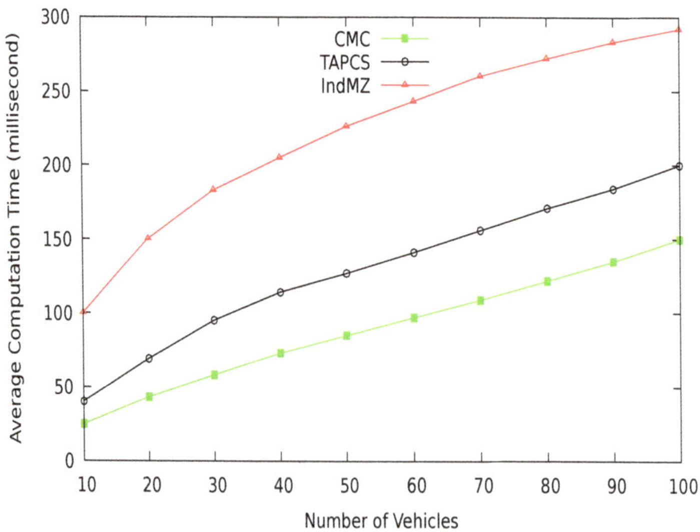

Figure 21. Average computation cost with a various number of vehicles.

9.4. Privacy and Traffic Optimization

The majority of the research conducted for location privacy protection in VANETs is based on the pseudonyms' changing process. The design of the privacy scheme may have an impact on the performance of other fields of the VANETs. One consideration is the impact on communication protocols, high frequency of changing pseudonym improves privacy but creates complications for the routing protocols. It degrades the performance of communication protocols that may impact the packet delivery ratio. The impact of the pseudonym changing process discussed in [67] on the packet delivery ratio is as follows: if vehicles (nodes) change their pseudonyms after every five seconds, the packet delivery ratio is about 60% to the destination. If pseudonyms are changed after every

10 s, the successful packet delivery ratio is about above 75%, while increasing duration to 30 s successful delivery ratio becomes above 85%. However, the increasing duration for pseudonym changes will reduce the level of privacy. Thus, there should be a reasonable balance between privacy protection and traffic optimization.

10. Conclusions

This paper takes on the problem of location privacy in a vehicular network. We have proposed a new Crowd-based Mix Context (CMC) scheme for location privacy preservation in the vehicular network. CMC employs vehicle speed, direction, and traffic density for the pseudonym changing process. Based on these parameters, the vehicles update pseudonyms simultaneously, which creates confusion for an adversary to break the pseudonyms of vehicles at different location spots. We formally model and analyze the proposed scheme using HLPN. The evaluation results show that CMC improves the anonymisation of vehicles compared with existing schemes IndMZ and TAPCS at various traffic densities. This prevents the adversary from linking pseudonyms of vehicles and identifies a target vehicle in the road region. The proposed scheme reduced the computation burden on vehicles for generating fake pseudonyms in the existing methods. The CMC also minimizes the impact of anonymisation on safety applications by managing road context information. In the future, we are eager to do more experiments on the vehicle high speed and low traffic density and will determine a robust privacy preservation method in such a dynamic road network condition.

Author Contributions: Conceptualization, I.U. and M.A.S.; methodology, A.K.; software, I.U.; validation, A.W., M.A.S., G.J. and A.K.; formal analysis, I.U., G.J. and A.W.; investigation, C.M.; resources, C.M.; data curation, A.K.; writing—original draft preparation, I.U.; writing—review and editing, M.A.S., G.J. and A.W.; visualization, G.J., I.U. and A.W.; supervision, M.A.S.; project administration, A.K. and C.M.; funding acquisition, C.M. All authors have read and agreed to the published version of the manuscript.

Funding: Ikram Ullah wants to thank the Higher Education Commission Pakistan for supporting PhD studies. Maple would like to acknowledge the support of UKRI through the grants EP/R007195/1 (Academic Centre of Excellence in Cyber Security Research—University of Warwick), EP/N510129/1 (The Alan Turing Institute), EP/R029563/1 (Autotrust), and EP/S035362/1 (PETRAS, the National Centre of Excellence for IoT Systems Cybersecurity).

Institutional Review Board Statement: Not applicable.

Informed Consent Statement: Not applicable.

Data Availability Statement: Not applicable.

Acknowledgments: Thanks to Higher Education Commission, Pakistan for supporting academic studies.

Conflicts of Interest: The authors declare no conflict of interest.

References

1. Cunha, F.; Villas, L.; Boukerche, A.; Maia, G.; Viana, A.; Mini, R.A.; Loureiro, A.A. Data communication in VANETs: Protocols, applications and challenges. *Ad Hoc Netw.* **2016**, *44*, 90–103. [CrossRef]
2. Liang, W.; Li, Z.; Zhang, H.; Wang, S.; Bie, R. Vehicular ad hoc networks: Architectures, research issues, methodologies, challenges, and trends. *Int. J. Distrib. Sens. Netw.* **2015**, *11*, 745303. [CrossRef]
3. Cunha, F.D.D.; Boukerche, Z.; Villas, L.; Viana, A.C.; Loureiro, A.A.F. *Data Communication in VANETs: A Survey, Challenges and Applications*; [Research Report] RR-8498; INRIA Saclay: Palaiseau, France, 2014; pp. 1–26.
4. Qu, F.; Wu, Z.; Wang, F.Y.; Cho, W. A security and privacy review of VANETs. *IEEE Trans. Intell. Transp. Syst.* **2015**, *16*, 2985–2996. [CrossRef]
5. Engoulou, R.G.; Bellaïche, M.; Pierre, S.; Quintero, A. VANET security surveys. *Comput. Commun.* **2014**, *44*, 1–13. [CrossRef]
6. Omar, H.A.; Lu, N.; Zhuang, W. Wireless access technologies for vehicular network safety applications. *IEEE Netw.* **2016**, *30*, 22–26. [CrossRef]
7. Al-Sultan, S.; Al-Doori, M.M.; Al-Bayatti, A.H.; Zedan, H. A comprehensive survey on vehicular Ad Hoc network. *J. Netw. Comput. Appl.* **2014**, *37*, 380–392. [CrossRef]

8. Kenney, J.B. Dedicated short-range communications (DSRC) standards in the United States. *Proc. IEEE* **2011**, *99*, 1162–1182. [CrossRef]
9. Wang, S.; Yao, N.; Gong, N.; Gao, Z. A trigger-based pseudonym exchange scheme for location privacy preserving in VANETs. *Peer-to-Peer Netw. Appl.* **2018**, *11*, 548–560. [CrossRef]
10. Ullah, I.; Wahid, A.; Shah, M.A.; Waheed, A. VBPC: Velocity based pseudonym changing strategy to protect location privacy of vehicles in VANET. In Proceedings of the 2017 International Conference on Communication Technologies (Comtech), Rawalpindi, Pakistan, 19–21 April 2017; pp. 132–137.
11. Amro, B. Protecting Privacy in VANETs Using Mix Zones with Virtual Pseudonym Change. *Int. J. Netw. Secur. Its Appl.* **2018**, *10*, 11–21. [CrossRef]
12. Memon, I.; Chen, L.; Arain, Q.A.; Memon, H.; Chen, G. Pseudonym changing strategy with multiple mix zones for trajectory privacy protection in road networks. *Int. J. Commun. Syst.* **2018**, *31*, 1–44. [CrossRef]
13. Lu, R.; Lin, X.; Luan, T. H.; Liang, X.; Shen, X. Pseudonym changing at social spots: An effective strategy for location privacy in vanets. *IEEE Trans. Veh. Technol.* **2011**, *61*, 86–96. [CrossRef]
14. Boualouache, A.; Senouci, S.M.; Moussaoui, S. Vlpz: The vehicular location privacy zone. *Procedia Comput. Sci.* **2016**, *83*, 369–376. [CrossRef]
15. Boualouache, A.; Moussaoui, S. Urban pseudonym changing strategy for location privacy in VANETs. *Int. J. Ad Hoc Ubiquitous Comput.* 2017, 24, 49–64. [CrossRef]
16. Ali, Q.; Zhongliang, A.; Imran, D. Location Privacy with Dynamic Pseudonym-Based Multiple Mix-Zones Generation over Road Networks. *Wirel. Pers. Commun.* **2017**, *97*, 3645–3671.
17. Yu, R.; Kang, J.; Huang, X.; Xie, S.; Zhang, Y.; Gjessing, S. MixGroup: Accumulative pseudonym exchanging for location privacy enhancement in vehicular social networks. *IEEE Trans. Dependable Secur. Comput.* **2015**, *13*, 93–105. [CrossRef]
18. Zhang, L. OTIBAAGKA: A new security tool for cryptographic mix-zone establishment in vehicular ad hoc networks. *IEEE Trans. Inf. Forensics Secur.* **2017**, *12*, 2998–3010 [CrossRef]
19. Mei, Y.; Jiang, G.; Zhang, W.; Cui, Y. A collaboratively hidden location privacy scheme for VANETs. *Int. J. Distrib. Sens. Netw.* **2014**, *10*, 473151. [CrossRef]
20. Wasef, A.; Shen, X.S. REP: Location privacy for VANETs using random encryption periods. *Mob. Netw. Appl.* **2010**, *15*, 172–185. [CrossRef]
21. Sampigethaya, K.; Huang, L.; Li, M.; Poovendran, R.; Matsuura, K.; Sezaki, K. *CARAVAN: Providing Location Privacy for VANET*; Department of Electrical Engineering, University of Washington: Seattle, WA, USA, 2005.
22. Khacheba, I.; Yagoubi, M.B.; Lagraa, N.; Lakas, A. Location privacy scheme for VANETs. In Proceedings of the 2017 International Conference on Selected Topics in Mobile and Wireless Networking (MoWNeT), Avignon, France, 17–19 May 2017; pp. 1–6.
23. Amro, B.; Saygin, Y.; Levi, A. Enhancing privacy in collaborative traffic-monitoring systems using autonomous location update. *IET Intell. Transp. Syst.* **2013**, *7*, 388–395. [CrossRef]
24. Guo, N.; Ma, L.; Gao, T. Independent Mix Zone for Location Privacy in Vehicular Networks. *IEEE Access* **2018**, *6*, 16842–16850. [CrossRef]
25. Boualouache, A.; Moussaoui, S. TAPCS: Traffic-aware pseudonym changing strategy for VANETs. *Peer-to-Peer Netw. Appl.* **2017**, *10*, 1008–1020. [CrossRef]
26. Zidani, F.; Semchedine, F.; Ayaida, M. Estimation of Neighbors Position privacy scheme with an Adaptive Beaconing approach for location privacy in VANETs. *Comput. Electr. Eng.* **2018**, *71*, 359–371. [CrossRef]
27. Sampigethaya, K.; Li, M.; Huang, L.; Poovendran, R. AMOEBA: Robust location privacy scheme for VANET. *IEEE J. Sel. Areas Commun.* **2007**, *25*, 1569–1589. [CrossRef]
28. Buttyán, L.; Holczer, T.; Weimerskirch, A.; Whyte, W. Slow: A practical pseudonym changing scheme for location privacy in vanets. In Proceedings of the 2009 IEEE Vehicular Networking Conference (VNC), Tokyo, Japan, 28–30 October 2009; pp. 1–8.
29. Wei, Y.-C.; Chen, Y.-M. Safe Distance Based Location Privacy in Vehicular Networks. In Proceedings of the 2010 IEEE 71st Vehicular Technology Conference, Taipei, Taiwan, 16–19 May 2010; pp. 1–5.
30. Emara, K.; Woerndl, W.; Schlichter, J. CAPS: Context-aware Privacy Scheme for VANET Safety Applications. In Proceedings of the 8th ACM Conference on Security & Privacy in Wireless and Mobile Networks, New York, NY, USA, 22–26 June 2015; pp. 21:1–21:12.
31. Khacheba, I.; Yagoubi, M.B.; Lagraa, N.; Lakas, A. CLPS: Context-based location privacy scheme for VANETs. *Int. J. Ad Hoc Ubiquitous Comput.* **2018**, *29*, 141–159. [CrossRef]
32. Singh, P.K.; Gowtham, S.N.; Tamilselvan, S.; Nandi, S. CPESP: Cooperative pseudonym exchange and scheme permutation to preserve location privacy in VANETs. *Veh. Commun.* **2019**, *20*, 100183. [CrossRef]
33. Arain, Q.A.; Memon, I.; Deng, Z.; Memon, M.H.; Mangi, F.A.; Zubedi, A. Location monitoring approach: Multiple mix-zones with location privacy protection based on traffic flow over road networks. *Multimed. Tools Appl.* **2018**, *77*, 5563–5607. [CrossRef]
34. Deng, X.; Xin, X.; Gao, T. A location privacy protection scheme based on random encryption period for VSNs. *J. Ambient. Intell. Humaniz. Comput.* **2020**, *11*, 1351–1359. [CrossRef]
35. Weerasinghe, H.; Fu, H.; Leng, S.; Zhu, Y. Enhancing unlinkability in vehicular ad hoc networks. In Proceedings of the 2011 IEEE International Conference on Intelligence and Security Informatics, Beijing, China, 10–12 July 2011.

36. Ni, J.; Lin, X.; Shen, X. Privacy-preserving data forwarding in VANETs: A personal-social behavior based approach. In Proceedings of the GLOBECOM 2017—2017 IEEE Global Communications Conference, Singapore, 4–8 December 2017; pp. 1–6.
37. Huanguo, Z.; Yi, M. *A Privacy Protection Scheme in VANETs Based on Group Signature*; Springer: Singapore, 2019; Volume 10.
38. Ullah, I.; Shah, M.A.; Khan, A.; Maple, C.; Waheed, A. Virtual Pseudonym-Changing and Dynamic Grouping Policy for Privacy Preservation in VANETs. *Sensors* **2021**, *21*, 3077. [CrossRef] [PubMed]
39. Hoh, B.; Gruteser, M. Protecting location privacy through path confusion. In Proceedings of the First International Conference on Security and Privacy for Emerging Areas in Communications Networks (SECURECOMM'05), Athens, Greece, 5–9 September 2005; pp. 194–205.
40. Karim, E. Location privacy in vehicular networks. In Proceedings of the 2013 IEEE 14th International Symposium on "A World of Wireless, Mobile and Multimedia Networks" (WoWMoM), Madrid, Spain, 4–7 June 2013.
41. Chaudhary, B.; Singh, K. A Dummy Location Generation Model for Location Privacy in Vehicular Ad hoc Networks. In Proceedings of the International Conference on Innovative Computing and Communications, New Delhi, India, 20–21 February 2021; Springer, Singapore, 2021; pp. 1–10.
42. Cui, J.; Wen, J.; Han, S.; Zhong, H. Efficient Privacy-preserving Scheme for Real-time Location Data in Vehicular Ad-hoc Network. *IEEE Internet Things J.* **2018**, *4662*, 1–8. [CrossRef]
43. Yu, H.; Li, G.; Wu, J.; Ren, X.; Cao, J. A location-based path privacy protection scheme in internet of vehicles. In Proceedings of the IEEE INFOCOM 2020—IEEE Conference on Computer Communications Workshops (INFOCOM WKSHPS), Toronto, ON, Canada, 6–9 July 2020; pp. 665–670.
44. Zhong, H.; Ni, J.; Cui, J.; Zhang, J.; Liu, L. Personalized Location Privacy Protection Based on Vehicle Movement Regularity in Vehicular Networks. *IEEE Syst. J.* **2021**. [CrossRef]
45. Ullah, I.; Shah, M.A.; Khan, A.; Jeon, G. Privacy-preserving multilevel obfuscation scheme for vehicular network. *Trans. Emerg. Telecommun. Technol.* **2021**, *32*, e4204. [CrossRef]
46. Chen, Z.; Bao, X.; Ying, Z.; Liu, X.; Zhong, H. Differentially private location protection with continuous time stamps for VANETs. In Proceedings of the International Conference on Algorithms and Architectures for Parallel Processing, Guangzhou, China, 15–17 November 2018; Springer: Cham, Switzerland, 2018.
47. Wang, W.; Min, M.; Xiao, L.; Chen, Y.; Dai, H. Protecting Semantic Trajectory Privacy for VANET with Reinforcement Learning. In Proceedings of the ICC 2019–2019 IEEE International Conference on Communications (ICC), Shanghai, China, 20–24 May 2019; pp. 1–5.
48. Benarous, L.; Kadri, B.; Boudjit, S. Alloyed Pseudonym Change Strategy for Location Privacy in VANETs. In Proceedings of the 2020 IEEE 17th Annual Consumer Communications & Networking Conference (CCNC), Las Vegas, NV, USA, 10–13 January 2020; pp. 1–6.
49. Chen, Y.; Lo, T.; Lee, C.; Pang, A. Efficient Pseudonym Changing Schemes for Location Privacy Protection in VANETs. In Proceedings of the 2013 International Conference on Connected Vehicles and Expo (ICCVE), Las Vegas, NV, USA, 2–6 December 2013; pp. 937–938.
50. Corser, G.; Fu, H.; Shu, T. Endpoint Protection Zone (EPZ): Protecting LBS User Location Privacy Against Deanonymisation and Collusion in Vehicular Networks. In Proceedings of the 2013 International Conference on Connected Vehicles and Expo (ICCVE), Las Vegas, NV, USA, 2–6 December 2013; pp. 369–374.
51. Corser, G.; Fu, H.; Shu, T.; D'Errico, P.; Ma, W.; Leng, S.; Zhu, Y. Privacy-by-decoy: Protecting location privacy against collusion and deanonymisation in vehicular location based services. In Proceedings of the 2014 IEEE Intelligent Vehicles Symposium Proceedings, Dearborn, MI, USA, 8–11 June 2014; pp. 1030–1036.
52. Gerlach, M.; Felix, G. Privacy in VANETs using Changing Pseudonyms—Ideal and Real. In Proceedings of the IEEE 65th Vehicular Technology Conference-VTC2007-Spring, Dublin, Ireland, 22–25 April 2007; pp. 2521–2525.
53. Freudiger, J.; Raya, M.; Félegyházi, M.; Papadimitratos, P.; Hubaux, J.-P. Mix-Zones for Location Privacy in Vehicular Networks. In Proceedings of the Association for Computing Machinery (ACM) Workshop on Wireless Networking for Intelligent Transportation Systems (WiN-ITS), Vancouver, BC, Canada, 14–17 August 2007; Volume 51, pp. 1–7.
54. Carianha, A.M.; Barreto, L.P.; Lima, G. Improving Location Privacy in Mix-Zones for VANETs. In Proceedings of the 30th IEEE International Performance Computing and Communications Conference, Orlando, FL, USA, 17–19 November 2011; pp. 1–6.
55. Ying, B.; Makrakis, D.; Mouftah, H.T. Dynamic mix-zone for location privacy in vehicular networks. *IEEE Commun. Lett.* **2013**, *17*, 1524–1527. [CrossRef]
56. Ying, B. Pseudonym Changes Scheme based on Candidate- Location-List in Vehicular Networks. In Proceedings of the 2015 IEEE International Conference on Communications (ICC), London, UK, 8–12 June 2015; pp. 7292–7297.
57. Boualouache, A.; Moussaoui, S. S2si: A practical pseudonym changing strategy for location privacy in vanets. In Proceedings of the 2014 International Conference on Advanced Networking Distributed Systems and Applications, Bejaia, Algeria, 17–19 June 2014; pp. 70–75.
58. Ying, B.; Makrakis, D.; Hou, Z. Motivation for protecting selfish vehicles' location privacy in vehicular networks. *IEEE Trans. Veh. Technol.* **2015**, *64*, 5631–5641. [CrossRef]
59. Bidi, Y.; Makrakis, D. Reputation-based Pseudonym Change for Location Privacy in Vehicular Networks. In Proceedings of the 2015 IEEE International Conference on Communications (ICC), London, UK, 8–12 June 2015; pp. 7041–7046.

60. Boualouache, A.; Senouci, S.; Moussaoui, S. Towards an Effi cient Pseudonym Management and Changing Scheme for Vehicular Ad-Hoc Networks. In Proceedings of the 2016 IEEE Global Communications Conference (GLOBECOM), Washington, DC, USA, 4–8 December 2016; pp. 1–7.
61. Zuberi, R.S.; Ahmad, S.N. Secure Mix-Zones for Privacy Protection of Road Network Location Based Services Users. *J. Comput. Netw. Commun.* **2016**, *2016*. [CrossRef]
62. Guo, N.; Ma, L.; Gao, T. A Location Privacy-Preserving Scheme for VANETs Based on Virtual Mix Zone. 2017; Volume 3, pp. 1–8.
63. Memon, I.; Mirza, H.T.; Arain, Q.A.; Memon, H. Multiple mix zones de-correlation trajectory privacy model for road network. *Telecommun. Syst.* **2019**, *70*, 557–582. [CrossRef]
64. Kalaiarasy, C.; Sreenath, N.; Amuthan, A. An effective variant ring signature-based pseudonym changing mechanism for privacy preservation in mixed zones of vehicular networks. *J. Ambient Intell. Humaniz. Comput.* **2020**, *11*, 1669–1681. [CrossRef]
65. Wahid, A.; Yasmeen, H.; Shah, M.A.; Alam, M. Holistic approach for coupling privacy with safety in VANETs. *Comput. Netw.* **2019**, *148*, 214–230. [CrossRef]
66. Sajjad, H.; Kanwal, T.; Anjum, A.; Malik, R. An efficient privacy preserving protocol for. *Comput. Secur.* **2019**, *86*, 358–371. [CrossRef]
67. Schoch, E.; Kargl, F.; Leinmüller, T.; Schlott, S.; Papadimitratos, P. Impact of pseudonym changes on geographic routing in vanets. In *European Workshop on Security in Ad-Hoc and Sensor Networks*; Springer: Berlin/Heidelberg, Germany, 2006; pp. 43–57.

sustainability

Case Report

Sustainable Urban Development for Older Australians: Understanding the Formation of Naturally Occurring Retirement Communities in the Greater Brisbane Region

Jiaxuan E [1], Bo Xia [1,*], Laurie Buys [2] and Tan Yigitcanlar [1]

[1] Faculty of Engineering, School of Architecture and Built Environment, Queensland University of Technology, 2 George Street, Brisbane, QLD 4001, Australia; jiaxuan.e@hdr.qut.edu.au (J.E.); tan.yigitcanlar@qut.edu.au (T.Y.)

[2] Faculty of Health and Behavioural Sciences, The University of Queensland, St Lucia, QLD 4072, Australia; l.buys@uq.edu.au

* Correspondence: paul.xia@qut.edu.au

Abstract: As most older Australians prefer to age-in-place, providing sustainable and age-friendly communities poses a significant challenge to urban policymakers. The naturally occurring retirement communities (NORCs) have organically emerged as a collaborative model of care to support older adults to age-in-place, but neither academic research nor government policies recognise this housing option for older Australians. This paper aims to analyse the distributions and temporal patterns of NORCs in the Greater Brisbane Region, Australia, to understand the formation and development of NORCs. The geovisualisation method was employed to identify the distribution changes of NORCs between 2006 and 2016. The Global Moran's I and Local Moran's I measures were utilised to analyse the spatial correlation and the clusters of NORCs. The results show that NORCs increased significantly from 2006 to 2016, and their distribution was mainly clustered or co-located along the coastline and Brisbane River areas. The evolvement of NORCs reflected the change of aggregation pattern of older population between 2006 and 2016. Understanding the distribution trend of NORCs informs government policy and decisions in addressing issues of service delivery and community cooperation, and eventually leads to sustainable urban development and successful ageing in place for older Australians.

Keywords: age-in-place; ageing communities; naturally occurring retirement communities; sustainable urban development; age-friendly cities; older population; Brisbane; Australia

Citation: E, J.; Xia, B.; Buys, L.; Yigitcanlar, T. Sustainable Urban Development for Older Australians: Understanding the Formation of Naturally Occurring Retirement Communities in the Greater Brisbane Region. *Sustainability* **2021**, *13*, 9853. https://doi.org/10.3390/su13179853

Academic Editor: Boris A. Portnov

Received: 3 August 2021
Accepted: 30 August 2021
Published: 2 September 2021

1. Introduction

Most of the developed countries are ageing societies, and catering for this large proportion of the older population is becoming an important part of urban policy [1,2]. For most of the older people, ageing-in-place (remaining in their own home in the community as long as possible) is the preferred housing option, which has a positive impact on many aspects of daily life, such as feeling safe and comfortable, maintaining physical and mental wellbeing and receiving social support [3,4]. However, ageing-in-place can be challenging for older people due to various issues, such as reduced physical capabilities, the rising cost of long-term care, increased risk of social isolation, and an unprepared community environment [5]. As a result, many older Australians do not have a range of options other than to relocate into retirement villages or aged care facilities [6].

A retirement village is an older people-based community that provides a variety of accommodation, services and facilities to meet their unique requirements [7]. It is an institution and needs rules, regulations, programmes and staff to govern its residents' daily life [8]. An aged care is the health institution that provides services to meet the needs of help with day-to-day tasks or health care for older people in Australia. In some cases, the best way to receive help and support can be by living in an aged care home (also known as nursing home) either on a permanent basis or for a short stay. However, although both of the options above provide protection for older people with respect to health and housing, more than 90% of older Australians wish to age-in-place and perceive the benefits of staying at home as a secure, familiar environment with the local community they know and love [9].

To support older people's ageing in place, urban planners and local governments need to provide an age-friendly urban environment, both in the community and city levels, so that older people's daily needs can be fully met [10]. The 'naturally occurring retirement communities' (NORCs), a concept originating from the US in the 1980s, has emerged as a collaborative living and care model to support older adults to age-in-place and avoid moving to more restrictive settings [11]. The concept of NORC was first proposed by Hunt and Gunter-Hunt [12], defining NORCs as neighbourhoods or building complexes that are not originally designed for older adults, in which 50% of the residents are 50 years or older and have aged in their homes. The criteria of NORCs varied over time. For instance, Hunt and Ross [13] identified a NORC with at least half of the residents aged 60 or older. Likewise, it is also defined as a housing community where at least 65% of residents are over 50 years old [14]. Moreover, the U.S. Congress Senate (2006) defined NORCs as communities in which at least 40% of the heads of households are older individuals.

Although the original purpose was not to help the older people to age in the community, NORCs have developed naturally and provide a way for older adults to live independently as long as possible. [15]. Because of the high density of the older population in the same geographical locations, communities can effectively serve older adults and support formal and informal collaborations between residents, communities, service providers and public sectors. Hunt and Gunter-Hunt [12] showed that neighbourhood services (which support older people's need and capabilities) were a major attraction to NORCs and increase the residents' satisfaction with such community-based housing.

In the U.S., most of the NORCs occur throughout the southern third of the country (Sun Belt) and some of the denser cities, such as New York and Boston, because of their ample public transportation links and amenities. Canada is considered to have an ageing environment similar to that in the U.S, and thus NORCs can be identified in some of the densely populated areas, such as the Ontario province [16]. In some other countries of the world, the term NORC is not applied, but communities sharing similar features are still recognised, which facilitate independent living for the older residents. AARP International reported on piloting various housing solutions for the ageing population in Europe, and Japan is widely regarded as a hospitable ageing country [17].

In response to the NORC demographic cluster, NORC supportive service programmes (NORC-SSPs) were developed to serve their senior residents by providing social and health care services tailored to their specific needs. The first NORC programme was established in 1986 at Penn South Houses, in New York City. Since then, the NORC programme model has been broadly replicated in more than 25 states across the U.S. Masotti, Fick [18] proposed the idea of healthy NORCs—that some NORC environments are healthier or more attractive than others for seniors, so that older people are intentionally moving together and the health benefits within healthy NORCs are higher where physical and social environments facilitate greater activity and promote feelings of well-being. The NORC is, therefore, considered a positive model for ageing-in-place to promote the health and mental well-being of older individuals [19,20].

Although NORCs are well recognised in the U.S., they are much less so in Australia. Currently, only two 'virtual retirement communities' (one in Sydney and another in Perth) claimed to be inspired by the NORC movement in the U.S., where online local networks are provided to support older people living independently in their own homes and with access to local services [20]. Meanwhile, though, this pattern of ageing is happening naturally in Australia, given that an increasing number of communities in Australia have more than 40% of residents over 65 years old. For example, 40.6% of the residents of Queensland's Bribie

Island are 65+ years old [21]. It is thus clear that NORCs are an existing but unnoticed social phenomenon in Australia. In other words, we have already witnessed an Australian version of NORCs, but neither academic research nor government policies recognise this housing option for older Australians. Currently, there is very limited, if any, knowledge about the formation and development of NORCs.

This paper, hence, aims to understand the formation and development of NORCs in the Greater Brisbane Region (a geographical area with a much higher ageing rate than the national average of 15%), mainly in terms of their spatial and temporal changes over a 10-year period, from 2006 to 2016. The study is innovative in its spatial analysis of Australian NORCs, enabling various stakeholders, especially policymakers, to identify the formation trend of NORCs, with the ultimate purpose of developing age-friendly communities and cities to support older Australians' ageing-in-place.

2. Materials and Methods

A spatial and temporal data analysis, including a spatial autocorrelation analysis, a cluster and outlier analysis and a hotspot/coldspot analysis, upon a geographic information system (GIS), were applied to identify the distribution pattern of older population and NORCs on the basis of geographical boundary and census characteristics in Australia. The census statistics of the residential areas of the households with different characteristics can reflect the geographical patterns that naturally form with the development of time.

2.1. Data

The Australian Bureau of Statistics (ABS) Census data of 2006, 2011 and 2016, which include the distribution of usual residents in the households by different ages, were used. Given that regions with a rapid population growth, such as the Greater Brisbane Region, are especially sensitive to population ageing, which will necessarily involve substantial increases in public expenditure on health and aged care, the Greater Brisbane Region was chosen as the case locality for the data analysis. In particular, this region has become the popular retirement destination for older Australians [21], with some areas such as the Bribie Island with the oldest median age (60.6 years) in the capital city region. The Greater Brisbane Region comprises eight local government areas and covers a total area of around 15,800 square kilometres, or 1% of Queensland's total area, but was home to 2.27 million people, almost half of Queensland's population, which is about 4.7 million [22].

2.2. Unit of Analysis

An appropriate unit of analysis is critical in geography-based analyses. Every five years, ABS counts every person and home in Australia, in the Census of Population and Housing. For the 2011 and 2016 Census, Statistical Area Level 1 (SA1) was designed for a detailed spatial analysis of the Census data. This research used these areas to examine how Census characteristics vary at a neighbourhood scale within larger areas [23]. For the 2006 Census, the Census Collection District (CD) was designed as the smallest unit for data collection and processing—including about 200 dwellings. CD was the basis of output for most data and served as the basic building block in the Australian Standard Geographical Classification (ASGC) [24].

It should be noted that although CD in 2006 and SA1 in 2011 and 2016 were the basic census units of the same level for the data analysis in this study, the number of census units for SA1 were higher than that of CD because of the increase of population and immigrants. Nevertheless, as the purpose of this study is to understand the formation trend of NORCs over time, which focused on the population changes and the distribution features of the older population (65+), the slight differences in the number and scale between SA1 and CD units do not affect the overall result.

2.3. Analysis

The definition of NORC varies in different places and different periods. In the place of origin, the US Federal Government, through Title IV of the Older Americans Act, recognised NORCs as 'communities in which at least 40% of the heads of households are older individuals' [25]. According to prior studies, 'resident', 'head of household', 'old adult' and 'house owners and renters' are the most frequently used concepts (and their datasets) in identifying NORCs [13,14,26]. The most widely cited definition of NORCs is 'communities in which at least 40% of the heads of households are older individuals'. In Australian statistical data, older people are defined as people aged 65 years or more [27]; however, the concept of 'head of household' was not applied in the 2006, 2011 and 2016 ABS Census data. Alternatively, this research employed the concept and dataset of all household members who usually reside in the private dwellings rather than the head of household to define NORCs. It also excluded older residents living in nursing homes or aged care facilities. Therefore, this study adopted the combined circumscription of NORC as the community with 40% or more members of households aged 65 years and older, which has excluded holiday visitors and persons who have moved to nursing homes.

Four approaches of spatial analyses were conducted to identify the formation and development of NORCs. These were: (i) geovisualisation, (ii) spatial autocorrelation (global Moran's I), (iii) cluster and outlier analysis (local Moran's I), and (iv) hotspot and coldspot analysis (Getis-Ord Gi *). These are among the frequently applied spatial analysis techniques in the age and healthcare studies [28]. The software of ESRI ArcMap Version 10.8.1 was applied for data analysis.

A geovisualisation was conducted to identify the distribution of NORCs in 2006, 2011 and 2016. When working with spatially referenced data, geovisualisation is helpful to recognise patterns across large geographical regions [29,30]. Choropleth maps, in this study, are used to display different classes of proportion of older people in each census unit and recognise the possible NORCs that meet the criteria.

Global Moran's I was used to measure the spatial autocorrelation of NORCs based on both locations and the proportion of older population simultaneously. Given a set of features such as location, areas and population, it evaluated whether the distribution pattern of NORCs was clustered, dispersed or random. Global Moran's I supports geovisualisation by statistically distinguishing the level of the spatial structure, which qualitatively improves the reliability of the interpreted geovisualised information [31]. In this research, global Moran's I statistic for the proportion of household members in the greater Brisbane region aged 65 years and over in each census unit has value ranges from −1 to 1. A negative value reveals that farther census units are more related than closer ones, a positive value of I reveals that the closer census units are more connected than farther ones, and 0 informs no spatial autocorrelation between them. [32]. A Z-score and p-value were used for the statistical significance test to verify the result: when z-score < -1.65 or $>+1.65$ associated with p-value < 0.10, the confidence level is 90%; when z-score < -1.96 or $>+1.96$ associated with p-value < 0.05, the confidence level is 95%; when z-score < -2.58 or $>+2.58$ associated with p-value < 0.01, the confidence level is 99%.

Local Moran's I, 'a local spatial autocorrelation statistic' [33], was employed to identify local clusters or outliers of NORCs to understand their contribution to the 'global' cluster statistic. It assesses each feature of census units within the context of neighbouring features and compares the local circumstance to the overall situation. Local Moran's I could be utilised to detect the clusters of census units with 40% or more of the members of a household aged 65 years and over among nearby census units in the greater Brisbane region.

Getis-Ord Gi * (Gi *) was used to verify a statistically significant spatial cluster of high values (hotspots) or low values (coldspots) of proportion of older household members within a distance [34]. A Gi * analysis was employed to reveal whether a high or low proportion of older household members is concentrated over the greater Brisbane region at the different statistically significant levels, which means that to be a statistically significant hotspot, a census unit will have a high value and be surrounded by other census units with high values as well. The visualisation of hotspots and coldspots of older population is the enhanced tool of cluster statistic in Local Moran's I, where hotspot identification not only describes the state of aggregation of older population at the moment but also predicts a trend of cluster for some neighbourhoods with low proportions of older population located in the contexts of high values. Meanwhile, coldspot areas are indicated as the aggregation of suburbs with a low proportion of older population, with the neighbours co-locating with low values as well. This research focused on finding hotspots of highly proportioned older household members including NORCs.

3. Results

As shown in Figure 1, the proportion of older Australians increased obviously across the census units in the greater Brisbane region from 2006 to 2011 to 2016. The total number of household members aged 65 and over in the greater Brisbane region were 185,490, 215,149 and 267,281 in 2006, 2011 and 2016, respectively. Meanwhile, the proportion of older household members (aged 65+) increased from 10.68% in 2006 to 11.15% in 2011 to 12.76% in 2016 in the greater Brisbane region. This indicated a significant growth rate (44.1%) of older household members (65+), much higher than the growth rate of the total household members (20.6%) in the greater Brisbane region from 2006 to 2016. In addition, in 2006, 57.9% of the census units had less than 10% of older household members (those with 65+ years old), while in 2016 it dropped to 48.9%. By contrast, in 2006, only 7.6% of the census units had an older population of more than 20%, and this number increased to 10.7% in 2016.

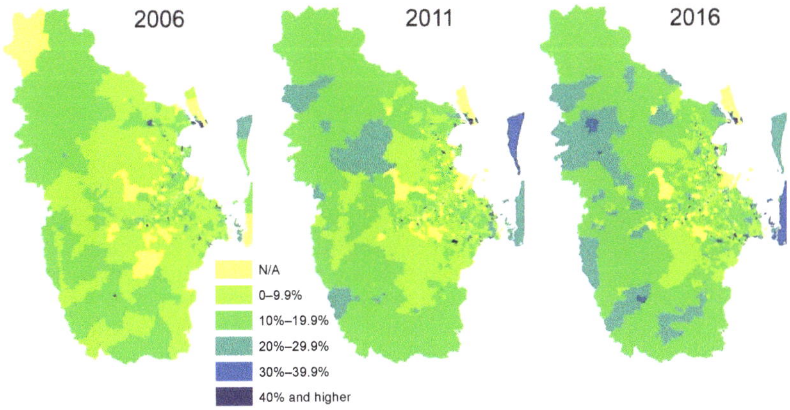

Figure 1. Map of census units with the percentage of older household members (age 65+) in the Greater Brisbane Region in 2006, 2011 and 2016. The colours are classified by N/A, 0–9.9%, 10–19.9%, 20–29.9%, 30–39.9%, 40% and higher.

Figure 2 shows the locations of NORCs in 2006, 2011 and 2016. In 2006, 25 (0.8%) out of 3236 census units were identified as NORCs (with the proportion of 65+ years old more than 40%) in the greater Brisbane region. These numbers increased to 65 (1.3%) out of 5164 census units in 2011 and 92 (1.7%) out of 5373 census units in 2016, which indicated a rapid growth of NORCs over the 10-year span. In addition, the older household members (65+) living in NORCs accounted for 3.4% of the total older household members in 2006, and it increased to 6.1% in 2011 and 7.2% in 2016. According to Figure 2, NORCs were distributed mostly along the coastline and Brisbane River. Especially on the Bribie Island, there already existed four NORCs accommodating 886 older household members (65+) in 2006, and this number increased rapidly to 10 NORCs with 1961 older household members (65+) in 2011 and 15 NORCs with 2963 older household members (65+) in 2011. Currently, the Bribie Island has the oldest median ages (60.6 years) in Queensland.

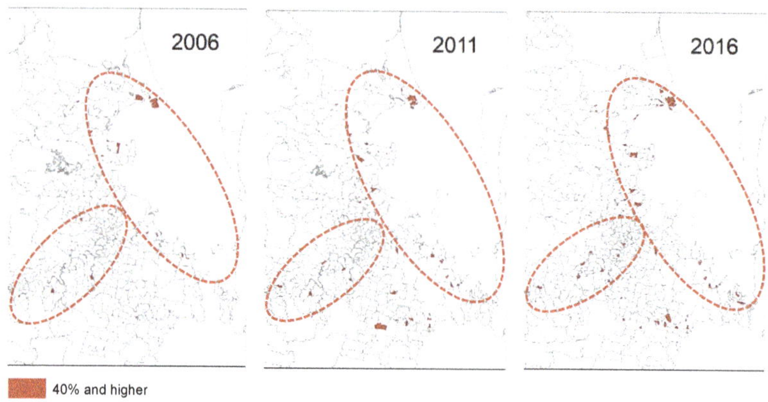

Figure 2. Map of naturally occurring retirement communities (NORCs) with suburb boundaries in the Greater Brisbane Region in 2006, 2011 and 2016. The distributions along the Brisbane River and coastal areas are marked out. Bribie Island has the greatest number of NORCs and oldest median ages.

As indicated by the NORCs' distribution pattern, Global Moran's I was 0.352423 (z = 33.740511) in 2006, 0.252765 (z = 30.630196) in 2011 and 0.282963 (z = 35.100097) in 2016. For the Global Moran's I statistic, the null hypothesis states that the NORCs are randomly distributed in the study area. This result indicated that the distributions of NORCs were spatially autocorrelated or considered as not randomly distributed (rejecting the null hypothesis), which means that the NORCs or census units with a high proportion of older household members (65+) tended to get close to similar ones. Similarly, census units with low proportions of older household members were close to similar ones as well. Given the z-score of 33.74, 30.63 and 35.10, there was less than 1% likelihood that this clustered pattern could be the result of random chance.

Local Moran's I was used to detect the clusters, outliers and hotspots of NORCs. Figure 3 shows that the high-high clusters of census units (i.e., the census units with high proportions of older household members were co-located or clustered together, with a red colour in the red dot-line circle in Figure 3) were located in some specific areas, such as the Bribie Island, Cleveland and Victoria Point, most of which were distributed along the coastal line and expanded their cluster areas rapidly from 2006 to 2011 and 2016. By contrast, the low-low cluster (where census units with a low proportion of older household members (65+) were co-located together) were mainly located in the mountainous areas. Especially those blue areas in the black dot-line circle were disappearing from 2006 to 2011 and 2016 primarily due to the increase of the older population. For high-low and low-high outliers, no obvious patterns exist in this study.

Figure 4 shows the optimised hotspot and coldspot living areas of older household members in 2006, 2011 and 2016. The optimised hotspot map shows that the statistically significant hotspot suburbs (with high proportions of older household members) are distributed along the coastline, and these areas were expanding rapidly from 2006 to 2016. The total number of older household members (65+) living in those optimised hotspot areas were 69,326, 60,341 and 69,595 in 2006, 2011 and 2016, respectively, accounting for 37.4%, 28.0% and 26.0% of the total older household members (65+). Meanwhile, some of the inland regions became hotspots as well, although the number of older household members (65+) was small. Take the Laidley, for example. Only 493 (18.9% of its total household members), 516 (16.1% of its total household members) and 619 (18.9% of its Total household members) older household members (65+) were living there in 2006, 2011 and 2016, respectively. By contrast, the coldspots (with low proportions of older household members) in the inner-city areas and urban suburbs were increasing along with the decrease of the hotspots in these areas during the 10-year period.

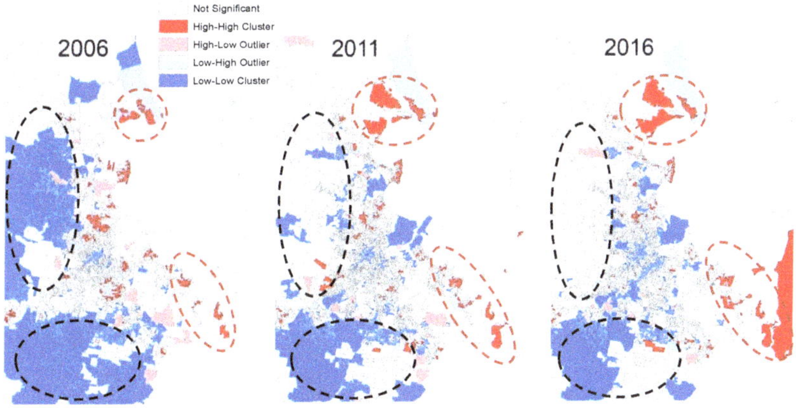

Figure 3. Cluster and outlier analysis of census units with a proportion of older household members (age 65+) in the Greater Brisbane Region in 2006, 2011 and 2016. Different colours indicate different meanings, including not significant, high-high cluster, low-low cluster, high-low outlier and low-high outlier. Significant high-high clusters and low-low clusters are marked out.

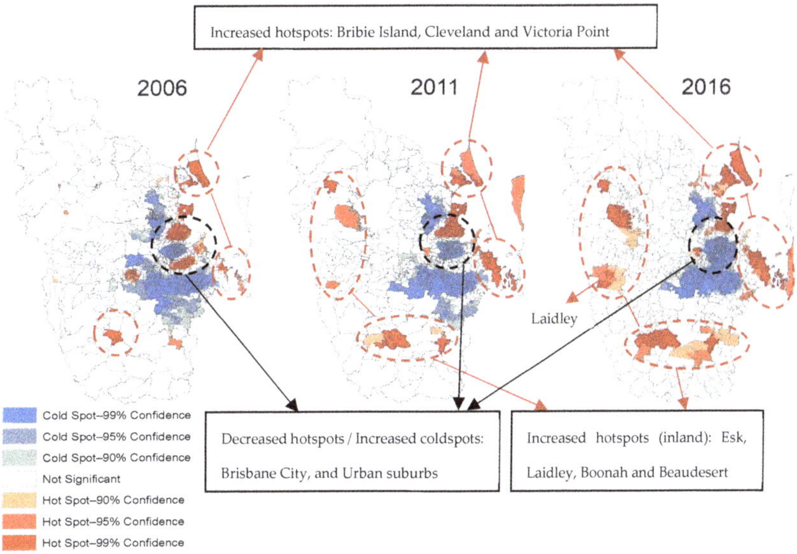

Figure 4. Optimised hotspot analysis of census units with the proportion of older household members (age 65+) in the Greater Brisbane Region in 2006, 2011 and 2016. Different colours indicate 90%, 95% and 99% confidence of hotspots and coldspots. The red circles mark out the increased hotspot areas from 2006 to 2011 to 2016, and the black circles indicate the increased coldspot and decreased hotspot areas from 2006 to 2011 to 2016.

4. Discussion

Our analysis results reveal that NORCs have developed rapidly over the 10-year period in the greater Brisbane region. The cluster of older population may generate both positive and negative impacts on active ageing. Firstly, in NORCs, service providers and multilevel collaborations may serve the older people more efficiently because of the spatially proximate locations for NORC residents living within a reachable range. It is a great opportunity for the government to provide home and community care in NORCs where the older people can enjoy professional care with comparatively low cost due to the economies of scale.

According to the Queensland Department of Communities, Housing and Digital Economy [35], the distribution of retirement villages in the greater Brisbane area is highly consistent with the distribution of older people, where the areas with high density of older residents have one or more retirement villages located around them. This implies that the retirement village developers have realised the advantage of the economic scales brought by the concentration of the older population. Similarly, one or more nursing homes can also be found in these popular areas with older people. NORC supportive service programmes as the solution of ageing-in-place provide the older people with another choice of ageing, which is different from the retirement village or nursing home, and the NORC participants can enjoy some of the services that the institutional care system could not provide, such as exercise and dance classes, trips and cultural events [36]. For the government, NORC programmes play a positive role in reducing the pressure of resource scarcity in the healthcare system [37]. However, it must be clarified that NORCs are different from NORC Service Support Programmes where multilevel collaborations, services and supports are provided to those NORCs with a high proportion of older people. Governments should provide NORC service support programmes to existing NORCs, with the ultimate purpose of providing an age-friendly urban environment to facilitate older Australians' ageing-in-place.

Nevertheless, neither academic research nor government policies recognise this housing option for older Australians. Although the Advisory Taskforce on Residential Transition for Ageing Queenslanders [38] recommended that the Queensland Government and the local governments provide incentives for the development of NORCs as one of the diverse age-friendly housing solutions, no relevant initiatives are available to date, most likely as NORCs do not fit the definition of a retirement community. However, NORCs present a huge opportunity for governments because supporting ageing-in-place can ultimately result in significant public expenditure savings, particularly given the fiscal demands of institutional aged care and healthcare.

In addition, a high density of older population may lead to diverse social networks in later life, which are related to more positive outcomes, including better cognitive functioning [39], lower levels of depressive symptomatology [40], better outcomes in serious physical illnesses [41], reduced likelihood of loneliness and anxiety and increased likelihood of happiness [42]. Studies have found that in maintaining a positive attitude and preventing depression, the existence of a network of friends is as important as, or even more important than, the existence of family members [43]. The Robert Woods Johnson Foundation also emphasises that communities that provide social services and material support have a positive impact on the health of their residents [44]. Moreover, Cornwell, Laumann [45] found that an older adult's larger socialisation network of friends and the frequency of that socialisation positively impact their longevity. However, it should also be noted that while ageing-in-place and ageing-in-community may have the advantages of familiarity and maintaining one's connections, it may also set older adults up for social isolation, particularly if they have limited mobility or access to other people. Social isolation and loneliness have been shown to have negative effects on health and well-being [46].

The temporal distribution of NORCs shows that the formation of NORCs closely aligns with spatial distributions of older population, which could be caused by the in-migration (move into a certain area for permanent living inside or outside the state), out-migration (move out of a certain area to a more suitable place for permanent living) and ageing of existing populations [47]. The increase of high-high clusters and hotspots along the coastal areas, where most of the NORCs are located, might be very likely due to the in-migration of older household members (65+), given that many Australians would like to live in an environment with seaside scenery, convenient seacoast facilities for activities and extending social networks. These people are known colloquially as 'sea-changers' [48].

Likewise, some of the rural residential suburbs and their neighbouring suburbs, such as Esk, Laidley, Boonah and Beaudesert, became ageing hotspots after a 10-year change of demographic mainly because of the out-migration of youth and tree-change in-migration of older people. The proportion of local young generations leaving their rural homelands was high—more than half of the cohort population in some cases [49]. Meanwhile, tree-change as a relatively nascent migration trend of urban residents moving towards inland regional areas of Australia becomes another option for some older people [50]. Clearly, NORCs in rural areas have the great potential to accommodate 'tree-changers', even though current facilities and services can be sparsely located/provided in these areas.

Compared with the increase of the above-mentioned hotspots for older people, the increase of coldspots in the inner city areas are mainly due to young people moving in for job opportunities, along with older people choosing out-migration for retirement [51]. On the contrary, the decrease of low-low clusters in inland areas, caused by the increasing proportion of older population, is mainly due to the natural ageing and younger generations moving out for job opportunities in capital cities or regional centres. Currently, how to better support older people who prefer to remain at home and independent in inland areas can be a challenge for the Australian society.

Understanding the distribution of hot and coldspots associated with NORCs may be useful in guiding governments in resource allocation and optimising investment to promote financial efficiency. As NORCs are clustered in the same hotspot areas, collaborations between industry and local governments across different NORCs, such as sharing the existing infrastructure and services (e.g., transportation and healthcare), will have the great potential to maximise their value to older people due to the increased economic scale. Since NORCs are based on the existing resources of the community, they may become a supplementary part of the local service delivery system. NORCs have become an opportunity to develop and/or strengthen community partnerships to improve how the community responds to the changing needs of residents as they age-in-place [11].

The growth of NORCs in the greater Brisbane region aligns with the phenomenon of population ageing not only in Australia but also in all developed countries. Today, more than 530 cities and communities from 37 countries are participating in the WHO's Global Network of Age-Friendly Cities and Communities; however, government structures are often isolated and lack flexibility to formulate and implement effective age-friendly programmes covering multiple disciplines. Working with local NGOs, citizen groups, service providers and other private sector entities, NORC supportive service programmes will enable local governments to integrate the vision of the age-friendly programme with the interests of older people, coordinate stakeholder priorities and leverage existing resources. With the identification of NORCs and their formation patterns, researchers, communities and local governments can collaborate to understand the local history, economic activities and experience of older residents in those NORCs to enrich policy planning. As the most popular interstate destination for people aged 65+ years (with 31% of interstate arrivals in 2015) [52], Queensland has launched the programme Queensland: An Age-friendly Community since 2016, aiming to ensure older people are free from age-related barriers that prevent community participation. The findings of this study provide a valuable data support this programme.

5. Conclusions

To support older Australians' preference for ageing-in-place, it is essential to provide a smart, sustainable and age-friendly urban environment [53,54].This study aimed to understand the distribution and formation of NORCs with temporal changes for the first time in Australia. As emerging age-friendly communisis, NORCs evolved rapidly in the greater Brisbane region, accommodating 7.2% of the older members of households, although accounting only for 1.7% (92) of the total census units in 2016. In addition, NORCs were not developed randomly, but are mainly clustered or co-located along the Brisbane River and the coastline areas, which attract an increasing number of older people moving into these areas for retirement living.

This study provides both knowledge and practical implications for NORCs and urban development. First, the spatial analytic approach made the geographical reference data more accessible and comprehensive. It visualised and described the distribution patterns, temporal trends and geographic locations of NORCs at the city level. Second, a multilevel cooperation among local governments, industries and agencies can be developed to provide better services and urban environment for the older population in these areas. Finally, this research offers an approach to data visualisation to interpret the temporal and spatial distribution of NORCs.

This study has some limitations. First, although the geographical factors affect the clusters or outliers of NORCs, the development of NORCs may be driven by multiple factors, such as built environment, natural scenery, walkability, accessibility and health and social services, the information of which was not available in the Census data. The second limitation in the spatial analysis is the mutative unit, which was modified from CD in 2006 to SA1 in 2011 and 2016. Different units of analysis may change the findings, even though the changing may follow the life pattern of the local population. Thirdly, the distribution of nursing homes and aged care facilities may likely affect the growth and development of NORCs, whose pattern of influence needs to be explored in future studies. Fourthly, among the census units with a high proportion of older household members (65+), some of them (accounting for less than 0.5% of the total census units, and many located in national parks or remote usual areas) have a small number of total household members (less than 200). As a result, the proportion of senior residents in these census units can be easily affected by the relocations of older people, even in a very small number. Therefore, whether these census units can be identified as NORCs needs to be further investigated according to the actual living environment and neighbourhood relationship.

The occurrence mechanisms of NORCs requires further investigation in the future. An ageing policy and related services regarding NORCs may be promoted to provide better ageing-in-place for older Australians.

Author Contributions: Conceptualisation, B.X.; methodology, J.E.; formal analysis, J.E.; writing—original draft preparation, B.X. and J.E.; writing—review and editing, L.B. and T.Y. All authors have read and agreed to the published version of the manuscript.

Funding: This research received no external funding.

Institutional Review Board Statement: Ethical review and approval were waived for this study, due to the data set this search using was from the Australian Bureau of Statistics Census Data which is the open resources database can be visited publicly.

Informed Consent Statement: Patient consent was waived due to no specific patient information was used in this study because the data set this search using was from the Australian Bureau of Statistics Census Data which is the open resources database can be visited publicly.

Data Availability Statement: Data are available in a publicly accessible repository that does not issue DOIs. Publicly available datasets were analysed in this study. These data can be found here: [www.abs.gov.au] accessed on 5 July 2021, Census TableBuilder, 2006, 2011 and 2016 Census-Counting Persons, Place of Usual Residence.

Conflicts of Interest: The authors declare no conflict of interest.

References

1. Carbonaro, G.; Leanza, E.; McCann, P.; Medda, F. Demographic Decline, Population Aging, and Modern Financial Approaches to Urban Policy. *Int. Reg. Sci. Rev.* **2018**, *41*, 210–232. [CrossRef]
2. Chan, S.; Ellen, I.G. Housing for an Aging Population. *Hous. Policy Debate* **2017**, *27*, 167–192. [CrossRef]
3. Smith, A.E. Ageing in urban neighbourhoods. In *Place Attachment and Social Exclusion*; Policy Press: Bristol, UK, 2009.
4. Buys, L.; Miller, E. Residential satisfaction in inner urban higher-density Brisbane, Australia: Role of dwelling design, neighbourhood and neighbours. *J. Environ. Plan. Manag.* **2012**, *55*, 319–338. [CrossRef]

5. Hu, X.; Xia, B.; Chong, H.-Y.; Skitmore, M.; Buys, L. Improving the sustainable retirement village framework: From theory to practice. *J. Clean. Prod.* **2020**, *248*, 119290. [CrossRef]
6. Stimson, R.J.; McCrea, R. A push–pull framework for modelling the relocation of retirees to a retirement village: The Australian experience. *Environ. Plan. A* **2004**, *36*, 1451–1470. [CrossRef]
7. Hu, X.; Xia, B.; Skitmore, M.; Buys, L.; Zuo, J. Retirement villages in Australia: A literature review. *Pac. Rim Prop. Res. J.* **2017**, *23*, 101–122. [CrossRef]
8. McDonald, J. Community participation in an Australian retirement village. *Aust. J. Ageing* **1996**, *15*, 167–171. [CrossRef]
9. Australian Institute of Health and Welfare. *The Desire to Age in Place among Older Australians*; Bulletin no. 114. Cat. no. AUS 169; AIHW: Canberra, Australia, 2013.
10. Van Hoof, J.; Kazak, J.K.; Perek-Białas, J.M.; Peek, S. The challenges of urban ageing: Making cities age-friendly in Europe. *Int J. Environ. Res. Public Health* **2018**, *15*, 2473. [CrossRef]
11. Ivery, J.M.; Akstein-Kahan, D. The Naturally Occurring Retirement Community (NORC) Initiative in Georgia: Developing and Managing Collaborative Partnerships to Support Older Adults. *Adm. Soc. Work* **2010**, *34*, 329–343. [CrossRef]
12. Hunt, M.E.; Gunter-Hunt, G. Naturally Occurring Retirement Communities. *J. Hous. Elderly* **1986**, *3*, 3–22. [CrossRef]
13. Hunt, M.E.; Ross, L.E. Naturally occurring retirement communities: A multiattribute examination of desirability factors. *Gerontologist* **1990**, *30*, 667–674. [CrossRef]
14. Lyons, B.P.; Magai, C. Reducing Health Risks and Psychological Distress Among Older Black Residents of Naturally Occurring Retirement Communities. *J. Gerontol. Soc. Work* **2001**, *35*, 53–69. [CrossRef]
15. Bradley, H.W.; Martin, J. *Planning Aging—Supportive Communities*; American Planning Association: Chicago, IL, USA, 1 June 2015; p. 579.
16. Donnelly, C. Beyond Long-Term Care: The Benefits of Seniors' Communities that Evolve on Their Own the Conversation: The Conversation. 2020. Available online: https://theconversation.com/beyond-long-term-care-the-benefits-of-seniors-communities-that-evolve-on-their-own-144269 (accessed on 15 May 2021).
17. Analytics AaF. *The Aging Readiness & Competitiveness Report—Community Social Infrastructure*; AARP International: Washington, DC, USA, 2018.
18. Masotti, P.J.; Fick, R.; Johnson-Masotti, A.; MacLeod, S. Healthy Naturally Occurring Retirement Communities: A Low-Cost Approach to Facilitating Healthy Aging. *Am. J. Public Health* **2006**, *96*, 1164–1170. [CrossRef] [PubMed]
19. Guo, K.L.; Castillo, R.J. The US long term care system: Development and expansion of naturally occurring retirement communities as an innovative model for aging in place. *Ageing Int.* **2012**, *37*, 210–227. [CrossRef]
20. The Weekly Source. Important: NORCs Arrive in Australia with WA Govt Sponsored over-55s Community Club. 2018. Available online: https://www.theweeklysource.com.au/important-norcs-arrive-in-australia-with-wa-govt-sponsored-over-55s-community-club/ (accessed on 5 April 2021).
21. Australia Bureau of Statistics. Twenty Years of Population Change. 2020. Available online: https://www.abs.gov.au/articles/twenty-years-population-change (accessed on 1 March 2021).
22. Statistics ABo. Regional Population by Age and Sex, Australia, 2017. Cat. No. 3235.0. Available online: https://www.abs.gov.au/AUSSTATS/abs@.nsf/Lookup/3235.0Explanatory%20Notes12017?OpenDocument (accessed on 30 May 2021).
23. Statistics ABo. 1270.0.55.001—Australian Statistical Geography Standard (ASGS): Volume 1—Main Structure and Greater Capital City Statistical Areas July 2016. Available online: https://www.abs.gov.au/ausstats/abs@.nsf/Lookup/by%20Subject/1270.0.55.001~{}July%202016~{}Main%20Features~{}Statistical%20Area%20Level%201%20(SA1)~{}10013 (accessed on 4 March 2021).
24. Statistics ABo. 2901.0—Census Dictionary, 2006 (Reissue). Available online: https://www.abs.gov.au/AUSSTATS/abs@.nsf/2f762f95845417aeca25706c00834efa/413876f3bae9cc70ca25720a000c428b!OpenDocument (accessed on 4 March 2021).
25. Senate USC. Older Americans Act of 1965. In *America SaHoRotUSo*; Amended Through, P.L., Ed.; United States Congress: Washington, DC, USA; p. 109.
26. Lanspery, S.C.; Callahan, J.J.; Schwartz, H. *Naturally Occurring Retirement Communities: A Report Prepared for the Pew Charitable Trusts*; Brandeis University: Waltham, MA, USA, 1994.
27. Statistics ABo. Disability, Ageing and Carers, Australia: Summary of Findings. 2019. Available online: https://www.abs.gov.au/statistics/health/disability/disability-ageing-and-carers-australia-summary-findings/2018#older-people (accessed on 4 March 2021).
28. Higgs, G. A literature review of the use of GIS-based measures of access to health care services. *Health Serv. Outcomes Res. Methodol.* **2004**, *5*, 119–139. [CrossRef]
29. Cutter, S.L.; Boruff, B.J.; Shirley, W.L. Social vulnerability to environmental hazards. *Soc. Sci. Q.* **2003**, *84*, 242–261. [CrossRef]
30. Goldman, B.A. *The Truth about Where You Live: An Atlas for Action on Toxins and Mortality*; Three Rivers Press: New York City, NY, USA, 1991.
31. Rivera-Hernandez, M.; Yamashita, T.; Kinney, J.M. Identifying naturally occurring retirement communities: A spatial analysis. *J. Gerontol. Ser. B Psychol. Sci. Soc. Sci.* **2015**, *70*, 619–627. [CrossRef] [PubMed]
32. Mitchell, A. *The ESRI Guide to GIS Analysis*; Esri Press: Redlands, CA, USA, 2005; Volume 2.
33. Anselin, L. Local indicators of spatial association—LISA. *Geogr. Anal.* **1995**, *27*, 93–115. [CrossRef]
34. Esri. How Hot Spot Analysis (Getis-Ord Gi*) Works. Available online: https://pro.arcgis.com/en/pro-app/latest/tool-reference/spatial-statistics/h-how-hot-spot-analysis-getis-ord-gi-spatial-stati.htm (accessed on 21 March 2021).

35. 2020–2021 Retirement Villages Register. 2021. Available online: https://www.data.qld.gov.au/dataset/retirement-village-schemes-registered-with-department-of-communities-housing-digital-economy (accessed on 20 May 2021).
36. Ormond, B.A.; Black, K.J.; Tilly, J.; Thomas, S. *Supportive Services Programs in Naturally Occurring Retirement Communities*; Assistant Secretary for Planning and Evaluation: Washington, DC, USA, 2004.
37. Kloseck, M.; Crilly, R.G.; Gutman, G.M. Naturally Occurring Retirement Communities: Untapped Resources to Enable Optimal Aging at Home. *J. Hous. Elder.* **2010**, *24*, 392–412. [CrossRef]
38. Buys, L. Residential Transition for Older Queenslanders. *Technology* **2016**, *54*, 3–8.
39. Ellwardt, L.; Van Tilburg, T.G.; Aartsen, M.J. The mix matters: Complex personal networks relate to higher cognitive functioning in old age. *Soc. Sci. Med.* **2015**, *125*, 107–115. [CrossRef] [PubMed]
40. Fiori, K.L.; Antonucci, T.C.; Cortina, K.S. Social network typologies and mental health among older adults. *J. Gerontol. Ser. B Psychol. Sci. Soc. Sci.* **2006**, *61*, P25–P32. [CrossRef]
41. Cohen, S.; Janicki-Deverts, D. Can we improve our physical health by altering our social networks? *Perspect. Psychol. Sci.* **2009**, *4*, 375–378. [CrossRef] [PubMed]
42. Litwin, H.; Shiovitz-Ezra, S. Social network type and subjective well-being in a national sample of older Americans. *Gerontologist* **2011**, *51*, 379–388. [CrossRef] [PubMed]
43. Fiori, K.L.; Antonucci, T.C.; Akiyama, H. Profiles of social relations among older adults: A cross-cultural approach. *Ageing Soc.* **2008**, *28*, 203–231. [CrossRef]
44. Cubbin, C.; Egerter, S.; Braveman, P.; Pedregon, V. *Where We Live Matters for our Health: Neighborhoods and Health*; Robert Wood Johnson Foundation: Princeton, NJ, USA, 2008.
45. Cornwell, B.; Laumann, E.O.; Schumm, L.P. The social connectedness of older adults: A national profile. *Am. Sociol. Rev.* **2008**, *73*, 185–203. [CrossRef]
46. de Jong Gierveld, J.; Van Tilburg, T.; Dykstra, P.A. Loneliness and social isolation. In *Cambridge Handbook of Personal Relationships*; Cambridge University Press: Cambridge, UK, 2006; pp. 485–500.
47. Davies, A.; James, A. *Geographies of Ageing: Social Processes and the Spatial Unevenness of Population Ageing*; Ashgate Publishing, Ltd.: Farnham, UK, 2011.
48. Bartlett, H.; Carroll, M. Ageing in place down under. *Glob. Ageing Issues Act.* **2011**, *7*, 25–34.
49. Argent, N.; Walmsley, J. Rural youth migration trends in Australia: An overview of recent trends and two inland case studies. *Geogr. Res.* **2008**, *46*, 139–152. [CrossRef]
50. Buckle, C.; Drozdzewski, D. Urban perceptions of tree-change migration. *Rural. Soc.* **2018**, *27*, 192–207. [CrossRef]
51. Mohnen, P. *The Impact of the Retirement Slowdown on the Us Youth Labor Market*; Mimeo: Memphis, TN, USA, 2019.
52. 3101.0-Australian Demographic Statistics, March 2016. Available online: https://www.abs.gov.au/AUSSTATS/abs@.nsf/featurearticlesbyCatalogue/81A7C0DAEADBC3C6CA257A850013DEFE?OpenDocument (accessed on 2 April 2021).
53. Tan, Y. *Sustainable Urban and Regional Infrastructure Development: Technologies, Applications and Management*; IGI Global: Hershey, PA, USA, 2010; 416p.
54. Yigitcanlar, T.; Han, H.; Kamruzzaman, M.; Ioppolo, G.; Sabatini-Marques, J. The making of smart cities: Are Songdo, Masdar, Amsterdam, San Francisco and Brisbane the best we could build? *Land Use Policy* **2019**, *88*, 104187. [CrossRef]

sustainability MDPI

Case Report

Understanding and Acceptance of Smart City Policies: Practitioners' Perspectives on the Malaysian Smart City Framework

Seng Boon Lim [1], Jalaluddin Abdul Malek [1], Md Farabi Yussoff Md Yussoff [2] and Tan Yigitcanlar [3,4,*]

1 Center for Research in Development, Social and Environment (SEEDS), Faculty of Social Sciences and Humanities, University Kebangsaan Malaysia, Bangi 43600, Malaysia; lims@ukm.edu.my (S.B.L.); jbam@ukm.edu.my (J.A.M.)
2 Federal Government Administrative Centre, Federal Department of Town and Country Planning, Ministry of Housing and Local Government, Putrajaya 62675, Malaysia; farabi@planmalaysia.gov.my
3 School of Architecture and Built Environment, Queensland University of Technology, Brisbane, QLD 4000, Australia
4 School of Technology, Federal University of Santa Catarina, Campus Universitario, Florianópolis 88040-900, SC, Brazil
* Correspondence: tan.yigitcanlar@qut.edu.au or tan.yigitcanlar@ufsc.br; Tel.: +61-7-3138-2418

Abstract: Whilst a plethora of research exists on the smart cities and project performance evaluations, only few studies have focused on the smart city policy evaluation from the perspective of its acceptance by practitioners. This paper aims to generate insights by evaluating the smart city policy through a developing country case study—i.e., Malaysia. This study employed a questionnaire survey method for data collection and analyzed the data by using Fuzzy Delphi analysis. A group of 40 practitioners was gathered in a focus group discussion through purposive sampling. The main objectives of this survey were to identify the understanding and acceptance levels of the seven smart city domains and respective strategies that are outlined in the Malaysian Smart City Framework. The results disclosed that the practitioners possessed divergent levels of understanding and acceptance in terms of smart city domains. The study participant practitioners accepted all understanding and acceptance objectives of smart economy, living, people, and governance domains (expert agreement 75–92% and threshold d value 0.123–0.188), but rejected all objectives for both smart environment and digital infrastructure domains (expert agreement 55–74% and threshold d value 0.150–0.212). Along with this, acceptance of smart mobility was also rejected (expert agreement 56% and threshold d value 0.245). The findings reveal that considering all opinions expressing dissensus is essential when building more inclusive smart city strategies. This study contributes to the smart city discourse as being one of the first in capturing professional practitioners' understanding and acceptance on a national level smart city policy by applying the Delphi method in the smart city context. Most importantly, the study informs urban policymakers on how to capture the voices and perspectives of the general public on national and local smart city strategy and initiatives.

Keywords: Europe; Fuzzy Delphi method; Hong Kong; India; Malaysia; smart cities; smart city policy; smart urbanization; urban policy; policy evaluation

Citation: Lim, S.B.; Malek, J.A.; Yussoff, M.F.Y.M.; Yigitcanlar, T. Understanding and Acceptance of Smart City Policies: Practitioners' Perspectives on the Malaysian Smart City Framework. *Sustainability* **2021**, *13*, 9559. https://doi.org/10.3390/su13179559

Received: 5 August 2021
Accepted: 23 August 2021
Published: 25 August 2021

1. Introduction

Since the early 2000s, smart city development has been gaining global momentum. Thus, many models or concepts have been formed, adopted, and evaluated [1]. For example, the seminal smart city concept by [2] laid the basis for the formation of six smart city domains (i.e., smart economy, people, governance, mobility, environment, and living) and emphasized activities that would cultivate independent citizens. Since then, many models have been adopted and adapted from the concept of [2], such as the smart cities wheel by [3], the initiative framework of the smart city by [4], the alternative framework for smart city governance by [5], the conceptual framework for defining the smart city by [6], and the Unified Smart City Model by [7]. On the other hand, top-down smart policies that have been adopted and adapted from the work of [2] include the Hong Kong Smart City Blueprint [8] and the Malaysian Smart City Framework (MSCF) [9].

Furthermore, many studies have evaluated smart city performance. For instance, Ref. [2] developed the European medium-sized (smart) city indicators and ranking; Ref. [10] used the analytic network process (ANP) to investigate the relations between smart city domains, actors (i.e., government, industry, university, and civil society), and strategies; Ref. [11] examined the Malaysian smart city domains through the AHP; Ref. [12] developed a smart city descriptor scoring table to qualitatively compare smart city domains performance in Singapore, Korea, and Malaysia; Ref. [13] developed a smart city sharable framework to evaluate 17 smart cities in China; Ref. [14] developed a fuzzy synthetic evaluation of the challenges facing smart city development in developing countries; Ref. [15] developed a typology of smart city assessment tools and evaluated 122 cities; Ref. [16] developed the smart city index and ranking; and ref. [17] recently developed a smart city measurement framework for inclusive growth.

Nevertheless, far less research has been conducted on evaluating the smart city policy, with the exception of scholars such as [18], who made a general evaluation of the smart city policy and the challenges facing five UK cities. It is crucial to evaluate each planned top-down policy, especially from the public perspective. With just internal assessments by the authorities and departments, actual situations and shortfalls may be overlooked. This might result in overall failure and wasted investment and resources. Taking the case of the MSCF, launched in 2019, to date there have been no evaluation reports on the strategies being planned. Furthermore, the period from 2021 to 2022 has been scheduled as the time to implement smart initiatives nationwide [9]. Many local authorities lack suitable references and benchmarking on the details of the smart city domains and strategies to be adopted [19]. Without reference to evaluation, authorities or officers on the ground tend to believe that a blueprint is perfect and will follow it to the letter. Thus, in this research, and given the practical knowledge gaps, the authors intend to answer the following questions:

- What level of understanding do practitioners have of the smart city domains stated in MSCF?
- What level of acceptance do practitioners have of the smart city domains stated in MSCF?

Based on these research questions, this study aims to evaluate the understanding and acceptance of practitioners from various sectors who are involved in smart city development in developing countries (using Malaysia as a case study). Knowing the levels of public understanding and acceptance was intended to be the output of this study, which would thus provide guidance to governments and policymakers to improve the smart city strategies and policies so that more smart and inclusive living is available to their citizens.

2. Literature Background

Understanding the basic smart city domains is mainly influenced by the six domains outlined by [2], namely the smart economy, living, environment, people, governance, and mobility.

According to [2], the smart economy component is characterized by competitiveness. Among the sub-components of the smart economy (in the case of medium-sized European city rankings) are an innovative spirit, entrepreneurship, an economic image and trademark, productivity, labor market flexibility, and international embeddedness. As the economy is a broad concept and its strategies are context-based, many scholars and agencies have suggested measuring specific components, including nineteen economic attributes in the case of India, as stated by [20]. These include promoting balanced and sustainable economic growth, making strategic investments on strategic assets, and knowing that all forms of economics function at the local level. In another case, the smart economy domain of the

Hong Kong Smart City Blueprint [8] promotes sharing economy, fintech, smart tourism, and re-industrialization.

In the case of Malaysia, the components stated in MSCF are to intensify the application of technology and digitalization in core business functions, enhance the usage of e-payment, attract investment in high value-added industries, create a workforce to match the jobs in these industries, provide technology labs and collaborative platforms, establish incubators and accelerators, and leverage existing government assistance and funding. Supporting literature can be found in Table 1.

Table 1. Smart economy domain.

Smart Economy Strategy	Reference
Intensify technology application and digitalization in core business functions	[2,20–22]
Enhance the usage of e-payment	[23–25]
Attract investment in high value-added industries	[26,27]
Create workforce to match jobs in high value-added industries	[4,28,29]
Provide technology labs and collaborative platforms	[22,30,31]
Establish incubators and accelerators	[32,33]
Leverage on existing government assistance and funding	[20,30]

High value-added activities refer to the major contribution of a private industry or government sector to overall gross domestic product (GDP) [34]. Contributions to GDP include higher wages and compensation for employees, taxes on production, lower import subsidies, and a gross operating surplus [34]. The Hong Kong labor market is an example of a concentration of high value-added service industries, with 25.9% of employees working in public administration or in the social and personal services industry in 2014 [35]. However, it is challenging to transition from low to high value-added industries in developing countries. This is the case in Indonesia, where low value-added industries such as textiles are desperately fighting rising wages and seeking protection from international competition. High value-added sectors largely utilize technology in various activities, including designing products, delivering products, processing customer orders, and improving product quality [27]. Nevertheless, according to MSCF, technology disruptors in Malaysia, such as robotics and analytics, are shifting traditional services towards value-adding and non-traditional service areas. However, the authors observed that MSCF did not refer to the issues of wages and imbalanced urban-rural development. Correspondingly, the smart city policy has offered opportunities within the Fourth Industrial Revolution (Industry 4.0) mostly in developed states and urban areas, while less-developed states and rural areas, such as Sabah, are mentioned far less.

The second domain of smart living is characterized by the quality of life. Among the sub-components found in the smart living concept outlined by [2] are cultural facilities, health conditions, individual safety, housing quality, educational facilities, touristic attractivity, and social cohesion. In the Indian case, [20] scoped smart living into 14 attributes, including promoting shared values in society, celebrating local history and culture, and opening highly accessible public spaces. In the case of Hong Kong, their strategies are in building a Wi-Fi-connected city, developing faster digital payment systems, providing free electronic identity (eID) citizenship for government and commercial online transactions, and launching a $1 billion funding scheme to support the procurement of technological products by elderly and rehabilitation service units [8].

In Malaysia, the MSCF strategies are to enhance safety and security, promote the provision of quality housing, optimize emergency responses, enhance the quality of healthcare services through digital technology and encourage urban farming for better living. Supporting literature can be found in Table 2.

Table 2. Smart living domain.

Smart Living Strategy	Reference
Enhance safety and security	[20,36,37]
Promote quality housing	[2,38]
Optimize emergency response	[20,39]
Enhance quality of healthcare services through digital technology	[40–43]
Encourage urban farming for better living	[23,44]

Concerning the element of enhancing safety and security, one key initiative in Malaysia is the focus on crime reduction [36,45]. For example, under the safe city initiative through the Ministry of Housing and Local Government, a safer city can be created using several strategies, such as crime prevention through environmental design (CPTED) and crime prevention through social design (CSPD) [46]. With CPTED, information and communication technology (ICT), and mechanical surveillance design initiatives are popular, including the installation of closed-circuit television (CCTV) in public spaces, IoT (internet-of-things) lighting, safety (panic button) alarms, and establishing GIS (geographic information system) mapping for crime detection [36]. In the case of the capital city, Kuala Lumpur, crime is always an important issue for the citizens and the city authorities. Research has shown that the challenges to making Kuala Lumpur a safe city can be mitigated by enhancing the role of guardians (i.e., the authorities); promoting CPTED and CSPD activities; and assisting victims and offenders with psychological, financial, and family assistance [47].

The idea behind the third domain, smart environment, centers on preserving natural resources. The smart environment sub-components outlined by [2] are the attractivity of natural conditions, pollution, environmental protection efforts, and sustainable resource management. Another source of reference from India, Vinod Kumar [20], presented 22 attributes to describe the smart environment, which included protecting nature; managing water resources, water supply systems, floods, and inundations effectively; encouraging neighborliness and a spirit of community; upgrading urban resilience to the impacts of climate change; and creating a low-carbon environment based on energy efficiency, renewable energy, and the like. In the Hong Kong case, the strategies are focused on reducing the carbon intensity; promoting energy efficiency and conservation in the community, with a particular focus on green and intelligent buildings; reducing waste; and monitoring the air pollution and cleanliness of public spaces [8].

In Malaysia, MSCF smart environment strategies include the need to preserve green areas and enhance the management of trees in public parks; strengthen the integrated and sustainable solid waste management; strengthen the solid waste laws and policies; improve the air quality and its monitoring system; improve the water quality and its monitoring system; increase energy efficiency and promote renewable energy sources in the community; enhance disaster risk management by adopting advanced technology applications; enhance the non-revenue water management; and encourage the development of a low-carbon city concept that can be adopted at the local level. Supporting literature can be found in Table 3.

Table 3. Smart environment domain.

Smart Environment Strategy	Reference
Preserve green area and enhance the management of trees in public parks	[2,48]
Strengthen the integrated and sustainable solid waste management	[2,48]
Strengthen the solid waste laws and policies	[49,50]
Improve the air quality and its monitoring system	[50,51]
Improve the water quality and its monitoring system	[2,50]
Increase energy efficiency and promote renewable energy sources in community	[2,20,37]
Enhance the disaster risk management by adopting advanced technology application	[52,53]
Enhance the non-revenue water management	[2,54]
Encourage the development of low carbon city concept to be adopted at local level	[48,55]

In terms of park and green area management, the reduction in size of reserved forest and the preservation of green space in development plans are continual issues in Malaysia. Although forest land may have been gazetted, new development plans have always resulted in excuses to degazette forest reserves in favor of mixed-use development. For example, the Selangor State Government has recently granted a mixed development project on 931 hectares of the Kuala Langat North Forest Reserve, which is largely a move to rescind the protected status of the remnants of a once-sprawling peat forest that has been home to four indigenous Temuan settlements. The project also threatens wildlife [56]. This is one case that demonstrates the image of the Malaysian government, which can easily override gazetted land protection with the introduction of new plans under political influence and with profitable intentions, despite concerns for the public good of civil society, climate change, and the overall environment.

In terms of community attitudes to environmental protection, much change is required in Malaysia, especially within the authority-dependence mindset. The study on the Iskandar territory, Johor, Malaysia, Ref. [57] showed that residents are conscious of the need for environmental cleanliness; however, their mindsets were hindered by the belief that the cleanliness of public space is mainly the responsibility of the authorities. Thus, Ref. [57] reaffirmed that the involvement and accountability of all parties are much needed in caring for the natural environment.

The fourth domain of smart people is characterized by social and human capital [2]. The indicators for the case of Europe include the level of qualification, affinity with lifelong learning, social and ethnic plurality, flexibility, creativity, cosmopolitanism, open-mindedness, and participation in public life. In the case of India, 'smart people' are proposed as being the fundamental building block of a smart city system because, without people's active participation, a smart city system would not function effectively (Figure 1). Thus, Ref. [20] proposed eleven attributes of smart people by including the need to be actively involved in the city's sustainable development; excel in creativity and finding unique solutions to challenging issues; opt for lifelong learning and use e-learning models; and be cosmopolitan and open-minded and hold a multicultural perspective. In the case of Hong Kong, this focuses on nurturing young talent, innovation, and entrepreneurial culture [8].

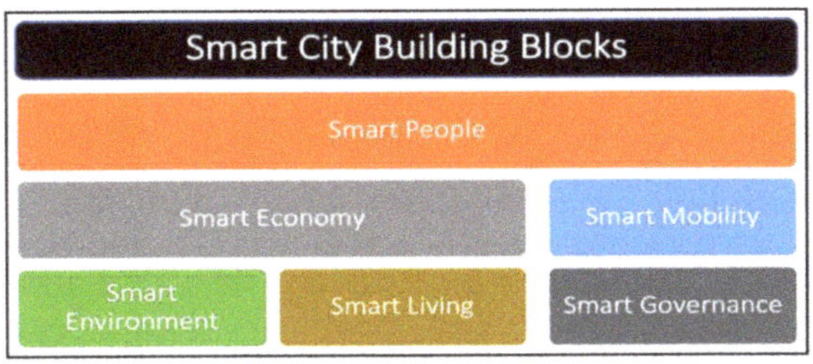

Figure 1. Smart city system building blocks, adapted from [20].

In the case of Malaysia, the strategies are to improve moral education in schools; enhance public awareness in practicing good moral and civic duties; increase skilled and talented human capital at every level; enhance public participation and community empowerment initiatives; improve gender sensitization and the inclusivity of vulnerable groups; and increase public willingness to adapt to emerging technologies. Supporting literature can be found in Table 4.

Table 4. Smart people domain.

Smart People Strategy	Reference
Improve moral education in schools	[58,59]
Enhance public awareness in practicing good moral and civic	[59,60]
Increase skilled and talented human capital at every level	[8,20]
Enhance public participation and community empowerment initiatives	[20,61–63]
Improve gender sensitization and inclusivity of vulnerable groups	[52,64]
Increase the public willingness to adapt with emerging technologies	[8,20,65]

The element of cultivating skilled and talented human capital is particularly crucial, as Malaysia is determined to adopt the National Fourth Industrial Revolution Policy (Malaysian Industry 4.0 Policy), which was launched recently on 1 July 2021 [66]. This Industry 4.0 policy was launched with the purpose of transforming Malaysia into a high-income state through technology and digitalization. Five fundamental technologies of the Industry 4.0 policy include artificial intelligence, the internet of things, blockchain, cloud computing and big data analytics, and advanced materials and technologies [67]. For the young generation to master these Industry 4.0 skills, it is crucial to plan every level of education properly. The Industry 4.0 policy is aligned with the Shared Prosperity Vision 2030, launched in 2019. The aim is to drive Malaysia towards developed nation status by 2030.

The moral and spiritual education element is considered appropriate for the majority Muslim society in Malaysia. The moral element of cultivating smart people is comparatively silent in most western European smart societies (refer to [2,68]). Since the early 1980s, Royal Professor Ungku Abdul Aziz bin Ungku Abdul Hamid, a well-known academician in Malaysia, has creatively interpreted a religious and moral form of development, which represents a balance between the spiritual and material world and is geared towards the needs of the local Muslim community [59]. The emphasis on the moral and spiritual element adopted in the MSCF will further strengthen the quality of Malaysian citizenship by developing a more peaceful and caring society.

Citizen participation and community empowerment are often identified as important elements in realizing a citizen-centric smart city [20,62]. However, this attention should never be blinded by political actions that assume that tokenism and non-participation (refer to [61]) satisfy this type of participation. On the contrary, it is vital to involve citizens in decision making and agenda setting in the smart city initiatives [69].

The core value of the fifth domain of smart governance is political participation. From the European perspective, Ref. [2] described smart governance using the components of participation in decision making, public and social services, and transparent governance. The systematic literature review by [70] summarized six attributes for building a smart governance system. It should be based on ICT, external collaboration and participation, internal coordination, decision-making processes, e-administration, and outcomes. Prior research also suggests that the main outcome of smart city governance is the production of a wide range of public values through innovative collaborations [70].

From the Indian perspective, Ref. [71] suggested 12 steps to convert existing e-governance to smart governance, including an increase in city expenditure on ICT; the ease of access to e-services such as lodge complaints, claims and rights to information; and the promotion of e-democracy through e-decision making and e-voting. From the Hong Kong perspective, smart governance is promoted through using open data for smart city innovations; building smarter city infrastructure, such as the fifth generation (5G) mobile network; building a new big data analytics platform; data sharing among government departments; and adopting building information modelling (BIM) for major government capital work projects [8].

From the MSCF perspective, the components include increasing the scope of e-government services, increasing the quality of e-government services, elevating the use of data sharing platforms across government agencies, and promoting information disclosure and open data from the Government. Table 5 shows the smart governance strategies in MSCF and the related citations.

Table 5. Smart governance domain.

Smart Governance Strategy	Reference
Increase the scope of e-government services	[64,71,72]
Increase the quality of e-government services	[2,40,71,73]
Elevate the use of data sharing platform across government agencies	[26,70,74,75]
Promote information disclosure and open data from government	[8,76–79]

It is crucial to be aware of the component of elevating the use of data sharing platforms across government agencies, as the isolated performance of government agencies was identified by the former prime minister as hindering the performance and services of government agencies [80]. In fact, this lack of efficiency, which is due to excessive bureaucracy, the reluctance of public servants to share data, and other factors, is not a new issue in the delivery of the Malaysian government system [81,82].

Concerning the sixth domain, smart mobility, the main concerns outlined by [2] were transport and ICT. The sub-components of [2] include local accessibility; (inter)national accessibility; the availability of an ICT infrastructure; and sustainable, innovative, and safe transport systems. In the case of India, Ref. [20] described smart mobility in terms of ten attributes, such as a focus on the mobility of people but not vehicles; advocating walkability and cycling; balanced transportation options such as a mass rapid transit system; and seamless mobility for differently abled people. In the Hong Kong case, the strategies are to focus on intelligent transport systems and traffic management; public transport interchanges/ bus stops and parking; environmental friendliness in transport; and smart airports with facial biometric technology. These features should offer a hassle-free travel experience [8].

In the Malaysian case, the smart mobility strategies address the need to establish intelligent transport management; enhance data sharing and digital mobility platforms; establish demand-based ridesharing services; utilize AI and the sensor-based predictive maintenance of a public transport fleet and infrastructure; enhance the dynamic smart parking infrastructure; establish an electric vehicle revolution; enhance collaboration with academia on research and development (R&D) into, and the commercialization of, EVs and next-generation automobiles; and promote the usage of public transport applications. Table 6 shows the smart mobility strategies in MSCF and the related citations.

Table 6. Smart mobility domain.

Smart Mobility Strategy	Reference
Establish intelligent transport management	[2,8,20]
Enhance data sharing and digital mobility platform	[83,84]
Establish demand-based ride sharing services	[8,20,85]
Utilize AI and sensor-based predictive maintenance of public transport fleet and infrastructure	[2,22,43,85,86]
Enhance dynamic smart parking infrastructure	[8,43,83]
Establish electric vehicle revolution	[85,87]
Enhance collaboration with academia on R&D and commercialization on EVs and next-generation automobile	[83,85]
Promote the usage of public transport application	[8,83–85]

In general, all the components and strategies in various countries discussed above indicate that smart mobility is universal, regardless of whether it is introduced in the global north or south. The common item is the promotion of people-centric (rather than vehicle-centric) [83] and environmentally friendly (rather than utility convenient) transportation means [84]. The measures involved include opting to cycle and walk and to take public transport in the city rather than using a personal vehicle that produces greenhouse gas, carbon emissions, and pollution. This is predominantly important in many Asian cities; for example, Kuala Lumpur is characterized by heavy car dependence, leading to traffic congestion and delays [85]. Planning for future mobility must focus less on building more highways and being car-dependent but rather on alternative ways of thinking about environmentally friendly mobility means and adoption. Considering the need for environmental protection and the preference for connecting two destination points via electronic platforms/communication, the actual physical cost of travelling could be reduced.

In addition to the above six basic domains, the authors would like to discuss another emerging domain, that of smart digital infrastructure. This domain did not appear as an individual domain in [2,8,20]. Giffinger et al. [2] explicitly merged this element into the smart mobility domain. Meanwhile, in the case of Hong Kong, this digital infrastructure is explained/inserted in the smart government domain. As digital infrastructure is a frequent practice in Western and developed countries in Europe and North America, it is quite ready and more embedded into other domains. Under the New York Smart and Equitable City Plan 2015, digital infrastructure was embedded in the domains of smart buildings and infrastructure; smart transport and mobility; smart energy and environment; smart public health and safety; and smart government and community [88]. All the sectors and strategies within the smart cities concept center on ICT infrastructure, a point on which the authors and the majority of smart city scholars agree (Figure 2).

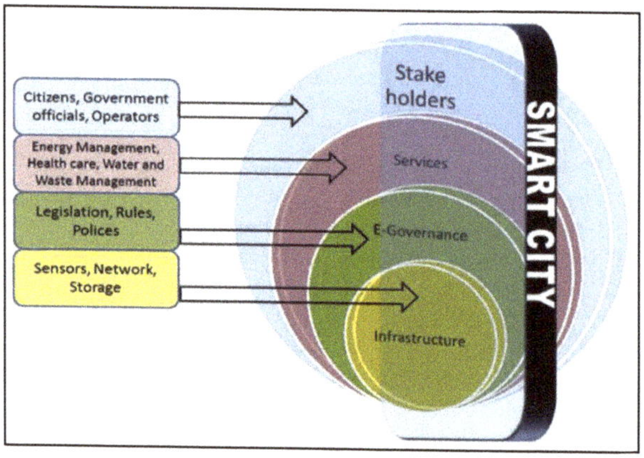

Figure 2. Digital infrastructure is the heart of smart city development [89].

However, in most private sectors conceptions, due to the propagation and sale of their latest technologies, this digital infrastructure element is explicitly highlighted. In the case of Frost and Sullivan, it is even divided into two different domains: smart technology and smart infrastructure (Figure 3).

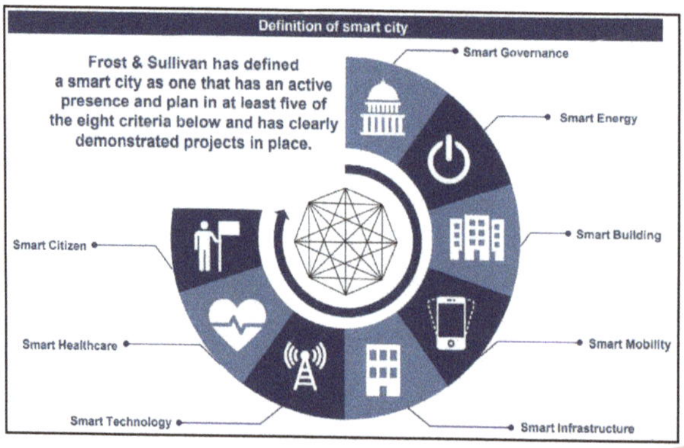

Figure 3. Smart city domains [90].

In MSCF, smart digital infrastructure has been designated as a separate seventh domain. The smart digital infrastructure strategies include the need to enhance the roles of service providers in developing digital infrastructure; enhance internet speed and connectivity; enhance the government's role in facilitating the development of communication infrastructure; enhance indoor and outdoor network coverage; strengthen policies related to personal data protection; and strengthen policies related to cybersecurity. Table 7 illustrates the strategies of the smart digital infrastructure domain and its related citations.

Table 7. Smart digital infrastructure domain.

Smart Digital Infrastructure Strategy	Reference
Enhance service provider's role in developing digital infrastructure	[8,37,90,91]
Enhance internet speed and connectivity	[90,92]
Enhance government's role in facilitating the development of communication infrastructure	[8,70,93]
Enhance indoor and outdoor network coverage	[2,94,95]
Strengthen policies related to personal data protection	[96–98]
Strengthen policies related to cybersecurity	[8,37,89,99,100]

One form of digital infrastructure to attract attention in smart city development is the IoT. Using the internet, the IoT is a network that interconnects ordinary physical objects, such as smartphones, with identifiable addresses to provide intelligent services [101]. In 2021, 35 billion IoT devices were expected to be installed and there were 46 billion connected devices around the world [102]. These numbers, in total, represent more than ten times the size of the world population. Therefore, it could be imagined that it is crucial to tackle the cybersecurity issues that relate to using IoT machines and to address the need for personal data protection as part of living in smart cities. In Malaysia, cybersecurity cases rose by 82.5% between 18 March and 7 April 2020 (838 cases), compared to the same timeframe in 2019 (459 cases) [103]. These cases include some form of cyberbullying; fraud or intruding into an unauthorized system such as phishing and email scams; data breaches and distributed denial of service (DDoS) attacks on local businesses; and hacking into private video conferencing chats and harassing the participants during the COVID-19 movement control period.

To tackle these cybersecurity problems in combination with promoting IoT adoption, in 2015, the National IoT Strategic Roadmap was launched by the Ministry of Science, Technology, and Innovation, with the national applied R&D center MIMOS Bhd. as the implementation secretariat and with the support of agencies such as Cybersecurity Malaysia [104]. This roadmap targeted the contribution of RM 9.3 billion (about USD 2.2 billion) to the gross national income and the creation of more than 14,000 highly skilled employment opportunities by 2020. In addition, other policies have been initiated, such as the National Industry 4.0 policy, the National Cyber Security Policy, and the Malaysia Personal Data Protection Act 2010. MSCF mentioned the need to review and enforce stronger laws, as well as upgrade security systems and procedures in the public and private sectors. In this context, cybersecurity has been identified as a policy to be strengthened in the smart city context.

3. Methodology

After outlining the smart city domains and examples of smart city policies worldwide, this methodology section explains the MSCF case study, the samples of respondents, the data collection, the questionnaire design, and the data analysis method.

3.1. The Case of Malaysian Smart City Framework

MSCF is the first top-down document to formulate the direction of smart city development in Malaysia. The document was launched in September 2019 and drafted by the Ministry of Housing and Local Government. In MSCF, smart cities are defined as "cities that use ICT and technological advancement to address urban issues, including to improve quality of life, promote economic growth, develop a sustainable and safe environment and encourage efficient urban management practices" [9]. This definition, in practice, aims to achieve the vision of "quality and smart living" [9].

In terms of planning, the implementation of smart cities nationwide is divided into three phases from 2019 to 2025. These are phase 1 (the foundation stage), from 2019 to 2020 (two years); phase 2 (the development stage), from 2021 to 2022 (two years); and phase 3 (the advanced development and monitoring stage), from 2023 to 2025 (three years). To ensure its effective implementation, understanding, and acceptance among the people must also be investigated, especially for urban residents. Hence, this study focuses on the understanding and acceptance among professionals of MSCF. The study has been developed to identify the appropriateness of the outlined strategies.

As discussed previously, various components/strategies can be found in academic and grey literature under the grouping of each smart city domain, all of which depends on managing problems and challenges in local contexts. The case is the same in Malaysia, where the government had customized the domains, components, and strategies according to the local challenges. Based on the planned domains, 90 questionnaire items were designed (see Appendix A) and face validated by two smart city experts.

3.2. Sampling and Data Collection

This study employed a quantitative survey via the Fuzzy Delphi method. In obtaining expert opinions using the Fuzzy Delphi method, the ideal sample size is between 10 and 50 respondents [105,106]. Therefore, the authors decided to sample 40 smart city practitioners from the Kuala Lumpur Greater Valley area, including the city of Kuala Lumpur, Putrajaya, and Cyberjaya (Table 8).

Table 8. Informant sampling.

Characteristic	Quantity (N = 40)	Percentage (%)
Gender		
Male	19	47.5
Female	21	52.5
Age		
23–30 years old	1	2.5
31–40	24	60.0
41–50	13	32.5
51 years old and above	2	5.0
Race		
Malay	36	90.0
Chinese	2	5.0
Bumiputera Sabah and Sarawak	2	5.0
Academic qualification		
Bachelor's degree	26	65.0
Master's degree	8	20.0
PhD	6	15.0
Employment sector		
Government	32	80.0
Private	4	10.0
Self-employed	4	10.0
Work experience		
5 to 8 years	4	10.0
9 to 10 years	8	20.0
11 to 15 years	13	32.5
16 to 20 years	8	20.0
21 years and above	7	17.5
Job Position		
Director/CEO	12	25.5
Assistant director/Senior officer	10	21.3
Executive officer	8	17.0
Engineer/Planner/Architect	14	29.8
Technician	3	6.4

As Table 8 shows, this group of practitioners consisted of those in the government, private, and self-employed sectors. They represented the middle class and various professional job roles, such as director/CEO, assistant director/senior officer, executive officer, engineer/planner/architect, and technician. Since the 1970s, the middle class has emerged as a significant group contributing to the urbanization process in major cities in Malaysia [107]. Thus, the selection of professionals as respondents was significant given the composition of this group, the majority of whom lived in urban areas. The professionals were selected based on various criteria: they had to have a minimum of five-year work experience; possess at least a bachelor's degree; and be primarily involved in the planning, design, delivery, and management of cities and their development.

The success of the Fuzzy Delphi method depends on the insights and information supplied by experts. Thus, a panel of experts/respondents was identified through a purposive sampling and nomination process, rather than random selection. Later, a focus group discussion was organized, and data were collected.

3.3. Questionnaire Design

Through a structured questionnaire, a survey strategy of enquiry was conducted. Three sections were used in the questionnaire to obtain information from the respondents. Section one was designed to determine the respondent's background. Section two focused on their understanding, while section three focused on their acceptance of the MSCF's domains. The questionnaire adopted a closed-ended design. The respondents were asked to rate the 90 variables based on their level of significance using a five-point Likert scale, with five being Strongly Agree and one being Strongly Disagree. For the details of the survey items, see Appendix A. Aghimien et al. [14] adopted a similar approach in their study that evaluated the challenges facing smart cities.

3.4. Data Analysis

In addition, the Fuzzy Delphi method was chosen as the analysis technique to obtain the agreement of experts, namely the professionals, based on the study objectives. The Fuzzy Delphi method is a Delphi method performed to obtain information regarding consensus on measurement variables or factors from a group of experts [108,109]. The Delphi Method has been shown to be effective in publishing the best ideas/views through collective responses from expert informants [110]. With the principle of "more minds are better than a single mind", the Fuzzy Delphi method is designed as a forecasting tool to gather the ideas of structured groups, which are said to be more accurate than unstructured predictions [111]. This technique allows experts to coordinate their actions systematically in addressing a particular problem or difficulty and reach a consensus.

In this study, expert consensus was evaluated based on the seven MSCF domains, namely the smart economy, smart living, smart environment, smart people, smart government, smart mobility, and smart digital infrastructure. Each of these domains has its own strategic initiatives to enable cities in Malaysia to achieve smart city status. Respondents' understanding and acceptance were analyzed to achieve the objectives of the study.

Questionnaire data obtained from the focus group feedback of professionals were analyzed using a formulated Microsoft Excel worksheet by [106]. The experts' score inputs were evaluated in stages. Mathematical scores—the Likert scale and the triangular fuzzy scale scores for each item—were obtained (Table 9) and converted into mean values. Later, the threshold value (d), the percentage of expert agreement and the "defuzzification" process of the fuzzy score with α-cut value were calculated. Finally, based on the above three criteria, the ranking positions of the consensus items accepted/rejected by the expert panel were analyzed.

Table 9. Triangular fuzzy number scale [106].

	Strongly Disagree			Disagree			Moderately Agree			Agree			Strongly Agree		
Likert scale	1			2			3			4			5		
Triangular fuzzy Delphi scale	0.0	0.0	0.2	0.0	0.2	0.4	0.2	0.4	0.6	0.4	0.6	0.8	0.6	0.8	1.0

In detail, let us say the item "I am ready to use e-payment in my daily affairs" was scored 5 (strongly agree) by an expert. The score is converted into the minimum, most plausible, and maximum values of 0.6, 0.8, and 1.0 fuzzy scores. It indicated the expert is agreeable to the item is 60%, 80%, and 100%, respectively. Then, the fuzzy scale of (0.6, 0.8, 1.0) is converted into mean value $(\overline{m})$ among the 40 responds.

Next, according to [112], the calculation of the threshold (d) value performed was as follows:

$$d(\overline{m},\overline{n}) = \sqrt{\frac{1}{3}[(m_1 - n_1)^2 + (m_2 - n_2)^2 + (m_3 - n_3)^2]} \quad (1)$$

where,

d = the threshold value,
m_1 = the smallest mean value of a fuzzy number,
m_2 = the most plausible mean value of a fuzzy number,
m_3 = the maximum mean value of a fuzzy number,
n_1 = the smallest value of a fuzzy number,
n_2 = the most plausible value of a fuzzy number, and
n_3 = the maximum value of a fuzzy number.

The value of 'd' (the threshold value) for all items of the questionnaire indicates expert consensus agreement for each item. According to [112], the value of 'd' must be greater than or equal to 0.2 to indicate consensus agreement for each item.

For the expert agreement/consensus percentage, if the expert consensus exceeded 75%, it was considered accepted [113,114]. Then, through the process of defuzzification or the process of determining the scores, the ranking positions of each item were determined. The formula used to determine the ranking/score for an item was as follows:

$$A_{max} = \frac{1}{3}(m_1 + m_2 + m_3) \quad (2)$$

After an assessment was made, if the fuzzy (A_{max}) score or α-cut value was equal to or exceeded 0.5, this indicated expert consensus to accept the item [115].

The Delphi method is a widely accepted, efficient, and effective way of bringing together experts to discuss, debate, and organize a body of information in order to develop a validated instrument, reach agreement on an issue, uncover common factors, or forecast trends [116,117]. This method is deemed particularly highly reliable when more than ten experts in the given field were employed [105,106]. Additionally, to minimize the bias, it is important to involve experts in a study that possess extensive experience, high qualifications, and knowledge in the field or the subject matter [118]. Evidently, this study meets these requirements as it has involved 40 experts with a minimum of five-year work experience, possessed at least a bachelor's degree, and involved intensively in the planning and management of smart cities in the context of Malaysia (Table 8). Hence, we did not employ an additional validation mechanism for the generated results of the Delphi study.

4. Results

In general, the understanding and acceptance of the targeted group of experts in this study were contested. This shows that the community has different perceptions of the smart city domains stated in the MSCF. This divergent phenomenon can be described in two ways. Firstly, from the domain perspective, the majority of domains (i.e., smart economy, living, people, and governance) were accepted, two domains (i.e., smart environment and digital infrastructure) were rejected, while the smart mobility domain was partially accepted. Secondly, from the objective perspective, more than half of the domains were accepted (Table 10).

Table 10. Results of Fuzzy Delphi analysis by smart city domains.

Domain	Threshold (d) Value		Expert Agreement (%)		Average of Fuzzy Score (A_{max})		Result		Ranking by the Fuzzy Score	
	(U)	(A)	(U)	(A)	(U)	(A)	(U)	(A)	(U)	(A)
Smart Economy	0.142	0.139	76%	89%	0.725	0.731	Accepted	Accepted	2	2
Smart Living	0.132	0.171	75%	91%	0.719	0.700	Accepted	Accepted	4	4
Smart Environment	0.189	0.212	57%	55%	0.654	0.639	Rejected	Rejected	7	7
Smart People	0.123	0.128	80%	83%	0.745	0.743	Accepted	Accepted	1	1
Smart Government	0.188	0.184	92%	91%	0.704	0.698	Accepted	Accepted	5	5
Smart Mobility	0.164	0.245	83%	56%	0.724	0.654	Accepted	Rejected	3	6
Smart Digital Infrastructure	0.204	0.150	72%	74%	0.670	0.725	Rejected	Rejected	6	3

Note: U stands for Understanding, A stands for Acceptance. Three conditions to accept an item: threshold value (d) ≤ 0.2, percentage of experts' consensus $\geq 75\%$, and average fuzzy score $(A_{max}) \geq \alpha -$ cut value = 0.5.

To accept the criteria of the Fuzzy Delphi analysis, the results must meet three conditions: (a) threshold value, $d \leq 0.2$, (b) expert agreement percentage $\geq 75\%$, and (c) average fuzzy score $(A_{max}) \geq \alpha -$ cut value = 0.5. Overall, all the domains fulfilled the third criteria, with fuzzy scores equal to or exceeding 0.5. Meanwhile, the threshold value and expert agreement showed mixed results.

To provide more detail on the item results, as shown in Table 11, the smart economy and living had a 100% acceptance rate for the objective of Acceptance, hinting that these two domains can be implemented directly at ground level with little modification. On the other hand, the smart environment scored the lowest acceptance rates, 22.22% for the Understanding objective and 33.33% for the Acceptance objective. This result indicates that the smart environment domain has experienced great public dissensus and more refinement is needed before its implementation to avoid later failures.

Table 11. Results of Fuzzy Delphi analysis by objectives.

Objective	Domain	Item	Accepted Item	% of Acceptance	Rejected Item	% of Rejection	Fuzzy Score Interval
Understanding	Economy	7	5	71.43	2	28.57	0.775 − 0.655 = 0.120
	Living	5	4	75.00	1	25.00	0.775 − 0.620 = 0.155
	Environment	9	2	22.22	7	77.78	0.745 − 0.572 = 0.173
	People	6	4	66.67	2	33.33	0.765 − 0.715 = 0.050
	Government	4	3	75.00	1	25.00	0.725 − 0.693 = 0.032
	Mobility	8	6	75.00	2	25.00	0.755 − 0.685 = 0.070
	Digital Infrastructure	6	3	50.00	3	50.00	0.735 − 0.523 = 0.212
Acceptance	Economy	7	7	100.00	0	0.00	0.770 − 0.710 = 0.060
	Living	5	5	100.00	0	0.00	0.725 − 0.670 = 0.055
	Environment	9	3	33.33	6	66.67	0.720 − 0.557 = 0.163
	People	6	4	66.67	2	33.33	0.770 − 0.720 = 0.050
	Government	4	3	75.00	1	25.00	0.715 − 0.680 = 0.035
	Mobility	8	3	37.50	5	62.50	0.700 − 0.563 = 0.137
	Digital Infrastructure	6	4	66.67	2	33.33	0.760 − 0.655 = 0.105
Total		90	54	61.36	34	38.64	

Note: Refer Appendix B for detailed calculations.

In general, the results of the analysis on the smart economy, living, people, and governance domains met all three conditions of the Fuzzy Delphi method in terms of Understanding and Acceptance. However, some item details must be addressed (refer to Appendix B).

First, for the Understanding objective of the smart economy, the two rejected items were items 3 (high value-added industry investment, with threshold value $d = 0.21$, and expert agreement at only 33%) and 7 (assistance to business operations, with 73% expert agreement). For the Acceptance objective of the smart economy, all the items were accepted. For the high value-added industry investment, the respondents did not arrive at a consensus. Some thought that the authorities should focus on the manufacturing sector, especially in suburban and rural areas, instead of prioritizing high value-added industry, which would accelerate the existing urbanization issues in metropolitan Malaysia, such as in Kuala Lumpur and the Klang Valley area.

Second, under smart living, the only problematic Understanding item was item 1 (crime reduction). Respondents were less able to comprehend why Malaysia was stated as having a high, instead of moderate, crime rate, since most of them lived in peaceful environments. Meanwhile, they were inclined to accept that the MSCF would be able to reduce the crime rate effectively through ICT applications, such as the installation of CCTV in public areas.

Third, for the understanding and acceptance of smart people, all four rejected items were due to the 70% to 73% expert agreement. For item 3, the acceptance of the education policy for human capital development, respondents were not fully confident that the restructuring of education at the tertiary level would produce innovative graduates. One respondent commented that the current graduate market indicated that graduates were able to perform at routine levels while lacking innovative thinking and solution-creation skills.

Fourth, for the understanding and acceptance of smart governance, item 3—inter-governmental data sharing—was the only item rejected as the threshold value $d = 0.224$ and 0.202. Respondent feedback suggested that they did not understand how inter-governmental data could be shared in practice, as some were still experiencing issues such as the separate performance of departments, the redundancy of providing data to particular departments, and the inability to receive valid and complete data through a single department enquiry. For example, the Department of Statistics does not provide open demographic data by city or district level so one needs to go to the particular local authorities.

The major focus of this study should be the smart environment and digital infrastructure domains because both were rejected in terms of the understanding and acceptance objectives. In general, for the environment, its threshold (d) construct for Acceptance (0.212) was more than 0.2 while both values of expert agreement (57% for Understanding and 55% for Acceptance) were less than 75%. For digital infrastructure, its threshold (d) construct for Understanding (0.204) was also more than 0.2 while both values of expert agreement (72% for Understanding and 74% for Acceptance) were also less than 75%. These negative results show that the public remain less likely to understand and accept the components planned in these two domains, smart environment, and digital infrastructure.

In detail, for the smart environment, the three lowest-ranked Understanding items related to items 1 (park and green area management), 8 (non-revenue water management and reporting), and 9 (low-carbon city and carbon emissions). Meanwhile, the three lowest-ranked Acceptance items related to items 7 (readiness towards disaster-resilient cities), 4 (air quality monitoring) and 2 (waste segregation and recycling). From the overall perspective, the environment-related issues worrying the public are broad in scope and a cause for grave alarm. The smart environment domain facings major public understanding and acceptance issues and the authorities should prioritize improvements in this domain.

For the smart digital infrastructure, two items of interest in terms of Understanding are items 6 (cybersecurity) and 5 (personal data protection); for Acceptance, they are items 1 (roles of service providers) and 2 (internet speed). It seems that respondents lacked confidence in the authority's online system security and personal data protection, and felt they were vulnerable to cyber-attacks and personal data leaks. Attention should also be given to the respondents who did not fully accept that private service providers were solely responsible and thought that the government was too. Another important issue involved rural areas with low internet speeds of 4G and below.

For the smart mobility, the result was accepted for Understanding but rejected for Acceptance. The acceptance of respondents was rejected since the threshold (d) conduct was 0.245, which is over the 0.2 required; furthermore, the expert agreement of 56% was much less than the 75% required.

Clearly, the rejection phenomenon identified for the Acceptance objective needs attention. A low level of expert agreement was observed for items 6 (electric vehicle), 1 (smart traffic management), 8 (public transport application), and 5 (smart parking infrastructure). These results showed that the respondents were worried about the traffic planning presented in the MSCF and were unconvinced by the solutions related to the issues stated above.

5. Discussion

5.1. Voicing Dissensus Opinions for Building a More Inclusive Smart City Blueprint

The findings indicate divergent expert perceptions. The different job roles and employment sectors of the respondents could be expected to produce diverse results. Figure 4 summarizes the occurrence of three conditions.

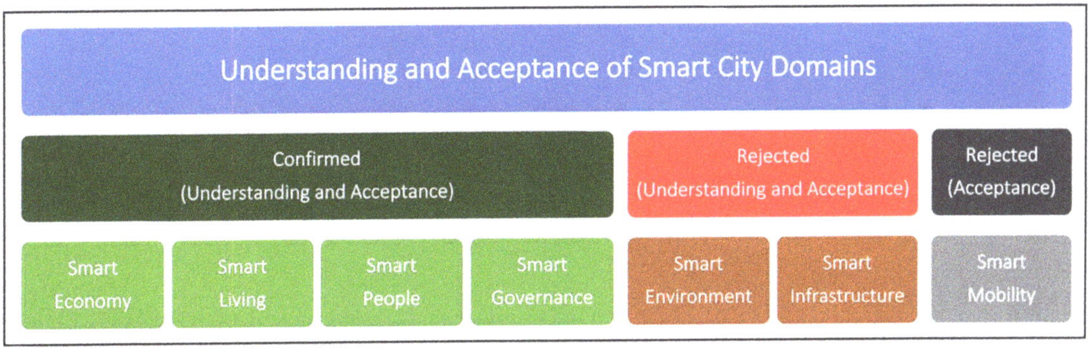

Figure 4. Dissensus opinions on the understanding and acceptance of smart city domains.

The dissensus in the results is a finding that leaders and policymakers should be aware of. They should accept this reality and include those opinions that do not always favor the majority. For example, people were dissatisfied with the smart environment domains; for example, they did not comprehend the ineffectiveness of the authorities' park and green area management and preservation, nor did they accept that the authorities had done enough pertaining to this matter. Evidence from [2,20] showed that the smart environment covers a wide range of natural resources preservation and the resilience of human actions to the impact of climate change; thus, it is crucial to contemplate the divergent opinions from the respondents.

For smart digital infrastructure, respondents did not recognize that their personal data are protected by the Malaysia Personal Data Protect Act 2010; they felt unsafe from cyber-attacks and were also unconvinced that the authorities would reduce cyber threats and have a positive impact by protecting online users.

For smart mobility, although most understood that this domain is important, they showed disagreement in accepting the implemented initiatives. For example, they did not express confidence in the promotion of autonomous, and electric or green vehicles, smart traffic light functioning, and public transport applications. This finding is interesting from a global perspective as these measures are working elsewhere in other countries, meaning they were understood by the respondents. However, it is clear that the authorities should improve these matters to bring about greater local acceptance and avoid wasted investment. The following subsections provide various ideas for reconsidering the implementation of smart city domains in MSCF.

In the broader topic of smart city's understanding and acceptance, the above result reflected a developing country's context and dynamic in practice [1,14]. The administrators should be responsive and improve the smart city domains and strategies from time to time. As for the specific scientific field, the smart environment needs more attention as climate change is real [2,20,87] and applying smart digital infrastructure with higher security [21,22,37] to counter this global issue is urgently needed to be addressed.

5.2. Rethinking the Viability of Smart City Domains and Strategies

A smart economy tends to possess high value-added industries, so it is proposed that high value-added industrial investment promotion initiatives be reconsidered geographically as the distribution of secondary industry is unbalanced and currently heavily favors the Peninsula and urban areas [119]. Sabah and Sarawak are still heavily dependent on primary products (i.e., timber, oil, and LNG). Targeting the relocation of manufacturing sectors to less-developed areas, which would create new urban growth centers or smart cities, needs far more attention, rather than targeting high-value investment in the already mature urban and metropolitan areas. Furthermore, wages in less-developed areas need to be improved since continuing to invest in high-value industries in urban areas will further exacerbate the urbanization issues. Malaysia could learn from India in promoting balanced and sustainable economic growth and ensuring all economic activities work well at the local level (refer to [20]). Furthermore, Malaysia could also learn from Hong Kong in placing greater focus on sharing economic activities among regions and in the re-industrialization of the necessary supporting primary and secondary sectors (refer to [8]).

Another point to consider is the potential of e-commerce. Online transaction expansion initiatives that are gaining a place in the hearts of consumers can be created in line with the increasingly busy lifestyle of urban citizens. A study of online purchasing practices in Malaysia by [120] found that buying online was chosen because it is a convenient and easy way to shop for necessities while avoiding long queues at the counter. Online shopping is a trend in modern society since internet usage has increased in the last decade. It has accelerated under the stay-at-home new normality caused by the COVID-19 pandemic threat. Thus, considering online shopping initiatives as part of the smart economy initiative should enable improvements in the economic status of urban residents, either as traders or customers, which would facilitate the lives of both parties [25].

For smart living, the experts rejected the understanding that the crime rate in Malaysia remains high compared to other countries. The respondents thought that the crime rate was under control level. This opinion matches the findings of [121], whose local studies in Kuala Lumpur showed that city residents are comfortable with the crime situation. Additionally, Ref. [121] found that the perception that the crime rate was high in Malaysia actually did exist in the foreign discourse. Thus, the authorities could consider all such perspectives, turning their focus to the means of adaptation to the fear of crime, the omnipresence of police in public spaces, and assistance to prevent criminal acts in community areas [47,121].

Furthermore, in terms of smart living, voluntary and more active community involvement initiatives related to the safety, educational, and health aspects of the local community can be added to reduce the extent of the dependence on government resources. According to [122], community involvement ensures that the needs and aspirations of the community are not neglected; the result is that community members will be educated and subsequently empowered. This shows that the role of the community can resolve local issues more effectively.

In terms of smart people, referring to skilled and talented human capital, the government must rethink tertiary education and determine how to actively produce digitally talented innovative graduates to suit the value-added industry in the Industry 4.0 era. More structured and holistic learning opportunities within the areas of IoT devices development, telecommunications, middleware, big data analytics, and artificial intelligence are needed. This is because engineering students currently focus mainly on hardware and connectivity aspects while computer science students learn middleware and big data analytics separately [104]. To adopt Industry 4.0 technologies in Malaysian smart city society, a radical paradigm shift in educating graduates so they transform into talented human capital should be the priority, a notion that was reaffirmed by the Prime Minister [67]. Thus, the 2013 national education policy is somewhat outdated, so rethinking how to enhance it through the Industry 4.0 perspective is crucial. Questions such as how to nurture young people so they master the fundamental Industry 4.0 technologies in stages, from primary, secondary, and tertiary education up to life-long learning for the elderly community members, should become the central aim in formulating a new national education policy.

Next, moral and ethical development, as mentioned in the MSCF, is considered a good move for developing countries like Malaysia, as many Western developed countries have resolved this moral element (refer to [2]). As mentioned in the literature, Ungku Aziz's 1980s ideas are still considered fundamental and remain relevant enough to be adopted in the current smart city development in Malaysia. Although the values of education related to moral development were stated in the National Education Blueprint 2013, the latest education plan could be enhanced based on the five principles of Maqasid Al-Syariah, namely caring for religion, caring for life, caring for intellect, caring for one's offspring, and caring for property. Maqasid Al-Syariah refers to the noble purpose of Islamic law, which is based on the principle of Maslahat and which mankind could universally obtain through the text or authority of Islamic law [123]. This universal concept is seen by all as practicable.

Another aspect to enhance in terms of smart people in Malaysia is the level of civic participation in local authority decision-making and programs [73]. Participation in decision-making differs from community empowerment: the former involves the level of citizen power and can influence agenda setting, while the latter refers to the tokenism level of service delivery stages [61]. Furthermore, in the former, people are active in decision-making and co-creating with the authorities, whereas, in the latter, people tend to be in a weaker, beneficiary, or reactive position when they are deemed 'empowered' by the authorities. Contemplating the lower level of participation in decision-making in Malaysia, the authors argue that the implementation of the MSCF could enhance the extent of this form of participation. Although it may face dissensus of opinion, in the long term, this move will help in building more democratic spaces and independent citizenship for Malaysian nation-building [124].

On the governance issue, excessive bureaucracy, delays in approving applications and licenses, as well as a lack of information on new policies and regulations are among the main problems plaguing the government's delivery system [81,82]. Inter-governmental data sharing is another challenge due to the separate departmental practices in Malaysia. To address this, Hong Kong's initiatives can be adopted, such as building a new big data analytics platform; adopting public cloud services, which would enable real-time data transmission and sharing among government departments; and enhancing security features so that government departments can deliver efficient and agile e-services [8]. Future smart city governance should make effective use of their data assets to secure outcomes that are appropriate to citizens' needs. Investment by agencies in system-wide data capture, integration, and analytics capabilities [75] is a crucial aspect to develop.

Apart from data sharing, smart governance ultimately aims to produce public values for citizens, such as from the perspective of asset management and financial and economic sustainability [93]. To realize such public values, e-democracy must be upheld through active e-voting and e-decision making [71], which is a major topic for Malaysian smart development advocates to deliberate. According to the Democracy Index 2020, out of 167 countries, Malaysia (ranked 39) and India (ranked 53) fell into the category of flawed democracies. Meanwhile, Hong Kong ranked 87 due to its hybrid regime of flawed democracy and authoritarian control [125]. In terms of the purpose of building independent citizens [2] within the conception that smart cities are democratic ecologies [126], Malaysia and similar places must actually strive further to achieve higher transparency and open governance. As suggested by [127], 'good enough governance' for smart city societies in Malaysia should consider the cultural context of the Muslim majority, prioritize governance content that allows more scope for political participation and free speech, and cultivate the imagination and unselfishness of children.

Furthermore, it was found that the understanding and acceptance of initiatives in the smart environment is the most critical among all the domains. In this regard, announcements on smart environment initiatives must be intensified and expanded to ensure the sustainability of the existing environment. Most importantly, the authors' view is that environmental accountability initiatives must be added to this component to enable each party to understand the concept and play their respective roles in caring for the environment. In the case of maintaining a clean environment, Ref. [57] found that all stakeholders should take responsibility, not solely the authorities. Efforts to maintain and preserve the environmental space relate to the question of community awareness and attitude, which, if sufficiently high, would ensure that the environment is always clean, healthy, and sustainable.

In terms of preserving parks and green spaces in urban areas, Malaysia's development control guidelines set a minimum of 10% green and open space reservation, which is considered relatively low. In comparison, the city of Wuhan, China, has launched its Wuhan Low-Carbon Urban Development Plan 2013, which reserves 28% for green areas in the city [48]. Therefore, a rethink is suggested that would impose a greater green space allocation in new development plans and, together with agencies such as PLAN-Malaysia, the MSCF could incorporate this higher green space allocation as one of its smart environment initiatives.

In terms of smart mobility and, in particular, electrical vehicles (EVs) in Malaysia, Putrajaya city bought 150 electric buses (each costing RM 1.5 million). They operate in Putrajaya and the vicinity, the aim being to cut carbon emissions, noise pollution, and traffic congestion while improving public transport and parking systems [128]. The operation of the electric buses is calibrated by battery capacity and charging facilities and has been found to outperform conventional bus operations [129]. However, cases of the inefficiency of public transport management were identified, whereby the electric buses were found abandoned at the Depoh Putrajaya. Bus breakdowns are frequent due to lack of maintenance, unreliable and delayed bus arrival times, and reductions in bus routes [130]. Thus, although electric cars are efficient in costs and energy saving with long-term usage [84], the adoption of an EV ecosystem is required, involving features such as efficient management and the availability of efficient power charging stations.

As for privately-owned electric cars, it has been found that the understanding and acceptance of the community is still low. The respondents in this study felt the costs involved in owning and maintaining a private electric vehicle were higher than those of a typical vehicle. It is true that research has shown that the cost of electric vehicle ownership in Malaysia is not yet as competitive as typical internal combustion vehicles [131]. This shows the market and society acceptance of electric cars remains still low. Those involved in the MSCF should rethink the issues of EVs, along with the latest National Automotive Policy 2020, in promoting affordable new technologies. For example, incentives and funding are available under the National Automotive Policy 2020 to develop the technology and engineering required for NxGV (next-generation vehicles), autonomous vehicles, MaaS (mobility-as-a-service), and Industry 4.0 [132]. Thus, MSCF initiatives such as promoting collaboration with the private sector in developing affordable EVs could be implemented.

Last but not least, for the smart digital infrastructure, in terms of the cybersecurity and personal data protection issues, it seems that the MSCF did not provide clear direction on how to strengthen the necessary cybersecurity and personal data protection. The ranking of Malaysia as eighth out of 194 countries in the Global Cybersecurity Index 2021 [133] seems to contradict the results of this study's finding. Recently reported cyber intrusion cases [103] were over the targeted 9000 to 10,000 per year [9], and the assessment of the National IoT Strategic Roadmap was also ambiguous [104]. All the supporting agencies, such as Cybersecurity Malaysia, the Department of Personal Data Protection, and the Malaysia Administrative Modernization and Management Planning Unit must work more closely together and actively provide improvements or amendments to the policies, especially more strict enforcement of the Malaysian Personal Data Protection Act 2010.

As for the issues of low internet speed and digital infrastructure coverage in less-developed states and rural areas, more MSCF initiatives could be planned in conjunction with the latest National Digital Infrastructure Plan (JENDELA), the Malaysian Industry 4.0 Policy, and the Malaysia Digital Economy Blueprint (MyDigital). For instance, the current wireless broadband coverage in Malaysia is 96.7% for 2G, 95.3% for 3G, and 91.8% for 4G coverage in populated areas, with 25 Mbps speed [134]. Therefore, MSCF initiatives could plan to achieve 100% 4G coverage in populated areas and a speed of 100 Mbps by adopting 5G.

6. Conclusions

First, Malaysia's experience in smart city development dates back to 1996 Multimedia Super Corridor Malaysia initiative and the later efforts in developing research universities and integrating them with the city they are located [135,136]—through knowledge-based urban development principles to make space and place for smart urban communities [137]. Today, with its new smart city framework, Malaysia aims to transform its cities and societies into smarter ones. This paper aims to generate insights into how this framework is perceived with professional practitioners. In order to do so, this study conducted an empirical investigation concerning the seven smart city domains planned as part of a top-down national policy of the Malaysian Smart City Framework (MSCF). The findings disclosed that smart environment and digital infrastructure require the most attention, followed by smart mobility, governance, living, economy, and people.

Second, this study has contributed to the smart city discourse and literature particularly by examining the levels of understanding and acceptance from the multi-perspectives of practitioners from various sectors. The study is unique as it is one of the first in capturing professional practitioners' voices and perspectives on a national level smart city policy that impacts a large portion of the population. This finding is an important insight added to the literature investigating, in detail, smart city domains in practice. The divergent and dissensus opinions from the ground are valuable references for leaders and policymakers to consider in building a more inclusive and smarter city blueprint. Furthermore, applying the Fuzzy Delphi method in smart city studies is rather new. It has great potential to be explored and expanded into urban studies and planning disciplines as this method is popular in education, business, and management studies [138].

Last, the limitations of this study are the selection of purposive sampling for the Fuzzy Delphi analysis and the formulation of questionnaire items from the broader scopes of the smart city domains. Thus, based on the smart city domains and after designing two objectives of understanding and acceptance, future studies could explore other qualitative or quantitative methods to justify the results in this study. Other studies that evaluate the implementation of the smart city domain objectives could be conducted, such as using structural equation modelling to assess the implementation of smart city strategies in Greece [139] and acceptance of smart meters in Malaysia [37]. Moreover, future studies could be expanded to capture the voices and perspectives of the general public on national and local smart city strategy and initiatives. This will be the focus of our prospective study.

Author Contributions: Conceptualization and methodology, writing—original draft preparation, software, formal analysis, investigation, and data curation, S.B.L.; supervision, validation, resources, funding acquisition, project administration, review and editing, J.A.M., M.F.Y.M.Y. and T.Y. All authors have read and agreed to the published version of the manuscript.

Funding: The study received funding support from the Malaysian Ministry of Higher Education (grant number FRGS/1/2019/SS06/UKM/02/2). The funder was not involved in the planning, execution, write-up, or other contents of this article.

Institutional Review Board Statement: Not applicable.

Informed Consent Statement: Informed consent was obtained from all participants involved in the study.

Data Availability Statement: Data are contained within the article and appendixes.

Acknowledgments: The authors wish to thank study participants, managing editor and three anonymous reviewers for their invaluable comments and constructive critiques.

Conflicts of Interest: The authors declare no conflict of interest and have no financial or proprietary interests in any material presented/discussed in this article.

Appendix A. Survey Items

Table A1. Domain 1: Smart Economy.

Item	Understanding	Acceptance
Intensify technology application and digitalization in core business functions	I am aware that the use of technology in the services sector needs to be expanded and intensified in order to be able to compete with the use of technology in the manufacturing sector.	I am sure the application of technology and digitization in core business functions can be implemented quickly.
Enhance the usage of e-payment	I understand that the widespread use of debit and credit cards catalyzes e-payment.	I am ready to use e-payment in my daily affairs.
Attract investment in high value-added industries	I understand that investment promotion activities have been restructured to target high value-added industry investors.	I am confident that the attractiveness of high value-added investments can be increased from time to time.
Create a workforce to match jobs in high value-added industries	I am aware that computer science skills and critical thinking need to be widely disseminated as the high value-added industry sector requires creative and innovative employees.	I am sure the matching of high-income work with high value-added industries can be more efficient.
Provide technology labs and collaborative platforms	I understand that the strengthening and establishment of technology laboratories can help entrepreneurs to become more efficient in penetrating a wider market.	I am confident that the establishment of technology laboratories and collaborative platforms can enhance knowledge exchange in various fields.
Establish incubators and accelerators	I understand that incubators and drivers need to work more closely to meet market needs.	I am sure the creation of incubators and drivers can help realize new ideas to produce more competitive businesses that impact the local community and society.
Leverage existing government assistance and funding	I understand that the assistance provided by various government agencies should be used optimally to improve business operations.	I am confident that the optimal use of facilities provided by the Government can help improve business operations.

Table A2. Domain 2: Smart Living.

Item	Understanding	Acceptance
Enhance safety and security through perspectives of crime	I understand the crime rate in Malaysia is still high compared to other countries.	I am sure the installation of analytical proactive surveillance, such as CCTV, can reduce the crime rate in Malaysia.
Promote quality housing	I understand the promotion of the smart home can improve the quality of life.	I believe that the adoption of smart home applications, such as facial recognition systems and IoT lighting, could provide better quality housing.
Optimize emergency response	I understand that an emergency call center can help in cases of emergency, such as fire.	I believe that the adoption of an emergency call center with a real-time mobile rescue application is crucial when an emergency occurs.
Enhance the quality of healthcare services	I believe good healthcare can improve the quality of life.	I believe that the Smart City Framework is able to facilitate public health services.
Encourage urban farming for better living	I believe that urban farming activities can enhance the relationship between neighbors and their sense of belonging.	I am confident that through urban farming, a community can work together more closely and appreciate each other.

Table A3. Domain 3: Smart Environment.

Item	Understanding	Acceptance
Preserve green areas and enhance the management of trees in public parks	I understand that local authorities have improved the management efficiency of public parks, such as adapting the use of smart management systems to preserve green areas (i.e., the use of RFID technology to inventory existing trees).	I gain positive effects from the use of well-managed public parks and green areas by local authorities, such as increased social interaction, peace of mind, and stress reduction.
Strengthen integrated and sustainable solid waste management	I am aware that waste segregation at source and the recycling of waste are the best solid waste management methods to maintain environmental sustainability.	I always practice waste segregation at source and the recycling of waste items at home and work.
Strengthen the solid waste laws and policies	I am of the view that the government has provided adequate laws and policies to improve solid waste management in the country.	I am confident that the laws and policies formulated by the government can improve the sustainability of the country's solid waste management.
Improve the air quality and its monitoring system	I am aware of the importance of using public transport as an initiative to reduce carbon emissions that can affect the environmental air quality.	I strive to increase my use of modes of public transportation in day-to-day affairs to help reduce carbon emissions.
Improve the water quality and its monitoring system	I understand that the government has, over time, improved the efficiency of the water monitoring system technology.	I am confident that the improvement of water monitoring system technology in urban areas by the government will provide a high-quality, clean water source for residents.
Increase energy efficiency and promote renewable energy sources	I find that efforts are being made by the government and the private sector to increase the use of renewable energy in the community.	I have applied the use of energy-efficient appliances, such as LED lighting, at home and at work to reduce the use of electricity generated from fossil fuels.
Enhance disaster risk management by adopting advanced technology applications	I understand that the government has implemented disaster risk management through the use of the latest technology, such as warning delivery systems, to facilitate the delivery of information to the public.	I participate in disaster management awareness programs organized at the community level so that I can be better prepared in the event of a disaster.
Enhance Non- Revenue Water Management	I am of the view that public understanding of the importance of non-revenue water management in water resources management has increased.	I immediately make an online report to the agency or responsible party when faced with incidents such as a burst pipe.
Encourage the development of the low-carbon city concept to be adopted at the local level	I understand that the implementation of initiatives to reduce carbon emissions from buildings and vehicles has been implemented in tandem with the urbanization process.	I believe the development of low-carbon cities by private developers is a positive step towards reducing the carbon footprint of urban areas.

Table A4. Domain 4: Smart People.

Item	Understanding	Acceptance
Improve moral education in schools	I agree that the element of moral education among the younger generation is important as it is an initial step in the formation of an ethical society.	I welcome the government's intention to improve moral education and prioritize it in the early stages of schooling.
Enhance public awareness in practicing good morals and civics	I understand and realize that the concept of a moral and ethical society is an important element in building a smart city culture.	I agree that the building of a smart city community culture can be achieved through civic awareness programs on public facilities, the environment, and the importance of community living.

Table A4. *Cont.*

Item	Understanding	Acceptance
Increase the volume of skilled and talented human capital at every level	I think that the formation of an educated and highly skilled generation is an important aspect of building a knowledgeable society as part of the construction of smart cities.	I am confident that the restructuring of the education policy at every level in the fields of research, science and technology will produce a generation of highly-educated, skilled and innovative people.
Enhance public participation and community empowerment initiatives	I realize that community participation, community engagement, and community empowerment are highly important in every policy formation of a country and lead to the well-being of the people.	I agree the community needs to be directly involved in the making of every government policy and initiative through a simple and fast digital platform.
Improve gender sensitization and the inclusivity of vulnerable groups	I am aware that the interests and needs of women and people with disabilities (OKU) must be taken into account in every aspect of urban development planning.	I support the idea that facilities are provided in every urban development, which through a digital medium, take into account the needs and safety of all groups, especially women and people with disabilities (OKU).
Increase public willingness to adapt to emerging technologies	I understand that the concept of the smart city formation will be formed from a skilled and efficient society and with the use of IT.	I agree that it is time for digital skills to be learned earlier in childhood and subsequently introduced into continuous learning in the community through digital billboards placed in public spaces.

Table A5. Domain 5: Smart Government.

Item	Understanding	Acceptance
Increase the scope of e-government services	I understand that through the Smart City Framework, the government can widen the scope of government services to the community.	I believe that through the Smart City Framework, a wider range of government services will be available to the community.
Increase the quality of e-government services	I understand that the use of the Smart City Framework can improve the quality of e-government services.	I believe the Smart City Framework can improve e-government services to the community.
Elevate the use of data sharing platforms across government agencies	I am confident that if inter-governmental data sharing works well, there will be fewer community complaints and better-quality government/private services.	Through inter-governmental data sharing, I have received valid and accurate data/information from government/private organizations.
Promote the use of information disclosure and open data on behalf of the government	I understand that the dissemination of open data and authentic information can expedite the transparency of governmental services.	I agree that the accessibility of open data and information dissemination would benefit all.

Table A6. Domain 6: Smart Mobility.

Item	Understanding	Acceptance
Establish intelligent transport management	I understand the importance of smart transportation management, such as the use of smart traffic lights, the use of sensors for traffic management, and pollution tracking.	I am satisfied with the way smart transportation management functions, such as with the use of smart traffic lights, the use of sensors for traffic management, and pollution tracking.

Table A6. *Cont.*

Item	Understanding	Acceptance
Enhance the use of data sharing and digital mobility platforms	I understand the importance of data sharing and digital mobility platforms.	I am willing to use data sharing and digital mobility platforms.
Establish demand-based ridesharing services	I know about on-demand ridesharing service applications for vans or shuttle buses, trains, Grab, or SOCAR.	I use on-demand ridesharing service applications for vans or shuttle buses, trains, Grab, or SOCAR services.
Utilize AI and sensor-based predictive maintenance for the public transport fleet and infrastructure	I understand the use of AI (Artificial Intelligence) and sensor-based maintenance forecasting for the public transportation infrastructure and traffic.	I agree that AI and sensor-based forecast maintenance for the public transport infrastructure is required so that forecast maintenance can take place before damage and disruption occurs.
Enhance the dynamic smart parking infrastructure	I know about dynamic smart parking infrastructures, like smart parking meters and apps that provide real-time parking vacancy information.	I use smart parking infrastructure, such as smart parking meters and apps that provide real-time parking vacancy information.
Establish an electric vehicle revolution	I understand the importance and necessity of the electric vehicle revolution.	I have used electric cars/green vehicles/energy-efficient vehicles/electric buses.
Enhance collaboration with academia on R&D into, and the commercialization of, EVs and next-generation automobile	I understand the importance of collaborating with academics and the private sector in R&D into, and the commercialization of, next-generation electric vehicles and cars.	I am willing to work with academics and the private sector on the framework, testing, and regulation of autonomous vehicles/long-term transit planning.
Promote the usage of public transport applications	I know about applications regarding travel on public transport services such as buses, trains, or taxis.	I use applications regarding travel on public transport services such as buses, trains or taxis.

Table A7. Domain 7: Smart Digital Infrastructure.

Item	Understanding	Acceptance
Enhance the roles of service providers in developing digital infrastructure	I am confident that the infrastructure sharing policy among service providers will provide better high-speed internet services.	I understand that the role of completing the communication infrastructure of a new development project is the responsibility of the developer.
Enhance internet speed and connectivity	I know that the government will enforce minimum internet speed standards in stages.	I am aware that most major cities in Malaysia are equipped with 4G high-speed internet facilities.
Enhance the government's role in facilitating the development of communication infrastructure	I understand that the government always assists service providers in facilitating the development of communication infrastructure.	I believe that the Malaysian Commission of Communications and Multimedia (MCMC) should enforce the appropriate standards for network services.
Enhance indoor and outdoor network coverage	I agree that development companies need to equip new development projects with fiber optic lines to support the Smart City policy.	I agree that new buildings are equipped with in-building fiber optic network access facilities.
Strengthen policies related to personal data protection	I am confident that the personal information of internet users is protected by the Personal Protection Act 2010.	I am sure that the reduction of cyber threats will have a positive impact on the government, companies and individuals.
Strengthen policies related to cybersecurity	I understand that online systems and information are safe from cyber-attacks.	I feel that policies and laws related to cybersecurity and personal data need to be updated periodically to protect consumers.

Appendix B. Fuzzy Delphi Analysis Results

Table A8. Domain 1: Smart Economy (Understanding).

No.	Triangular Fuzzy Number		Defuzzification Process				Result	Ranking according to the Fuzzy Score
	Threshold (d) Value	Expert Agreement (%)	m_1	m_2	m_3	Average of Fuzzy Score		
1	0.067	88%	0.575	0.775	0.975	0.775	Accepted	1
2	0.107	78%	0.555	0.755	0.955	0.755	Accepted	2
3	0.210	33%	0.455	0.655	0.855	0.655	Rejected	7
4	0.130	78%	0.545	0.745	0.945	0.745	Accepted	3
5	0.193	88%	0.490	0.690	0.890	0.690	Accepted	6
6	0.156	98%	0.515	0.715	0.915	0.715	Accepted	5
7	0.133	73%	0.540	0.740	0.940	0.740	Rejected	4

Table A9. Domain 1: Smart Economy (Acceptance).

No.	Triangular Fuzzy Number		Defuzzification Process				Result	Ranking according to the Fuzzy Score
	Threshold (d) Value	Expert Agreement (%)	m_1	m_2	m_3	Average of Fuzzy Score		
1	0.080	88%	0.570	0.770	0.970	0.770	Accepted	1
2	0.130	78%	0.545	0.745	0.945	0.745	Accepted	3
3	0.179	90%	0.510	0.710	0.910	0.710	Accepted	7
4	0.156	98%	0.515	0.715	0.915	0.715	Accepted	4
5	0.158	98%	0.510	0.710	0.910	0.710	Accepted	6
6	0.156	98%	0.515	0.715	0.915	0.715	Accepted	4
7	0.115	75%	0.550	0.750	0.950	0.750	Accepted	2

Table A10. Domain 2: Smart Living (Understanding).

No.	Triangular Fuzzy Number		Defuzzification Process				Result	Ranking according to the Fuzzy Score
	Threshold (d) Value	Expert Agreement (%)	m_1	m_2	m_3	Average of Fuzzy Score		
1	0.206	35%	0.420	0.620	0.820	0.620	Rejected	5
2	0.101	83%	0.560	0.760	0.960	0.760	Accepted	2
3	0.171	81%	0.41	0.78	0.87	0.687	Accepted	4
4	0.113	83%	0.555	0.755	0.955	0.755	Accepted	3
5	0.069	90%	0.575	0.775	0.975	0.775	Accepted	1

Table A11. Domain 2: Smart Living (Acceptance).

No.	Triangular Fuzzy Number		Defuzzification Process				Result	Ranking according to the Fuzzy Score
	Threshold (d) Value	Expert Agreement (%)	m_1	m_2	m_3	Average of Fuzzy Score		
1	0.179	90%	0.470	0.670	0.870	0.670	Accepted	5
2	0.183	90%	0.500	0.700	0.900	0.700	Accepted	3
3	0.165	85%	0.492	0.824	0.741	0.686	Accepted	4
4	0.155	95%	0.525	0.725	0.925	0.725	Accepted	1
5	0.172	93%	0.510	0.710	0.910	0.710	Accepted	2

Table A12. Domain 3: Smart Environment (Understanding).

No.	Triangular Fuzzy Number		Defuzzification Process				Result	Ranking according to the Fuzzy Score
	Threshold (d) Value	Expert Agreement (%)	m_1	m_2	m_3	Average of Fuzzy Score		
1	0.242	35%	0.375	0.570	0.770	0.572	Rejected	9
2	0.128	70%	0.540	0.740	0.940	0.740	Rejected	2
3	0.214	35%	0.425	0.625	0.825	0.625	Rejected	5
4	0.122	73%	0.545	0.745	0.945	0.745	Rejected	1
5	0.155	93%	0.530	0.730	0.930	0.730	Accepted	3
6	0.193	90%	0.490	0.690	0.890	0.690	Accepted	4
7	0.244	23%	0.400	0.600	0.800	0.600	Rejected	6
8	0.201	48%	0.390	0.585	0.785	0.587	Rejected	8
9	0.204	45%	0.400	0.590	0.790	0.593	Rejected	7

Table A13. Domain 3: Smart Environment (Acceptance).

No.	Triangular Fuzzy Number		Defuzzification Process				Result	Ranking according to the Fuzzy Score
	Threshold (*d*) Value	Expert Agreement (%)	m_1	m_2	m_3	Average of Fuzzy Score		
1	0.174	50%	0.425	0.625	0.825	0.625	Rejected	6
2	0.233	33%	0.395	0.590	0.790	0.592	Rejected	7
3	0.219	35%	0.445	0.640	0.840	0.642	Rejected	4
4	0.260	38%	0.380	0.570	0.770	0.573	Rejected	8
5	0.189	88%	0.505	0.705	0.905	0.705	Accepted	2
6	0.181	90%	0.505	0.705	0.905	0.705	Accepted	2
7	0.256	38%	0.360	0.555	0.755	0.557	Rejected	9
8	0.246	30%	0.435	0.630	0.830	0.632	Rejected	5
9	0.153	98%	0.520	0.720	0.920	0.720	Accepted	1

Table A14. Domain 4: Smart People (Understanding).

No.	Triangular Fuzzy Number		Defuzzification Process				Result	Ranking according to the Fuzzy Score
	Threshold (*d*) Value	Expert Agreement (%)	m_1	m_2	m_3	Average of Fuzzy Score		
1	0.088	83%	0.565	0.765	0.965	0.765	Accepted	1
2	0.107	78%	0.555	0.755	0.955	0.755	Accepted	2
3	0.133	73%	0.540	0.740	0.940	0.740	Rejected	5
4	0.122	73%	0.545	0.745	0.945	0.745	Rejected	4
5	0.122	80%	0.550	0.750	0.950	0.750	Accepted	3
6	0.169	93%	0.515	0.715	0.915	0.715	Accepted	6

Table A15. Domain 4: Smart People (Acceptance).

No.	Triangular Fuzzy Number		Defuzzification Process				Result	Ranking according to the Fuzzy Score
	Threshold (*d*) Value	Expert Agreement (%)	m_1	m_2	m_3	Average of Fuzzy Score		
1	0.110	80%	0.555	0.755	0.955	0.755	Accepted	2
2	0.080	88%	0.570	0.770	0.970	0.770	Accepted	1
3	0.128	70%	0.540	0.740	0.940	0.740	Rejected	3
4	0.133	73%	0.540	0.740	0.940	0.740	Rejected	3
5	0.150	95%	0.530	0.730	0.930	0.730	Accepted	5
6	0.165	93%	0.520	0.720	0.920	0.720	Accepted	6

Table A16. Domain 5: Smart Government (Understanding).

No.	Triangular Fuzzy Number		Defuzzification Process				Result	Ranking according to the Fuzzy Score
	Threshold (*d*) Value	Expert Agreement (%)	m_1	m_2	m_3	Average of Fuzzy Score		
1	0.182	95%	0.505	0.700	0.900	0.702	Accepted	2
2	0.192	90%	0.495	0.695	0.895	0.695	Accepted	3
3	0.224	88%	0.500	0.690	0.890	0.693	Rejected	4
4	0.155	95%	0.525	0.725	0.925	0.725	Accepted	1

Table A17. Domain 5: Smart Government (Acceptance).

No.	Triangular Fuzzy Number		Defuzzification Process				Result	Ranking according to the Fuzzy Score
	Threshold (*d*) Value	Expert Agreement (%)	m_1	m_2	m_3	Average of Fuzzy Score		
1	0.176	95%	0.495	0.695	0.895	0.695	Accepted	3
2	0.169	93%	0.515	0.715	0.915	0.715	Accepted	1
3	0.202	85%	0.480	0.680	0.880	0.680	Rejected	4
4	0.191	90%	0.500	0.700	0.900	0.700	Accepted	2

Table A18. Domain 6: Smart Mobility (Understanding).

No.	Triangular Fuzzy Number		Defuzzification Process				Result	Ranking according to the Fuzzy Score
	Threshold (*d*) Value	Expert Agreement (%)	m_1	m_2	m_3	Average of Fuzzy Score		
1	0.137	75%	0.540	0.740	0.940	0.740	Accepted	3
2	0.130	78%	0.545	0.745	0.945	0.745	Accepted	2
3	0.113	83%	0.555	0.755	0.955	0.755	Accepted	1
4	0.209	85%	0.495	0.695	0.895	0.695	Rejected	7
5	0.147	80%	0.540	0.740	0.940	0.740	Accepted	3
6	0.220	85%	0.485	0.685	0.885	0.685	Rejected	8
7	0.175	90%	0.515	0.715	0.915	0.715	Accepted	6
8	0.177	90%	0.520	0.720	0.920	0.720	Accepted	5

Table A19. Domain 6: Smart Mobility (Acceptance).

No.	Triangular Fuzzy Number		Defuzzification Process				Result	Ranking according to the Fuzzy Score
	Threshold (*d*) Value	Expert Agreement (%)	m_1	m_2	m_3	Average of Fuzzy Score		
1	0.311	20%	0.390	0.590	0.790	0.590	Rejected	7
2	0.184	90%	0.485	0.685	0.885	0.685	Accepted	4
3	0.227	80%	0.490	0.690	0.890	0.690	Rejected	3
4	0.176	93%	0.500	0.700	0.900	0.700	Accepted	1
5	0.245	25%	0.465	0.660	0.860	0.662	Rejected	5
6	0.346	25%	0.380	0.555	0.755	0.563	Rejected	8
7	0.200	93%	0.495	0.690	0.890	0.692	Accepted	2
8	0.273	20%	0.460	0.650	0.850	0.653	Rejected	6

Table A20. Domain 7: Smart Digital Infrastructure (Understanding).

No.	Triangular Fuzzy Number		Defuzzification Process				Result	Ranking according to the Fuzzy Score
	Threshold (*d*) Value	Expert Agreement (%)	m_1	m_2	m_3	Average of Fuzzy Score		
1	0.165	95%	0.530	0.725	0.925	0.727	Accepted	2
2	0.217	85%	0.495	0.690	0.890	0.692	Accepted	3
3	0.191	93%	0.485	0.680	0.880	0.682	Accepted	4
4	0.144	73%	0.535	0.735	0.935	0.735	Rejected	1
5	0.214	30%	0.460	0.660	0.860	0.660	Rejected	5
6	0.291	58%	0.330	0.520	0.720	0.523	Rejected	6

Table A21. Domain 7: Smart Digital Infrastructure (Acceptance).

No.	Triangular Fuzzy Number		Defuzzification Process				Result	Ranking according to the Fuzzy Score
	Threshold (*d*) Value	Expert Agreement (%)	m_1	m_2	m_3	Average of Fuzzy Score		
1	0.221	33%	0.455	0.655	0.855	0.655	Rejected	6
2	0.202	90%	0.490	0.690	0.890	0.690	Rejected	5
3	0.110	80%	0.555	0.755	0.955	0.755	Accepted	2
4	0.101	83%	0.560	0.760	0.960	0.760	Accepted	1
5	0.134	80%	0.545	0.745	0.945	0.745	Accepted	3
6	0.134	80%	0.545	0.745	0.945	0.745	Accepted	3

Note: Three conditions to accept an item: threshold value (d) ≤ 0.2, percentage of experts' consensus $\geq 75\%$, and average fuzzy score $(A_{max}) \geq \alpha -$ cut value = 0.5.

References

1. De Souza, K.C.; Hunter, M.; Jacob, B.; Yigitcanlar, T. Pathways to the Making of Prosperous Smart Cities: An Exploratory Study on the Best Practice. *J. Urban Technol.* **2020**, *27*, 3–32. [CrossRef]
2. Giffinger, R.; Fertner, C.; Kramar, H.; Kalasek, R.; Pichler, N.; Meijers, E. *Smart Cities: Ranking of European Medium-Sized Cities*; TU Vienna: Wien, Astria, 2007.
3. Cohen, B. What Exactly Is a Smart City? Available online: https://www.fastcodesign.com/1680538/what-exactly-is-a-smart-city (accessed on 8 August 2017).

4. Chourabi, H.; Nam, T.; Walker, S.; Gil-Garcia, J.R.; Mellouli, S.; Nahon, K.; Pardo, T.A.; Scholl, H.J. Understanding smart cities: An integrative framework. In Proceedings of the 45th Hawaii International Conference on System Sciences Understanding (ICSS), Maui, HI, USA, 4–7 January 2012; IEEE: Piscataway, NJ, USA, 2012; pp. 2289–2297.
5. Alonso, R.G.; Castro, S.L. Technology helps, people make: A smart city governance framework grounded in deliberative democracy. In *Smarter as the New Urban Agenda;* A Comprehensive View of the 21st Century City; Gil-Garcia, J.R., Pardo, T.A., Nam, T., Eds.; Springer: Cham, Switzerland, 2016; pp. 333–347.
6. Mosannenzadeh, F.; Vettorato, D. Defining Smart City. A Conceptual Framework Based on Keyword Analysis. *TeMA J. Land Use Mobil. Environ.* **2014**, *6*, 683–694. [CrossRef]
7. Anthopoulos, L.; Janssen, M.; Weerakkody, V. A Unified Smart City Model (USCM) for Smart City Conceptualization and Benchmarking. *Int. J. Electron. Gov. Res.* **2016**, *12*, 77–93. [CrossRef]
8. Hong Kong. Hong Kong Smart City Blueprint. Available online: https://www.smartcity.gov.hk/doc/HongKongSmartCityBlueprint(EN).pdf (accessed on 2 February 2020).
9. Ministry of Housing and Local Government. *Malaysia Smart City Framework;* Ministry of Housing and Local Government: Putrajaya, Malaysia, 2019.
10. Lombardi, P.; Giordano, S.; Farouh, H.; Yousef, W. Modelling the smart city performance. *Innov. Eur. J. Soc. Sci. Res.* **2012**, *25*, 137–149. [CrossRef]
11. Tahir, Z.; Malek, J.A. Main criteria in the development of smart cities determined using analytical method. *Plan. Malays. J.* **2016**, *14*, 1–14. [CrossRef]
12. Yasmin, M.A.; Hasniyati, H.; Melasutra, M.D.; Md Nasir, D.; Anuar, A. An initiatives-based framework for assessing smart city. *Plan Malays.* **2016**, *14*, 13–22.
13. Li, C.; Dai, Z.; Liu, X.; Sun, W. Evaluation system: Evaluation of smart city shareable framework and its applications in China. *Sustainability* **2020**, *12*, 2957. [CrossRef]
14. Aghimien, D.O.; Aigbavboa, C.; Edwards, D.J.; Mahamadu, A.-M.; Olomolaiye, P.; Nash, H.; Onyia, M. A fuzzy synthetic evaluation of the challenges of smart city development in developing countries. *Smart Sustain. Built Environ.* **2020**. [CrossRef]
15. Sharifi, A. A typology of smart city assessment tools and indicator sets. *Sustain. Cities Soc.* **2020**, *53*, 101936. [CrossRef]
16. Institute for Management Development (IMD). *Smart City Index;* Institute for Management Development and Singapore University of Technology and Design: Singapore, 2019.
17. OECD. *Measuring Smart Cities' Performance: Do Smart Cities Benefit Everyone?* Organisation for Economic Co-operation and Development: Paris, France, 2020.
18. Caird, S.P.; Hallett, S.H. Towards evaluation design for smart city development. *J. Urban Des.* **2019**, *24*, 188–209. [CrossRef]
19. Lim, S.B.; Malek, J.A.; Hashim, N. Implementing the smart city concept in Malaysia: Contemporary challenges, strategies and opportunities in the COVID-19 era. *Malays. Townplan J.* **2021**, 1–20, forthcoming.
20. Vinod Kumar, T.M.; Dahiya, B. Smart economy in smart cities. In *Smart Economy in Smart Cities, Advances in 21st Century Human Settlements;* Vinod Kumar, T.M., Ed.; Springer: Singapore, 2017; pp. 3–76.
21. Deng, T.; Zhang, K.; Shen, Z.J. A systematic review of a digital twin city: A new pattern of urban governance toward smart cities. *J. Manag. Sci. Eng.* **2021**, *6*, 125–134. [CrossRef]
22. Yigitcanlar, T.; Cugurullo, F. The Sustainability of Artificial Intelligence: An Urbanistic Viewpoint from the Lens of Smart and Sustainable Cities. *Sustainability* **2020**, *12*, 8548. [CrossRef]
23. British High Commission Kuala Lumpur. *Smart City Handbook Malaysia: How Technology and Data Are Shaping the Future of Malaysian Cities;* British High Commission Kuala Lumpur: Kuala Lumpur, Malaysia, 2021.
24. Centre for Liveable Cities. *ASEAN Smart Cities Network;* Centre for Liveable Cities: Singapore, 2018.
25. Low, S.; Ullah, F.; Shirowzhan, S.; Sepasgozar, S.M.E.; Lee, C.L. Smart Digital Marketing Capabilities for Sustainable Property Development: A Case of Malaysia. *Sustainability* **2020**, *12*, 5402. [CrossRef]
26. Hasibuan, A.; Sulaiman, O.K. Smart city, konsep kota cerdas sebagai alternatif penyelesaian masalah perkotaan kabupat-en/kota. *Bul. Utama Tek.* **2019**, *14*, 127–135.
27. Oxford Business Group. Going Strong: The Shift toward Higher-Value-Added Activities Is Creating New Challenges as well as Prospects. Available online: https://oxfordbusinessgroup.com/overview/going-strong-shift-toward-higher-value-added-activities-creating-new-challenges-well-prospects (accessed on 27 July 2021).
28. Rana, N.P.; Luthra, S.; Mangla, S.K.; Islam, R.; Roderick, S.; Dwivedi, Y.K. Barriers to the development of smart cities in Indian context. *Inf. Syst. Front.* **2019**, *21*, 503–525. [CrossRef]
29. Firmansyah, H.S.; Supangkat, S.H.; Arman, A.A. Studi tentang model pengembangan kota cerdas. In Proceedings of the e-Indonesia Initiatives (eII-Forum), Bandung, Indonesia, 15–16 October 2015; Institut Teknologi Bandung: Bandung, Indonesia, 2015; pp. 42–47.
30. Eden Strategy Institute. *Top 50 Smart City Governments;* Eden Strategy Institute and Ong&Ong Pte Ltd.: Singapore, 2018.
31. Diez, T. Personal Fabrication: Fab Labs as Platforms for Citizen-Based Innovation, from Microcontrollers to Cities. *Nexus Netw. J.* **2012**, *14*, 457–468. [CrossRef]
32. Monzon, A. Smart cities concept and challenges: Bases for the assessment of smart city projects. In *Smartgreens 2015 and Vehits 2015, Communications in Computer and Information Science 579;* Helfert, M., Krempels, K.H., Klein, C., Donellan, B., Guiskhin, O., Eds.; Springer: Cham, Switzerland, 2015; pp. 17–31.

33. Katz, B.; Wagner, J. *The Rise of Innovation Districts: A New Geography of Innovation in America*; Brookings Metropolitan Policy Program: Washington, DC, USA, 2014.
34. BEA. What Is Industry Value Added? Bureau of Economic Analysis, US Department of Commerce. Available online: https://www.bea.gov/faq/index.cfm?faq_id=1842006 (accessed on 24 July 2021).
35. Govada, S.S.; Spruijt, W.; Rodgers, T. Introduction to Hong Kong's Development. In *Smart Economy in Smart Cities*; Vinod Kumar, T.M., Dahiya, B., Eds.; Springer: Singapore, 2017; pp. 171–186.
36. Lim, S.B.; Yong, C.K.; Malek, J.A.; Jali, M.F.M.; Awang, A.H.; Tahir, Z. Effectiveness of Fear and Crime Prevention Strategy for Sustainability of Safe City. *Sustainability* **2020**, *12*, 593. [CrossRef]
37. Alkawsi, G.A.; Ali, N.; Mustafa, A.S.; Baashar, Y.; Alhussian, H.; Alkahtani, A.; Tiong, S.K.; Ekanayake, J. A hybrid SEM-neural network method for identifying acceptance factors of the smart meters in Malaysia: Challenges perspective. *Alex. Eng. J.* **2021**, *60*, 227–240. [CrossRef]
38. Hollands, R.G. Beyond the corporate smart city? Glimpses of other possibilities of smartness. In *Smart Urbanism: Utopian Vision or False Dawn?* Marvin, S., Luque-Ayala, A., McFarlane, C., Eds.; Routledge: London, UK, 2016; pp. 168–184.
39. Anttiroiko, A.-V. Successful government responses to the pandemic: Contextualizing national and urban responses to the COVID-19 outbreak in East and West. *Int. J. E-Plan. Res.* **2021**, *10*, 1–17. [CrossRef]
40. Choi, C.; Choi, J.; Kim, C.; Lee, D.; Baek, U.; Sim, Y. The Smart City Evolution in South Korea: Findings from Big Data Analytics. *J. Asian Financ. Econ. Bus.* **2020**, *7*, 301–311. [CrossRef]
41. Kashif, M.; Samsi, S.Z.M.; Awang, Z.; Mohamad, M. EXQ: Measurement of healthcare experience quality in Malaysian settings. A contextualist perspective. *Int. J. Pharm. Healthc. Mark.* **2016**, *10*, 27–47. [CrossRef]
42. Petrova-Antonova, D.; Ilieva, S. Smart cities evaluation—A survey of performance and sustainability indicators. In Proceedings of the 44th Euromicro Conference on Software Engineering and Advanced Application (SEAA), Prague, Czech Republic, 29–31 August 2018; IEEE: Piscataway, NJ, USA, 2018; pp. 486–493.
43. Yigitcanlar, T.; De Souza, K.; Butler, L.; Roozkhosh, F. Contributions and Risks of Artificial Intelligence (AI) in Building Smarter Cities: Insights from a Systematic Review of the Literature. *Energies* **2020**, *13*, 1473. [CrossRef]
44. Smart Selangor Delivery Unit (SSDU). *Smart Selangor Action Plan to 2025 (SSAP 2025)*; Menteri Besar Selangor Inc.: Shah Alam, Malaysia, 2020.
45. PLANMalaysia. *Bandar Selamat 2010*; PLANMalaysia (Jabatan Perancangan Bandar dan Desa): Kuala Lumpur, Malaysia, 2010.
46. PLANMalaysia. *Reka Bentuk Bandar Selamat: Panduan Pelaksanaan. Crime Prevention through Environmental Design (CPTED) Implementation Guide*; PLANMalaysia (Jabatan Perancangan Bandar dan Desa): Kuala Lumpur, Malaysia, 2010.
47. Lim, S.B.; Yong, C.K.; Rashid, M.F.A.; Malek, J.A. A framework of challenges facing the safe city programme in Kuala Lumpur. *Plan. Malays.* **2020**, *18*, 47–61.
48. Tan-Mullins, M.; Cheshmehzangi, A.; Chien, S.-S.; Xie, L. *Smart-Eco Cities in China: Trends and City Profiles 2016*; University of Exeter (SMART-ECO Project): Exerter, UK, 2017.
49. Valencia, S.C.; Simon, D.; Croese, S.; Nordqvist, J.; Oloko, M.; Sharma, T.; Buck, N.T.; Versace, I. Adapting the Sustainable Development Goals and the New Urban Agenda to the city level: Initial reflections from a comparative research project. *Int. J. Urban Sustain. Dev.* **2019**, *11*, 4–23. [CrossRef]
50. Kundu, D.; Sietchiping, R.; Kinyanjui, M. (Eds.) *Developing National Urban Policies: Ways Forward to Green and Smart Cities*; Springer: Singapore, 2020.
51. Yigitcanlar, T.; Butler, L.; Windle, E.; DeSouza, K.; Mehmood, R.; Corchado, J. Can Building "Artificially Intelligent Cities" Safeguard Humanity from Natural Disasters, Pandemics, and Other Catastrophes? An Urban Scholar's Perspective. *Sensors* **2020**, *20*, 2988. [CrossRef] [PubMed]
52. UN-Habitat. *World Cities Report 2020: The Value of Sustainable Urbanization*; United Nations Human Settlements Programme: Nairobi, Kenya, 2020.
53. Connolly, C. From resilience to multi-species flourishing: (Re)imagining urban-environmental governance in Penang, Malaysia. *Urban Stud.* **2019**, *57*, 1485–1501. [CrossRef]
54. Fong, V. Smart Selangor Blueprint: Selangor Smart Cities? Available online: http://www.smartcitiesasia.com/blueprint-smart-city-selangor/ (accessed on 1 February 2018).
55. Yigitcanlar, T.; Teriman, S. Rethinking sustainable urban development: Towards an integrated planning and development process. *Int. J. Environ. Sci. Technol.* **2015**, *12*, 341–352. [CrossRef]
56. Qureshi, F. In Malaysia, the Fate of a Peat Forest Relies on a Powerful State Official. Available online: https://www.eco-business.com/news/in-malaysia-the-fate-of-a-peat-forest-relies-on-a-powerful-state-official/ (accessed on 25 July 2021).
57. Abdullah, M.Y.H.; Pawanteh, L.; Abdullah, S.M.S.; Mustaffa, N. Kesedaran dan sikap komuniti dalam pengurusan perseki-taran di Wilayah Iskandar, Johor. *J. Melayu.* **2010**, *5*, 71–86.
58. Hare, R.M. A Philosophical Autobiography. *Utilitas* **2002**, *14*, 269–305. [CrossRef]
59. Borhannuddin, M.S. Ungku Aziz's perspective on "development". *Turk. J. Islam. Econ.* **2015**, *2*, 1–15. [CrossRef]
60. Tegos, S. Civility and politeness in early modern thought. In *Encyclopedia of Early Modern Philosophy and the Sciences*; Jalobeanu, D., Wolfe, C.T., Eds.; Springer: Cham, Switzerland, 2021; pp. 1–8.
61. Arnstein, S.R. A Ladder of Citizen Participation. *J. Am. Inst. Plan.* **1969**, *35*, 216–224. [CrossRef]

62. Malek, J.A.; Lim, S.B.; Yigitcanlar, T. Social Inclusion Indicators for Building Citizen-Centric Smart Cities: A Systematic Literature Review. *Sustainability* **2021**, *13*, 376. [CrossRef]
63. Rotta, M.J.R.; Sell, D.; dos Santos Pacheco, R.C.; Yigitcanlar, T. Digital Commons and Citizen Coproduction in Smart Cities: Assessment of Brazilian Municipal E-Government Platforms. *Energies* **2019**, *12*, 2813. [CrossRef]
64. United Nations (UN). *E-Government Survey 2020*; Department of Economic and Social Affairs, United Nations: New York, NY, USA, 2020.
65. Yigitcanlar, T.; Kankanamge, N.; Vella, K. How are smart city concepts and technologies perceived and utilized? A systematic geo-Twitter analysis of smart cities in Australia. *J. Urban Technol.* **2021**, *28*, 135–154. [CrossRef]
66. Bedi, R.S. Malaysia Embarks on Digitalisation Journey with National 4IR Policy Launch. Available online: https://www.thestar.com.my/news/nation/2021/07/01/malaysia-embarks-on-digitalisation-journey-with-national-4ir-policy-launch (accessed on 26 July 2021).
67. Economic Planning Unit. *National Fourth Industrial Revolution (4IR) Policy*; Economic Planning Unit, Prime Minister's Department: Putrajaya, Malaysia, 2021.
68. Caragliu, A.; del Bo, C.F.; Nijkamp, P. Smart Cities in Europe. *J. Urban Technol.* **2011**, *18*, 65–82. [CrossRef]
69. Lim, S.B.; Malek, J.A.; Hussain, M.Y.; Tahir, Z.; Saman, N.H.M. SDGs, smart urbanisation, and politics: Stakeholder partner-ships and environmental cases in Malaysia. *J. Sustain. Sci. Manag.* **2021**, *16*, 190–219. [CrossRef]
70. Meijer, A.; Bolívar, M.P.R. Governing the smart city: A review of the literature on smart urban governance. *Int. Rev. Adm. Sci.* **2016**, *82*, 392–408. [CrossRef]
71. Vinod Kumar, T.M. (Ed.) E-Governance for Smart Cities. In *E-Governance for Smart Cities, Advances in 21st Century Human Settlements*; Springer: Singapore, 2015; pp. 1–46.
72. Johnson, P.A.; Acedo, A.; Robinson, P.J. Canadian smart cities: Are we wiring new citizen-local government interactions? *Can. Geogr.* **2020**, *64*, 402–415. [CrossRef]
73. Lim, S.B.; Malek, J.A.; Hussain, M.Y.; Tahir, Z. Participation in e-government services and smart city programs: A case study of malaysian local authority. *Plan. Malays. J.* **2020**, *18*, 300–312. [CrossRef]
74. Meijer, A. Datapolis: A public governance perspective on "smart cities". *Perspect. Public Manag. Gov.* **2018**, *1*, 195–206. [CrossRef]
75. Urban Tide. *Overview of the Smart Cities Maturity Model: Joining the Dots of Smart Cities*; Urban Tide and Scottish Government: Edinburgh, UK, 2014.
76. Komninos, N.; Kakderi, C.; Panori, A.; Tsarchopoulos, P. Smart City Planning from an Evolutionary Perspective. *J. Urban Technol.* **2019**, *26*, 3–20. [CrossRef]
77. Purwanto, A.; Zuiderwijk, A.; Janssen, M. Citizen engagement with open government data: A systematic literature review of drivers and inhibitors. *Int. J. Electron. Gov. Res.* **2020**, *16*, 1–25. [CrossRef]
78. Yigitcanlar, T.; Han, H.; Kamruzzaman, M.; Ioppolo, G.; Sabatini-Marques, J. The making of smart cities: Are Songdo, Masdar, Amsterdam, San Francisco and Brisbane the best we could build? *Land Use Policy* **2019**, *88*, 104187. [CrossRef]
79. Yigitcanlar, T. Smart city policies revisited: Considerations for a truly smart and sustainable urbanism practice. *World Technop. Rev.* **2018**, *7*, 97–112.
80. Thestar. Smart City Players Working in Silos Hindering Development as a Whole, Says Dr M. Available online: https://www.thestar.com.my/news/nation/2019/09/23/smart-city-players-working-in-silos-hindering-development-as-a-whole-says-dr-m (accessed on 2 February 2021).
81. Idrus, N.A.; Ismail, S.; Sanusi, F.A. Delays in Malaysian government projects: Learning from project management failure. *Jurutera* **2019**, *10*, 13–19.
82. Ramli, H. Birokrasi Jejas Sistem Penyampaian. Available online: https://www.mef.org.my/MEFITN/utusan070209a.pdf (accessed on 20 July 2021).
83. Trombin, M.; Pinna, R.; Musso, M.; Magnaghi, E.; De Marco, M. Mobility management: From traditional to people-centric approach in the smart city. In *Emerging Technologies for Connected Internet of Vehicles and Intelligent Transportation System Networks, Studies in Systems, Decision and Control 242*; Elhoseny, M., Hassanien, E., Eds.; Springer: Cham, Switzerland, 2020; pp. 165–182.
84. Yigitcanlar, T.; Kamruzzaman, M. Smart Cities and Mobility: Does the Smartness of Australian Cities Lead to Sustainable Commuting Patterns? *J. Urban Technol.* **2019**, *26*, 21–46. [CrossRef]
85. Asirvatham, D.; Brohi, S.N.; Xion, T.E.; Theng, T.G.; Fei, N.J.; Kaur, S.; Pillai, T.R.; Sukumaran, S.; Yue, W.S.; Wei, G.; et al. *Smart Mobility Cities: Connecting Bristol and Kuala Lumpur Project Report*; University of Bristol: Bristol, UK, 2018.
86. D'Amico, G.; L'Abbate, P.; Liao, W.; Yigitcanlar, T.; Ioppolo, G. Understanding Sensor Cities: Insights from Technology Giant Company Driven Smart Urbanism Practices. *Sensors* **2020**, *20*, 4391. [CrossRef] [PubMed]
87. Walker, A. Safe Streets Are the Best Tool We Have to Combat Climate Change and We Need to Act Now. Available online: https://www.curbed.com/2018/10/10/17957532/climate-change-street-design-vision-zero (accessed on 14 March 2020).
88. NYC Mayor's Office. *Building a Smart + Equitable City*; NYC Mayor's Office of Tech + Innovation: New York, NY, USA, 2015.
89. Sarkar, A. Smart Cities: A Futuristic Vision. Available online: https://www.thesmartcityjournal.com/en/articles/1333-smart-cities-futuristic-vision (accessed on 26 July 2021).
90. Frost and Sullivan. Smart Cities—Frost & Sullivan Value Proposition. Available online: https://www.frost.com/wp-content/uploads/2019/01/SmartCities.pdf (accessed on 5 August 2021).

91. Dirks, S.; Keeling, M. *A Vision of Smarter Cities: How Cities Can Lead Way into a Prosperous and Sustainable Future*; International Business Machines Corporation: New York, NY, USA, 2009.
92. Praharaj, S.; Han, J.H.; Hawken, S. Innovative Civic Engagement and Digital Urban Infrastructure: Lessons from 100 Smart Cities Mission in India. *Procedia Eng.* **2017**, *180*, 1423–1432. [CrossRef]
93. Castelnovo, W.; Misuraca, G.; Savoldelli, A. Smart cities governance: The need for a holistic approach to assessing urban participatory policy making. *Soc. Sci. Comput. Rev.* **2016**, *34*, 724–739. [CrossRef]
94. Wilson, B.; Chakraborty, A. Planning Smart(er) Cities: The Promise of Civic Technology. *J. Urban Technol.* **2019**, *26*, 29–51. [CrossRef]
95. Caprotti, F.; Springer, C.; Harmer, N. 'Eco' For Whom? Envisioning Eco-urbanism in the Sino-Singapore Tianjin Eco-city, China. *Int. J. Urban Reg. Res.* **2015**, *39*, 495–517. [CrossRef]
96. Breuer, J.; Pierson, J. The right to the city and data protection: Complementary for developing citizen-centric digital cities. *Inf. Commun. Soc.* **2021**, *24*, 797–812. [CrossRef]
97. Gjermundrød, H.; Dionysiou, I. A conceptual framework for configurable privacy-awareness in a citizen-centric eGovernment. *Electron. Gov. Int. J.* **2015**, *11*, 258. [CrossRef]
98. Banisar, D. *The Right to Information and Privacy: Balancing Rights and Managing Conflicts*; The World Bank: Washington, DC, USA, 2011.
99. Nautiyal, L.; Malik, P.; Agarwal, A. Cybersecurity System: An Essential Pillar of Smart Cities. In *Smart Cities, Computer Communications and Networks*; Mahmood, Z., Ed.; Springer: Cham, Switzerland, 2018; pp. 25–50.
100. Goodwin, C.F.; Nicholas, J.P. *Developing a City Strategy for Cybersecurity: A Seven-Step Guide for Local Governments*; Microsoft: Washington, DC, USA, 2014.
101. Ma, H.D. Internet of Things: Objectives and Scientific Challenges. *J. Comput. Sci. Technol.* **2011**, *26*, 919–924. [CrossRef]
102. Nick, G. How Many IoT Devices Are There in 2021? Available online: https://techjury.net/blog/how-many-iot-devices-are-there/#gref (accessed on 27 July 2021).
103. Devanesan, J. Cybersecurity Is Top Concern, as Online Threats Mount in Malaysia by 82.5%. Available online: https://techwireasia.com/2020/04/cybersecurity-is-top-concern-as-online-threats-mount-in-malaysia-by-82-5/ (accessed on 27 July 2021).
104. Yun, T.Z. Moving beyond Awareness to Adoption of IoT. Available online: https://www.theedgemarkets.com/article/cover-story-moving-beyond-awareness-adoption-iot (accessed on 27 July 2021).
105. Jones, H.; Twiss, B.L. *Forecasting Technology for Planning Decisions*; Macmillan: New York, NY, USA, 1978.
106. Jamil, M.R.M.; Noh, N.M. *Kepelbagaian Metodologi Dalam Penyelidikan: Rekab Bentuk dan Pembangunan*; Qaisar Prestige Resources: Shah Alam, Malaysia, 2020.
107. Embong, A.R. *State-Led Modernization and the New Middle Class in Malaysia*; Palgrave: New York, NY, USA, 2002.
108. Dalkey, N.; Helmer, O. An Experimental Application of the DELPHI Method to the Use of Experts. *Manag. Sci.* **1963**, *9*, 458–467. [CrossRef]
109. Lateh, N.; Yaacob, S.E.; Md Rejab, S.N. Applying the fuzzy delphi method (FDM) to analyze the expert consensus values for instrument of shariah-compliant gold investment. *Pertanika J. Soc. Sci. Humanit.* **2017**, *25*, 165–178.
110. Ulschak, F. *Human Resource Development: The Theory and Practice of Need Assessment*; Reston Publishing Company: Reston, VA, USA, 1983.
111. Grime, M.M.; Wright, G. Delphi Method. In *Wiley StatsRef Statistics Reference Online*; Balakrishnan, N., Colton, T., Everitt, B., Piegorsch, W., Ruggeri, F., Teugels, J.L., Eds.; Wiley: Hoboken, NJ, USA, 2016; pp. 1–6.
112. Cheng, C.-H.; Lin, Y. Evaluating the best main battle tank using fuzzy decision theory with linguistic criteria evaluation. *Eur. J. Oper. Res.* **2002**, *142*, 174–186. [CrossRef]
113. Chu, H.-C.; Hwang, G.-J. A Delphi-based approach to developing expert systems with the cooperation of multiple experts. *Expert Syst. Appl.* **2008**, *34*, 2826–2840. [CrossRef]
114. Murry, J.W.; Hammons, J.O. Delphi: A Versatile Methodology for Conducting Qualitative Research. *Rev. High. Educ.* **1995**, *18*, 423–436. [CrossRef]
115. Tang, C.-W.; Wu, C.-T. Obtaining a picture of undergraduate education quality: A voice from inside the university. *High. Educ.* **2010**, *60*, 269–286. [CrossRef]
116. Esmaeilpoorarabi, N.; Yigitcanlar, T.; Guaralda, M.; Kamruzzaman, M. Evaluating place quality in innovation districts: A Delphic hierarchy process approach. *Land Use Policy* **2018**, *76*, 471–486. [CrossRef]
117. Perveen, S.; Kamruzzaman, M.; Yigitcanlar, T. Developing policy scenarios for sustainable urban growth management: A Delphi approach. *Sustainability* **2017**, *9*, 1787. [CrossRef]
118. Sulaiman, H.F.; Ismail, R.; Yusoff, H.M.; Anuar, N.; Jamil, M.R.M.; Daud, F. Validation of occupational zoonotic disease questionnaire using fuzzy Delphi method. *J. Agromed.* **2020**, *25*, 166–172. [CrossRef] [PubMed]
119. Drabble, J. The Economic History of Malaysia. Available online: http://eh.net/encyclopedia/economic-history-of-malaysia/ (accessed on 24 July 2021).
120. Mat Arisah, F.; Zainal Badari, S.A.; Hashim, A.H. Amalan pembelian secara atas talian dan faktor-faktor mempengaruhi. *Malays. J. Soc. Sci. Humanit.* **2016**, *1*, 111–123.
121. Lim, S.B.; Malek, J.A.; Kong, Y.C.; Tahir, Z.; Hernowo, B. Lessons from differences in the global and local discourse views on the safe city status of Kuala Lumpur, Malaysia. *Geogr. Malays. J. Soc. Sp.* **2021**, *17*, 143–158.
122. Malek, J.A.; Lim, S.B.; Tahir, Z. Understanding the issues of citizen participation. *J. Nusant. Stud.* **2019**, *4*, 1–22. [CrossRef]

123. Aziz, S.A. Maqasid Al-Syariah dalam perlembagaan persekutuan: Suatu perbahasan awal. *Kanun J. Undang. Malays.* **2016**, *28*, 278–312.
124. Lim, S.B.; Malek, J.A.; Mohd Yusof, H.; Tahir, Z. Malaysia Smart City Framework: A trusted framework for shaping smart Malaysian citizenship? In *Handbook of Smart Cities*; Augusto, J.C., Ed.; Springer: Cham, Switzerland, 2021; pp. 515–538.
125. Economist Intelligence Unit (EIU). *Democracy Index 2020: In Sickness and in Health?* Economist Intelligence Unit: London, UK, 2020.
126. Araya, D. (Ed.) *Smart Cities as Democratic Ecologies*; Palgrave Macmillan: New York, NY, USA; London, UK, 2015.
127. Lim, S.; Malek, J.; Yigitcanlar, T. Post-Materialist Values of Smart City Societies: International Comparison of Public Values for Good Enough Governance. *Future Internet* **2021**, *13*, 201. [CrossRef]
128. Majid, N.A. 150 Electric Buses for Putrajaya by 2025. Available online: https://www.nst.com.my/news/nation/2017/10/290305/150-electric-buses-putrajaya-2025 (accessed on 25 July 2021).
129. Teoh, L.E.; Khoo, H.L.; Goh, S.Y.; Chong, L.M. Scenario-based electric bus operation: A case study of Putrajaya, Malaysia. *Int. J. Transp. Sci. Technol.* **2018**, *7*, 10–25. [CrossRef]
130. Alias, W.N.H.W. Bas Awam Putrajaya Tersadai, "Merangkak". Available online: https://www.bharian.com.my/berita/nasional/2018/08/460658/bas-awam-putrajaya-tersadai-merangkak (accessed on 25 July 2017).
131. Mustapa, S.; Ayodele, B.; Ishak, W.M.; Ayodele, F. Evaluation of Cost Competitiveness of Electric Vehicles in Malaysia Using Life Cycle Cost Analysis Approach. *Sustainability* **2020**, *12*, 5303. [CrossRef]
132. MITI. *National Automotive Policy 2020*; Ministry of International Trade and Industry: Putrajaya, Malaysia, 2020.
133. Culpan, T. A Quiet Leader: Why Malaysia Ranks High in Global List of Cybersecurity. Available online: https://www.business-standard.com/article/international/malaysia-among-top-ten-countries-in-cybersecurity-due-to-global-cooperation-121070200403_1.html (accessed on 27 July 2021).
134. MCMC. *National Digital Infrastructure Lab (NDIL). Report 3 September 2020*; Malaysian Communications and Multimedia Commission: Putrajaya, Malaysia, 2020.
135. Yigitcanlar, T.; Sarimin, M. Multimedia super corridor, Malaysia: Knowledge-based urban development lessons from an emerging economy. *Vine* **2015**, *45*, 126–147. [CrossRef]
136. Yigitcanlar, T.; Sarimin, M. The Role of Universities in Building Prosperous Knowledge Cities: The Malaysian Experience. *Built Environ.* **2011**, *37*, 260–280. [CrossRef]
137. Yigitcanlar, T.; Dur, F. Making space and place for knowledge communities: Lessons for Australian practice. *Australas. J. Reg. Stud.* **2013**, *19*, 36–63.
138. Perveen., S.; Kamruzzaman, M.; Yigitcanlar, T. What to assess to model the transport impacts of urban growth? A Delphi approach to examine the space–time suitability of transport indicators. *Int. J. Sustain. Transp.* **2019**, *13*, 597–613. [CrossRef]
139. Siokas, G.; Tsakanikas, A.; Siokas, E. Implementing smart city strategies in Greece: Appetite for success. *Cities* **2021**, *108*, 102938. [CrossRef]

sustainability

MDPI

Article

Risk, Recessions, and Resilience: Towards Sustainable Local Labor Markets through Employment Portfolio Analysis

Bharman Gulati [1,*] and Stephan Weiler [1,2]

[1] Department of Economics, Colorado State University, Fort Collins, CO 80523, USA; Stephan.Weiler@colostate.edu
[2] Regional Economic Development Institute (REDI@CSU), Colorado State University, Fort Collins, CO 80523, USA
* Correspondence: b.gulati927@gmail.com

Abstract: This paper explores the role of local labor market dynamics on the survival of new businesses. The characteristics of the local labor market are likely to influence the survival of new businesses, the level of entrepreneurship, and the resilience of the regional economy. We apply portfolio theory to evaluate employment-based and income-based measures of risk-and-return trade-offs in local labor markets on new business survival in the United States. Our results show that volatility in local labor markets has a positive impact on new business survival, especially in Metropolitan Statistical Areas. The results are robust across different timeframes, including during economic downturns, thus highlighting the contribution of new businesses in developing the resilience of the local economy, and further promoting sustainable regional economic development.

Keywords: business survival; economic resilience; employment portfolio; risk–return tradeoff

Citation: Gulati, B.; Weiler, S. Risk, Recessions, and Resilience: Towards Sustainable Local Labor Markets through Employment Portfolio Analysis. *Sustainability* **2021**, *13*, 7926. https://doi.org/10.3390/su13147926

Academic Editor: Tan Yigitcanlar

Received: 31 May 2021
Accepted: 9 July 2021
Published: 15 July 2021

1. Introduction

Factors that impact the survival of a firm can be broadly categorized into firm-specific, industry-specific, and region-specific. Among region-specific factors, a significant factor is the local labor market environment. The local labor market's characteristics are likely to significantly impact the level of entrepreneurship and the success achieved by enterprises in that region.

Recent empirical research has identified the employment portfolio of a local market as a critical determinant influencing entrepreneurship in the region [1–4]. Borrowing from portfolio theory [5] in financial economics, the employment portfolio of the local labor market is now being used to evaluate the prospects of a business enterprise in that location. Portfolio theory helps identify the frontier whereat a combination of financial instruments will yield the optimal return, given their risk profile. Similarly, empirical investigation of the employment portfolio in a region has been shown to create a U-shaped frontier in the local labor market, given the risk–return profile in that market. The return in the labor market, on the horizontal axis, represents the growth in jobs, and the risk, on the vertical axis, represents the volatility in the growth in jobs in that local market.

The trade-off between employment-based risk and return measures in the local labor market is crucial for prospective entrepreneurs to estimate the likelihood of survival of their establishment in the future. Considering the interrelationship between risk and return in a local labor market, we hypothesize that this interrelationship, captured by the employment portfolio, can impact the survival of a new business operating in that region. We are interested in investigating whether the employment portfolio of a region influences the likelihood of survival of a new business in the next period. Is there any heterogeneity in the impact of employment portfolio on new business survival in metro and non-metro regions? How significant is this impact on new business survival in different regions?

Answers to these questions are likely to help design customized policies that promote and support existing and new business establishments in the region. This may also help local governments efficiently allocate financial resources for local economic growth rather than spending on blanket programs.

There is a gap in the existing literature highlighting the impact of local labor market dynamics on new business survival. This paper attempts to fill that gap. Notable investigations in this domain have focused on the role of human capital in the workforce [6], the impact of employment density on labor productivity [7], and the education level of the employer [8]. In turn, we measure the impact of changes in the local labor market environment on new business survival in three steps. First, we begin by identifying the existence of an optimal frontier, represented by risk and return trade-off in the local labor market (at county-level), by using the stochastic frontier estimation technique [1,4]. Second, we investigate the impacts of changes in the employment portfolios in the counties on the new business survival rate in 2010. These business establishments were born in 2005–2006. We find that the local employment-based measures significantly impact the survival of new businesses operating in metro counties. We then use income-based measures to validate our results and find that the latter also have a consistent and significant influence on new businesses, again on those operating in metro counties. Third, we narrow our sample to new businesses operating in Metropolitan Statistical Areas (MSAs) only, using both employment-based and income-based measures, testing the hypothesized relationship with new business survival rates in 2010. We also replicate the third step to investigate the impact on the new business survival rate in 2005. This carries dual benefits: one, it tests for the robustness of our initial findings, and second, it tests for any effects of exogenous shocks on the local labor market environment and their consequent impact on new business survival. The 2001 recession and the Great Recession are two major economic shocks that happened during the study period. By replicating the empirical exercise in different periods, we wish to isolate and study each exogenous event's impact on local labor market dynamics and subsequently on new business survival.

We find that volatility in the local labor market consistently impacts the likelihood of survival of new businesses in MSAs. We also find that this impact is stronger during recessions. Our results suggest that the continued survival of new businesses despite volatility in local labor markets contributes to local economic growth, and to the resilience of the local economy during economic downturns. We acknowledge the clear limitations of this research given the probability of reverse causality/endogeneity, and suggest trajectories for future work given these likely inter-relationships.

Our endeavor also addresses sustainability; in this case in the form of sustainable economic systems, identified through the viability and opportunities created by the local labor market. We find that over the long-term, the dynamics of local labor markets likely influence the behavior of employees, which further enhances the resilience of new businesses. This resilience displayed by new businesses, especially during recessions, promotes sustainable local economic development.

The following sections exhibit the details of this research endeavor. Section 2 reviews the new firm survival literature, the literature on the application of modern portfolio theory in regional economics, and how the application of this approach helps in identifying the impacts of changes in local labor markets on new business survival. Section 3 highlights the methodology and data used for the research. It presents the theoretical framework and the empirical model used in this attempt. The results of this exercise are discussed in Section 4. The final section concludes this paper, highlighting the impact of local labor market dynamics on new business survival, and discusses potential applications to policies aimed at local economic development.

2. Literature Review, Research Niche, and Theoretical Structure

Empirical research into the role of regional factors in firm survival has broadly focused on the impact of agglomeration, scale economies, and institutional legitimacy [6,9–11].

Investigations into the impact of local labor on firm survival have concentrated primarily on the availability and quality of human capital. In this regard, most of the research focuses on the role of entrepreneur-centric human capital in determining firm survival. The level of human capital possessed by the entrepreneur is typically measured in terms of entrepreneurs' personal characteristics. Ref. [6] cites the role of specific characteristics, namely, the level of education [12], entrepreneur's experience in similar roles [13–16], psychological factors such as motivation and ambition [17], preparation, and time spent on the operations [16]. Interestingly, another strand of related literature found no evidence between such personal characteristics and business survival [18,19].

Ref. [6] found a positive relationship between regional human capital and new firm survival in the growth period (1993–1995), and a not so strong relationship during the recession period (1990–1992) when studying the labor market areas (LMAs) in the U.S. They found that "high school dropout rates are negatively associated with firm survival rate for both periods, and college-degree-share is positively related to firm survival in the growth period." Ref. [7] investigated the variation in labor productivity across states in the U.S. They found that doubling employment density in a county results in a 6% increase in average labor productivity. Ref. [8] found that in Finland, the likelihood of survival of firms established by highly educated employers is greater in the recessionary period than in the growth period. They conclude that the general labor market conditions determine the likelihood of firms surviving in the market.

2.1. Theoretical Structure

Modern portfolio theory, espoused by [5], is used in identifying the unique investment portfolio(s) that yield the highest return for the level of risk tolerance of the investor. The portfolio(s) that meet the risk–return trade-off lie on the border, which is commonly referred as the efficient frontier. Since it was first applied by [20] to find the efficient trade-off between incentives to offer and the desired industrial mix in the region, regional economists have increasingly used portfolio theory to capture the risk–return trade-off with relevant variables of interest. Ref. [3] applied portfolio theory and found that luring businesses by offering economic incentives led to increased volatility in growth rates in metropolitan areas. Ref. [1] found an efficient frontier when mapping economic growth and instability for states in the United States. The paper suggests that economic diversity and instability appear to be negatively correlated.

Ref. [4] pioneered the application of portfolio theory to investigate the impacts of local employment portfolios on the entrepreneurship levels. Focusing on commuting zones and counties, they found that entrepreneurship may be an attractive avenue in areas having high employment risks and low returns. They suggest that self-employment can be an attractive alternative income mechanism to wage and salary jobs in such regional units.

The closure of a business establishment results not only in the loss of jobs or potential losses of tax revenue for the local authorities, but also in a unique externality. Business births and deaths represent business dynamism in the region, which potential entrepreneurs closely watch. Low survival rates of new businesses may have a snowball effect in discouraging potential entrepreneurs from investing in the region. The information spillover from local business birth and death rates is found to have a significant impact on subsequent entrepreneurship and job creation [21].

The number of jobs available, the growth in new jobs, and the risks associated with job growth are some of the factors that define the dynamic character of the local labor market. The interaction of these factors in the light of macroeconomic conditions and the unique economic–cultural milieu of the regional unit create opportunities (expected returns) and threats (risks) for the workforce in the region. The impact of exogenous shocks in the local labor market influences the risk–return trade-off of the worker, which affects their economic behavior.

We attempt to capture these dynamics in the local labor market through the representative labor market portfolio. The portfolio approach helps one to analyze the interplay of numerous idiosyncratic but significant labor market undercurrents, in the context of the prevalent macroeconomic environment, when evaluating the likely impact on the survival of new businesses in the regional unit.

We begin by selecting the county as the regional unit of analysis. The resource base and socio-economic environment unique to the county affect a business's growth; hence, the regional unit or location is vital in new business survival.

2.2. How Does Survival of New Businesses Determine the Resilience of the Region?

Recent empirical investigations have highlighted the contribution made by entrepreneurs and local businesses to the recovery of the local economy from exogenous shocks [22–24]. Entrepreneurship provides local policymakers with an appropriate tool to remove path dependency and the threat of lock-in in the local economy [25]. In the aftermath of the Great Recession, local authorities in many states in the U.S. initiated policy measures to promote entrepreneurship and business ventures in sunrise sectors, with the motivation of stimulating economic growth and reducing dependency on historically dominant sectors.

New businesses provide diversity to the local economy [26]. Given their small size, for all the challenges borne by new businesses, they respond faster to external shocks than their larger counterparts. Not only are they able to adapt to the changes, but they are also able to innovate to respond to the repercussions of the shock [23].

We find that risk in local labor markets helps in the survival of new businesses in metro counties and MSAs. Risk manifested through employment-based and income-based portfolios increases the likelihood of survival of new businesses. This impact has been observed to be stronger during the recent recessions in these regional units.

Economists are increasingly using the lens of resilience to investigate the contribution of new businesses and young establishments when evaluating the responses of regional economies to recent recessions. This ability to absorb, endure and recover from an exogenous shock is commonly referred to as resilience.

Resilience is commonly defined from engineering, ecological, and adaptive perspectives. From an engineering perspective, resilience refers to the displacement of an entity from its equilibrium, caused by an exogenous shock, and the subsequent return to the original equilibrium [27,28]. From an ecological perspective, resilience refers to a subsequent shift towards a new equilibrium, in response to displacement caused by the exogenous shock [27,29]. From an adaptive perspective, resilience follows an evolutionary approach, whereby an entity evolves over time to create new sustainable paths by adapting to changes occurring in its environment [30,31].

In this paper, we employ the definition of resilience from an ecological perspective. We find that the economic downturn highlights the undercurrents operating in the local labor market. The economic agents absorb and endure the economic shock, and eventually the economy recovers to achieve a new steady state after a period of time. This process of absorbing, enduring, and recovering over a period of time to reach a new steady state is significantly determined by the dynamics prevalent in the local labor market besides the existing stock of economic resources, local amenities, and level of aid received.

The ability to survive in the face of adverse economic downturns also concerns the human element, as the competencies and skills of both employees and the entrepreneur are entwined with the business. The success and failure of a business are determined, to a large extent, by the fortitude, perseverance, patience, and ingenuity displayed by the employees during challenging times. The resilience displayed by new businesses is hence related to the resilience displayed by the employees, which further determines the resilience of the community.

2.3. Why Does Risk in Labor Markets Affect Workers?

We identify risk through both employment-based and income-based measures. Employment-based risk (wage and salary employment risk, WSE-Risk) represents the volatility in the growth of employment opportunities in a local county. Firms often freeze hiring or lay off workers in response to economic downturns. WSE-Risk is likely to have a profound impact in a recession. Even after the recession, recovery in local employment lags behind recovery observed in other economic indicators. Consequently, the negative impact, economic and social, is more painful and lasts longer for the unemployed.

On the other hand, income-based risk (wage and salary income risk, WSI-Risk) represents the volatility in growth in income from wages and salaries. Wage income makes up for a significant proportion of the total income of the bottom 20 percentile of households [32]. Volatility in wage income thus has a substantial impact on the living standards of the majority of households. Empirical research highlights the divergence between growth in wages and growth in productivity. From 1973 to 2017, net productivity increased by 77%, while real wages increased by only 12.4% [33,34]. A closer look at the wage growth across different segments of labor force shows that the hourly wage of the middle-wage worker grew only 6% during 1979–2013. Wages were more or less stagnant during the 1980s, 1990s, and 2000s, except for a short period during late 1990s when wages grew due to a tight labor market [32]. The rise in unemployment, observed during recessions, also leads to the suppression of wages.

The overall trend indicates stagnation in wages, which translates into higher income risk for an average worker. The divergence between wage and productivity growth, and near stagnant wage growth, have been identified as major contributors to increases in income inequality in the U.S. With a higher cost of living and divergence in household income, the impact of WSI-Risk is likely to manifest more strongly in urban areas than in rural areas.

A common feature of recession is increased turmoil in the local labor market. This turmoil is apparent through the increased number of mass layoffs, significant and rapid increases in unemployment, a decline in new jobs, increased furloughs, reduced work hours, stagnation, or falls in wage and salary income for the employed.

The Bureau of Labor Statistics (BLS) in the March 2013 survey on Jobs Openings and Labor Turnover highlighted that the number of job openings in the private sector fell sharply, from 3.8 million in December 2007 to a low of 1.9 million in July 2009. During the same period, the number of quits (voluntary separation initiated by the employee) also declined from 2.7 million in December 2007 to 1.5 million in September 2009. Since the end of the recession, the numbers of both job openings and quits increased by 81% and 34%, respectively, in March 2013.

The increased volatility in the local labor market, demonstrated through a sharp decline in job openings and new hiring, is likely to discourage rational employees from taking the risk of quitting their current employment. Besides the risk of loss of income from quitting, an experienced employee also faces the risk of loss of skills and becoming less competitive in the job market. Ref. [35] found that such workers face a higher risk of skill loss, and thus accept a lower wage in exchange for job security. This decision results in lower employee turnover and a higher retention rate for existing business establishments. As evident in Figure 1, corresponding to an increase in layoffs and discharges, voluntary quits by employees declined significantly during both recessions.

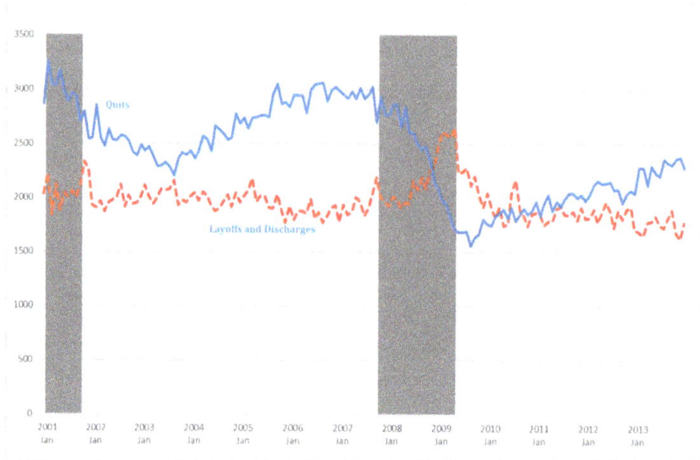

Figure 1. Quits, and layoffs and discharges, 2000–2014 (Seasonally Adjusted, 2000s). Source: Bureau of Labor Statistics, Jobs Openings and Labor Turnover Survey dataset.

3. Methodology

This research begins by investigating the impact of the employment portfolio in a county on the survival rates of new business establishments in that county. Growth in employment in the county is measured through the average annual growth in wage and salary employment. Risk in employment in the county represents the standard deviation in wage and salary employment during the same period.

Before it can be established that risk–return trade-off in a county affects business survival rates, it is imperative to establish the existence of a risk–return relationship. Investigation into risk–return trade-off has shown a U-shaped relationship for states, metro areas, and commuting zones [2–4]. Plotting the risk and growth variables for all counties for the 1996–2005 timeframe suggests that a U-shaped relationship does exist (Figure 2).

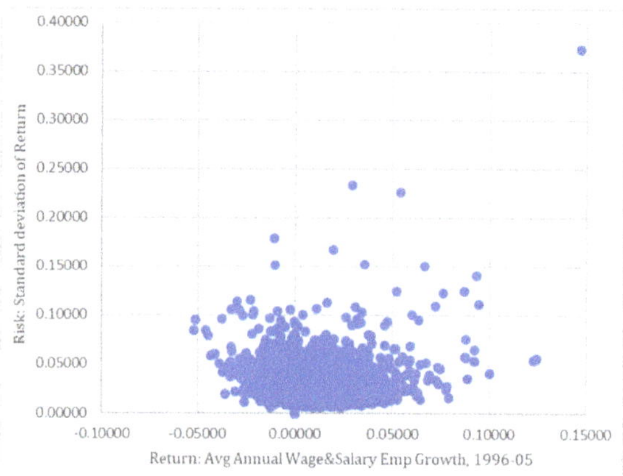

Figure 2. County WSE Risk and Return, 1996–2005.

The relationship between risk and return appears to be nonlinear. As returns increase, the risk declines. However, beyond a certain level of growth, the risk starts rising again. This indicates that the growth in returns (or job growth) has a dual effect on the risk, hence quadratic variation in the growth variable, growth-squared, is also included as an additional parameter in the model.

Visual observation suggests that the relationship between risk and returns is nonlinear. To confirm this inference, this relationship is tested econometrically. A widely used technique for this purpose is stochastic frontier estimation [1,4]. This technique may help in identifying the shape of the frontier and the parameters that define this shape. A nonlinear estimator, the maximum likelihood estimator (MLE), is used to estimate the model. The proposed model to be estimated is as follows:

$$\sigma_i = \alpha_i + \beta_i G_i + \beta_2 G^2{}_i + \varepsilon_i \quad (1)$$

where

σ_i = the standard deviation of WSE for region i
G_i = the annual rate of growth of WSE for region i
β_1 = parameter 1 to be estimated
β_2 = parameter 2 to be estimated

The results displayed in Table 1 confirm that the risk–return trade-off is indeed U-shaped. The risk–return profile can now be used to estimate whether the employment portfolio in a county influences the survival rate of new businesses in that county.

Table 1. Test for U-shaped risk–return trade-off in counties.

	1996–2005		2000–2008	
	Estimate	**Std Error**	**Estimate**	**Std Error**
Constant	0.028 ***	0.0003	0.025 ***	0.0003
Growth	−0.355 ***	0.0226	−0.219 ***	0.0205
Growth Squared	10.5 ***	0.412	11.5 ***	0.397
n	3024		3024	
Log-likelihood	8264.07		8283.7	

*** $p < 0.01$.

We test the hypothesized relationship using the model developed by [21]. We use the following reduced form equation to identify the labor market dynamics, manifested through the portfolio, in the context of new business survival (S_i) in the selected regional unit. Broadly, new business survival depends on the regional labor market portfolio (L_i), amenity score (A_i), demand shock (D_i), local workforce education measures (EE_i), bank deposits (B_i), regional housing market variables (H_i), regional income variables (I_i), market access (M_i), and regional employment variables (E_i).

$$S_i = f(L_i, A_i, D_i, EE_i, B_i, H_i, I_i, M_i, E_i) \quad (2)$$

The above reduced -form model evolves into the following empirical model:

$$\begin{aligned} Survival\ Rate_i = {} & \beta_0 + \beta_1 Local\ Employment\ Portfolio_i + \beta_2 Median\ Home\ Value, 2000_i \\ & + \beta_3 Owner\ Occupied\ Houses, 2000_i + \beta_4 BartikShock_i \\ & + \beta_5 CountyIncome\ per\ capita, 2000_i + \beta_6 Demand\ Shocks_i + \beta_7 Distance\ to\ Metro_i \\ & + \beta_8 Density_i + \beta_9 Amenities_i + \beta_{10} County\ Employment \\ & + \beta_{11} CountyIncome\ per\ capita\ growth\ rate, 2000-07_i \\ & + \beta_{12}\ Share\ of\ pop\ living\ in\ Rural, 2000_i \\ & + \beta_{13}\ County\ Employment\ growth\ rate,\ 2000-07_i\ + \beta_{14}\ Self\ Employment\ rate_i \\ & + \beta_{15}\ Population\ growth_i\ + \beta_{16}\ Share\ of\ Population\ with\ HighSchool, 2000_i \\ & + \beta_{17}\ Share\ of\ Population\ with\ BA+, 2000_i\ + \beta_{18}\ Bank\ Deposits\ per\ capita, 2005_i \end{aligned} \quad (3)$$

3.1. Data

Data for this research have been collected from the Bureau of Economic Affairs' (BEA) Regional Economic Information System (1969–2016). This dataset has comprehensive information about local labor markets at the county level for key variables such as wage and salary employment, self-employment (both in farm and non-farm sectors), personal income, and population. Data for variables representing regional characteristics such as education level, homeownership, and median house values are collected from the U.S. Census Bureau, decennial census.

Data for the survival rates of establishments are taken from the National Establishment Time Series database of U.S. establishments. The survival rate of new businesses in the counties is the dependent variable. Metropolitan counties are those with core populations exceeding 50,000 residents, while micropolitan counties have core populations of between 10,000 and 50,000. Town counties are based on core urban areas of less than 10,000 residents. Non-metropolitan counties are defined as micropolitan counties plus town counties. Descriptive statistics are detailed in the Appendix A.

3.1.1. Local Employment Portfolio—The Core Variables of Interest

Average wage and salary employment is selected as the measure of risk–return trade-off in the county because, as a frequently used employment-based measure, it follows trends in incomes as well as in population and tax revenue in the county [4]. This makes wage and salary employment an appropriate variable for investigating risk and return trade-off.

Risk is measured, based on the conventional definition, as the standard deviation of annual wage and salary employment growth (WSE-Risk) during the selected timeframe. The data for annual wage and salary employment growth are obtained from the Regional Economic Information System (REIS) at the Bureau of Economic Analysis (BEA), U.S. Department of Commerce, 1969–2016.

Growth is defined as the average annual wage and salary employment growth (WSE-Growth) in the county during the selected timeframe(s).

Growth squared is defined as the average annual wage and salary employment growth squared (WSE-Growth Squared) in the county during the selected timeframe(s).

The composition and character of the local labor market takes several years to change, and subsequently influence the local businesses in an average county. The timeframes selected in the study were around ten years on average (longer in a few instances), in order to encapsulate these changes and their subsequent manifestations in the survival of local businesses.

3.1.2. Regional Control Variables

The location of the establishment is likely to have an impact on its survival rate. To capture the impact of the local economic environment, commonly used measures of regional characteristics from the literature are used. Besides regional controls, financial capital and human capital controls are also used to evaluate their impact on establishment survival rate.

To partially address the issue of the potential endogeneity of some of these measures with the dependent variable, in the spirit of the econometric technique commonly followed in similar studies, lagged values of some of these variables are used in the econometric analysis. The key regional control variables used in the model and their respective data sources are provided in Appendix A.

No multicollinearity is found in the model. The variance inflation factor (VIF) for each regional control variable is less than 5.

4. Empirical Results

The empirical exercise is conducted by testing the impact of the local labor market portfolio, based on wage and salary employment data, on the new business survival rate in 2010 in county$_i$. Following the convention in the employment portfolio literature, employment-based measures of risk and return trade-off are used [1,4]. We begin by looking at the role of employment-based portfolios across the spectrum of the chosen regional unit: counties in the U.S., categorized as all, metro, nonmetro, towns, and micro.

WSE portfolios from 1996 to 2005 and from 1997 to 2006 are found to have a significant impact on the survival rate of new businesses in 2010 in metro counties. These are establishments that were born in 2005–2006. All three measures, WSE-Risk, WSE-Growth, and WSE-Growth Squared, were found to be significant, though to varying degrees (Table 2).

Table 2. Summary results of the WSE portfolio for new business survival rates in 2010.

	Survival Rate 2010				
	All	**Metro**	**Nonmetro**	**Towns**	**Micro**
WSE-Risk, 1996–2005	0.151 (0.1070)	0.365 ** (0.1410)	0.0292 (0.1380)	−0.127 (0.1670)	0.411 (0.2500)
WSE-Growth, 1996–2005	0.279 * (0.1560)	0.435 * (0.2370)	0.313 (0.1910)	0.254 (0.2290)	0.324 (0.3440)
WSE-Growth Squared 1996–2005	−2.466 (2.4300)	−5.491 * (2.6680)	−5.013 (3.9680)	−3.966 (4.7190)	−9.631 (7.3330)
WSE-Risk, 1997–2006	0.103 (0.1030)	0.375 ** (0.1200)	−0.056 (0.1380)	−0.238 (0.1680)	0.439 (0.2420)
WSE-Growth, 1997–2006	0.343 * (0.1660)	0.780 *** (0.2200)	0.306 (0.2090)	0.305 (0.2470)	0.0807 (0.3910)
WSE-Growth Squared, 1997–2006	−3.132 (2.6610)	−12.37 *** (2.8730)	−1.621 (3.9020)	−0.551 (4.7150)	−9.305 (6.8790)

* $p < 0.1$, ** $p < 0.05$, *** $p < 0.01$; Standard errors are in parenthesis.

We validate these results by using another potential measure of risk and return trade-off in the local labor market, wage and salary income data. A likely criticism of using this measure is the possible influence on the results because of differences in the cost of living in the regional units. Though challenging, the use of this measure is expected to facilitate the validation of the results found earlier using employment-based measures.

Conducting the empirical exercise with the income-based risk and return measures for different timeframes yields interesting results. Similar to the employment-based portfolio, this portfolio comprises wage and salary income growth (WSI-Growth; defined as the average annual wage and salary income growth in the county during the selected timeframe(s)), wage and salary income growth squared (WSI-Growth Squared; defined as the average annual wage and salary income growth squared in the county during the selected timeframe(s)), and wage and salary income risk (WSI-Risk; the standard deviation of annual wage and salary income growth during the selected timeframe). The WSI portfolios from 1996–2005, 1995–2005, 1991–2005, 1991–2004, 1992–2005, 1993–2005, and 1997–2006 were found to be significant for new business survival in 2010. Similar to before, the observed impact was consistently visible in the metro counties, and all three measures, WSI-Risk, WSI-Growth, and WSI-Growth Squared, were found significant.

The results indicate that risk and return trade-off in local labor markets, identified through employment-based and income-based measures, is instrumental in the survival of new businesses. This impact was found to be strong, especially in metro counties, which encouraged us to focus the investigation in Metropolitan Statistical Areas (MSAs; identified using the crosswalk file from NBER). The existence of a non-linear relationship between risk and growth variables was also verified in MSAs. Similar to counties, the existence of a U-shaped relationship was found in MSAs also (See Table A1 in Appendix A, with results of the MLE tests in MSAs).

The validity of these results could potentially be challenged due to the impact of the Great Recession on the local economy. The influence of this macroeconomic shock cannot be ignored, though considerable heterogeneity was observed in its impact on regions across the country.

A widely used practice in econometric analysis is to replicate the empirical exercise in a different period. To validate the results, the exercise was replicated to determine the impact of local labor market dynamics on the new business survival rate in 2005. A similarity that encouraged the selection of new business survival rate in 2005 was found in the fact that the new businesses that operated during this period also had to go through the recession in 2001, just like those that survived in 2010.

To test the robustness of the results, the empirical exercise was replicated in MSAs only, using both employment-based and income-based risk and return portfolios separately to gauge their impact on new business survival rates in 2005 (three regional control variables (Bartik shock, county/MSA income growth 2000–2007, county/MSA employment growth 2000–2007) were dropped when running regressions, as historical data were not available for them for this timeframe) and 2010.

Similar to the results found for counties, income-based risk and return (WSI) measures were consistently found to have a significant impact on new business survival in 2010 in the MSAs as well. WSI-Risk and WSI-Growth from 1991–2001, 1991–2005, 1995–2005, 1996–2005, 1996–2006, 1997–2006, 2001–2010, 2000–2009, 1997–2007, 2000–2008, 2002–2007, and 1998–2007 were found to be significant (Table A3 in the Appendix A). WSI-Growth Squared was found to be significant for all these timeframes, except in 2000–2009 and 2000–2008.

Next, we replicate this exercise in order to specifically investigate the impacts of employment-based and income-based measures on new business survival rates in MSAs in 2005. The regression results reveal that the WSE-Risk related to employment-based measures in 1991–1999, 1991–2000, 1992–2000, 1993–2000, 1991–2001, 1991–2004, 1991–2005, 1993–2005, 1993–2003, and 1993–2004 had a consistent and significant impact on the new business survival rates in 2005 (Table A4 in the Appendix A). WSE-Growth and WSE-Growth Squared were not found to have any impact during these periods. Income-based risk and return measures were also found to have no impact on the new business survival in 2005.

5. Discussion of Key Results

The analysis of the roles of wage and salary employment (WSE) and wage and salary income (WSI) portfolios in new business survival in MSAs revealed some interesting results. WSE-Risk was found to have a significant impact on new business survival in 2005, and the aggregate WSI portfolio on new business survival in 2010. Risk, manifested both in WSE and WSI, was consistently found to affect the likelihood of new business survival when investigated during different timeframes. Reverse causality or endogeneity may well be a factor in these results, and future work would need to incorporate robust identification strategies to untangle these relationships.

5.1. New Business Survival Rate 2010

Wage and salary income risk–return (WSI) portfolios were found to play a significant role in the survival of new businesses in 2010 in MSAs. These are the businesses that were born in 2005–2006.

Similar to the wage and salary employment risk and return (WSE) portfolio, the WSI portfolio also showed a U-shaped relationship between WSI Growth and WSI Risk, when plotted for all the counties in different timeframes (See Figure A1 in the Appendix).

The wage and salary income risk and return portfolios of 1996–2005, 1996–2006, 1995–2005, 1991–2005, 1991–2001, 1997–2006, 2001–2010, 2000–2009, 1997–2007, 2000–2008, 2002–2007, and 1998–2007 were selected for this analysis. The impact of the WSI risk–return portfolio on the new business survival rate in 2010 was found to be robust across these timeframes in the MSAs. We offer the following speculative scenarios to help understand these results.

A severe macroeconomic event was found while analyzing new business survival in the years preceding 2010, namely, the Great Recession. This recession was a distinctive event, comparable only to the Great Depression. It was deep, and compared to recent recessions, lasted for a longer period—from December 2007 to June 2009. The severity of this recession can be gauged from the fact that the national unemployment rate rose to almost 10% in late 2009. The average expenditures per household declined from USD 52,203 in 2007 to USD 48,109 in 2010 (The Recession of 2007–2009, Bureau of Labor Statistics, 2012, https://www.bls.gov/spotlight/2012/recession/, accessed on 3 December 2019) [36]. Spending declined in all major categories except healthcare during this period. Labor productivity increased marginally, while output and number of hours worked dropped significantly [36]. The wages and salaries of employees in the private sector grew by only 1.3% in December 2009, compared to an increase of 3.6% in March 2007.

Wage and salary income growth was found to have a positive effect, and is strongly significant in this analysis. This suggests that with increased growth in WSI, the survival of businesses increased. The Great Recession was deep, and lasted for a longer period compared to other recessions in the recent past. New businesses that rewarded existing employees, who stayed with them during this unprecedented event, with small but significant increases in wages and salaries were likely to have encouraged these employees to prolong their stay with them. Sharing the economic rewards with their employees is likely to help them not only survive in challenging times, but also to achieve a quick turnaround.

WSI-Risk was found to have a positive effect, and was significant in new businesses' survival in MSAs during the studied timeframes.

MSAs with a higher percentage of the population residing in rural areas in 2000 had a higher business survival rate in 2010 during the selected timeframes. Rural areas witnessed relatively sharper declines in jobs and prolonged unemployment in the local labor market. More job opportunities and higher wages are key factors in attracting younger workers to cities and urban areas. Older workers are comparatively less mobile and able to switch occupations when faced with an economic crisis, which creates challenges when it comes to making structural changes in the local economy [37,38]. The migration of a semi-skilled and skilled labor force contributed to the bigger pool of potential employees to choose from for the employers in MSAs.

"Median Household Value 2000" was found to be negative and significant during the selected timeframes. The Great Recession originated in the housing sector, and subsequently spread to other sectors of the economy. It is no surprise that the bursting of the housing bubble resulted in a sharp decline in housing values. This decline in housing prices resulted in negative home-equity for many homeowners. The unprecedented decline in personal wealth and the unfavorable shift in the labor market likely encouraged employees to remain in their current employment, which had a positive impact on the survival of business establishments.

Both "Population with BA+, 2000", and "Population with High School degree but No BA+, 2000", were found to have a positive and consistent impact on new business survival in 2010 during the selected timeframes. It is not surprising that MSAs with a higher share of both highly skilled labor force and semi-skilled labor force had higher survival rates. Higher human capital endowment is often associated with higher levels of experience, transferable knowledge, and skills, which helps alleviate the negative impacts of recessions in MSAs [39]. Such employees are highly valued assets, and firms prefer to retain them even when rationing jobs during recessions. Large cities and urban areas have a higher share of labor force with a college education. Ref. [40] shows that places with a higher share of college graduates suffered smaller effects from the crisis.

MSAs with a higher share of skilled labor force attract high-tech jobs. Ref. [41] found that the job-multiplier in commuting zones ranges from 1.7 to 2.9, depending upon the existence of high-tech clusters. This means that for every 100 new jobs created in high-tech sectors, 70 to 190 additional local jobs (including semi-skilled and unskilled) would be created in the regional unit. The survival of new businesses in the high-tech sectors in MSAs saved not only jobs in their units, but also several jobs that they helped create.

5.2. New Business Survival Rate 2005

We found that WSE-Risk had a consistent impact on the survival of new businesses in 2005. These businesses were born in 2000–2001. The analysis was conducted over different timeframes.

WSE-Growth and WSE-Growth Squared were not found to have any impact. However, WSE-Risk was consistently found to be positive and have a significant impact on business survival in MSAs in 2005. The impact of WSE-Risk was found to be robust across all selected timeframes. Again, we offer the following speculative scenarios to help understand these results.

The distinctive event in the lives of new businesses was the 2001 recession. This recession was unique compared to the recessions preceding it as it lasted for eight months, which is less than the average age of eleven months of previous recessions [42], and largely saw a technology, telecommunications, and tourism downturn. Comparatively, it was a milder and shorter recession. The unemployment rate increased by 2.10 percentage points, non-farm employment declined by 1.34 percentage points, while real output dropped by 1.6% from first quarter to the second quarter of 2001 [42]. Despite the decline in output and increase in unemployment, the 2001 recession was also unique because, surprisingly, the U.S. economy observed increased spending on consumer durables, new residential housing, and a significant increase of 2.2% in labor productivity. The latter is broadly attributed to the increased spending on information technology and computer hardware by corporations to combat the threat of Y2K in the late 1990s. The causes of this recession are the dot-com bubble burst, the decline in international trade, and the terrorist attacks of 11 September, all of which resulted in a decline in business spending and eventually the gross output. The 2001 recession severely affected the Technology sector compared to other sectors of the economy.

"Median Household Value, 2000" was found to have a positive and significant impact on new business survival in 2005 during the selected timeframes. An increase in median household value in MSAs is likely to affect business survival through dual channels. Ref. [42] found that expenditure on new housing and consumer durables grew in the years before the 2001 recession and continued to grow throughout the recession. The long-term interest rates peaked ten months before the March 2001 business cycle peak and continued to decline subsequently. Low-interest rate regimes before and during the 2001 recession fueled the growth in the housing market, which resulted in a boom in this sector in subsequent years. Many cities across the country witnessed unprecedented growth in housing prices. Ref. [43] estimated the price elasticities of housing stock at the MSA level, and found that in areas with high elasticity of housing supplies, the number of new houses increased, and areas with low elasticity saw sharp increases in housing prices. With increased home prices, existing homeowners cashed out their dormant home equity. A significant portion of this extra cash was spent on buying consumer durables and big-ticket items. This additional spending is likely to have trickled down as increased revenue for many existing and newly born establishments. The cashed-out home equity also became an additional and untapped source of funds for the owners of new establishments to invest in business and support existing operations, as well as to fuel growth.

"Population with High School degree but No BA+" was found to have a negative and significant impact on business survival in 2005. A distinctive feature of the 2001 recession was the sharp rise of 2.2 percentage points in non-farm labor productivity [42]. The late 1990s witnessed a significant investment in computer equipment and software by corporations to combat the threat of Y2K, and also the adoption of new technology, which likely resulted in increased labor productivity. Growth in labor productivity resulted in real disposable income growth (0.37%). The increased investments in IT and computer hardware by corporations may not have improved the productivity and the resulting disposable income of semi-skilled and unskilled labor force. Regions with less skilled labor force are likely to have missed these benefits, which would have affected the survival of businesses there. This impact is likely to have been stronger, especially during the recession.

5.3. Role of Risk in New Business Survival in MSAs

Risk is the key factor consistently influencing new business survival in MSAs, through both wage and salary employment and wage and salary income portfolios. The positive impact of volatility in the local labor market on new business survival is found to be stronger during recessions. As previously, we offer the following speculative scenarios to help explain these results.

The role of risk in the birth of enterprises is well documented [4]. Our results indicate that the risk displayed in the local labor market significantly increased the likelihood of new business survival. The increased turmoil in the labor market manifests through mass-layoffs, higher unemployment, stagnation in wage and salary income, etc. Empirical investigations highlight the fall in labor productivity, a secular decline in demand for semi-skilled workers, increases in low-wage and insecure jobs, and decreases in union membership as some of the long-term trends observed in the U.S. labor market [44]. Ref. [45] show that the growth rate in productivity picked up in the mid-1990s, slowed in the mid-2000s, and remained flat going into the Great Recession. The response of a rational worker aware of both the impact of exogenous macroeconomic shock and the ongoing dynamic changes in the local labor market would be to continue with their existing employment. This decision results in a lower employee turnover and a higher retention rate for newly established businesses.

A low employee turnover rate is beneficial for establishments. Excessive employee turnover has both monetary and non-monetary costs that can have a significant and far-reaching effect on the economic and operational efficiency of the establishment. In extreme cases, high employee turnover may have detrimental impacts on new business survival. Ref. [46], citing a nationwide survey, found that the average internal cost-per-hire for an engineer was USD 4901, a computer programmer USD 2500, a secretary USD 1000, and a retail sales associate USD 350 in the 1990s. The direct monetary costs can be attributed to the costs of advertising, recruitment, candidate travel, selection, hiring, assignment, orientation, training, signing bonus, and relocation expenses for the new employee(s). These costs would be significantly higher in current dollars. The non-monetary costs include the breakdown of customer relations, the disruption of workflow, declines in morale of the remaining employees, and additional costs incurred till the newly hired employee acquires necessary job skills and can work at the desired level of efficiency [46].

Highly talented and competent employees generally leave for better opportunities or get poached by competitors. The monetary and non-monetary costs from the loss of such employees are likely to have a substantial impact on the survival of new establishments. "County/MSA Employment growth, 2000–2007" was found to have a negative impact on new business survival rate in 2010 during the 2000–2009 and 2000–2008 timeframes. The sign of this variable indicates an inverse relationship with new business survival rate, which reinforces the impact of high employee turnover on the survival of new businesses. Broadly, new businesses operating in MSAs that witnessed high employment growth are also likely to experience high employee turnover, which would result in significant monetary and non-monetary costs. The aggregate consequences of these costs for new businesses, operating on shoe-string budgets, could be attributed to the closure of some of these businesses. The shift in employee preference to current employment vis-a-vis the potential gains derived from a risky switch during a recession helps a new establishment save significant economic resources. Additionally, the prolonged employment of such employees during a recession facilitates a continued focus on innovation, and on maximizing efficiency, which are the comparative advantages of a new business and are the bedrock of their survival.

An interesting result found in the empirical exercise is that risk only consistently manifested on survival rate via wage and salary employment in 2005, and via wage and salary income in 2010. Our analysis of new business survival also highlights the role of the salient structural changes observed in the labor market over the years, the impact of which became strongly visible during both recessions. The labor market witnessed a significant export of jobs to low-wage countries, beginning with manufacturing jobs going to China and Vietnam, and subsequently low-end service sector jobs going to India and the Philippines. The former was facilitated through increased global trade, and the latter accelerated through the technological improvements adopted by corporations in the late 1990s and early 2000s. Ref. [47] found that increased Chinese imports resulted in reductions in both employment and wage levels in the manufacturing sector. They also found that increased transfer payments made through multiple federal and state programs masked the loss in average earnings of the affected households during this period. Ref. [48] showed that globalization affected wages by pushing workers out of the manufacturing sector into low-paying jobs elsewhere. These subtle structural changes over the prolonged period changed the composition and quantity of jobs available in the key sectors, which first became evident during the 2001 recession. The Great Recession resulted in a sharp increase in unemployment across all sectors of the economy, thereby strongly demonstrating the impact of the structural changes that had begun earlier in the 1980s and 1990s. Workers across all sectors experienced a significant and prolonged decline in or loss of wage and salary income, which manifested through income-based measures in our investigation.

As noted above, the 2001 recession significantly impacted the technology and tourism sectors. Regions that specialized in these two sectors were most affected by this recession. States in the Mideast Census Region were least affected by the 2001 recession [49].

The Great Recession originated in the housing sector in urban centers, then transmitted to the financial sector, and eventually engulfed the national economy, before spreading to other countries. The origin of this economic shock and its transmission created an initial perception that its impact would create severe economic challenges in large cities and urban centers. However, investigations into the responses of regional economies to the Great Recession have shown that the impact has been more severe, and the recovery has been slower in smaller towns and rural areas.

Empirical research has shown that urban areas performed better during the 2001 and 2007–2009 recessions [49–51]. Not only were MSAs more able to endure these recessions, but they were also able to recover faster than non-MSAs. Surviving businesses contribute valuable tax dollars, and at the same time help reduce the additional expenditure of state and local governments in the form of transfer payments, such as unemployment benefits. These contributions significantly enhance the resilience of the community, especially during economic downturns. New businesses that were able to survive during these recessions sustained the local economy during them and accelerated economic growth in the recovery phase.

6. Conclusions

The findings in this paper highlight that risk in the local labor market improves the survival of new businesses, which provides much needed resilience to the local economy. This dynamic was evident in MSAs when investigating the responses of new businesses during the recent recessions of 2001 and 2007–2009.

The impact of macroeconomic shock on the local labor market, captured through volatility in employment-based and income-based measures, likely influences employees' decisions when evaluating the risk–return trade-off involved in quitting existing employment. The turmoil in the local labor market encourages employees to continue with their present employment and postpone any career-related decisions (until the economic outlook gets better). This shift in employee behavior, even if temporary, is beneficial for new businesses, as it shields the business from monetary and non-monetary costs associated with high employee turnover. Many new businesses struggle with numerous challenges, which compound exponentially due to the exogenous shock. The savings on such costs and the continued contribution of valued employees in business operations are nothing short of a lifeline for a new business.

This paper contributes to the literature focusing on the role of human capital in regional economic growth. Research on the role of human capital in firm survival is predominantly anchored in the entrepreneur's personal characteristics. Our endeavor contributes to this literature by uniquely applying portfolio theory in the context of employee-centric risk and return trade-off in the regional unit, also highlighting the contribution of employees to new business survival.

The findings from this paper may help policymakers design economic development policies tailor-made to the needs of the regional unit. The targeted approach may help in the more efficient use of tax dollars compared to the effect from applying a blanket policy. The Paycheck Protection Program (PPP) implemented by the Federal Government in response to the economic downturn caused by the COVID-19 pandemic is a good example of using a targeted approach to support local businesses. Not only does this help businesses survive the downturn and keep their employees on the payroll, but it also ensures business establishments are ready to participate in and contribute to a faster economic recovery. In the absence of this program, many more business establishments would have closed down, and consequently, the post-pandemic recovery would have been more painful and elongated.

We find encouraging results to support our hypothesis, and expect potential opportunities for further research on this topic. Our analysis is limited to the 2001 and 2007–2009 recessions, while countries all across the world are experiencing severe economic shock due to the widespread pandemic. Although the COVID-19 pandemic is a very different type of economic downturn, similar analyses of the COVID recession should provide promising opportunities to investigate its impact on the dynamics of the local labor market, and further, on business survival.

We use counties as a regional unit to capture the dynamics of the local labor market for our analysis. The selection of counties for the analysis can be questioned on the ground that they are not perfect reflections of the prevalent dynamics in the local labor market. For further research, we plan to incorporate granular data from both MSAs and commuting zones into our analyses. As discussed above, reverse causality or endogeneity is likely to be present, so future work needs to incorporate an appropriate identification structure for addressing these confounding relationships.

Author Contributions: B.G.: formal analysis, investigation, resources, methodology, data curation, original draft preparation; S.W.: conceptualization, supervision, writing—review and editing. Both authors have read and agreed to the published version of the manuscript.

Funding: This research received no external funding.

Institutional Review Board Statement: Not applicable in this study.

Informed Consent Statement: Not Applicable.

Data Availability Statement: Data presented in this study are available on request from the corresponding author. Detailed information about the data sources is provided in the Appendix A (Table A2).

Acknowledgments: The authors would like to acknowledge the valuable contribution of faculty members at Department of Economics, CSU, and external stakeholders at various stages of this endeavor.

Conflicts of Interest: The authors declare no conflict of interest.

Appendix A

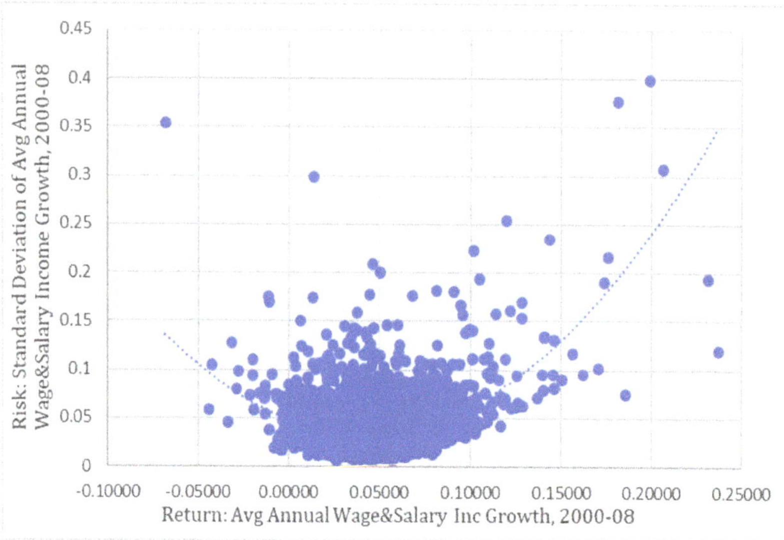

Figure A1. County WSI Risk and Return, 2000–2008.

Table A1. Test of risk–return trade-off in MSAs.

	Metropolitan Statistical Areas	
	Estimate	Std Error
Constant	0.018 ***	0.0005
Growth	−0.122 **	0.0599
Growth Squared	4.992 ***	1.367
n	374	
Log-likelihood	1326.19	

*** $p < 0.01$, ** $p < 0.05$.

Table A2. Key regional variables and data source (Regional Economic Information Systems (REIS), Bureau of Economic Analysis (BEA); United States Department of Agriculture (USDA); United States Census Bureau (USCB)).

Regional Variable	Source
Self Employment Rate	*REIS/BEA*
Amenity Score	*McGranahan Index, USDA Economic Research*
Share of Pop. Living in Rural, 2000	*USCB, Decennial Census 2000*
Share of pop. with BA+, 2000	*USCB, Decennial Census 2000*
Share of pop. With High School but No BA+, 2000	*USCB, Decennial Census 2001*
Bank Deposits per capita 2005	*Federal Deposit Insurance Corp (FDIC) & BEA*
Median Home Value, 2000	*USCB, Decennial Census 2000*
Owner Occupied Houses, 2000	*USCB, Decennial Census 2000*
County Income percapita, 2000	*USCB, Decennial Census 2000*
County Employment, 2000	*USCB, Decennial Census 2000*
County Income per capita growth, 2000–2007	*USCB*
County Employment growth, 2000–2007	*USCB*
Distance to nearest Metro	*USDA*
Employment-pop ratio, 2000	*REIS/BEA*

Table A3. Results from WSI portfolios from different timeframes as regards SR2010 in MSAs only. (Selected timeframes are displayed due to space constraints).

	Survival Rate, 2010					
	1991–2001	**1995–2005**	**1996–2005**	**1996–2006**	**1997–2006**	**2000–2009**
WSI-Risk	0.426 * (0.2230)	0.438 * (0.2240)	0.482 ** (0.2100)	0.494 ** (0.2150)	0.484 ** (0.2020)	0.401 * (0.2060)
WSI-Growth	2.080 *** (0.6290)	1.912 *** (0.6740)	1.722 *** (0.6490)	2.056 *** (0.6740)	2.017 *** (0.6800)	1.385 ** (0.5030)
WSI-Growth Squared	−12.90 ** (4.6010)	−12.44 ** (5.4620)	−11.26 ** (5.2460)	−12.33 ** −5.534	−12.15 ** (5.5920)	1.89 (5.5990)
Bartik Shock, 2005	−0.0000796 (0.0010)	−0.000047 (0.0010)	−0.00000365 (0.0010)	−0.000104 (0.0010)	−0.0000941 (0.0010)	−0.000975 (0.0009)
Dynamism: Centered	−0.00213 (0.0016)	−0.00191 (0.0016)	−0.00187 (0.0016)	−0.00193 (0.0015)	−0.00188 (0.0015)	−0.00253 * (0.0015)
County Income per capita, 2000	−0.000000924 (0.0000)	−0.0000012 (0.0000)	−0.0000013 (0.0000)	−0.0000012 (0.0000)	−0.0000012 (0.0000)	−0.0000010 (0.0000)
County Inc per capita growth rate, 2000–2007	0.000389 (0.0003)	0.000174 (0.0003)	0.000154 (0.0003)	0.000057 (0.0003)	0.00004 (0.0003)	−0.000683 (0.0004)
County Employment, 2000	−0.00000001 (0.0000)	−0.00000001 (0.0000)	−0.00000001 (0.0000)	−0.00000001 (0.0000)	−0.00000001 (0.0000)	−0.00000001 (0.0000)
County Employment growth rate, 2000–2007	−0.000198 (0.0003)	−0.000284 (0.0003)	−0.000259 (0.0003)	−0.000444 (0.0003)	−0.000431 (0.0003)	−0.00111 *** (0.0003)
Amenity Score	0.00089 (0.0013)	0.00025 (0.0013)	0.000294 (0.0013)	0.0000129 (0.0013)	0.0000737 (0.0013)	−0.0000789 (0.0013)
Deposits per capita, 2005	−0.0000004 (0.0000)	−0.0000004 (0.0000)	−0.0000004 (0.0000)	−0.0000005 (0.0000)	−0.0000005 (0.0000)	−0.0000004 (0.0000)
Share of Homes Owner Occupied	0.0456 (0.0510)	0.0551 (0.0510)	0.0554 (0.0510)	0.0594 (0.0507)	0.0617 (0.0508)	0.0835 * (0.0500)
Median HH Value, 2000	−0.000000314 ** (0.0000)	−0.000000285 ** (0.0000)	−0.000000282 ** (0.0000)	−0.000000285 ** (0.0000)	−0.000000284 ** (0.0000)	−0.000000273 ** (0.0000)
Share of pop living in Rural, 2000	0.0595 *** (0.0172)	0.0557 ** (0.0175)	0.0545 ** (0.0176)	0.0535 ** (0.0175)	0.0539 ** (0.0175)	0.0585 *** (0.0165)
County Self-Employment Rate, 2000	−0.00158 (0.0039)	−0.00232 (0.0039)	−0.00237 (0.0039)	−0.00231 (0.0039)	−0.00221 (0.0039)	−0.00188 (0.0039)
Population with BA+, 2000	0.375 *** (0.0874)	0.329 *** (0.0860)	0.325 *** (0.0858)	0.322 *** (0.0855)	0.324 *** (0.0857)	0.344 *** (0.0865)
Population with HS degree but NO BA+, 2000	0.121 * (0.0548)	0.125 * (0.0549)	0.118 * (0.0547)	0.128 * (0.0545)	0.126 * (0.0545)	0.170 ** (0.0552)
Employment-Pop Ratio, 2000	0.0173 (0.0379)	0.0267 (0.0373)	0.0276 (0.0372)	0.0236 (0.0370)	0.0241 (0.0370)	0.0403 (0.0361)
Distance to the Nearest MSA	−0.0145 (0.0083)	−0.0132 (0.0083)	−0.0137 (0.0083)	−0.0136 (0.0082)	−0.0137 (0.0083)	−0.01669 ** (0.0081)
Constant	0.452 *** (0.0619)	0.465 *** (0.0622)	0.476 *** (0.0620)	0.460 *** (0.0618)	0.461 *** (0.0619)	0.435 *** (0.0592)
R-squared	0.258	0.253	0.254	0.263	0.263	0.293
Observations	374	374	374	374	374	373

* $p < 0.1$, ** $p < 0.05$, *** $p < 0.01$. Standard errors are in parenthesis.

Table A4. Results from WSE portfolios from different timeframes as regards SR2005 in MSAs only. (Selected timeframes are displayed due to space constraints).

	Survival Rate 2005				
	1991–1999	**1991–2001**	**1992–2000**	**1993–2000**	**1993–2003**
WSE-Risk	0.494 ** (0.2340)	0.500 * (0.2160)	0.436 * (0.2350)	0.489 ** (0.2360)	0.534 * (0.2370)
WSE-Growth	0.188 (0.3100)	0.147 (0.3640)	0.387 (0.3390)	0.299 (0.3320)	0.0172 −0.365
WSE-Growth Squared	−4.762 (4.3560)	−5.011 (5.3860)	−6.324 (4.4020)	−4.872 (4.2690)	−2.658 (5.8880)
Dynamism Centered	0.0019 (0.0024)	0.00192 (0.0023)	0.00177 (0.0023)	0.00175 (0.0023)	0.00146 (0.0023)
County Inc per capita, 2000	−0.00000116 (0.0000)	−0.00000114 (0.0000)	−0.00000111 (0.0000)	−0.00000111 (0.0000)	−0.0000009 (0.0000)
County Employment, 2000	0.00000001 (0.0000)	0.00000001 (0.0000)	0.00000001 (0.0000)	0.00000001 (0.0000)	0.00000001 (0.0000)

Table A4. *Cont.*

	Survival Rate 2005				
	1991–1999	1991–2001	1992–2000	1993–2000	1993–2003
Amenity Score	−0.00311	−0.00298	−0.00305	−0.00286	−0.00301 *
	(0.0019)	(0.0019)	(0.0019)	(0.0019)	(0.0018)
Deposits per capita, 2000	0.000000469	0.000000453	0.000000426	0.000000447	0.00000048
	(0.0000)	(0.0000)	(0.0000)	(0.0000)	(0.0000)
Share of Homes Owner Occupied	−0.205 **	−0.209 **	−0.206 **	−0.200 **	−0.201 **
	(0.0723)	(0.0722)	(0.0724)	(0.0725)	(0.0717)
Median HH Value, 2000	0.000000430 **	0.000000431 **	0.000000420 **	0.000000415 **	0.000000424 **
	(0.0000)	(0.0000)	(0.0000)	(0.0000)	(0.0000)
Share of pop Residing in Rural, 2000	0.0162	0.018	0.0153	0.0169	0.0205
	(0.0233)	(0.0232)	(0.0233)	(0.0232)	(0.0228)
County Self-Employment Rate, 2000	−0.0058	−0.00537	−0.00568	−0.00591	−0.00552
	(0.0055)	(0.0055)	(0.0056)	(0.0055)	(0.0055)
Population with BA+, 2000	−0.152	−0.15	−0.151	−0.145	−0.153
	(0.1180)	(0.1170)	(0.1180)	(0.1180)	(0.1160)
Population with HS degree but NO BA+, 2000	−0.269 ***	−0.273 ***	−0.278 ***	−0.269 ***	−0.262 ***
	(0.0761)	(0.0762)	(0.0763)	(0.0761)	(0.0744)
Employment-Pop Ratio, 2000	−0.0414	−0.0399	−0.0398	−0.0399	−0.0347
	(0.0524)	(0.0523)	(0.0526)	(0.0526)	(0.0513)
Distance to the Nearest MSA	−0.0116	−0.0116	−0.0112	−0.0111	−0.0119
	(0.0118)	(0.0118)	(0.0118)	(0.0118)	(0.0116)
Constant	0.900 ***	0.903 ***	0.905 ***	0.895 ***	0.879 ***
	(0.0820)	(0.0817)	(0.0822)	(0.0824)	(0.0782)
R-squared	0.102	0.105	0.101	0.103	0.103
Observations	370	370	370	370	370

* $p < 0.1$, ** $p < 0.05$, *** $p < 0.01$; Standard errors are in parenthesis.

Table A5. Summary Statistics for All and Metro Counties.

	All				Metro			
	n = 3024				*n* = 808			
	Mean	Std. Dev.	Min	Max	Mean	Std. Dev.	Min	Max
Survival Rate 2010	0.676	0.091	0.000	1.000	0.644	0.060	0.429	0.870
WSE-Risk, 1996–2005	0.029	0.017	0.006	0.373	0.022	0.017	0.006	0.373
WSE-Growth, 1996–2005	0.009	0.017	−0.052	0.147	0.017	0.019	−0.037	0.147
WSE-Growth Squared, 1996–2005	0.000	0.001	0.000	0.022	0.001	0.001	0.000	0.022
Bartik Shock, 2005	2.2	4.4	−21.9	84.5	3.0	3.4	−10.5	23.0
Business Dynamism	1.1	3.1	−30.1	72.0	0.9	2.0	−3.8	26.4
County Income per capita, 2000	27,929	7075	3395	87,711	33,717	7898	17,104	87,711
County Income per capita growth	27.4	17.0	−42.5	341.2	24.0	10.6	−25.7	107.4
County Employment, 2000	53,286	186,743	294	5,404,010	165,187	336,265	2106	5,404,010
County Employment growth	6.46	14.46	−37.51	170.11	13.20	16.28	−37.51	145.26
Amenity Score	0.04	2.28	−6.40	11.17	0.27	2.41	−5.40	11.17
Bank Deposits per capita, 2005	14,509	11,259	390	241,738	15,026	16,387	2220	241,738
Share of Owner Occupied Houses, 2000	0.86	0.09	0.23	0.98	0.92	0.05	0.46	0.98
Median House Value, 2000	83,422	44,159	20,100	750,000	114,614	49,067	47,700	514,600
Share of pop residing in rural, 2000	0.60	0.30	0.00	1.00	0.31	0.24	0.00	1.00
Self-Employment Rate, 2000	0.73	3.10	0.00	81.51	0.27	0.93	0.00	15.55
Share of pop with BA+, 2000	0.11	0.05	0.03	0.44	0.14	0.06	0.03	0.44
Share of pop with HS, 2000	0.40	0.06	0.14	0.57	0.38	0.05	0.20	0.51
Employment-to-pop Ratio, 2000	0.52	0.15	0.13	2.79	0.54	0.16	0.21	1.79
Distance to nearest metro	0.81	0.65	0.00	4.34	0.26	0.23	0.00	1.74
Nonmetro*Amenityscore	−0.04	1.91	−6.40	11.15	0.00	0.00	0.00	0.00

Table A6. Summary Statistics for Nonmetro, Towns, and Micro Counties.

	Nonmetro				Towns				Micro			
	n = 2216				*n* = 1581				*n* = 635			
	Mean	Std. Dev.	Min	Max	Mean	Std. Dev.	Min	Max	Mean	Std. Dev.	Min	Max
Survival Rate 2010	0.688	0.097	0.000	1.000	0.692	0.102	0.000	1.000	0.679	0.080	0.250	1.000
WSE-Risk, 1996–2005	0.031	0.017	0.006	0.226	0.033	0.017	0.007	0.226	0.027	0.014	0.006	0.123
WSE-Growth, 1996–2005	0.005	0.015	0.0520	0.1228	0.005	0.016	0.0520	0.1228	0.007	0.014	0.0427	0.0921
WSE-Growth Squared, 1996–2005	0.000	0.001	0.000	0.015	0.000	0.001	0.000	0.015	0.000	0.001	0.000	0.008
Bartik Shock, 2005	1.90	4.65	−21.94	84.53	2.00	4.88	−21.94	84.53	1.66	3.99	−16.96	12.50
Business Dynamism	1.21	3.48	−30.08	72.04	1.33	3.82	−19.30	72.04	0.91	2.40	−30.08	18.60
County Income per capita, 2000	25,818	5376	3395	83,631	24,868	5310	3395	57,187	28,184	4781	12,301	3631
County Income per capita growth	28.65	18.6	−42.54	341.19	29.56	20.26	−42.54	341.19	26.39	13.40	−12.70	134.04
County Employment, 2000	12,485	12,802	294	95,004	8634	9660	294	89,957	22,073	14,525	436	95,004
County Employment growth	4.00	12.89	−35.86	170.11	3.45	13.18	−35.86	170.11	5.36	12.05	−24.68	94.77
Amenity Score	−0.05	2.23	−6.40	11.15	−0.10	2.17	−6.40	8.27	0.06	2.38	−6.10	11.15
Bank Deposits per capita,2005	14,320	8662	390	219,273	14,451	6957	390	60,109	13,994	11,890	1694	219,273
Share of Owner Occupied Houses, 2000	0.84	0.10	0.23	0.97	0.82	0.10	0.23	0.97	0.87	0.08	0.38	0.97
Median House Value, 2000	72,063	36,067	20,100	750,000	68,784	37,091	20,100	750,000	80,228	31,976	24,900	369,100
Share of pop residing in rural, 2000	0.71	0.25	0.06	1.00	0.78	0.23	0.06	1.00	0.53	0.22	0.07	1.00
Self-Employment Rate, 2000	0.89	3.56	0.00	81.51	1.05	3.86	0.01	81.51	0.52	2.64	0.00	49.38
Share of pop with BA+, 2000	0.09	0.04	0.03	0.44	0.09	0.04	0.03	0.44	0.10	0.04	0.04	0.32
Share of pop with HS, 2000	0.41	0.06	0.14	0.57	0.41	0.06	0.19	0.57	0.39	0.06	0.14	0.54
Employment-to-pop Ratio, 2000	0.51	0.14	0.13	2.79	0.50	0.14	0.13	2.04	0.54	0.15	0.19	2.79
Distance to nearest metro	1.00	0.64	0.24	4.34	1.03	0.68	0.25	4.34	0.94	0.54	0.24	3.82
Nonmetro*Amenityscore	−0.052	2.231	−6.40	11.15	−0.097	2.167	−6.40	8.27	0.060	2.380	−6.10	11.15

References

1. Chandra, S. A test of the regional growth-instability frontier using state data. *Land Econ.* **2002**, *78*, 442–462. [CrossRef]
2. Lande, P. Regional industrial structure and economic growth and instability. *J. Reg. Sci.* **1994**, *34*, 343–360. [CrossRef]
3. Spelman, W. Growth, stability, and the urban portfolio. *Econ. Dev. Q.* **2006**, *20*, 299–315. [CrossRef]
4. Low, S.A.; Weiler, S. Employment Risk, Returns, and Entrepreneurship. *Econ. Dev. Q.* **2012**, *26*, 238–251. [CrossRef]
5. Markowitz, H. Portfolio Selection. *J. Financ.* **1952**, *7*, 77–91.
6. Acs, Z.; Armington, C.; Zhang, T. The determinants of new-firm survival across regional economies: The role of human capital stock and knowledge spillover. *Pap. Reg. Sci.* **2007**, *86*, 367–391. [CrossRef]
7. Ciccone, C.; Hall, R.E. Productivity and the density of economic activity. *Am. Econ. Rev.* **1996**, *86*, 54–70.
8. Kangasharju, A.; Pekkala, S. The role of education in self-employment success in Finland. *Growth Chang.* **2002**, *33*, 216–237. [CrossRef]
9. Audretsch, D.B.; Mahmoon, T. New firm survival: New results using a hazard function. *Rev. Econ. Stat.* **1995**, *77*, 97–103. [CrossRef]
10. Audretsch, D.B.; Keilbach, M.C.; Lehman, E.E. *Entrepreneurship and Economic Growth*; Oxford University Press: Oxford, UK, 2006.
11. Christie, T.; Sjoquist, D. New Business Survival in Georgia: Exploring the Determinants of Survival Using Regional Level Data. *Growth Chang.* **2012**, *43*, 110–142. [CrossRef]
12. Bates, T. A comparison of franchise and independent small business survival rates. *Small Bus. Econ.* **1995**, *7*, 377–388. [CrossRef]
13. Cooper, A.C.; Woo, C.Y.; Dunkelberg, W.C. Entrepreneurship and the initial size of firms. *J. Bus. Ventur.* **1989**, *4*, 317–332. [CrossRef]
14. Bruderl, J.; Preisendorfer, P.; Ziegler, R. Survival chances of newly founded business organizations. *Am. Sociol. Rev.* **1992**, *57*, 227–242. [CrossRef]
15. Dyke, L.S.; Fisher, E.M.; Reuber, A.R. An inter-industry examination of the impact of owner experience on firm performance. *J. Small Bus. Manag.* **1992**, *30*, 72.
16. Peña, I. Business Incubation Centers and New Firm Growth in the Basque County. *Small Bus. Econ.* **2004**, *22*, 223–236. [CrossRef]
17. Keasy, K.; Watson, R. The state of the art of small business failure prediction: Achievements and prognosis. *Int. Small Bus. J.* **1991**, *9*, 11–29. [CrossRef]
18. Wicker, A.W.; King, J.C. Employment, ownership and survival in microbusiness: A study of new retail and service establishments. *Small Bus. Econ.* **1989**, *1*, 137–152. [CrossRef]
19. Keeble, D.; Walker, S. New firms, small firms and dead firms: Spatial patterns and determinants in the United Kingdom. *Reg. Stud.* **1994**, *28*, 411–427. [CrossRef]
20. Conroy, M. Alternative strategies for regional industrial diversification. *J. Reg. Sci.* **1974**, *14*, 31–46. [CrossRef]
21. Bunten, D.; Weiler, S.; Thompson, E.; Zahran, S. Entrepreneurship, Information, and Growth. *J. Reg. Sci.* **2015**, *55*, 560–584. [CrossRef] [PubMed]
22. Williams, N.; Vorley, T.; Keetikidis, P. Economic Resilience and Entrepreneurship: A case study of the Theessaloniki City Region. *Local Econ.* **2013**, *28*, 399–415. [CrossRef]
23. Williams, N.; Vorley, T. Economic Resilience and Entrepreneurship: Lessons from the Sheffield City Region. *Entrep. Reg. Dev.* **2014**, *26*, 257–281. [CrossRef]
24. Cowling, M.; Weixi, L.; Ledger, A.; Zhang, N. What really happens to small and medium-sized enterprises in a global economic recession? UK evidence on sales and job dynamics. *Int. Small Bus. J.* **2015**, *33*, 488–513. [CrossRef]

25. Martin, R.L. The Roepke lecture in economic geography—Rethinking regional path dependence: Beyond lock-in to evolution. *Econ. Geogr.* **2010**, *86*, 1–27. [CrossRef]
26. Glaeser, E. Reinventing Boston: 1630–2003. *J. Econ. Geogr.* **2005**, *5*, 119–153. [CrossRef]
27. Holling, C. Engineering resilience versus ecological resilience. In *Engineering within Ecological Constraints*; Schulze, P.C., Ed.; National Academies Press: Washington, DC, USA, 1996; pp. 31–43.
28. Hill, E.; Clair, T.; Wial, H.; Wolman, H.; Atkins, P.; Blumenthal, P.; Ficenec, S.; Fiedhoff, A. Economic shocks and regional economic resilience. In *Urban and Regional Policy and Its Effects: Building Resilient Regions*; Paper to Conference; Brookings Institution Press: Washington, DC, USA, 2010; pp. 20–21.
29. Martin, R. Regional economic resilience, hysteresis and recessionary shocks. *J. Econ. Geogr.* **2012**, *12*, 1–32. [CrossRef]
30. Bristow, G.; Healy, A. Regional resilience: An agency perspective. *Reg. Stud.* **2013**, *48*, 923–925. [CrossRef]
31. Martin, R.; Sunley, P. ON the notion of regional economic resilience: Conceptualization and explanation. *J. Econ. Geogr.* **2014**. [CrossRef]
32. Mishel, L.; Gould, E.; Bivens, J. *Wage Stagnation in Nine Charts*; Economic Policy Institute: Washington, DC, USA, 2015.
33. The Productivity-Pay Gap, August 2018. Economic Policy Institute. Available online: https://www.epi.org/productivity-pay-gap/ (accessed on 14 October 2019).
34. Alon, T.; David, B.; Dent, R.; Pugsley, B. Older and Slower: The Startup Deficit's Lasting Effects on Aggregate Productivity Growth. *J. Monet. Econ.* **2018**, *93*, 68–85. [CrossRef]
35. Fujita, S. Declining Labor Turnover and Turbulence. *J. Monet. Econ.* **2018**, *99*, 1–19. [CrossRef]
36. Bureau of Labor Statistics. The Recession of 2007–2009. Bureau of Labor Statistics. Available online: https://www.bls.gov/spotlight/2012/recession/pdf/recession_bls_spotlight.pdf (accessed on 23 October 2019).
37. Longhi, S.; Nijkamp, P.; Poot, H.J. Meta-Analysis of empirical evidence on the labour market: Impacts of immigration. *Rég. Dév.* **2008**, *27*, 161–191.
38. Robertson, A.; Tracy, S. Health and productivity of older workers. *Scand. J. Work Environ. Health* **1998**, *24*, 85–97. [CrossRef] [PubMed]
39. Lee, N. Grim down South? The determinants of unemployment increase in British cities in the 2008–2009 recession. *Reg. Stud.* **2014**, *48*, 1761–1778. [CrossRef]
40. Kitsos, A.; Bishop, P. Economic resilience in Great Britain: The crisis impact and its determining factors for local authority districts. *Ann. Reg. Sci.* **2018**, *60*, 329–347. [CrossRef]
41. Bartik, T.J.; Sotherland, N. *Realistic Local Job Multipliers*; Policy Brief; W.E. Upjohn Institute for Employment Research: Kalamazoo, MI, USA, 2019. [CrossRef]
42. Kliesen, K.L. *The 2001 Recession: How Was It Different and What Developments May Have Caused It?* Federal Reserve Bank of St. Louis Review: St. Louis, MO, USA, 2003; pp. 23–38. [CrossRef]
43. Saiz, A. The Geographic Determinants of Housing Supply. *Q. J. Econ.* **2010**, *125*, 1253–1296. [CrossRef]
44. Kalleberg, A.L.; Wachter, T.M. The U.S. Labor Market During and After the Great Recession: Continuities and Transformations. *RSF Russell Sage Found. J. Soc. Sci.* **2017**, *3*, 1–19.
45. Fernald, J.; Hall, R.; Stock, J.; Watson, M. *The Disappointing Recovery in U.S. Output after 2009*; FRBSF Economic Letter, Research from Federal Reserve Bank of San Francisco: San Francisco, CA, USA, 2018.
46. Abbasi, S.; Hollman, K. Turnover: The Real Bottom Line. *Public Pers. Manag.* **2000**, *29*, 333–342. [CrossRef]
47. Autor, D.H.; Dorn, D.; Hanson, G.H. The China Syndrome: Local Labor Market Effects of Import Competition in the United States. *Am. Econ. Rev.* **2013**, *103*, 2121–2168. [CrossRef]
48. Ebenstein, A.; Harrison, A.; McMillan, M.; Phillips, S. Estimating the Impact of Trade and Offshoring on American Workers Using the Current Population Surveys. *Rev. Econ. Stat.* **2014**, *96*, 581–595. [CrossRef]
49. Connaughton, J.E.; Madsen, R.A. Regional Implications of the 2001 recession. *Ann. Reg. Sci.* **2009**, *43*, 491–507. [CrossRef]
50. Capello, R.; Caragliu, A.; Fratesi, U. Spatial heterogeneity in the costs of the economic crisis in Europe: Are cities sources of regional resilience? *J. Econ. Geogr.* **2015**, *14*, 951–972. [CrossRef]
51. Connaughton, J.E.; Madsen, R.A. U.S. State and Regional Economic Impact of the 2008/2009 Recessio. *J. Reg. Anal. Policy.* **2012**, *42*, 177–187.

sustainability

MDPI

Article

Urban Morphological Feature Extraction and Multi-Dimensional Similarity Analysis Based on Deep Learning Approaches

Chenyi Cai [1,2,†], Zifeng Guo [2,†], Baizhou Zhang [1], Xiao Wang [1], Biao Li [1] and Peng Tang [1,*]

1 School of Architecture, Southeast University, 2 Sipailou, Nanjing 210096, China; caichenyi@seu.edu.cn or cai@arch.ethz.ch (C.C.); zhangbaizhou@seu.edu.cn (B.Z.); wangxiao_seu@seu.edu.cn (X.W.); jz_studio@seu.edu.cn (B.L.)

2 Department of Architecture, Swiss Federal Institute of Technology Zurich (ETHZ), Stefano-Franscini-Platz 1, 8093 Zürich, Switzerland; guo@arch.ethz.ch

* Correspondence: tangpeng@seu.edu.cn

† These authors contributed equally to this work.

Abstract: The study of urban morphology contributes to the evolution of cities and sustainable development. Urban morphological feature extraction and similarity analysis represents a practical framework in many studies to interpret and introduce the current built environment to aid in proposing novel designs. In conventional methods, morphological features are represented based on qualitative descriptions, symbolical interpretation, or manually selected indicators. However, these methods could cause subjective bias and limit the generalizability. This study proposes a hybrid data-driven approach to support quantitative morphological descriptions and multi-dimensional similarity analysis for urban design decision-making and to further morphology-related studies using information abundance via a deep-learning approach. We constructed a dataset of 3817 residential plots with geometrical and related infrastructure information. A deep convolutional neural network, GoogLeNet, was implemented with the plots' figure–ground images, by quantifying the morphological features into 2048-dimensional feature vectors. We conducted a similarity analysis of the plots by calculating the Euclidean distance between the high-dimensional feature vectors. Then, a comparison study was performed by retrieving cases based on the plot shape and plots with buildings separately. The proposed method considers the overall characteristics of the urban morphology and social infrastructure situations for similarity analysis. This method is flexible and effective. The proposed framework indicates the feasibility and potential of integrating task-oriented information to introduce custom and adequate references via deep learning methods, which could support decision making and association studies on morphology with urban consequences. This work could serve as a basis for further typo-morphology studies and other morphology-related ecological, social, and economic studies for sustainable built environments.

Keywords: urban morphology; deep learning; similarity analysis; cluster analysis; feature extraction

Citation: Cai, C.; Guo, Z.; Zhang, B.; Wang, X.; Li, B.; Tang, P. Urban Morphological Feature Extraction and Multi-Dimensional Similarity Analysis Based on Deep Learning Approaches. *Sustainability* **2021**, *13*, 6859. https://doi.org/10.3390/su13126859

Academic Editor: Tan Yigitcanlar

Received: 23 May 2021
Accepted: 11 June 2021
Published: 17 June 2021

1. Introduction

Urban morphology refers to the multidisciplinary study of urban forms regarding the physical environment, the cultural preservation process, and sustainable development [1,2]. The morphological approach provides the idea that morphology has the potential to be an animating force for urban design [3]. Many studies relate urban typo-morphology to studies of the citizens' lives, the social economy, and the energy system efficiency [4–6]. Therefore, urban morphology provides a valuable basis for urban planners and managers. Urban morphology is related to complex urban system analysis, such as energy performance [7], citizen behavior, and economic benefits [8].

On the other hand, in the urban renewal process, communications with the as-built urban fabric need to be considered in urban design. In a study of urban neighborhoods [9], evolutionary patterns related to sustainable urban neighborhoods were extracted based on the morphological classification and clustering of footprint patterns over time.

In addition to simulation-based urban morphological analysis [10], researchers focusd on data-driven approaches for urban morphological studies recently by considering the methods for constructing relations between the as-built and the to-be-built environment, such as multi-dimensional (e.g., the geometric dimension and social dimension) of urban forms. Technologies, such as 3D scanning, depth detection, multi-directional scanning, and simultaneous localization and mapping (SLAM) are becoming increasingly sophisticated. The abundance of web-based map data, such as AutoNavi, Baidu, and OpenStreet Map (OSM), allows architects to grasp data efficiently. In the information abundance, machine learning approaches support design by providing designers with previous cases for new design solutions based on the case-based reasoning (CBR) [11].

In data-driven urban design, learning from reality (in this case, urban morphology) helps decision makers in making comprehensive design decisions and researchers in the study of spatial form-related functionality and performance. On one hand, the suggestions for urban morphology design in the decision-making process take social, environmental, and economic factors into consideration. The suggestions serve as a reference as well as guidance for decision-makers.

Urban morphology is the physical carrier of quality of life. In the urban design process, the lack of a concrete understanding of urban morphology by non-professional decision-makers limits the space for discussion by designers [12]. Both developers and designers could develop discussions and ideas from the cases in similar situations, introducing the information and knowledge. On the other hand, the associations of urban morphology with functionality and further consequences supports researchers in morphology-related studies. The suggestions for urban cases could be based not only on the spatial form but also on the morphology-related traffic networks, energy performance, economic conditions, and so on, thus, supporting further scientific utility.

Effective morphological quantification methods for cases is crucial in terms of the data representation for a case retrieval system. However, cities are developed in complicated historical, economic, and behavioral contexts. Every city is unique in urban form [13]. There is not a clear-cut answer of the critical factors or the factor weights for city development or collapse simulation [14]; therefore, deduction and verification methods are challenging to apply in a comprehensive urban morphology study for urban design. For example, in the MApUCE tools chain study, a processing chain was proposed to calculate 64 standardized urban morphological indicators to represent the buildings, blocks, and spatial units [15].

More indicators could be extracted for more precise representation. However, each indicator's weight influences the calculation since it should be appropriate for reaching the global fitting of the instances, and there were still missing factors by selecting and calculating indicators. Moreover, the critical indicators varied from the cases from different cultural and historical contexts, which need to be studied and verified. Quantifying the morphology of a large number of cases with indicators would lead to a generalizability limitation. Therefore, a descriptive framework for urban support analysis is essential, and its effectiveness could be shown in the subsequent data analysis and visualization of urban case retrieval.

In developing an efficient approach for urban morphological quantification methods for similarity analysis, three facets of obstacles are observed: (1) the construction of multi-dimensional urban dataset that includes geometrical and social information, (2) the quantification of urban morphological features with concise but informative descriptions, and (3) similarity calculations of the extracted morphological features. According to the above discussion, an automatic data mining method would support the detailed information collection of the related infrastructures from various aspects. A fully automated feature extraction method may help to overcome the drawbacks of manually selecting indicators and balancing weights. A feature extraction method considering the statistical and overall characteristics of the instances could be introduced.

Machine learning approaches help to efficiently represent and retrieve cases from a considerable amount of data [16]. Deep learning is one of the branches of machine learning. Deep learning algorithms promote evolutionary methodologies for morphological analysis. This approach is robust with pictorial datasets because of the development in convolutional neural networks [17]. The convolutional methods support a fully automated feature extraction process among a large amount of data. This represents the samples' characteristics with concise and comprehensive information in feature vectors. Methods, such as image-data-based (RGB), numerical labeling, and semantic segmentation, are dedicated to feeding samples into neural networks with continuous and informative features. Cluster analysis supports comparison and similarity studies by data-deduction and distance calculating techniques, taking the feature extraction data as a basis.

We clarified our study scope on case retrieval to support design decision-making and serve as a basis for further scientific utility via deep learning, as, regarding the information abundance and complexity of cities, we need a solution space implying urban knowledge rather than certain answers. For example, the efficient similarity analysis of urban associations (e.g., the morphology, traffic, energy, and economy) represents the task-oriented retrieved cases for decision makers in certain applications and researchers for further scientific analysis, based on texts, images, models, and other representation media carrying concrete urban information.

Therefore, the construction of the spectrum of cases based on situation similarity is a promising way to efficiently introduce references from the information abundance for a wider discussion space for design decision making and precise association of morphology with urban consequences for scientific studies. The reference's effectiveness depends on the case quantification methods and the related social information (e.g., infrastructure, industrial distribution, and traffic conditions) of the cases. The method would also have potential for a general search engine in terms of urban morphology. Therefore, an effective morphological quantification approach and multi-dimension datasets, including related geometric and social information, would be needed for comprehensive urban design decision-making.

In this study, we propose a multi-dimensional similarity analysis approach for introducing cases in similar situations from the information abundance. The similarity analysis includes morphological similarity and social situation similarity. The proposed method combines data mining and cluster analysis via deep learning, taking the residential cases in Nanjing, China, for instance. In this study, a multi-dimension dataset, including geometrical information and infrastructure information, is constructed. The samples' morphological features are extracted into high-dimensional feature vectors (HDFV) via a deep convolutional neural network. This study further completes the case retrieval based on the HDFV. The architects can retrieve cases according to the plot-shape-similarity or building-distribution-similarity, along with the infrastructure information. The significance of the study is as follows:

- This study is an interdisciplinary study that integrates urban design with computer science and applies the latest deep learning techniques to the study of urban morphological case retrieval.
- The proposed approach provides a feasible method for quantifying the overall morphological characteristics automatically and informatively by pictorial-feature-mapping high-dimensional data, excluding the manual indicator selection process.
- Multiple cases ranking in similarity with the input and social information were achieved simultaneously, reducing the difficulty in deciding the weights of indicators.
- The proposed method is independent of the cases' morphological types by learning from the samples directly. Therefore, it has the flexibility to do clustering and retrieval for various urban morphologies, as long as the cases are feeding into the deep learning model.

- The framework could be flexible for integrating more environment-related datasets for image-based similarity analysis, introducing custom references to support the designer' decision-making towards sustainable development.
- This work could serve as a basis for further typo-morphology studies and integrate morphology-related ecological, social, and economic studies for the built environment.

2. Literature Review

In addition to a qualitative and conceptual description of urban morphology, recent studies have been performed for the quantitative representation of urban morphology, such as data discretization methods [18]. Researchers utilized various quantification methods for morphology-to-data transmission by selecting and adjusting indicators. Deep learning models were used to automatically learn features and classifiers at once by error backpropagation, adjusting the layers' importance depending on the problem. Then, similarity analysis techniques were used to construct an efficient case retrieval system based on the feeding samples [19].

2.1. Urban Morphological Quantification

Multiple quantitative approaches have been implemented for transmitting morphology to data. Researchers have used pre-proposed indicators (e.g., semantic indicators and geometrical indicators) to index buildings and urban relationships for the following studies [20]. For example, in the early stage application of case-based reasoning for design, ARCHIE is an intelligent case browsing system. The cases are represented as attribute-value pairs with 150 features, including the concept, text, actual number, and function [21]. In the study of urban typologies [22], block and street types were studied as a context-sensitive sample of types by describing the cases with semantic and geometric values, such as land use, length, area, and ground space index (GSI). Nahyun developed a model adopting case-based reasoning and a genetic algorithm to predict the maintenance costs for aging residential buildings [8].

On the other hand, hierarchical structures along with symbolic representations have been constructed to describe urban forms containing plots, buildings, and streets [23,24]. Based on the established hierarchical structure, measurements (e.g., distance and connectivity) were calculated for further analysis. For example, in the study of spatial design network analysis (sDNA), Crispin developed a tool for network analysis, which was implemented for the representation and calculation of the network nodes and link density [25].

Song proposed an access structure to symbolically describe and measure the relations between the fundamental elements: plots, buildings, and streets. They took eight cases for similarity analysis to validate the access structure [26]. Although discrete indicators and structural measurements help in morphology quantification, there are still features that are hard to describe numerically—especially the geometric information, such as the volume distribution, directions, shapes, and connections. This leads to a generalizability limitation.

2.2. Similarity Analysis of Urban Morphology

A further step based on the extracted features is to use regression models to cluster the samples and analyze their similarities. Researchers have used multiple methods, such as classification, clustering, or ranking the solution-instances to manage the database [27]. In recent decades, researchers emphasized typo-morphology-related, context-sensitive, and systematic urban form studies, such as socio-ecological spatial morphology [28]. Mathematical methods can also be used for similarity analysis. In the study of case-based design with 3D mesh architectural models [29], a TRAMMA (Topology Recognition and Aggregation of Mesh Models of Architecture) method was proposed for the clustering, retrieving, and reconstructing architectural elements for a new architectural model.

For the urban renewal process, the Roma urban renewal [30] study built a similarity searching system based on the block shapes to search for cases similar to the new block shape and adapt the urban fabric to the new site to preserve the historical context. The

Roman study investigated the case retrieval method for finding a case with similar block shapes to the block to be designed. They adjusted the pre-defined indicators related to the cases' geometric characteristics for input, until they obtained satisfactory retrieved results. The referred retrieval indicators in the Roman study included the block area, building density, the block-to-bounding-box ratio, the length-to-width ratio, and the function type. Then, they generated the flyover above the railway station for their design proposal [31]. The case similarity was detected based on the evaluation of these indicators. Then, one case that met the input conditions was retrieved from the dataset.

Since the regeneration study based on the retrieved cases required efforts from other aspects related to specific design tasks, it is important to make the interface for connecting the case retrieval and regeneration periods. For example, Hua regenerated new 3D mesh models by combing the parts of the cases based on the extracted floor elements and the corresponding walls. The extracted elements are the interface connecting the retrieved cases and the regenerated models [29].

In the Roman study, the retrieved case was arbitrarily applied to a new block automatically by simple geometrical operations (e.g., scaling and rotation), and architects can adjust the applied results manually based on a visual model software [31]. Therefore, the regeneration period could be realized by multiple techniques as long as the case retrieval process is highly automated. Therefore, in this study, we focused on the morphological quantification method's efficiency, the ability to bring more cases, and the flexibility for implementing new datasets for case retrieval.

2.3. Deep Learning for Morphological Analysis

Compared with conventional methods, the advantages of the deep learning method is that it is an end-to-end (e.g., image to segmentation label) process, with the potential of connecting task-oriented things [32]. Today, convolutional neural networks (CNNs) are one of the most prominent deep learning approaches for image processing and computer vision. In a convolution neural network (CNN), the feature mapping of the images extracts the input's underlying features by convolution kernels, promoting deep learning algorithms in recent decades [33]. Features of the input are continuously convolving into multi-dimensional vectors layer by layer. Neural networks have been implemented for solving problems in the field of architecture, such as the prediction of energy performance [34], pattern recognition of 2D images [35], as well as typological form-finding on 3D models [36].

New data analysis techniques have been proposed for the design process to address the abundance of variables. Neural network techniques, such as classification, prediction, and cluster analysis, provide technical support for regression models for data analysis. Cluster analysis is a powerful technique for urban morphological analysis, which extends the technical support of high efficiency and accuracy for classification and clustering analysis from a large amount of data.

In a study of typo-morphology in Lisbon [37], the block and street types were classified based on the prepared plans using the k-means clustering algorithm. The ETH team's "City of indexes" project interpreted urban morphological patterns based on the massive data of city images combined with personal preferences [38].

Concise but informative feature vectors extracted via deep learning were used to quantitatively characterize morphological features [39]. With the introduction of deep learning, architects need to focus on formulating inputs related to architectural problems, understanding the meaning of each layer's output, and finding the application sceneries. Since cluster analysis is an unsupervised technique, it excludes labeling data, which is time-consuming. It automatically features the images with high-dimensional data to support further analysis.

3. Materials and Methods

Figure 1 shows the general workflow of our study. First, we collected web map data, including geometrical information and Point of Interests (POI) in Nanjing, which were downloaded with the open source AutoNavi API. We filtered the collected data into a geometrical dataset and additional social information (distribution of related infrastructure) based on functionality with the ArcGIS software. We exported images for the case slices in terms of plots.

Second, the morphological features were automatically extracted through a deep convolutional neural network with inception-v3 modules into high dimensional feature vectors (HDFV). Third, the cluster analysis was visualized based on the t-SNE algorithm in a two-dimensional plane. The Euclidean distance was applied to calculate the similarity between the cases. Finally, the performance of the HDFV on the similarity analysis was verified by a comparison case retrieval study.

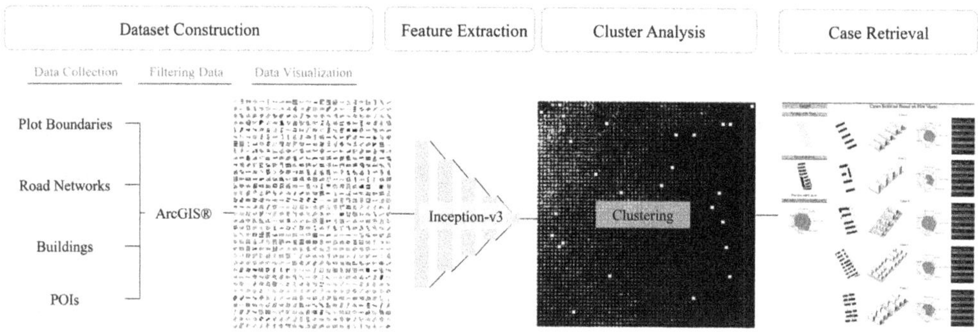

Figure 1. The general workflow of quantitative urban morphological analysis for case-based design via deep learning.

3.1. Dataset Construction

We chose Nanjing for our case study. The study scope focused on applying deep learning methods to the quantitative study of the urban fabric. The plot morphology is highly related to the land use of the plot. In Chinese cities, the land use of the plots is determined according to "GB 50137-2011 Code for Classification of Urban Land Use and Planning Standards of Development Land" [40], which is an important standard in city and town planning in China. According to the land sale data in Nanjing, the sold R2 residential plots were 58.9% in 2017, 42.5% in 2018, and 54.5% in 2019 of all sold land.

This indicates that residential plots occupy a large proportion of the urban plots, which helps to ensure a sufficient sample size. In addition, the residential plots showed no significant difference, which can ensure the similarity distribution of the datasets; therefore, the performance of the quantification method can be tested by making distinctions from similar morphologies. Therefore, by taking residential plots as the case study, a high-quality dataset can be obtained. We chose urban plots rather than blocks as research objects because the plot connects the buildings and the city. The buildings in one plot are usually planned as a whole. The plots with the same functions have similar design indicators, which implies the significance of morphological identification.

The AutoNavi open platform provides a web mapping service API in China, where we downloaded geometric information, navigation, and infrastructure information of a particular area. As the downloaded data includes all the information visualized on the web map, we need to filter out the specific data for our study direction according to the locations or the labels. For example, the Area of Interest (AOI) data includes various function types of areas, including residence areas, water areas, tourism areas, commercial areas, and education areas.

The building boundaries and road networks we downloaded from the platform involve all buildings and roads in Nanjing city. In addition to map-related geometry data, infrastructure information can be revealed by Point of Interests (POIs). The POI contains information, such as the ID, name, category, address, etc., and a series of related information (e.g., streetscape and user comments) can be obtained by keyword search. Based on the POIs from Nanjing city, we filtered out the residential-related POIs.

After downloading the data as shapefiles, we filtered out the plot boundaries according to the "residence" label of AOIs. Then, we filtered out the buildings and roads according to the locations compared to the filtered areas. These operations were based on ArcGIS. This study selected the infrastructure categories closely related to residential areas: restaurants, shopping, physical facilities, public services, medical facilities, education services, and public traffic. Figure 2 shows the distribution of the different types of POIs, taking one of the districts of Nanjing: Qinhuai District as an instance. The POIs information was visualized based on the tableau software.

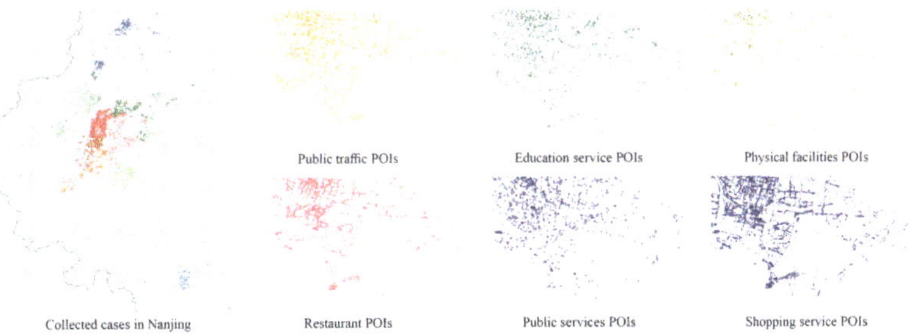

Figure 2. The distribution of the related infrastructure information around the residential cases (in Qinhuai District).

Finally, 4172 residential plots were filtered out from Nanjing. Some residential areas were still under construction or had no building information. Therefore, the dataset contained 3817 residential areas with valid data. For each residential case, there was infrastructure information, exported plot images, and geometric models, as well as semantic information attached as attribute values, including the name, address, map-ID, location, and site area.

Figure 3 shows the information included in the dataset after we processed the shapefiles and POIs downloaded from the AutoNavi platform. All the plots were exported to images as training data for the deep convolutional network model. The amount of infrastructure covered per residential area was calculated based on the service radius of the POI points and visualized by radar charts. We wrote the codes for drawing radar charts with JAVA language.

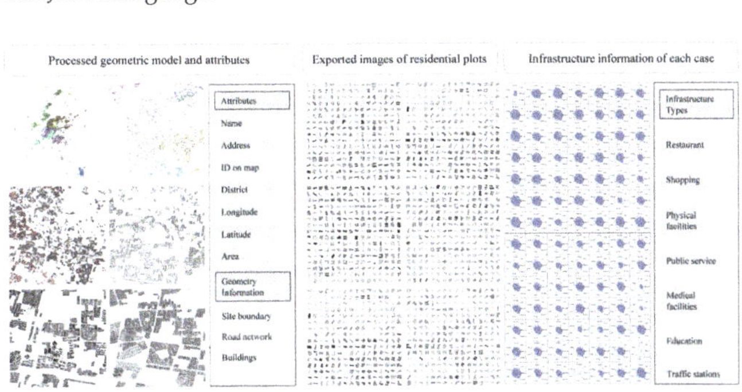

Figure 3. Part of the collected Nanjing residential cases with the attribute information shown on the ArcGIS platform.

3.2. Feature Extraction

The convolutional layers extract the overall features by sweeping the image pixels in a certain step sequence through the convolutional kernel, which is the feature mapping process (Figure 4). Each kernel is an n*n matrix containing weight values. The high dimensional feature vector output after multi-layer convolutional operations could represent the overall features of the input image because spatially adjacent pixels in an image have considerable correlations [41].

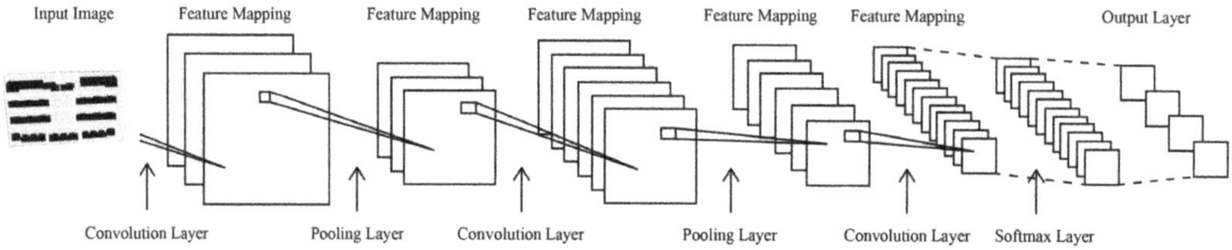

Figure 4. Schematic of a convolutional kernel sweeping over an image for feature mapping.

To quantify the urban morphological features, we used the deep convolutional neural network GoogLeNet, which includes the Inception-v3 module. This architecture of the Inception-v3 module contributes to the high performance of the deep convolutional neural networks on image classification because it is susceptible to the context of input images [42]. A pre-trained GoogLeNet could be implemented for various feature extraction tasks. The weights of the kernels were optimized by training the model with the ImageNet dataset [43]. In this way, the deep convolutional neural networks map the image features into high dimensional feature vectors by feature mapping. Figure 5 briefly illustrates the structure of the GoogLeNet.

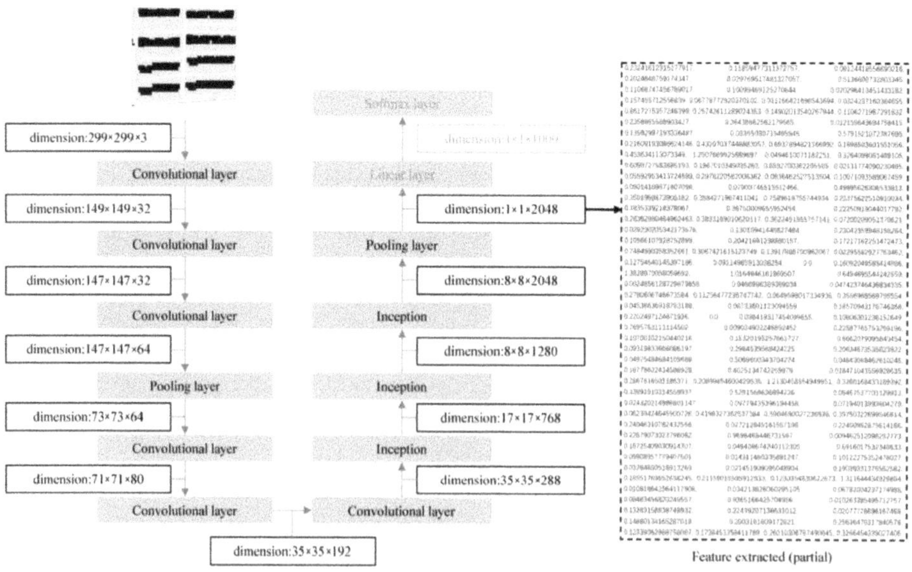

Figure 5. The structure of the GoogLeNet and the 2048-dimensional feature vector output of the bottleneck layer is taken out as high dimensional feature vectors (HDFV).

The process of passing an image through a trained convolutional neural network up to the bottleneck layer can be viewed as a feature extraction process for the image [44]. Typically, the final layer's output in a convolutional neural network is a number between 0 and 1 to represent the prediction of the categories. Before the linear layer is the so-called bottleneck layer, the output size of which is 1 × 1 × 2048. The bottleneck layer's output can be considered a more concise and representative feature vector of the image. This layer can represent the features learned by the neural network. Therefore, we take the penultimate layer's output, where the dimension of the input image increases to 1 × 1 × 2048. The output data were collected as HDFV for further comparison. We carried out a comparative study regarding the case retrieval performance on the plot shape and building distribution, focusing on the plot shape and plots with distributed buildings as independent inputs.

3.3. Cluster Analysis and Visualization

Using image data of the residence plots as the input for clustering is often impractical due to the gigantic size of the data matrix converted from the images. Therefore, a process of dimension reduction is often required. Currently, there are three mainstream techniques for data reduction: Principal Component Analysis, the t-SNE algorithm, and an Auto-Encoder. In this experiment, we used the t-SNE algorithm to map high-dimensional feature points to a two-dimensional plane without losing the information of the feature vectors. The samples with similar features were placed as neighbors in the cluster cloud (Figure 6). Figure 6 shows the spectrum of all the cases based on morphological similarity. The left picture represents the clustering results in terms of the cases' plot shapes, while the right picture shows the cases' plots with buildings. The more similar the cases are, the closer they are on the clustering map.

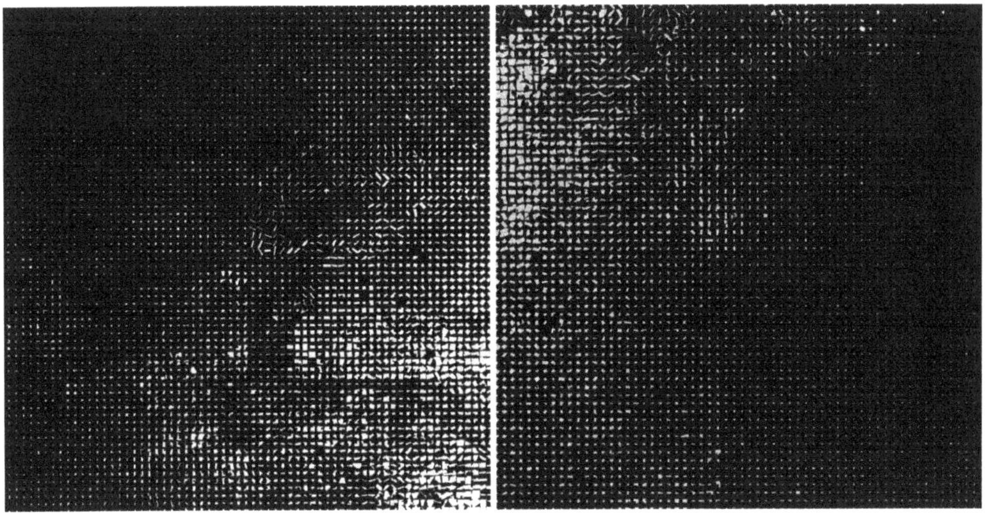

Figure 6. Atlas of the clustering analysis of 3817 residential plots in Nanjing, China.

To show the clustering map more clearly, we zoomed in to some parts and present them in Figure 7. The samples are shown on the same scale. Clusters of squaring, narrow, or irregular shapes can be intuitively seen in Figure 7a. The result is different in Figure 7b, where the distribution of buildings influences the clustering result. Different residential types could be observed, such as plots with few rows of buildings, closely spaced residential buildings, and loosely arranged villas, etc. We can intuitively see that the plots belonging to the same cluster have morphological features in common. The cluster analysis performed better in near-square plot shapes rather than irregular plot shapes, as more cases had square plots.

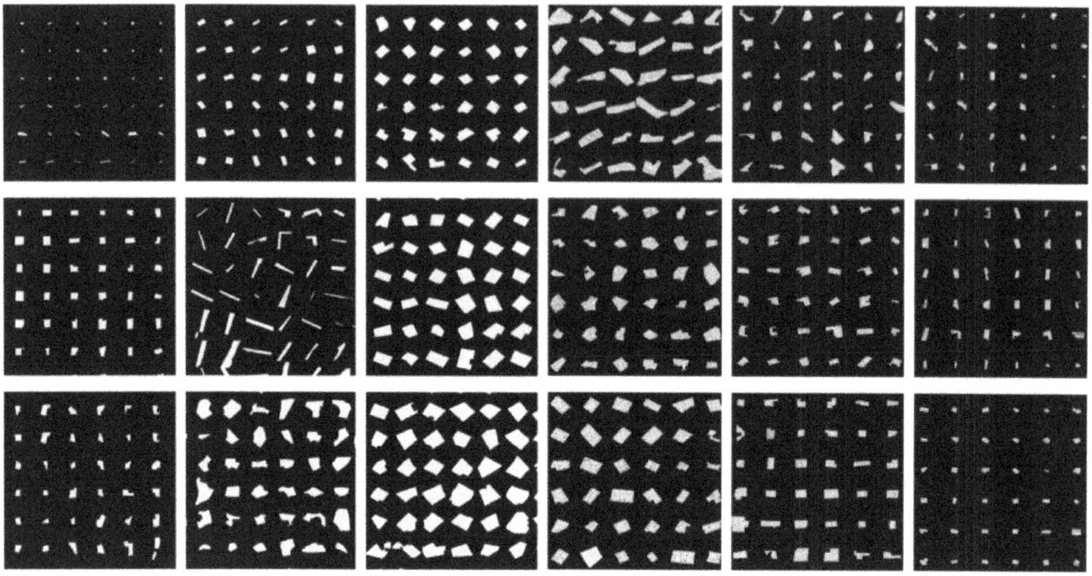

a. Plot shape clustering   b. Plot with buildings clustering

Figure 7. The visualization of the clustering based on plot shape and plots with buildings.

3.4. Similarity Analysis and Case Retrieval

The data of HDFV have the same contribution weights for the similarity analysis, as they reflect the overall morphological characteristics of each sample rather than the independent indicators. Therefore, the Euclidean distance was used to calculate the difference between the input images. The closer the distance was, the more similar the two cases were. The case retrieval system can rank one or more cases for the target based on the distance. We constructed a case retrieval system that realized case pairs according to the plot shape or plots with buildings. When integrated with other sample attributes, architects can choose the proper ones among the recommended cases according to the specific task.

4. Results and Discussion

After the feature extraction by the deep CNN, each image was assigned with HDFV and its corresponding attributes (semantic information and infrastructure information). Four cases were selected to explain the HDFV performance in representing the morphological similarity of the images. The urban fabric images and the distances between each HDFV are listed in Figure 8. We conclude that plots b and c, and a and d are pairs of similar morphological types according to the HDFV distances. Plots a and d are distributed with intensively lined-up buildings. These are aged residential areas built in the 20th century. Plots b and c were built in recent decades, also with lined-up buildings but with more sparse textures.

Plot Index	Euclidean Distance			
			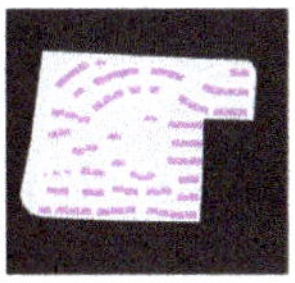	
a		15.032	14.667	12.963
b	15.032		13.306	15.778
c	14.667	13.306		15.106
d	12.963	15.778	15.106	

Figure 8. The Euclidean distance between the samples.

4.1. Clustering based on Euclidean Distance

To compare the HDFV performance featuring the figure–ground images, we selected five clusters with different characteristics and picked six samples from each. Figure 9 shows the nearest five cases to the targets according to the plot shape clustering and plot with building clustering. The five clusters show different characters. For example, cluster 1 has a square plot with lined buildings, while the buildings in cluster 3 are distributed intensively. The samples in cluster 2 are a relatively small plot with one or two buildings. The samples in cluster 4 and cluster 5 have linear plots and irregular plots. This result indicates that samples of different morphology types could be clustered automatically according to HDFV without the need to pre-define the morphology types.

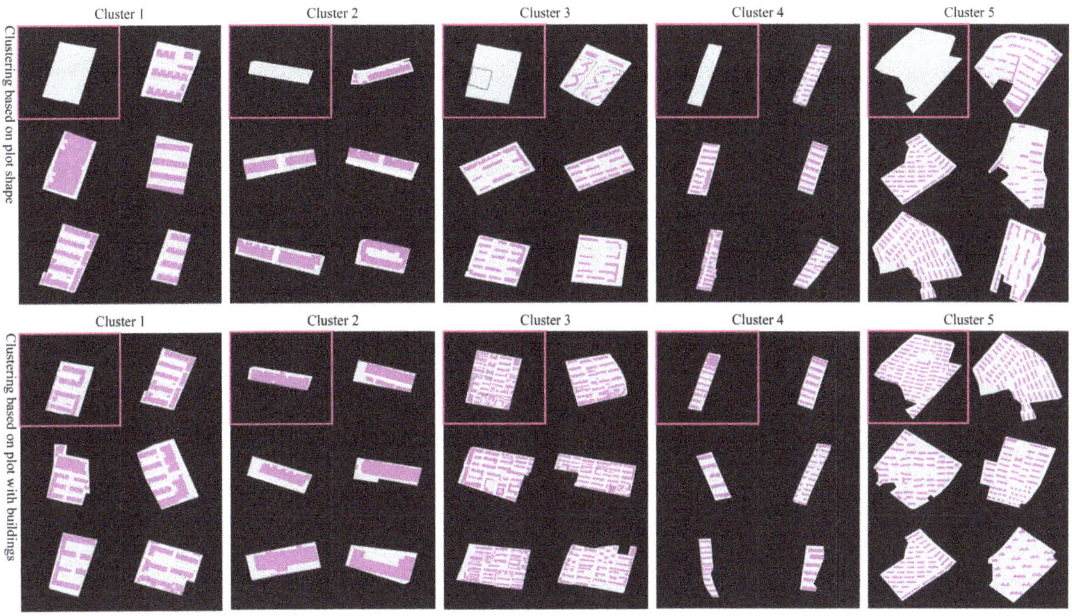

Figure 9. The plot shape most similar five cases and plot with buildings most similar five cases retrieved based Euclidean distance between HDFVs.

The samples in a cluster reflect the similarity in plot shape when clustering based on plot shape and in building distribution texture when clustering based on the plot with buildings. For example, in cluster 3, the target is an aged residential area in Nanjing. According to an architect's intuitive observation, cases retrieved based on the plot with buildings were similar to the target, with intensive building texture as the target, while some plot shapes included corners. On the contrary, for example, in cluster 2, cases retrieved based on the site shape varied in the building distribution but were similar in shape. We found potential indicating that the HDFV is sensitive to the urban fabric.

In the deep learning model, the HDFV was calculated by flattening the grayscale value matrix of the image, reflecting the distribution of $n \times n$ pixel matrix values over a $1 \times n \times n$ matrix. The HDFV reflects the morphology characteristic based on the distribution of pixels in an image. Moreover, the HDFV compressed the pixel distribution by increasing the impacts of effective pixels and decreasing the influence of ineffective pixels.

The clusters represent the similarity in the plot shape or the building texture. Different morphological characteristics, such as narrow plots with intensively distributed buildings and irregular plot shapes with multiple buildings, can be observed in terms of the clusters, consistent with an architect's intuitive observation. Therefore, the HDFV is sensitive to the urban morphology indicated by the pixel distribution by carrying the overall and informative features of the samples.

4.2. Examples of Case Retrieval

We took the case retrieval test as an example (Figure 10). The target plot was close to a rectangle, with intensively lined up buildings as well as vertical buildings. The distributions of the relevant infrastructures are shown with the radar chart. The case retrieval process was completed in seconds.

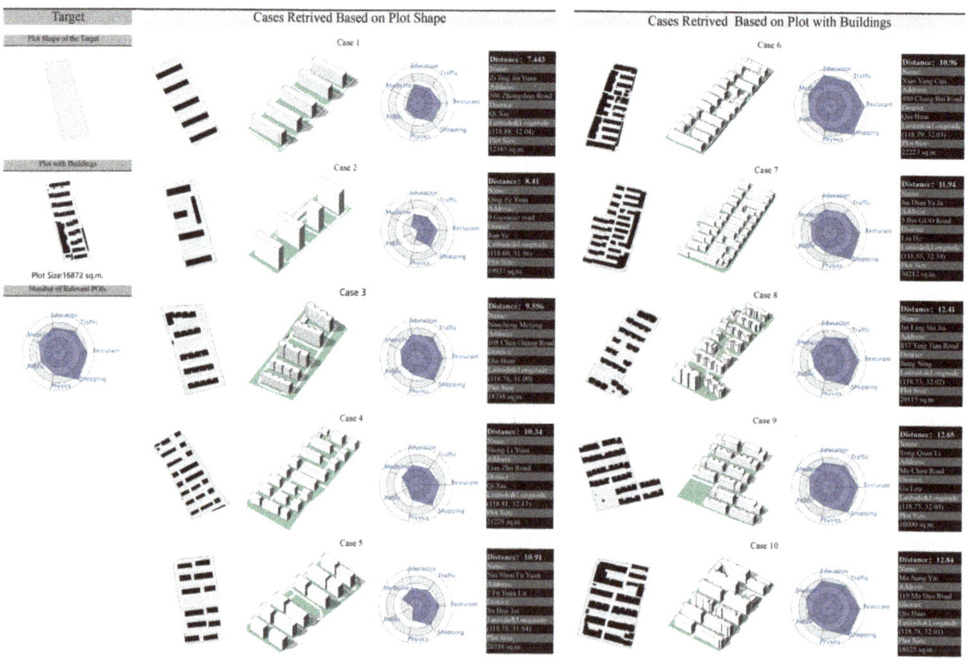

Figure 10. Comparison study of the case retrieval process by recommending cases based on plot shape and plots with buildings.

The building distribution influenced the results when the cases were retrieved based on plots with buildings. They were more similar in building texture than in plot shape with the target. For example, case 6 in target one was retrieved based on the plot with buildings. It had lined-up buildings and vertical ones, and the buildings were distributed intensively. The building distribution texture was similar to the target, which is like the aged Nanjing residential plots. Case 1 was retrieved based on the plot shape; therefore, it showed more similarity in the plot shape, while the building distribution texture was different from the target.

The results demonstrate the effectiveness of the case retrieval with similar morphologies based on the morphological quantification method. With the multi-dimensional dataset, the related information of the case (e.g., name and address) and the distribution of the related infrastructure could help decision makers and researchers obtain new information through similar cases to support decision making and further studies.

4.3. Discussions on the Method

The advantages of the proposed method regarding similarity analysis can be discussed by comparing the study using conventional methods, including the Roma urban renewal project mentioned in Section 2.2.

First, the HDFV had higher efficiency on the urban fabric quantification and similarity analysis. In the study of Roman, the authors extracted several geometric indicators to evaluate the block shape similarity. This took time for the extraction and verification process of selecting the indicators and balancing each weight. The HDFV carries comprehensive morphology characteristics by matrix operation of the sample pixels based on the deep learning model. It saves effort in addressing a large number of images since the whole process from feature extraction to case retrieval is performed automatically.

Secondly, the HDFV has more generalizability in applying the method to various urban morphologies. The Roman study had a limitation when evaluating the building distribution texture similarity. It required another extraction and verification process of selecting indicators to describe the building distribution characteristics, which is a more complicated and diverse process compared with describing the block shape. With the HDFV, we could perform the similarity investigation of plot shape or building distribution under the same framework. The proposed method goes beyond the morphology types from different historical contexts because it learns from the samples directly and clusters them based on the feature vector distance.

Thirdly, the proposed method has more flexibility regarding similarity-based case retrieval. In the study of Roman, the system recommended another case that was similar to the input. However, we could find a series of cases that are similar to the input from sufficient samples. The result space would be broader if there were more samples in the dataset. In this way, more references and information could be brought to designers.

What is more, the proposed method could broaden the similarity analysis because it takes related social information (e.g., the infrastructure) into consideration by integrating a multi-dimensional dataset. In addition to the morphology and the POI information studied in this experiment, the framework could be implemented for quantification and similarity analysis with more information related to the urban sustainability. For instance, the cultural background, the energy performance, the traffic conditions, etc. could be added to the dataset, depending on the design task. The image-based similarity analysis can be done via the deep learning model. This would provide precise references in similar situations to better support the designers' decision-making towards sustainable cities.

The limitations of this study involve that the highly automated process increases the difficulty of emphasizing the specialties from a particular aspect. There are three main limitations. First, the insufficient number of samples with similar plot sizes leads to some noise in the case retrieval results. The results would be more robust if there were around 1000 cases with similar plot sizes. This limitation could be overcome by simply adding more cases collected from cities around Nanjing based on AutoNavi.

Second, all 3D information (e.g., building height and building shape) is lost since the samples are represented as 2D images for learning. This drawback could be overcome by adding one more color channel to represent the building height or by using voxels instead of pixels to describe the plot in three dimensions. Third, the target needs to be trained together with the cases in the dataset. In other words, once a new target is introduced, the entire neural network has to be retrained.

5. Conclusions

This morphological similarity analysis represents a helpful analysis framework for many fields, such as typo-morphological, historical evolution, pre-design contextualization, and building energy performance. Finding cases in similar situations to the target could support designers in obtaining new information and knowledge resulting in better decision making and furthering scientific studies. Quantitative descriptions of urban morphology provide a baseline for in-depth urban fabric interpretation. This study aimed to develop a data-driven approach to quantitatively describe urban morphology and to develop a multi-dimensional case retrieval method for urban design decision-making in the early stage for association studies on morphology with specific social or economic aspects.

In this study, 3817 residential cases with geometrical and social service information from Nanjing, China, were filtered to construct the dataset. The data source was exported as figure–ground images for training the deep CNN GoogLeNet with the inception-v3 module, encoding the images into 2048-dimensional feature vectors based on grayscale values. The similarity analysis of the cases was verified by calculating the Euclidean distance between HDFV. A comparison study was conducted in the case retrieval process to integrate the morphological and infrastructural similarity.

This study demonstrated the feasibility and power of the deep learning network in urban morphological similarity analysis and multi-dimensional decision making. The deep learning algorithms provided a method to automatically extract/learn the intrinsic features from a large amount of data. The morphological features were represented by HDFV, which contained comprehensive information for the morphological characteristics.

The multi-dimensional case retrieval method can support comprehensive decision-making and morphology-related scientific studies by providing customers with many references in similar situations with the target based on the comprehensive and precise similarity analysis. This method is integrated with easy access to related infrastructure and social and economic information. Other information that is related to the specific task (e.g., culture, traffic, energy performance, and economic consequences) could be easily implemented under the same framework to support decision-making and further scientific studies regarding associations of morphology and other urban aspects.

Future work will focus on technological improvements and more application scenarios. Adding more dimensions to the data source, such as additional color values to indicate building heights, would be an effective improvement of the model performance. Approaches to using geometric spatial data as direct inputs to the neural network rather than figure–ground images will be explored for better computational efficiency and more precise case retrieval. More typo-morphology-related attributes could be added to the data sources according to specific scenes.

For example, energy performance, user testimonials, traffic conditions, industrial distributions, the natural environment, and so on could be introduced for more comprehensive similarity analysis to better support design decision making. In addition, the HDFV could serve as the interface for connecting retrieved cases and regeneration. For example, new design proposals could be generated derived from retrieved cases by implementing energy evaluation and optimization or rule-based generative design.

Author Contributions: Conceptualization, C.C.; Methodology, C.C., Z.G. and B.Z.; Software, C.C. and Z.G.; Validation, C.C. and Z.G.; Formal Analysis, C.C.; Resources, X.W.; Data Curation, X.W.; Writing—Original Draft Preparation, C.C.; Writing—Review and Editing, C.C., Z.G., P.T. and B.L.; Visualization, B.Z.; Supervision, B.L. and P.T.; Funding Acquisition, B.L., C.C. and Z.G. All authors have read and agreed to the published version of the manuscript.

Funding: This study was funded by the National Natural Science Foundation of China (NSFC) project (No. 51978139). This study was funded by China Scholarship Council (CSC) grant 201706090254 and 202006090151.

Conflicts of Interest: The authors declare no conflict of interest.

Abbreviations

The following abbreviations are used in this manuscript:

HDFV	High Dimensional Feature Vector
CBR	Case-based Reasoning
CNN	Convolutional Neural Network
POI	Point of Interest
AOI	Area of Interest

References

1. Kropf, K. Aspects of urban form. *Urban Morphol.* **2009**, *13*, 105.
2. Hemphill, L.; Berry, J.; McGreal, S. An indicator-based approach to measuring sustainable urban regeneration performance: Part 1, conceptual foundations and methodological framework. *Urban Stud.* **2004**, *41*, 725–755. [CrossRef]
3. Rossi, A.; Eisenman, P.; Ghirardo, D.Y.; Ockman, J. *The Architecture of the City*; MIT Press: Cambridge, MA, USA, 1982.
4. Chen, F. Urban Morphology and Citizens' Life. In *Encyclopedia of Quality of Life and Well-Being Research*; Springer: Dordrecht, The Netherlands, 2014; pp. 6850–6855.
5. Osman, T.; Divigalpitiya, P.; Osman, M.M.; Kenawy, E.; Salem, M.; Hamdy, O. Quantifying the relationship between the built environment attributes and urban sustainability potentials for housing areas. *Buildings* **2016**, *6*, 39. [CrossRef]
6. Sarralde, J.J.; Quinn, D.J.; Wiesmann, D.; Steemers, K. Solar energy and urban morphology: Scenarios for increasing the renewable energy potential of neighbourhoods in London. *Renew. Energy* **2015**, *73*, 10–17. [CrossRef]
7. Emmanuel, M.R. *An Urban Approach to Climate-Sensitive Design: Strategies for the Tropics*; Taylor & Francis: London, UK, 2005.
8. Kwon, N.; Song, K.; Ahn, Y.; Park, M.; Jang, Y. Maintenance cost prediction for aging residential buildings based on case-based reasoning and genetic algorithm. *J. Build. Eng.* **2020**, *28*, 101006. [CrossRef]
9. Ariga, T. Morphology, Sustainable Evolution of Inner-urban Neighborhoods in San Francisco. *J. Asian Archit. Build. Eng.* **2005**, *4*, 143–150. [CrossRef]
10. Hua, H.; Li, B.; Yu, M.; Hao, Z.; Tang, H. Fourth-grade Workshops on Computational Design. In *Proceedings of the 11th National Conference on Digital Technologies in Architectural Education*; China Architecture & Building Press: Beijing, China, 2016.
11. Watson, I.; Perera, S. Case-based design: A review and analysis of building design applications. *AI EDAM* **1997**, *11*, 59–87. [CrossRef]
12. Han, D. The Main Practice Forms of Urban Design in China at the Present Stage and Chanllenges it Faced. *Jiangsu Constr.* **2018**, 8–10.
13. Zhang, L.; Ding, W. Changing urban form in a planned economy: the case of Nanjing. *Urban Morphol.* **2018**, *22*, 15–34.
14. Bettencourt, L.M.; Lobo, J.; West, G.B. Why are large cities faster? Universal scaling and self-similarity in urban organization and dynamics. *Eur. Phys. J.* **2008**, *63*, 285–293. [CrossRef]
15. Bocher, E.; Petit, G.; Bernard, J.; Palominos, S. A geoprocessing framework to compute urban indicators: The MApUCE tools chain. *Urban Clim.* **2018**, *24*, 153–174. [CrossRef]
16. Hovestadt, L. *Beyond the Grid: Architecture and Information Technology: Applications of a Digital Architectonic*; Birkhauser: Basel, Switzerland, 2010.
17. Lauzon, F.Q. An introduction to deep learning. In Proceedings of the 2012 11th International Conference on Information Science, Signal Processing and their Applications (ISSPA), Montreal, QC, Canada, 2–5 July 2012; pp. 1438–1439.
18. Witten, I.H.; Frank, E. Data mining: Practical machine learning tools and techniques with Java implementations. *ACM Sigmod Record* **2002**, *31*, 76–77. [CrossRef]
19. Sönmez, N.O. A review of the use of examples for automating architectural design tasks. *Comput.-Aided Des.* **2018**, *96*, 13–30. [CrossRef]
20. Chen, H.C.; Han, Q.; de Vries, B. Urban morphology indicator analyzes for urban energy modeling. *Sustain. Cities Soc.* **2020**, *52*, 101863. [CrossRef]
21. Goel, A.; Kolodner, J.; Pearce, M.; Billington, R.; Zimring, C. *ARCHIE: A Case-Based Architectural Design System*; Research Report; College of Computing, Georgia Institute of Technology Atlanta: Atlanta, GA, USA, 1991.

22. Chen, F. Interpreting urban micromorphology in China: Case studies from Suzhou. *Urban Morphol.* **2012**, *16*, 133.
23. Hillier, B.; Hanson, J. *The Social Logic of Space*; Cambridge University Press: Cambridge, MA, USA, 1989.
24. Gil, J. Analyzing the configuration of multimodal urban networks. *Geogr. Anal.* **2014**, *46*, 368–391. [CrossRef]
25. Cooper, C.; Chiaradia, A.J. sDNA: How and why we reinvented Spatial Network Analysis for health, economics and active modes of transport. In Proceedings of the GIS Research UK (GISRUK) 2015, Leeds, UK, 15–17 April 2015; pp. 122–127.
26. Song, Y.; Zhang, Y.; Han, D. Access structure. *Environ. Plan. Urban Anal. City Sci.* **2021**. [CrossRef]
27. Liu, Y.; Stouffs, R.; Tablada, A. Rethinking the Urban Design Process from a Data Perspective. In Proceedings of the 35th International Conference on Education and Research in Computer Aided Architectural Design in Europe, Rome, Italy, 20–22 September 2017; Volume 1; pp.449–460.
28. Marcus, L.; Pont, M.B.; Barthel, S. Towards a socio-ecological spatial morphology: Integrating elements of urban morphology and landscape ecology. *Urban Morphol.* **2019**, *23*, 115–124.
29. Hua, H. A case-based design with 3D mesh models of architecture. *Comput.-Aided Des.* **2014**, *57*, 54–60. [CrossRef]
30. Tang, P.; Li, H.; Wang, X. Generative Design on Conservation and Inheritance of Traditional Architecture and Settlement Based on Machine Learning: A Case Study on the Urban Renewal Design of Roma Termini Railway Station. *Architect* **2019**, *1*, 100–105.
31. Xu, J.; Li, B. Application of Case-Based Methods and Information Technology in Urban Design-The Renewal Design of the urban region around Roma Railway Station. In Proceedings of the 24th CAADRIA Conference, Wellington, New Zealand, 15–18 April 2019; pp. 625–634.
32. Nogueira, K.; Penatti, O.A.; Dos Santos, J.A. Towards better exploiting convolutional neural networks for remote sensing scene classification. *Pattern Recognit.* **2017**, *61*, 539–556. [CrossRef]
33. Krizhevsky, A.; Sutskever, I.; Hinton, G.E. Imagenet classification with deep convolutional neural networks. *Commun. ACM* **2017**, *60*, 84–90. [CrossRef]
34. Mohammadiziazi, R.; Bilec, M.M. Application of machine learning for predicting building energy use at different temporal and spatial resolution under climate change in USA. *Buildings* **2020**, *10*, 139. [CrossRef]
35. Ochoa, K.S.; Guo, Z. A framework for the management of agricultural resources with automated aerial imagery detection. *Comput. Electron. Agric.* **2019**, *162*, 53–69. [CrossRef]
36. De Miguel, J.; Villafane, M.E.; Piskorec, L.; Sancho-Caparrini, F. Deep form finding-Using variational autoencoders for deep form finding of structural typologies. In Proceedings of the 37th Conference on Education and Research in Computer Aided Architectural Design in Europe (eCAADe) & 23rd Conference of the Iberoamerican Society Digital Graphics (SIGraDi), Porto, Portugal, 11–13 September 2019.
37. Gil, J.; Beirão, J.N.; Montenegro, N.; Duarte, J.P. On the discovery of urban typologies: Data mining the many dimensions of urban form. *Urban Morphol.* **2012**, *16*, 27.
38. Alvarez-Marin, D.; Ochoa, K.S. Indexical Cities: Articulating Personal Models of Urban Preference with Geotagged Data. *arXiv* **2020**, arXiv:2001.10615.
39. Dong, J.; Li, L.; Han, D. New Quantitative Approach for the Morphological Similarity Analysis of Urban Fabrics Based on a Convolutional Autoencoder. *IEEE Access* **2019**, *7*, 162–174. [CrossRef]
40. Wang, K.; Zhang, J.; Xu, Z.; Xu, Y. Coordinative And Transition-oriented Technical Regulations For Land-use Planning: An Analysis On The Code For Classification of Urban Land Use And Planning Standards of Development Land. *City Plan. Rev.* **2012**, *4*, 42–48.
41. Goodfellow, I.; Bengio, Y.; Courville, A.; Bengio, Y. *Deep Learning*; MIT Press Cambridge: Cambridge, MA, USA, 2016; Volume 1.
42. Szegedy, C.; Vanhoucke, V.; Ioffe, S.; Shlens, J.; Wojna, Z. Rethinking the inception architecture for computer vision. In Proceedings of the IEEE Conference on Computer Vision and Pattern Recognition, Las Vegas, NV, USA, 27–30 June 2016; pp. 2818–2826.
43. Szegedy, C.; Liu, W.; Jia, Y.; Sermanet, P.; Reed, S.; Anguelov, D.; Erhan, D.; Vanhoucke, V.; Rabinovich, A. Going deeper with convolutions. In Proceedings of the IEEE Conference on Computer Vision and Pattern Recognition, Boston, MA, USA, 7–12 June 2015; pp. 1–9.
44. Tsang, S.H. Review: Inception-v3—1st Runner Up (Image Classification) in ILSVRC 2015. Available online: https://bit.ly/2MKWI5k (accessed on 10 September 2018).

sustainability MDPI

Article

How to Sustain Sustainability Monitoring in Cities: Lessons from 49 Community Indicator Initiatives across 10 Latin American Countries

Ludger Niemann [1,2,*] and Thomas Hoppe [2]

1 Faculty of Public Management, Law and Safety (PLS), The Hague University of Applied Sciences, 2521 The Hague, EN, The Netherlands
2 Faculty of Technology, Policy and Management (TPM), Delft University of Technology, 2628 Delft, BX, The Netherlands; T.Hoppe@tudelft.nl
* Correspondence: l.h.h.niemann@hhs.nl

Abstract: Since the 1990s, many countries have witnessed the emergence of organizations publishing environmental, social, and quality-of-life indicators at a city level in order to promote public awareness, democratic participation, and sustainability policies. Many such initiatives are short-lived, however, and reasons for their success and failure under-researched. Using interviews, surveys, and documental data, we explored the survival rates, obstacles, and achievements of 49 initiatives in 10 Latin American countries. Contrary to those in other world regions, most initiatives have civil society stakeholders (notably universities, media, and businesses), excluding governments. Implementing citizen perception surveys proved effective to gain public attention. Several initiatives obtained name recognition and policy influence, which are significant achievements in megacities such as Bogotá, São Paulo, and Lima, where numerous NGOs vie for attention. Frequent obstacles include a lack of finances. After a seven-year period (2014–2021), 55% of the sampled initiatives remained active, ranging from 90% in Colombia to none in other countries. Organizational continuity appeared to be associated with network membership and discontinuity with diverging obstacles, including political pressures in some countries (e.g., Mexico), data scarcity in poorer ones (e.g., Bolivia), and a lack of sustained interest in relatively richer ones (e.g., Chile). Recent increases in socio-economic inequalities are strengthening the potential of community indicators.

Keywords: sustainability monitoring; sustainability indicators; community indicators; quality-of-life; transparency; accountability; participatory governance

Citation: Niemann, L.; Hoppe, T. How to Sustain Sustainability Monitoring in Cities: Lessons from 49 Community Indicator Initiatives across 10 Latin American Countries. *Sustainability* **2021**, *13*, 5133. https://doi.org/10.3390/su13095133

Academic Editor: Tan Yigitcanlar

Received: 2 April 2021
Accepted: 30 April 2021
Published: 4 May 2021

Publisher's Note: MDPI stays neutral with regard to jurisdictional claims in published maps and institutional affiliations.

1. Introduction

> *We made recommendations on adjustments to the city development plans, some of which were accepted. However, relations with the municipal administration have been a bit tense due to the poor results they obtained in our citizen perception surveys. This made them very upset.*
>
> (Quote by the representative of an indicator initiative in Colombia; informant CO#3).

To improve a city's liveability and democratic governance, a strategy with much appeal in this digital, data-driven age involves tracking numbers on issues that many citizens and elected officials care about [1]: How is our city or neighbourhood doing (and comparing to others) with regard to, say, air quality standards, quality-of-life, municipal expenditure, and crime rates? Organizations dedicated to compiling and publicly reporting on local wellbeing or sustainability indicators at regular intervals have been called "urban observatories" or "community indicators"; in line with other studies [2], we label them "community indicator initiatives" in this article. Pioneers such as *Sustainable Seattle* were founded in the 1990s [3] and hundreds of other local projects were active at the turn of the millennium [4].

In Bogotá in 1998, the Colombian capital's chamber of commerce, a reputed foundation, media firm, and university jointly created *Bogotá Cómo Vamos* ("Bogota, how are we doing" in Spanish). This organization's core mandate is to publicly report on the city's quality-of-life and sustainability on the basis of relevant indicator data obtained from official sources and supplemented with their citizen perception surveys. Inspired by this Colombian model, dozens of similar initiatives subsequently appeared across Latin America [5,6]. Many chose equivalent names (*Lima Cómo Vamos*, *Rio Como Vamos*, etc.) or variants often combining city names and plural first-person pronouns—in Brazil, *Nossa São Paulo* ("Our São Paulo" in Portuguese) was created in 2007 and emulated by namesake initiatives in Argentina such as *Nuestra Buenos Aires*. To exchange experiences and coordinate joint activities, they collectively created the *Latin American Network of Just and Sustainable Cities* in 2011 that comprised, in its heyday, some 60 like-minded initiatives from 10 countries [7].

Across Latin America, "community indicators" evidently proliferated in the previous decade, fuelled by substantial investments in time and money by civil society volunteers, private foundations, universities, journalists, and entrepreneurs. In several cities, international donors, such as the Inter-American Development Bank (IDB), co-financed the creation of a "civil society monitoring system" [8]. A quick internet search for key actors reveals that *Bogotá Cómo Vamos* (cf. www.bogotacomovamos.org (accessed on 1 April 2021)) and Colombian sister organizations continue to be active, just as *Nossa São Paulo* (www.nossasaopaulo.org.br (accessed on 1 April 2021)) and fellow initiatives in Brazil. However, *Rio Como Vamos* and *Nuestra Buenos Aires* no longer exist, the Latin American network became inactive in 2016, and in some countries, none of the original community indicator initiatives survived.

The rise (and fall) of community indicators in different countries triggers two fundamental questions: What do they achieve? Which contextual factors influence their achievements and survival? Various academic studies address the first question and point to several positive impacts, including gains in public awareness and sustainability-oriented decision-making [4]. In the literature on sustainability indicators at large (thus including their application at a national level, in industries, etc.), scholars have also identified the possibility of misuse (e.g., disinformation campaigns) but non-use appears to be the more prevalent risk [9]. Indeed, failure to achieve long-term usage seems to be a common ailment. According to a recent review by Wray, Stevens, and Holden [10] (p. 10), "not unlike the story in other realms of voluntary and community work, efforts in community indicators have been plagued by the short lifespan of many initiatives. All too often, the cycle is one of a burst of investment of, enthusiasm, dedication, skill, and resources, a hard slog to establish an initial reputation and reporting system, some small triumphs of media, community, and perhaps even political attention, followed by a series of disappointments in efforts to repeat, accelerate, or institutionalize the work, and ultimately by the decline or disappearance of the initiative".

In that view, fizzling out over time seems to be the normal fate of indicator projects; moreover, unrealistic expectations on the side of practitioners appear to be more important than either the local or national context or the specific choices each initiative makes about its activities and organizational set-up. With regards to internal governance, some researchers offer specific recommendations. Extrapolating from a case study in Australia, Davern et al. [11] (p. 571) posit that "all indicator systems should include these best practice principles in their development and operation", which include the prescription to "include a balanced mix of government, business and community representation". The last point is remarkable since, in Latin America, virtually all initiatives operate at an arm's length from governments and have governance arrangements that explicitly exclude elected officials.

Therefore, it is fair to state that there are both theoretical and empirical reasons to map and evaluate community indicator initiatives in Latin America. Theoretically, various assumptions about the effectiveness and "best practice" of community indicators can be put to test, in particular, with regards to prescriptions about organizational governance. As this article shows, community indicator initiatives differ widely in the number of stakeholders involved, their degree of cooperation with media firms, reliance on volunteers, indicator choices, dissemination methods, and other characteristics. Empirically, the Global South is under-researched, and a more representative selection of case studies is repeatedly called for by scholars of sustainability indicators [12]. Latin American initiatives operate in diverse environments, ranging from smaller towns to the world's largest cities, in countries showing differences in terms of public service levels, political violence, and access to information laws. This provides unique opportunities to open the "black box of contextual drivers" [13]. Responding to calls in the literature for comparative, longitudinal approaches, this study, therefore, sought to answer the following research question:

Which Design and Context Factors Are Associated with the Influence and Long-Term Viability of Community Indicator Initiatives in Latin American Cities?

Three research sub-questions help structure this article: How do city-level community indicator initiatives function in different Latin American countries? What do they perceive as their objectives, barriers, and achievements? Which contextual factors, in combination with organizational strategies, are associated with their success and failure?

Through our theoretical and empirical contributions, we aim to strengthen the global body of knowledge on indicator initiatives and to provide insights to practitioners; the latter include civil society activists, donors, and decision-makers involved in the design of national transparency and accountability policies. We further expect that the research frameworks elaborated for this study, including the typology of context and design factors as well as effects, will inform future studies. Our approach and objectives are summarised in Figure 1.

Figure 1. Outline of research issues, approach, and objectives.

This article is organized as follows. The next section contains an overview of essential literature on community indicators, with a view to guiding the elaboration of the conceptual model applied in this study. This is followed by a section describing the research population and methods and one presenting key results. The final section contains a discussion and conclusion. Further details and raw data are available as Supplementary Materials.

2. Community Indicators in the Context of Transparency and Accountability Initiatives

Community indicators function at the intersection of three issues of high standing on the global policy agenda and SDG framework [14]—namely, urban sustainability, transparency, and civic participation. As such, they relate to vast and overlapping academic disciplines, including policy studies, governance, communication, and management sciences [9]. To contextualize our comparative, longitudinal evaluation of Latin American initiatives, we reviewed the literature for key findings in three areas: activities of sustainability indicator projects, their organizational structure, and their achievements, including longevity; a transversal concern in each area is the influence of contextual factors.

Context differences are under-researched regarding community indicator initiatives yet are likely to play a major role in their success or failure. Informed by sectoral reviews, such as the one by Wray et al. [10], and the transparency and accountability literature, such as Grandvoinnet et al. [13], contextual factors with major relevance for community indicator initiatives include the political–legal regime, data availability and the presence of alternative, competing transparency and accountability actions at the country and city levels.

With regards to organizational set-up, a joint venture of governmental and non-governmental organizations is often taken for granted [15] or even put forward as the "best practice [11]. A comparative review of the longevity of 82 indicator initiatives assessed the participation of governments and academic institutions and also concluded that "a broad and cross-cutting alliance of stakeholders is the most promising, because it can best ensure ongoing demand for the product, collective accountability, [...] a robustness of funding" [2] (p. 25). The same study implicitly describes barriers impairing the long-term functioning of indicator initiatives, namely funding and sustained demand for indicator information. An organization's legitimacy is also mentioned yet rarely studied comparatively; this is no surprise, as it is highly contextual and hard to operationalize [16].

Research on indicator projects has been dominated for years by studies exploring how they select which indicators. One main distinction concerns bottom-up, participatory, and context-specific versus top-down, expert-led, standardized approaches, with each having advantages and advocates [17,18]. While a set of indicators can also be merged into one index, most initiatives forgo this option, as aggregation obscures meaning and transparency [19]. Sustainability monitoring typically involves outcome indicators, often clustered in the environmental, social, and economic domains [20]. According to a recent global analysis of 67 measurement initiatives, a city's unemployment rate and green urban space are the most widely used indicators [21]. For these two examples, indicator initiatives in rich countries can obtain relevant information from official statistics or they may just need to convert raw data into more meaningful indicators, such as "green space per capita". For subjective wellbeing, however, another indicator that is widely used among community initiatives internationally [2,22], and pioneers in Latin America [23,24], all data needs to be collected through surveys, representing an expensive undertaking. In many indicator systems, political processes only receive marginal attention [21], though exceptions have been documented, such as a multi-year, voluntary benchmark among Dutch municipalities exclusively tracking sustainability policies [25]. Further, in addition to reporting outcome indicators, many initiatives deal with political processes, either through monitoring policy-related indicators—for example, government expenditures—or active community engagement, for example, by organizing round tables, public events, and so forth.

Figure 2 visualizes this array through sample activities found in this study and the literature [4]. The set is non-exhaustive, but it illustrates the spectrum of activities involving indicator reporting (concerning outcomes in the city and processes or inputs by the local government) and actions for community engagement. The latter may also be conceptualized as targeting generic processes (e.g., via community events) as well as the outcomes of deliberations, such as formalized spatial development plans. This spectrum mirrors that of sustainability reporting by local governments; a study by Niemann and Hoppe concerning the practices of European pioneers found significant divergence in terms of content along outcome and process dimensions and various degrees of citizen engagement strategies [26]. A number of North American community indicator initiatives also attempt to monitor government performance at large [27]. On the other hand, process-focused activities may also be carried out by other organizations not working on sustainability. From this perspective, community indicator projects with their diverging scope of community engagement can be conceptualized as belonging to the broad field of transparency and accountability initiatives [13].

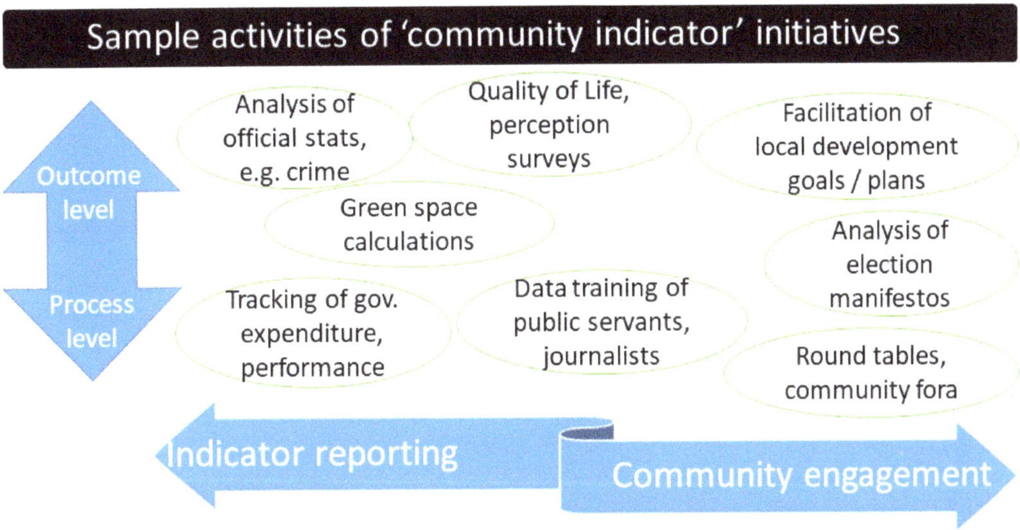

Figure 2. Sample activities of community indicator projects or urban observatories.

Assessments of effectiveness conceptually require taking objectives into account so one can measure progress against actual goals. Naturally, goals differ between indicator projects in different contexts and lifecycles [28]. Beyond the implicit notion of improving local governance and sustainability, however, initiatives often lack explicitly stated objectives [29]. To help assess outcomes in general terms, other studies about sustainability indicators [30] distinguished three clusters of use and influence: instrumental (i.e., bearing on decision and policymaking), conceptual (e.g., learning and capacity development among public servants), and political–symbolic (e.g., public discourse). These clusters overlap but help identify intended effects—such as instrumental influence on local government policies—as well as unintended uses and misuses.

A common challenge for research on the effects of "infomediaries" [13] and multi-stakeholder actions is identifying causality. As all community indicator projects operate in networks (of different data providers, users, target audiences, etc.), their societal impact cannot be studied in isolation but requires a systems perspective. The method of process tracing allows for the gathering of high-quality evidence; a study applying it to three American community indicator initiatives found positive impacts regarding agenda-setting and other dimensions [29]. Process tracing and similar methods, however, require complex methodological rigour, which cannot easily be applied at a large scale. This explains why large-N comparative studies typically resort to more superficial measures of success, such as organizational survival or "staying power" [2]. Figure 3 summarizes the main factors thus identified in this literature review—various contextual factors (at country and city levels) as well as design choices (in particular, about organizational set-up and actions) that have a bearing on what community indicator projects achieve in terms of uses and influence. A further variable of interest is organizational evolution, which also serves as a reminder that none of the other constructs are static but actually co-evolve over time.

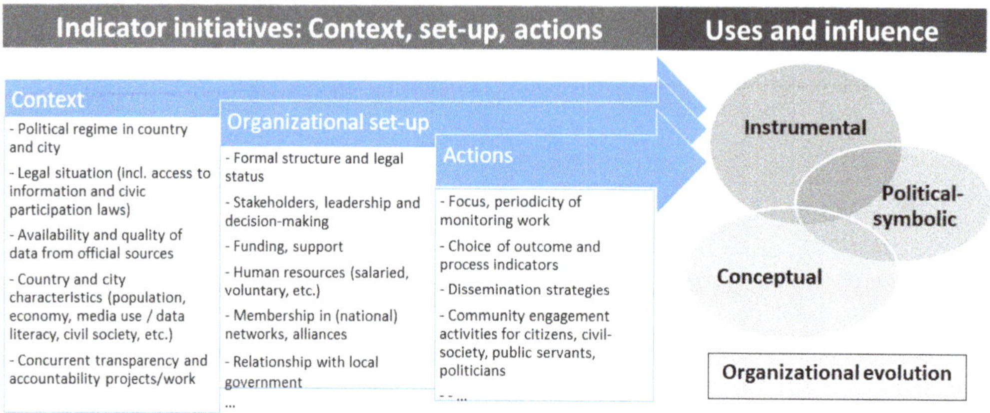

Figure 3. Conceptual framework based on the literature review.

3. Population and Methods

The world's "battle for sustainability will be won or lost in cities", the United Nations proclaimed [31]. According to this metaphor, Latin America is an important battleground inasmuch as it is the world's most urbanized region, where about 80% of the population lives in cities. In 2011, the Latin American Network for Fair, Democratic, and Sustainable Cities (In Spanish: *Red Latinoamericana por Ciudades y Territorios Justos, Democráticas y Sustentables*) was created by community indicator organizations from across the region. Some, for example, *Bogotá Como Vamos,* had already been working for years with salaried staff. In other countries (e.g., Ecuador), all initiatives were run by volunteers, without funding or legal status, and with just a Facebook page to begin with. In conferences and discussions, a shared purpose was agreed—namely, promoting equitable development and democracy by reporting on sustainability indicators and the local government's management—as well as the stipulation that members are civil society organizations. In terms of joint activities, priority was given to communication (websites, newsletters, etc.) and developing an indicator database for benchmarking. In the absence of their own financial resources, grants from philanthropic foundations (notably, AVINA) allowed the funding of a part-time communication officer in Lima, a secretary in Mexico, and consultants in Brazil developing a collective indicator database (For joint communication, the network used to run the website www.redciudades.net (accessed on 1 April 2021) that following the end of project funding, has been offline since 2016). Network representatives also decided to support this longitudinal research by sharing the (contact) details of member initiatives and volunteering to be interviewed. Importantly, key informants concurred that the network included virtually all relevant, city-based Latin American initiatives and that no alternative institutional model (such as sustainability reporting by local governments) were prevalent in any country. Therefore, membership, as assessed in 2014, can be considered representative of the population of Latin American community indicator initiatives at the time. The network counted 65 initiatives on its own website, but verification of individual internet sites showed 8 as actually inactive in 2014. As documented in the Supplementary File, these initiatives were discarded from this longitudinal study, just as 8 fledgling initiatives only started in 2013 (As also documented in the Supplementary File, at least 9 initiatives have been founded since 2014, including one in Guatemala, as an additional country). Therefore, the main sample consisted of 49 city-based community indicator initiatives started by 2012 and active in 2014.

For research on the practices, experiences, and long-term viability of these Latin American community indicator initiatives, the conceptual model derived from previous studies (cf. Figure 3) served as a starting point. For each of its main dimensions (context, set-up, actions, use and influence, evolution), we identified the key parameters and data collection methods, as presented in Table 1. To assess an initiative's context, we relied on "hard data", such as city population numbers and the UN's Human Development Index at the country level. To explore the political–legal landscape, we further mapped country ratings from four international indices (by the World Bank, Transparency International, Reporters Without Borders, RTI). Such governance indicators are methodologically contested [32,33], unsuited to reliably capture all discontinuities, such as those arising from radically different presidencies in Brazil and Mexico, and not necessarily valid sub-nationally. However, country differences plausibly have repercussions at the city level and comparative indices may be meaningful to show overall trends and country clusters. For further validation and information at the city level, we interviewed and surveyed representatives of indicator initiatives in Spanish (and in Portuguese for Brazil).

Table 1. Operationalization and data sources.

Construct	Factor	Parameters	Data Source
Context	Country characteristics	• Indices on human development, government effectiveness, corruption, press freedom, right of access to information.	• United Nations, World Bank, Transparency International, etc. • Interviews
	City characteristics	• Population size. • Availability/quality of sustainability data. • Prevalence of political pressure	• Gov. statistics • Survey; interviews
Organizational set-up	Formal structure	• Type of formal registration.	• Internet query; survey
	Stakeholders	• Categories of formal stakeholders; in particular, (local) governments.	• Internet query; survey
	Human resources	• Number of staff/volunteers.	• Survey
Actions	Focus of monitoring/indicators	• Broad set of (sustainability and quality-of-life) indicators and/or government policies.	• Internet query; survey
	Dissemination methods	• Internet channels (website, blog, social media).	• Internet query; survey
	Engagement activities	• Type of activities mentioned.	• Survey
Use and influence	Conceptual (→ capacity development)	• Knowledge, ideas, "data literacy" in public institutions and civil society. • Name recognition: among civil servants/technical staff.	• Survey; interviews
	Instrumental (→ Policy influence)	• Reference to reports and indicators in government plans. • Name recognition among decision-makers. • Evidence of policy changes.	• Survey; interviews
	Political and symbolic (→ public awareness)	• References in debates, publications. • Name recognition among media and the general public.	• Survey; interviews
Evolution	Organizational continuity	• Evidence of activity in the last 6 months (2021).	• Internet query

The main survey contained open and closed questions about each initiative's set-up, objectives, activities, and experiences (including political interference and quality of public data). It was sent electronically in 2014 and responded to by 44 initiatives, a 90% response rate. Most answers were given by a person with high knowledge of internal processes, such as the initiative's director, and included frank statements about sensitive topics, such as relations with local governments. In response to requests for confidentiality, the survey and interview data were anonymized and summarized at the country level without identifying individual cities. Additional interviews were held with 11 representatives of indicator initiatives in 2014, 2019, and 2020.

To assess current activities, the websites of all initiatives were screened in 2021. This allowed for reliably mapping whether an initiative publishes indicators (on sustainability, quality-of-life, or government performance) but not their inherently diffuse and less visible community engagement activities; the majority of initiatives lack annual reports. Uses and influences were assessed qualitatively, with estimations (in 2014) of name recognition among various target groups serving as one quantifiable proxy indicator.

An initiative's level of continuity in 2021 was assessed through the visibility of activities on the internet. If an initiative's main website no longer existed, alternatives were explored via search engines and social media sites. If the initiative showed substantial activity in any channel (beyond a simple "like" or retweet) in the last 6 months, it was classified as "active". If no sign of life was found in more than 12 months, the initiative was classified as "discontinued", and cases in between as "unclear" (see Supplementary File for details). Average continuity or "survival rates" were subsequently compared to other factors, such as city size and country context. Comparisons based on country averages, however, are only explorative in nature due to the small number of cases involved.

4. Results

To structure this section, we present our findings as responses to three research sub-questions. These are descriptive (about the functioning of community indicator initiatives), analytic (about their experiences), and tentatively predictive (exploring the relationship between continuity and context).

4.1. Sub-Question (1): How Do City-Level Community Indicator Initiatives Function in Different Latin American Countries?

The Latin American community indicator initiatives studied hail from a geographically vast area, as they work in 49 cities located in 10 countries between Mexico and Argentina. See the Supplementary Files for details on names, city population, year of establishment, main activities, and website. Table 2, below, summarizes, at the country level, the number of initiatives that existed in 2014, with certain characteristics in terms of organizational set-up and actions, and that continue to exist in 2021. As evident from this table and the Supplementary File, Brazil (15) and Colombia (10) have had the largest number of initiatives. Some were started in small municipalities of just 20,000 inhabitants but the majority in large cities. The average population size is close to 2 million, and from among the 10 largest Latin American cities, all except for Caracas were at some moment represented in the *Latin American Network for Fair, Democratic and Sustainable Cities*. In terms of organizational set-up, a large majority of initiatives are distinct civil society organizations. Exceptions include those where the community indicator work is run as another NGO's project, or entirely by volunteers, or as a public–private partnership led by a local government. In fact, local governments are only included as stakeholders in 6% of all initiatives. Significantly more prevalent are media firms (stakeholders in 50% of all initiatives), academic institutions such as universities (63%) and business institutions (e.g., chamber of commerce), and private foundations (69%). Whereas initiatives called *Cómo Vamos* are usually governed by three or four-member institutions, others opted for more fluid internal governance; *Nossa Sao Paolo* auto-identifies as a network with hundreds of stakeholders. In 2014, the number of salaried staff varied from 0 (in fledgeling initiatives or those organized purposefully as a "citizen collective") to 13; the median was 3.

Table 2. Number of initiatives per country in terms of characteristics and staying rate.

		Organizational Set-Up				Actions			Evolution
		Stakeholders (Beyond Civil Society):				Monitoring/Reporting on:			Continuity
Country	Total # of Initiatives (2014)	Academia	Media Firms	Business/Pı Foundations	Local Government	Various Sustain-Ability Indicators	Quality-of-Life/Percep Surveys	Local Government Policy	Evidence of Activity in the Past 6 Months (2021)
Argentina	**6**	2	2	4	1	6	2	4	2
Bolivia	**3**	2	2	3	/	3	2	/	1
Brazil	**15**	9	7	9	1	11	5	5	8
Chile	**2**	1	?	?	/	2	2	/	0
Colombia	**10**	10	10	10	/	10	10	1	9
Ecuador	**2**	/	/	/	/	1	2	/	1
Mexico	**5**	3	1	4	1	5	5	3	4
Paraguay	**1**	?	?	?	/	?	1	/	1
Peru	**3**	3	2	3	/	3	3	1	1
Uruguay	**2**	1	/	1	/	1	1	1	0
SUM:	**49**	**31 (63%)**	**24 (50%)**	**34 (69%)**	**3 (6%)**	**42 (86%)**	**33 (67%)**	**15 (31%)**	**27 (55%)**

In terms of activities assessed in the study period, about a third of initiatives dedicated explicit attention on their websites to their local government's policies, such as ordinances and compliance with election pledges. Most combined reporting about a set of sustainability-related indicators (obtained from official sources) with those of quality-of-life surveys implemented by the initiatives themselves. In Colombian cities, such subjective perception surveys are a trademark activity and implemented annually via professional pollsters contracted at a high cost. Other initiatives only have periodic surveys. One Ecuadorian, volunteer-run initiative used sociology students assigned by their university for polling. All initiatives also carry out a wide array of outreach activities not amenable to simple quantification. A case in point is capacity building; "the personnel of eight municipalities have been trained to fill out the files to collect the indicators. It's a permanent job", explained informant PE#3. Other examples include organizing public debates with mayoral candidates in the run-up to elections [34] and numerous thematic round tables with officials and experts about issues such as sustainable transport or child malnutrition. In terms of organizational continuity, the verification of internet sites showed that 55% of the 49 initiatives tracked since 2014 were assessed as being active in 2021.

4.2. Sub-Question (2): What Do Such Initiatives Perceive as Their Objectives, Barriers, and Achievements?

According to the charter of the *Latin American Network for Fair, Democratic and Sustainable Cities* (agreed upon in 2011), member initiatives have five common objectives: (i) to monitor a city's situation in terms of quality-of-life, social justice, democracy, and sustainability; (ii) to promote civic participation; (iii) to monitor and influence public policies; (iv) to promote space for dialogue between civil society, the private sector, and the state; (v) to disseminate information and knowledge for informed civic participation and decision-making processes. In terms of target groups and strategies, the sampled initiatives developed a range of priorities. Whereas virtually all (95% of survey respondents) considered elected officials a main target group, about 60% also targeted researchers and staff in public institutions. As a Peruvian respondent (PE#03) explained, "The work is in stages. We are giving priority to young people through actions and training. The media is getting information and workshops, just as the technical personnel of the institutions, the latter primarily on indicator issues. Social leaders are convened; we sign institutional agreements".

Since (indicator) data represent the lifeblood of initiatives, their availability and reliability are a main concern in some countries. In Bolivia, a respondent (BO#1) explained that "there's very little access to information. Normally we gather it from sources such as interviews and focal groups". In neighbouring Chile, "the official statistics generally have high fidelity" (CH#2). At an aggregate level, 7% of survey respondents in 2014 considered data obtained from national sources as having no or little reliability, and even 28% thought so of data from local sources. The former, however, are often not usable at the local level; as an Argentine respondent (AR#5) explained, "we were unable to access disaggregated data in our locality".

To explore barriers and bottlenecks, the survey contained questions about the prevalence of difficulties in several areas—out of these, finding suitable staff and media attention was not considered problematic by most respondents. Over 50% of them, however, reported significant difficulties in accessing indicator data and finding funding (for running costs and specific projects). Further, on a survey question about political pressure or interference (in relation to areas of work, research, positive and negative news, and involved personnel), 38% of initiatives indicated suffering from them "frequently" or "very frequently"; only a third did not experience pressures. Sometimes this was experienced as a matter of evolving trust—in the words of one Peruvian informant (PE#3), "At the beginning of our activities, there was mistrust of our institution and it was accused of having a political overtone—this on the part of the municipal authorities and also in some cases of civil society". In other instances, however, maturing initiatives felt increasing heat. A Brazilian informant (BR#5) stated, "We are going through a period of great political pressure from the current municipal public administration".

Achievements are largely intangible and hard to measure. To quantitatively assess to what extent initiatives were effective in achieving name recognition, the 2014 survey asked respondents to estimate which percentages of three target groups knew the initiative's work; the reported averages were 26% for the city's general population, 49% for relevant technical staff of public institutions, and 72% for decision-makers such as the mayor and councillors. Regarding the latter, individual responses ranged from 1% to 100% between fledging and mature organizations. It is worth mentioning that some indicator initiatives assess their name recognition via surveys, and also have reliable ways to monitor their interaction with office-bearers if they are invited to present quality-of-life survey results in, say, a municipal council meeting. As a Colombian informant (CO#3) explained, "The percentage of the population was measured in our citizen perception survey [...]. Among decision-makers, everyone knows us because we constantly interact with them, either requesting information or at different round tables".

Regarding other outcomes, key informants reported evidence of effects (in some cases backed up by detailed explanations in annual reports) in various dimensions. In terms of conceptual use and influence, 77% of survey respondents indicated having contributed to the development of capacities among public institutions. As a Peruvian respondent (PE#03) stated, "We have achieved that information-generating institutions, as well as public institutions, disseminate their information and try to update it [as] municipal public servants are finally understanding the importance of data". Regarding instrumental use and influence, 72% of initiatives (according to the survey) reported having had some—and 30%, even "very large"—achievements in influencing the design of public policies. According to a Colombian respondent (CO#8), "We made recommendations on the elaboration of the municipal development plans [and] 20% of our recommendations were accepted". A major achievement for many initiatives—especially in Brazil and Argentina—was also the successful lobby for new bylaws introducing a legal obligation for mayoral candidates to create action plans with targets for various sustainability indicators, about which the elected mayor then has to report back in public accountability meetings; such bylaws are known as "plan(o) de metas" in Spanish and Portuguese [35]. By nature, the cluster of political–symbolic uses and influences is the most intangible; one economic way to assess them is through subjective perceptions of key informants and proxy indicators such as media coverage. Over 90% of survey respondents reported that their initiative had contributed to the city's agenda, public discourse, and knowledge. As an Argentine informant (AR#3) observed, "if by impact we understand that the issues are debated or published, we have had important achievements since in general, the information that we produce is published by the main media outlets".

4.3. *Sub-Question (3): Which Contextual Factors in Combination with Organizational Strategies Are Associated with the Success and Failure of These Initiatives?*

The analysis of activity levels showed that 27 of 49 (i.e., 55%) of city initiatives sampled in 2014 were confirmed as active in 2021, but 13 were classified as "discontinued" and 10 as "unclear"; the latter category applies to initiatives with a functioning website but no evidence of recent activities (cf. Supplementary File). This leads to a conservative estimate of continuity since some of the initiatives classified as "unclear" continue to work as a less formalized citizen movement with rudimentary social media activity. Interestingly, the cities hosting each of these three groups (active, unclear, and discontinued) showed similar average population numbers. Therefore, city size does not appear to be a predictor of long-term viability. It deserves mentioning, though, that small, rural communities were overrepresented among the initiatives that had stopped by 2014 (cf. Section 3).

Chile, Ecuador, Paraguay, and Uruguay each had only one or two initiatives in the initial sample, which implies a limited base for generalizations. In these cases, the average "survival rates" thus need to be interpreted with much caution. Nonetheless, the number of initial initiatives per country appears to matter. As shown in Table 2, Brazil (53%), Colombia (90%), and Mexico (80%) showed the highest continuity rates. The first two also have active networks at the national level (www.redcomovamos.org (accessed on 1 April 2021) and www.cidadessustentaveis.org.br (accessed on 1 April 2021). (The Colombian network also registered *Cómo Vamos* as a national trademark; elsewhere, the brand is not protected and has been adopted by other actors such as a newspaper in Mexico City). With its 90% survival rate, the Colombian model, including its established brand and tight internal governance (centred around a small set of stakeholders including a media firm, university, and chamber of commerce), is ostensibly successful. According to the network's website, it now boasts 16 initiatives covering 60% of Colombia's urban population. A similar institutional *Cómo Vamos* model was successfully emulated in Lima but abandoned in Rio de Janeiro. In several countries, network-oriented models prevailed. As a radical example of a community indicator initiative of the non-institutionalized type, a "citizen collective" in the Ecuadorian city of Cuenca has now run successfully for 10 years with neither legal personality nor salaried staff. Both tight and loose, network-oriented models of internal governance have thus been shown to thrive. This is also evident from diverging advice given by key informants about institutional strategies: "Make sure to establish an organizational structure very early in order to receive resources and have professionals working full time" (BR#9) versus "be an inclusive movement and look for less institutionalized operating models" (BR#2).

To explore the relevance of contextual factors, Table 3 shows the number and percentage of surviving initiatives per country juxtaposed to the selected sample of socio-economical and governance indices. Due to methodological limitations, including the small number of cases and untested validity of country governance assessments, statistical analyses are not meaningful. However, the data suggest the presence of country clusters. Chile and Uruguay are, according to international comparisons (with Table 3 showing most recent data available), the region's top performers in terms of government effectiveness, press freedom, and low levels of perceived corruption. Bolivia, Paraguay, and Peru represent the other end of the spectrum, while Argentina holds a middle ground. In five of these six countries, the observed continuity rate is below 50%, whereas it is higher in the four countries occupying average positions in regional comparisons of human development (as measured by the HDI) and governance. (Paraguay had seen the emergence of five initiatives since 2013 that did not attain organizational continuity. The one established in its capital in 2010 and included in this study remains active.) At the city level, these observations match the prevalence of obstacles reported by key informants. In Chile and Uruguay, the quality of public data sources and political interference was no major concern. In Bolivia, Peru, and Paraguay, however, access to data was a frequently mentioned problem. Another dimension on which key informants reported diverging experiences is political interference and intimidation at the city level—these were high in Mexico and Brazil but virtually absent in other countries (According to the RTI, Paraguay, Bolivia, and Ecuador have poorly developed access to information laws, whereas Mexico belongs to the world's top performers. At the city level, key informants reported that progressive laws can be helpful yet miss immediate relevance for their work if public institutions do not collect sustainability data).

Table 3. Survival rate of community indicators in relation to country context.

	Study	**Country Statistics**				
	Active Initiatives (2021)	**Human Development Index (0–1)**	**Government Effectiveness (1–100%)**	**Corruption Perception Index (0–100)**	**Press Freedom Index (1–100)**	**Legal Right to Information (1–150)**
	Percent Compared to 2014	**United Nations, 2018**	**World Bank (WGI), 2019**	**Transparency International, 2020**	**Reporters Without Borders, 2020**	**RTI-Rating.org, 2020**
Argentina	2 (33%)	0.83	49	45	29	92
Bolivia	1 (33%)	0.70	25	31	35	70
Brazil	8 (53%)	0.76	44	37	34	108
Chile	0 (0%)	0.85	82	67	27	94
Colombia	9 (90%)	0.76	56	37	43	102
Ecuador	1 (50%)	0.76	37	38	33	74
Mexico	4 (80%)	0.77	46	29	46	136
Paraguay	1 (100%)	0.72	33	28	33	62
Peru	1 (33%)	0.76	50	37	31	93
Uruguay	0 (0%)	0.81	75	71	16	92

5. Discussion and Conclusions

In response to the research question "Which design and context factors are associated with the influence and long-term viability of community indicator initiatives in Latin American cities?", this study showed that 49 initiatives located in 10 countries had an average continuity or survival rate of 55% after seven years. This is a conservative estimate since some of the remaining 45% may be active in a different or less institutionalized form.

Our finding that many initiatives fizzled out over time complements evidence from other continents (e.g., [4]). Underneath the regional averages, we found significant differences at the country level. Some countries studied for this article—notably, Colombia—evidenced high continuity rates of 90%, with further institutional growth in recent years. Importantly, community indicator initiatives struggle in different contexts for different reasons. Our data suggest that in higher income and more democratic countries, maintaining institutional funding and sustained (media) attention are common difficulties, while the absence of immediate policy impacts constitutes a main reason for disillusionment among practitioners. In poorer or more violent contexts, however, scarce availability of reliable public data and political intimidation often represent additional existential threats for community indicator projects. This finding aligns with the growing body of literature about the fate of transparency and accountability initiatives [36] and makes a new contribution to community indicator research, which has hitherto been biased in favour of high-income countries, while ignoring developments in the Global South.

Regarding their organizational set-up, virtually all Latin American community indicator initiatives are civil society alliances excluding governmental stakeholders. This differs from other world regions and challenges "best practice" recommendations about the inclusion of governments [11]. Moreover, successful Latin American initiatives show diversity in their organizational set-up; whereas some are governed by a small set of founding members (notably, philanthropic foundations, media firms, and universities), others function as looser networks.

The implementation of citizen perception surveys proved an expensive yet effective way of gaining wide (media) attention. This mirrors similar trends found by scholars elsewhere [2,37]. Many Latin American initiatives also offer training in data literacy and indicator use to diverse target groups, such as public servants, journalists, and community leaders. According to key informants, and corroborated by evidence including representative household surveys, several initiatives also gained significant name recognition in their cities among the general public, technical staff of local governments, and political decision-makers, such as the mayor and councillors. Using their standing, some initiatives successfully influenced public policies such as municipal development plans; another achievement in several cities concerned the approval of new legal requirements for local governments to incorporate sustainability considerations and citizen-led accountability fora into their planning and reporting systems. For relatively small organizations with usually less than 10 staff members, these are significant achievements in megacities such as Bogotá, São Paulo, Lima, and La Paz, where numerous pressure groups vie for attention.

This study is exploratory in nature, which implies methodological limitations. Tracking dozens of initiatives over time required selecting a limited set of research constructs, and thereby disregarding the complex nature, evolution, and interaction between a country's and city's context and a networked community indicator initiative. However, to avoid blind spots, we supplemented the deductive analysis of constructs derived from previous research with qualitative, inductive feedback from key informants.

More research is needed. Our findings and the conceptual models developed (cf. Figure 3, Table 1) contain leads for further investigations. To facilitate follow-up research, a number of hypotheses (and the countries serving as relevant cases) were identified for each of the main dimensions of the conceptual model, as listed in Table 4. In terms of context, for example, we surmise that the demise of civil society-run community indicator initiatives in Chile and Uruguay may partially be attributable to the relatively better quality of government-run indicator systems in these countries (cf. [38]). In terms of organizational set-up, partnerships with media firms perceived as unbiased (where available) are hypothesized to foster organizational continuity. Further research on internal leadership and the choice of working as broad networks or via a small group of stakeholders will help elucidate the relative advantages of each strategy and how these play out over time. Such studies may be informed by findings in the literature that any (performance) indicator system requires continual reconsideration [39], and that citizen participation tools should be aligned to a local population's evolving "sustainability literacy" [40]. Regarding activities, we hypothesize (based on findings in various countries) that the implementation of surveys that tap citizens' satisfaction with mayors may bring attention but also accusations of partisan meddling. Beyond these hypotheses, further research is needed for several other issues, such as the precise nature of community engagement strategies, including cooperation with other urban movements [41], and various ways of using indicators and indices, as well as relating local monitoring efforts to international frameworks such as the UN's Sustainable Development Goals [42].

Table 4. Hypotheses for further research.

Main Dimension	Hypotheses for Future Studies	Relevant Case/Country in This Study
Context	The higher a country's development level and public provision of sustainability indicators, the smaller the niche and perceived added value of community indicator initiatives.	Chile, Uruguay
	In countries with too limited availability of reliable public data, community indicator projects are hard to be maintained.	Bolivia, Paraguay, Peru
	Progressive laws on access to information foster the functioning of community indicator projects.	Mexico, Brazil
	The local presence of philanthropic foundations (and supportive government policies), as well as universities, fosters the successful foundation of community indicator projects.	Various
Organizational set-up	Partnership with media organizations perceived as unbiased help community indicator projects thrive.	Colombia, Argentina
	When relations between governments, NGOs, and media firms are highly politicised, creating joint associations is not viable and work in looser networks more effective.	Ecuador
	Establishing community indicator initiatives in multiple cities (using a joint brand) fosters media attention (also via benchmarking of city data) and institutional stability.	Colombia
Actions	Implementing household surveys on quality-of-life perceptions is expensive yet effective for gaining political and media attention.	Colombia, Brazil, Mexico
	Monitoring citizen satisfaction with mayors via surveys ensures salience but increases the risk of being perceived as partisan.	Colombia, Peru
	Monitoring government performance and compliance with election pledges increases the effectiveness of government work.	Brazil

We conclude by positing that 20 years of sustainability indicator projects in Latin America offers a wealth of lessons for this and other world regions. Arguably, the "socio-ecological niche" and potential positive contribution of community indicator projects is globally increasing due to three major trends: (i) continued urbanization, (ii) increased data availability due to public investments, technological developments, and open data laws and (iii) many increases in socio-economic inequalities, recently exacerbated by the viral pandemic. Sadly, the more unequal a city, the more informative and newsworthy a localized comparison of relevant indicators. In this endeavour, collaboration involving researchers, practitioners, and policymakers is key, as is stamina. As an Argentine informant (AR#3) recommended to others wanting to start a community indicator project, "It is very important to have and guarantee the continuity of the work long-term. Keep in mind that we work with people and that the generation of trust is a central point for the success of the initiative. It is very important to achieve the most heterogeneous participation possible so that different sectors feel identified and represented".

Supplementary Materials: The following are available online at https://www.mdpi.com/article/10.3390/su13095133/s1.

Author Contributions: L.N.: Conceptualization, Investigation, Writing—Original Draft; T.H.: Supervision, Writing—Review and Editing. All authors have read and agreed to the published version of the manuscript.

Funding: This research received no external funding.

Institutional Review Board Statement: Not applicable.

Informed Consent Statement: Informed consent was obtained from all subjects involved in the study. As explained in the Methods Section, all respondents were anonymized.

Data Availability Statement: Cf Supplementary File for additional material.

Acknowledgments: We sincerely thank numerous activists and observers of the "Latin American Network of Just, Democratic and Sustainable Cities" for generously sharing their experiences. For feedback on the draft manuscript, we are very grateful to Hans de Bruijn, Senj Temple and three anonymous reviewers.

Conflicts of Interest: The authors declare no conflict of interest.

References

1. Kitchin, R.; Lauriault, T.P.; McArdle, G. Knowing and governing cities through urban indicators, city benchmarking and real-time dashboards. *Reg. Stud. Reg. Sci.* **2015**, *2*, 6–28. [CrossRef]
2. Barrington-Leigh, C. The role of subjective well-being as an organizing concept for community indicators. In *Community Quality-of-Life Indicators: Best Cases VII*; Holden, M., Phillips, R., Stevens, C., Eds.; Springer: Cham, Switzerland, 2017; pp. 19–33.
3. Holden, M. Revisiting the local impact of community indicators projects: Sustainable seattle as prophet in its own land. *Appl. Res. Qual. Life* **2007**, *1*, 253–277. [CrossRef]
4. Moreno Pires, S.; Magee, L.; Holden, M. Learning from community indicators movements: Towards a citizen-powered urban data revolution. *Environ. Plan. C Polit. Space* **2017**, *35*, 1304–1337. [CrossRef]
5. Pozzebon, M.; Mailhot, C. Citizens engaged to improve the sustainability and quality of life of their cities: The case of nossa sao paulo. *J. Chang. Manag.* **2012**, *12*, 301–321. [CrossRef]
6. Hevia, F.J. *Monitoreo Ciudadano En México. Participación Ciudadana Para Mejorar La Provisión de Servicios Públicos*; Centro de Investigación y Docencia Económicas: México City, Mexico, 2016.
7. Hernández Quiñones, A.; Flórez Herrera, J.H. *Accountability Social y Democracia: El Caso de La Red Latinoamericana Por Ciudades Justas, Democráticas y Sustentables*; Universidad de los Andes, CIDER: Bogotá, Colombia, 2013.
8. Inter-American Development Bank. *Evaluation of the IDB's Emerging and Sustainable Cities Initiative*; Office of Evaluation and Oversight: Washington, WA, USA, 2016.
9. Lyytimäki, J.; Tapio, P.; Varho, V.; Söderman, T. The use, non-use and misuse of indicators in sustainability assessment and communication. *Int. J. Sustain. Dev. World Ecol.* **2013**, *20*, 385–393. [CrossRef]
10. Wray, L.; Stevens, C.; Holden, M. The history, status and future of the community indicators movement. In *Community Quality-of-Life Indicators: Best Cases VII*; Holden, M., Phillips, R., Stevens, C., Eds.; Springer: Cham, Switzerland, 2017; pp. 1–16.
11. Davern, M.T.; Gunn, L.; Giles-Corti, B.; David, S. Best practice principles for community indicator systems and a case study analysis: How community indicators victoria is creating impact and bridging policy, practice and research. *Soc. Indic. Res.* **2016**, *131*, 567–586. [CrossRef]
12. Ramos, T.B. Sustainability assessment: Exploring the Frontiers and paradigms of indicator approaches. *Sustainability* **2019**, *11*, 824. [CrossRef]
13. Grandvoinnet, H.; Aslam, G.; Raha, S. *Opening the Black Box: The Contextual Drivers of Social Accountability*; The World Bank: Washington, WA, USA, 2015; ISBN 978-1-4648-0481-6.
14. Klopp, J.M.; Petretta, D.L. The urban sustainable development goal: Indicators, complexity and the politics of measuring cities. *Cities* **2017**, *63*, 92–97. [CrossRef]
15. Moreno-Pires, S.; Fidélis, T. A proposal to explore the role of sustainability indicators in local governance contexts: The case of Palmela, Portugal. *Ecol. Indic.* **2012**, *23*, 608–615. [CrossRef]
16. Holden, M. Sustainability indicator systems within urban governance: Usability analysis of sustainability indicator systems as boundary objects. *Ecol. Indic.* **2013**, *32*, 89–96. [CrossRef]
17. Fraser, E.D.; Dougill, A.J.; Mabee, W.E.; Reed, M.; McAlpine, P. Bottom up and top down: Analysis of participatory processes for sustainability indicator identification as a pathway to community empowerment and sustainable environmental management. *J. Environ. Manag.* **2006**, *78*, 114–127. [CrossRef]
18. Moreno Pires, S.; Fidélis, T.; Ramos, T.B. Measuring and comparing local sustainable development through common indicators: Constraints and Achievements in practice. *Cities* **2014**, *39*, 1–9. [CrossRef]
19. Barrington-Leigh, C.; Escande, A. Measuring progress and well-being: A comparative review of indicators. *Soc. Indic. Res.* **2016**. [CrossRef]
20. Tanguay, G.A.; Rajaonson, J.; Lefebvre, J.-F.; Lanoie, P. Measuring the sustainability of cities: An analysis of the use of local indicators. *Ecol. Indic.* **2010**, *10*, 407–418. [CrossRef]
21. Merino-Saum, A.; Halla, P.; Superti, V.; Boesch, A.; Binder, C.R. Indicators for urban sustainability: Key lessons from a systematic analysis of 67 measurement initiatives. *Ecol. Indic.* **2020**, *119*, 106879. [CrossRef]
22. Cox, D.; Frere, M.; West, S. Developing and using local community wellbeing indicators: Learning from the experience of community indicators victoria. *Aust. J. Soc. Issues* **2010**, *45*, 71–88. [CrossRef]
23. Lora, E. Using life satisfaction data to identify urban problems, prioritize local public expenditures and monitor the quality of urban life. In *Handbook of Happiness Research in Latin America*; Rojas, M., Ed.; Springer: Dordrecht, The Netherlands, 2016; pp. 273–294. ISBN 978-94-017-7202-0.

24. Martinez, J.; Mikkelsen, C.A.; Phillips, R. Introduction: Quality of life and sustainability, socio-spatial, and multidisciplinary perspectives. In *Handbook of Quality of Life and Sustainability*; Martinez, J., Mikkelsen, C.A., Phillips, R., Eds.; International Handbooks of Quality-of-Life; Springer International Publishing: Cham, Switzerland, 2021; pp. 1–14. ISBN 978-3-030-50539-4.
25. Niemann, L.; Hoppe, T.; Coenen, F. On the benefits of using process indicators in local sustainability monitoring: Lessons from a Dutch Municipal Ranking (1999–2014). *Environ. Policy Gov.* **2017**, *27*, 28–44. [CrossRef]
26. Niemann, L.; Hoppe, T. Sustainability reporting by local governments: A magic tool? Lessons on use and usefulness from European pioneers. *Public Manag. Rev.* **2018**, *20*, 201–223. [CrossRef]
27. de Lancer Julnes, P.; Broom, C.; Park, S. A suggested model for integrating community indicators with performance measurement. Challenges and opportunities. *Int. J. Community Well-Being* **2019**. [CrossRef]
28. Joss, S.; Tomozeiu, D.; Cowley, R. Eco-city indicators: Governance challenges. *WIT Trans. Ecol. Environ.* **2012**, *155*, 109–120. [CrossRef]
29. Wood, D.M. Community indicators and collective impact: Facilitating change. *Community Dev.* **2016**, *47*, 194–208. [CrossRef]
30. Hezri, A.A.; Dovers, S.R. Sustainability Indicators, policy and governance: Issues for ecological economics. *Ecol. Econ.* **2006**, *60*, 86–99. [CrossRef]
31. United Nations 'Our Struggle for Global Sustainability Will Be Won or Lost in Cities,' Says Secretary-General. Available online: http://www.un.org/press/en/2012/sgsm14249.doc.htm (accessed on 29 March 2014).
32. Høyland, B.; Moene, K.; Willumsen, F. The tyranny of international index rankings. *J. Dev. Econ.* **2012**, *97*, 1–14. [CrossRef]
33. Michener, G. Policy evaluation via composite indexes: Qualitative lessons from international transparency policy indexes. *World Dev.* **2015**, *74*, 184–196. [CrossRef]
34. Cabrera Paredes, M.; González Moscoso, E.; Vintimilla Carrasco, M.I. *Colectivo Cuenca Ciudad Para Vivir. Diez Años de Apuesta Para La Ciudanía Activa. Sistematización de La Experiencia*; CCCV: Cuenca, Ecuador, 2020.
35. Romanutti, V.; Cáceres, P. Innovación institucional y transformación democrática. Ocho años de implementación del plan de metas de gobierno en la ciudad de Córdoba. *Adm. Pública Soc.* **2020**, *10*, 252–264.
36. Fox, J.A. Social accountability: What does the evidence really say? *World Dev.* **2015**, *72*, 346–361. [CrossRef]
37. Rinne, J.; Lyytimäki, J.; Kautto, P. From sustainability to well-being: Lessons learned from the use of sustainable development indicators at national and eu level. *Ecol. Indic.* **2012**, *35*, 35–42. [CrossRef]
38. Orellana, A.; Vicuña, M. Requerimientos de indicadores y estándares urbanos para una descentralización regional efectiva. In *Desarrollo Territorial Colaborativo: Descentralizando Poder, Competencias Y Recursos*; von Baer, H., Bravo, N., Eds.; Ediciones Universidad de La Frontera: Temuco, Chile, 2019; pp. 557–570. ISBN 978-956-236-373-0.
39. Pollitt, C. Performance management 40 years on: A review. some key decisions and consequences. *Public Money Manag.* **2018**, *38*, 167–174. [CrossRef]
40. Cohen, M.; Wiek, A.; Kay, B.; Harlow, J. Aligning public participation to stakeholders' sustainability literacy—A case study on sustainable urban development in phoenix, arizona. *Sustainability* **2015**, *7*, 8709–8728. [CrossRef]
41. Friendly, A. Urban Policy, Social movements, and the right to the city in Brazil. *Lat. Am. Perspect.* **2017**, *44*, 132–148. [CrossRef]
42. Thomas, R.; Hsu, A.; Weinfurter, A. Sustainable and inclusive—Evaluating urban sustainability indicators' suitability for measuring progress towards SDG. *Environ. Plan. B Urban Anal. City Sci.* **2020**. [CrossRef]

sustainability

Article

Learning and Teaching Interdisciplinary Skills in Sustainable Urban Development—The Case of Tampere University, Finland

Jonathon Taylor [1,*,†], Salla Jokela [2,†], Markus Laine [3,†], Juho Rajaniemi [1], Pekka Jokinen [3], Liisa Häikiö [2] and Antti Lönnqvist [1]

1 Faculty of Built Environment, Tampere University, 33720 Tampere, Finland; juho.rajaniemi@tuni.fi (J.R.); antti.lonnqvist@tuni.fi (A.L.)
2 Faculty of Social Sciences, Tampere University, 33014 Tampere, Finland; salla.e.jokela@tuni.fi (S.J.); liisa.haikio@tuni.fi (L.H.)
3 Faculty of Management and Business, Tampere University, 33100 Tampere, Finland; markus.laine@tuni.fi (M.L.); pekka.jokinen@tuni.fi (P.J.)
* Correspondence: jonathon.taylor@tuni.fi
† J.T., S.J., and M.L. contributed equally to this work as joint first authors.

Abstract: Developing the economic, environmental, and social sustainability of urban environments is challenging due to the complex and interconnected nature of the context and objectives. In order to be successful in this challenging environment, professionals working in the urban development arena should have a holistic understanding of the different pillars of sustainable development, as well as various competencies and skills. This paper looks at sustainable urban development (SUD) from the perspective of the skills and competencies required and identifies effective pedagogic practices that could help educate future professionals. In particular, we explore interdisciplinary and transdisciplinary learning, reflective thinking, and experiential learning, which are needed for understanding various aspects of a complex phenomenon, collaborating with professionals from different fields and coming up with novel and constructive ways of solving complex problems. We first examine these through reviewing and analyzing relevant literature on education for sustainable development, with a focus on SUD. Then, we explore the application of these approaches in practice by describing and analyzing a newly introduced degree program at Tampere University, Finland.

Keywords: education for sustainable development; pedagogy; urban; multidisciplinary learning

Citation: Taylor, J.; Jokela, S.; Laine, M.; Rajaniemi, J.; Jokinen, P.; Häikiö, L.; Lönnqvist, A. Learning and Teaching Interdisciplinary Skills in Sustainable Urban Development—The Case of Tampere University, Finland. *Sustainability* **2021**, *13*, 1180. https://doi.org/10.3390/su13031180

Academic Editor: Manuel Duarte Pinheiro
Received: 22 December 2020
Accepted: 21 January 2021
Published: 23 January 2021

Publisher's Note: MDPI stays neutral with regard to jurisdictional claims in published maps and institutional affiliations.

1. Introduction

There is an urgent requirement to improve global environmental sustainability, including addressing critical challenges such as reducing greenhouse gas (GHG) emissions, water consumption, biodiversity loss, and accelerating land and resource use [1]. In parallel, a number of interrelated or unique social sustainability challenges need to be addressed, including inequity, social cohesion, inclusion, and justice [2]. These challenges must be faced within the economic and administrative constraints, resulting in social learning and new forms of social-ecological reflectivity and sustainability governance [3,4].

Urban areas are a natural focus for efforts to improve sustainability. They account for an increasing majority (54%) of the global population [5] and Gross Domestic Product (GDP) (85%); have a disproportionate share of global energy (60–80%) and resource consumption (75%); produce around 50% of global waste; and emit around 70% of global GHG emissions [6]. Urban development—defined here as the improvement or expansion of urban physical, social, administrative, and economic infrastructure—needs to be done in a sustainable manner in order to help avoid lock-in of unsustainable urban systems. There are also significant opportunities for sustainable urban development policies to provide co-benefits for population health and wellbeing [7–9]. However, to-date, very few cities have had success in meeting either local or global sustainability challenges (e.g., [10]).

Urban areas are highly complex systems, with feedbacks, interdependencies, and non-linear linkages between environmental, social, economic, and governance elements of the urban system. This complexity can lead to so-called wicked problems in planning and policy that are extremely difficult if not impossible to solve. Addressing complex societal challenges, such as those targeted for improvement in the Sustainable Development Goals (SDGs), cannot be achieved within narrow disciplinary silos, but require a shift in development practice, and technological, social, and administrative solutions and innovations. Hence, there is a growing need for integrated skills and knowledge to address the key challenges of the Anthropocene [11]. Transdisciplinarity as a practice processes complex, real-life problems by means of methodological cooperation between disciplines, and between researchers and practical actors, and thus enables integrated learning between the scientific community and society [12,13].

Education for Sustainable Development (ESD) has a critical role in providing transdisciplinary skills, with universities having an important role as drivers of sustainable change. However, in education, transdisciplinary teaching of Sustainable Urban Development (SUD) has not seen significant adoption into the established curricula so far [14] and there is limited experience on how to develop and facilitate transdisciplinary teaching of SUD in practice. Therefore, the aims of this paper are to:

(1) Identify the knowledge and skills required for future SUD professionals;
(2) Evaluate current pedagogical practices and limitations of sustainability teaching at the Higher Education Institution (HEI)—level; and
(3) Illustrate and critically reflect on the practical implementation of SUD education at the bachelor's degree level.

To achieve the first two aims, we review and summarize current literature in ESD and analyze how SUD education has been implemented in HEIs worldwide. To achieve the third, we describe as a case study the development of a new bachelor's degree program in SUD at Tampere University, Finland, which aims to equip future sustainable urban development technology, social, and administrative professionals with the skills and holistic understanding to enact sustainable change in a global context. Finally, we reflect on the degree program, illustrating how learning outcomes and pedagogical practices identified in the literature review are integrated into the curriculum, as well as exploring the perspectives of students and practical experiences of lecturers following the programs first implementation.

2. Methodological Approach

Our paper includes both a literature review and case study. Existing research into ESD and SUD education has explored the challenges and best practices for educating students about sustainability whilst providing them with the core competencies to be agents of sustainable change. However, the extant literature on the topic can be characterized as fragmented and it is not easy to form a coherent and comprehensive view of key issues related to ESD, particularly in the context of SUD education. Therefore, in order to frame current thinking in ESD, a literature search was conducted in Google Scholar and Scopus. Relevant studies were reviewed to answer the following questions:

- What are the key competencies students of SUD should acquire to make them capable of addressing key challenges?
- How best to teach SUD in a manner that crosses both disciplinary boundaries, as well as the boundaries between the HEI and the wider society? What are the key challenges in doing so? Further, what are the methods of teaching that can provide students with these key competencies?
- How has SUD has been taught in other HEIs, and what lessons can be learned from the various approaches?

We then conduct a case study on the development and implementation of a new bachelor's degree program, which has been developed using the theoretical basis described in the review. The purpose of the case study is not only to illustrate how theory is applied in practice, but also to highlight the context-specific nature of teaching and learning SUD, offering a situated account of a dynamic process through which theoretical ideas are translated into practice. Simultaneously, we add to existing academic discussion on SUD education by focusing on a less-studied geographical context and education level (bachelor's program). Our methods take inspiration from previous educational research with an ethnographic case study approach (e.g., [15,16]). This approach entails being immersed in natural environment of the research area and using an insider position to generate and analyze data. Our knowledge emerges both from previous literature and our own position as the developers and teachers of the bachelor's degree program. Data sources and methods include classroom observations, "field notes" by the lecturers, reflective discussions and exercises with SUD students assisted by the online mind map tool Flinga, and analysis of student learning diaries. When analyzing the data, we performed cross-checks to identify consistent themes and to resolve ambiguities that arose during some of the discussions. The combination of review and case study provides a useful way to obtain a comprehensive view of the complex phenomenon we are examining and provides answers to our research questions.

3. Review of Current Practices in SUD Education

3.1. Contextual Challenges for SUD Education in HEIs

There is a critical need for the knowledge, innovation, and skills to address significant challenges in sustainable development. HEIs have an important role to play in driving sustainable change—through the provision of education, skills, and research—and as such, universities globally have positioned themselves as critical actors in achieving sustainable development and the SDGs. In Finland this year, universities agreed on a number of theses intended to move sustainable development and responsibility "from words to actions".

However, teaching of sustainability poses a multitude of challenges to HEIs, as sustainability problems are inherently multidisciplinary, complex, and require a broad understanding of issues. Wiek et al. [17] claim that instead of focusing on isolated events, it is important to gain deep understanding of "sustainability problem constellations" in order to deal with the root causes of these events. This way of thinking directs attention to complex and dynamic processes that underlie individual, easily perceptible problems. These processes take place and interact across multiple scales and are dependent upon people's beliefs, habits, motives, and practices.

In addition, HEIs also face the challenge of needing to evolve to meet societal needs [18]. There is a need to change the ways that HEIs function and interact with wider society [19], including government and industry, in order to help develop the solutions to key environmental, social, and economic sustainability problems. Furthermore, HEI administrative structures are typically organized around rigid disciplinary frameworks, and integrating a broader, interdisciplinary subject such as sustainability into such a framework is challenging.

3.2. Key Competencies in SUD

Traditional teaching in HEIs has focused on knowledge specialization, reductionist thinking, and education within disciplinary silos [20]. However, there is a need to educate a new generation of professionals capable of analyzing and dealing with complex problems in sustainable development, and a narrow focus on knowledge acquisition limits the ability of graduates to gain a holistic, integrated understanding of the issues. In response, researchers have defined key competencies needed to develop a broad understanding of sustainability and solve real-life sustainability problems. Rather than focusing on cognitive learning, competencies focus on developing skills for solving diverse sets of problems, including effective communication, teamwork, and methododological competencies. The most widely used set of competencies for sustainable development is perhaps that of UNESCO [21] (Table 1), which is based Wiek et al.'s [22] synthesizing of key competencies in sustainability, and Rieckmann's [23] research on the key competencies that should be fostered through university teaching and learning:

Table 1. Key competencies for sustainability according to UNESCO [21].

Competency	UNESCO Definition
Systems thinking competency	The abilities to recognize and understand relationships; to analyze complex systems; to think of how systems are embedded within different domains and different scales; and to deal with uncertainty.
Anticipatory competency	The abilities to understand and evaluate multiple futures—possible, probable, and desirable; to create one's own visions for the future; to apply the precautionary principle; to assess the consequences of actions; and to deal with risks and changes.
Normative competency	The abilities to understand and reflect on the norms and values that underlie one's actions; and to negotiate sustainability values, principles, goals, and targets, in a context of conflicts of interests and trade-offs, uncertain knowledge, and contradictions.
Strategic competency	The abilities to collectively develop and implement innovative actions that further sustainability at the local level and further afield.
Collaboration competency	The abilities to learn from others; to understand and respect the needs, perspectives, and actions of others (empathy); to understand, relate to and be sensitive to others (empathic leadership); to deal with conflicts in a group; and to facilitate collaborative and participatory problem solving.
Critical thinking competency	The ability to question norms, practices, and opinions; to reflect on own one's values, perceptions, and actions; and to take a position in the sustainability discourse.
Self-awareness competency	The ability to reflect on one's own role in the local community and (global) society; to continually evaluate and further motivate one's actions; and to deal with one's feelings and desires.
Integrated problem-solving competency	The overarching ability to apply different problem-solving frameworks to complex sustainability problems and develop viable, inclusive, and equitable solution options that promote sustainable development, integrating the abovementioned competences.

> "Key competencies represent cross-cutting competencies that are necessary for all learners of all ages worldwide (developed at different age-appropriate levels). Key competencies can be understood as transversal, multifunctional and context-independent." [21]

All eight UNESCO competencies are highly necessary to tackle problems—wicked or otherwise—in sustainable development. In addition, all students need to be taught methodological competences like traditional project management techniques, planning and decision-making methods [24], skills to work across national, geographical, and cultural boundaries [25], and generic working life skills required for their future careers [26], as well as specialist skills and competencies related to their field of study. For example, technology graduates should have strong analytical and computer literacy skills. In social sciences, basic methodological, reading, writing and presentation skills are essential to understanding social scientific concepts and theories, as well as their relevance in making sense of everyday life [24]. These skills support the development of effective citizenship, enabling graduates to navigate in a politically polarized social climate and transform existing soci(et)al structures and patterns of behavior, including, inter alia, sexism, racism, and colonialism [25].

3.3. Key Themes in SUD Education

SUD is a problem- and solution-oriented field that operates in a complex environment of various communities of knowledge, non-academic stakeholders, and citizens. SUD education is not only about learning and producing knowledge, but also about goals, norms and visions of transition and transformation [27]. Therefore, we describe three interlinked themes that are vital in SUD education: transdisciplinary, reflective thinking and transformative learning. All these approaches demand the capability to think of problems, solutions, and their framing from multiple perspectives in various contexts.

3.3.1. Interdisciplinarity and Transdisciplinarity

Successful development of sustainable solutions in urban areas requires integrating knowledge and skills from multiple perspectives, and so a common theme for all competencies is the ability to think across disciplinary and institutional boundaries. The concepts of interdisciplinarity and transdisciplinarity are often referred to in ESD literature. Interdisciplinary education aims to teach students subjects from multiple perspectives, developing students' abilities to synthesize knowledge from different disciplines, change perspectives, and cope with complexity [28]. Transdisciplinary education extends this further, entailing solution-oriented collaboration between academics and non-academic actors [20,27,29] and bringing the world outside of the HEI into student education [20]. Lang et al. [27] define transdisciplinarity in broad terms as follows:

> "Transdisciplinarity is a reflexive, integrative, method-driven scientific principle aiming at the solution or transition of societal problems and concurrently of related scientific problems by differentiating and integrating knowledge from various scientific and societal bodies of knowledge."

As there is no single discipline that would cover the various aspects of SUD, interdisciplinary and transdisciplinary approaches—rather than monodisciplinary—are important (see, e.g., [30–32]). As a result, inter-/transdisciplinary education and research have started to become more commonplace although the monodisciplinary tradition still dominates HEI teaching practices. Indeed, a number of studies have discussed the importance of HEI teaching of sustainability aiming for transdisciplinary problem-based learning instead of the accumulation of discipline-based knowledge [33], and evidence suggests that it can improve course outcomes [34].

3.3.2. Reflective Thinking

Reflective or critical thinking refers to evaluating one's own thinking patterns whilst trying to solve problems in order to learn how to be able to think better in such situations [35,36]. It is an ability to think critically about practices, opinions, and norms, but also a capacity to reflect on your own values, perceptions, and actions [37] and understand external views. Pragmatist philosopher John Dewey defined reflective thinking:

> "Active, persistent and careful consideration of any belief or supposed form of knowledge in the light of the grounds that support it, and the further conclusions to which it tends, constitutes reflective thinking." [38]

Reflective thinking is hard to reduce to the definable learning outcomes traditionally used by HEIs. As Rodgers [39] has pointed out reflective thinking is not end itself, but rather a tool to transform an experience into meaning filled theory that is grounded in practice, informed by existing theory, and serves a bigger purpose of making society better. It is an iterative, progressive process from experience to theory and from theory to experience, with a similar form to scientific inquiry. While reflective thinking is central to HEI education, few educators actively teach students how to think critically.

3.3.3. Transformative Learning

The key challenges of the Anthropocene demand transformative changes to society. In the field of sustainable development one faces uncertainty, poorly defined circumstances, and conflicting interests and realities. Still, these unknowns of the future need to be coped with a creative transformative commitment, rather than inaction [40].

Transformative or "experiential learning", as UNESCO [41] terms it, aims to challenge the core assumptions and values that teachers, students and as society hold. The teaching of SUD in HEIs should be transformative [35,42], focusing on the topics that are needed the most for societal transformation [43]. It shifts attention from the teacher to the student and emphasizes immediate experiences and close interaction with local communities [44]. Experiential learning can be further divided to several distinct—and partly overlapping—approaches. Some of these emphasize students' contribution to meeting the needs of local communities (e.g., service learning), while others focus on bridging scientific knowledge and practice by developing evidence-based solutions to real-life problems (e.g., problem-based learning and solution-oriented learning), and still others focus on empowering students to acknowledge their capacity to make a difference (e.g., participatory action learning) [33,45–47]. Practical forms of learning and teaching associated with these approaches include project-based courses, studios, and workshops. The common denominator in these approaches is that they put significant emphasis on "how" to teach instead of focusing on "what" questions or the accumulation of knowledge related to SUD [19].

These forms of learning and teaching can be supported by teaching strategies that foster open-mindedness and open-ended trajectories, enabling and encouraging both teachers and students to confront hegemonic or conventional ideas, think about alternatives, and acknowledge different or opposing positions as well as underlying value. For example, interdisciplinary team-teaching brings together teachers from multiple faculties, schools, or disciplines to collaborate in planning and delivering courses [48]. This practice may be beneficial, not only because it encourages teachers to seek integration of different approaches and viewpoints, but also because it breaks existing boundaries. As Brewer [49] has noted, "the world has problems, but universities have departments". Thus, challenging or changing the institutional context in which knowledge is being (re)produced may ultimately enable universities to flexibly and effectively respond to the real-world sustainability problems.

3.4. Teaching of SUD in Practice

The above perspectives on competencies, teaching, and learning aim to equip students with the knowledge and skills to go beyond traditional empirical understanding ("what" and "why" questions) to knowledge synthesis, normative questions, and ethical analysis ("how and "should" questions).

Urban areas are at the center of many of the key sustainability challenges and are where the complexity of these challenges becomes manifest. As such, cities are prime examples of the multifaceted nature of sustainability challenges, which do not neatly follow the disciplinary boundaries of traditional HEI departments (see [49]). Instead, they call for various forms of collaboration that extend beyond existing institutional structures. Therefore, urban issues are often the focus in ESD curricula [45–47]. Urban areas offer abundant real-world learning settings where students can obtain first-hand observations and experiences of sustainability challenges, combine scientific knowledge with local knowledge, and engage in collaboration and co-creation of solutions with stakeholders and experts from various fields. These settings encourage active, student-centered learning, critical, inter- and transdisciplinary thinking and flexible application of different problem-solving frameworks, which are integral to acquiring key sustainability competencies and contributing to sustainability-oriented societal transformations [35,46].

As global interest in sustainable development has increased, so have the number of higher education institutions offering degree programs on SUD [28,50–53]. However, often, the teaching of sustainability is integrated into existing curricula without a suitable pedagogical framework based on ESD [19,54], and many existing programs—for example in urban studies and urban planning—are facing the challenge of embedding sustainability in their teaching [55]. While there is general agreement of the value of inter- and transdisciplinarity in addressing sustainability problems, integrating sustainability into new or existing courses and programs can be challenging and there can be multiple barriers [56]. For example, challenges remain as to how collaboration between disciplines can and should be fostered in practice. Coordinating teaching between different schools and disciplines while fostering interaction with the wider society entails balancing commitments to research, service, and teaching in various study programs, overcoming administrative and institutional challenges, and negotiating working definitions of "sustainability" and ways of assessing of student work [48,57].

Developing transdisciplinary curricula in practice is not without its significant challenges [58]. Teachers in HEIs have typically not been educated in an inter- or transdisciplinary manner themselves and lack the training to use such approaches in their teaching [59,60]. It requires close collaboration between academics from different disciplines, designing course learning objectives, curricula, lectures, and workshops. It also requires that academics have the time to learn about sustainability issues within their own fields, as well as an appreciation for the perspectives and knowledge of their academic collaborators. However, without embedded inter- or transdisciplinary approaches, sustainability is taught within disciplinary silos without broader perspectives.

In response to this, many institutions are opting to develop new interdisciplinary programs in SUD (e.g., [28,51–53]), while there are fewer but increasing numbers of transdisciplinary courses (e.g., [61]). This trend is also apparent in the field of urban studies, which has witnessed the emergence of new master's and bachelor's degree programs. So far, research on inter- and transdisciplinary learning and teaching in the urban context has mostly focused on individual university courses (e.g., [45,46,48,62]) or their institutional settings [19] instead of entirely new study programs in SUD. The majority of existing inter- and transdisciplinary SUD degree programs are at postgraduate level, and there are a limited number of undergraduate degree programs on sustainable urban development that can provide students with different perspectives on sustainable development prior to establishing their disciplinary expertise.

4. Case Study—SUD at Tampere University

The above review highlights the critical need for transdisciplinary in sustainability teaching at HEIs, and that—to be successful—transdisciplinarity needs to be embedded in the core of the teaching curricula. In addition, there are core skills that future sustainability professionals must have, and different pedagogical approaches to help students develop these skills. There are therefore opportunities to offer undergraduate degree programs on SUD that include transdisciplinarity and effective pedagogical practices at the very core of the program.

In response, Tampere University has developed a new SUD Bachelor's Degree Program, jointly delivered by the faculties of Social Sciences, Administrative Sciences, and the Built Environment. This program aims to provide students with an inter- and transdisciplinary education on sustainable urban development, with the opportunity to study three streams: administrative sciences, social sciences, and technology. While each stream has courses specific to the disciplines, all students participate in common courses on sustainable development. It is an international program, taught in English to local and overseas students, and with both a local and global perspective.

4.1. Program Development

The development of the SUD program began in 2018 when two universities in the city of Tampere—University of Tampere and Tampere University of Technology—were preparing a merger. There were many drivers for the merger but an important one was the belief that the new structure would provide better opportunities for interdisciplinary research and education. The merger was completed in 2019, and the merged university was named Tampere University. In order to support the merger process and facilitate interdisciplinary collaboration, pilot projects were identified and given financial resources. SUD education was selected as a pilot project because of the societal relevance, and because there was an existing research collaboration between the scholars of the merged universities.

The planning of the SUD program took two and a half years. During the first year, several seminar events were held, during which the theme and the idea of the program was discussed and developed by urban development practitioners and scholars. Based on these seminars, it was decided that the new program would be at bachelors' level, delivered by three collaborating faculties. The faculties and disciplines were chosen based on relevance to the theme, the existing research connections, and participants' engagement in planning.

The first year of program development was challenging because the new interdisciplinary and multi-faculty program did not fit existing institutional structures, and it was not clear how it would be connected to existing programs or their resources. It took time to develop a model that was acceptable to key stakeholders. A key decision was to develop an international English language bachelor's program—novel in contrast to the existing bachelor's programs at the participating faculties that were all in Finnish.

The second year of planning involved detailed discussions of the learning objectives and content, mainly by a team representing the three participating faculties, but with several events for faculty members and stakeholders, e.g., representatives of the City of Tampere, to provide feedback and ideas. Substantial iterations were required to develop a curriculum that would cover both the common SUD learning objectives, discipline-specific postgraduate eligibility, and various administrative and other requirements from the three faculties. At the end of the second year, three lecturers were recruited to take over the final planning and launch. The first intake of students was admitted during Spring 2020, and teaching began in August 2020. Detailed planning is still being undertaken as the new courses are being developed and delivered.

4.2. Theoretical Basis of the Tampere SUD Program

The development of the program has been framed by current research, outlined in the review above. It is designed to overcome some of the limitations of current ESD teaching by:

- Developing students core competencies, as framed by UNESCO, as well as additional competencies in intercultural communication and languages, and discipline-specific knowledge.
- Teaching sustainability at the undergraduate level, prior to the development of students' discipline identity and enabling them to learn transdisciplinary thinking at an early stage. This is in contrast to many of the current existing degree programs that are at master's level only.
- It is designed for interdisciplinary teaching of sustainability from the outset, rather than as an add-on to existing degree programs, and the structure allows students the opportunity to learn from different perspectives and interact with other students from different streams during workshops and projects throughout their studies.
- It provides opportunities for engagement with stakeholders outside of the university during transdisciplinary project-based courses.

The degree structure and curriculum can be seen in Figure 1. The degree program has learning outcomes for all students, as well as outcomes for specific streams. All students will graduate with a holistic understanding of the key concepts, interlinkages, actors, questions, conflicts, and solutions in the field of sustainable urban development globally, as well as the core discipline-specific skills that are required by employers or to continue on to masters level. Those in the technology stream will also gain a foundational knowledge of mathematics, physical sciences, and computer science as they relate to the urban environment. Students in administrative sciences will be able to understand the administrative, political, and economic phenomena in cities, as well as the key concepts in administrative sciences. Finally, students in the social science stream will understand the key theories, concepts and methods in social sciences, and the societal significance of social research. The curriculum has been designed with core courses on sustainability throughout the degree, bringing together the students in the separate streams for these common interdisciplinary and/or transdisciplinary courses in each year of study.

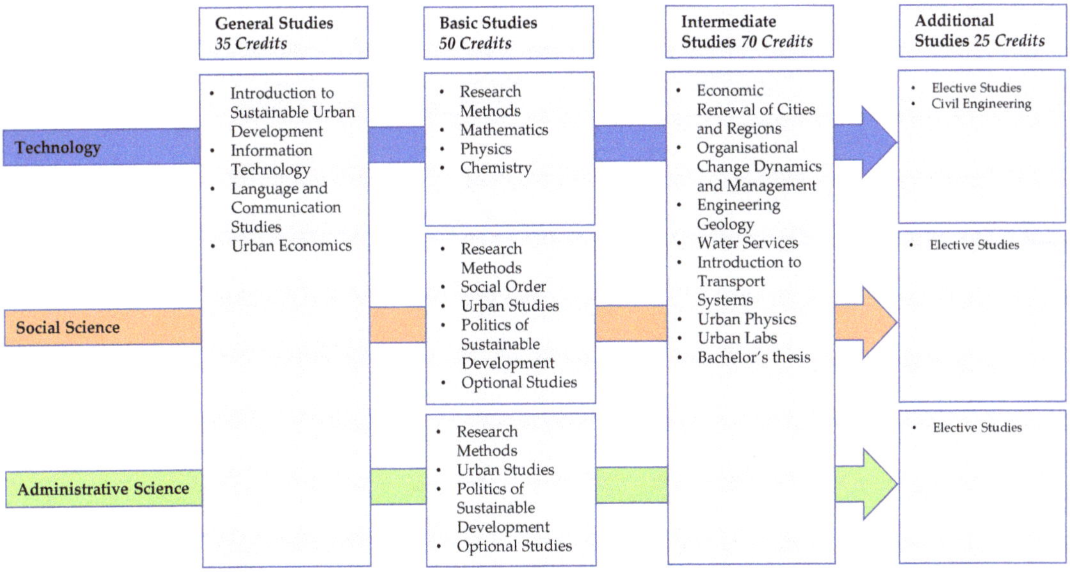

Figure 1. Degree structure and curriculum.

The degree program has resulted in the development of several new courses, common to all SUD students. These courses are intended to integrate across the degree curriculum, unifying concepts and providing interdisciplinary or transdisciplinary perspectives on common subject matters. This enables students to make connections and give broader perspectives on the issues they have been taught in more traditional subject-specific courses. Common courses have been developed and are delivered by three lecturers, one from each faculty, using team-based teaching to deliver integrated learning; these lecturers are provided the resources and support from the university to do so.

While the common integrating courses are intended to provide an interdisciplinary foundation, a unique element of the SUD degree program is the transdisciplinary teaching in the two Urban Lab courses. These courses will enable students to engage in real-life urban development projects in collaboration with public organizations such as the City of Tampere, or with commercial organizations. Currently, urban lab projects done in collaboration with Tampere city will make use of a future city district—Hiedanranta—as a real-life laboratory. Hiedanranta is being developed as a testbed for innovative projects related to smart and sustainable housing, energy, transportation, and business, providing ample opportunities for students to engage with actors outside of the university on real-world projects. These project-based courses also allow for students to collaborate between streams to get continued experience working with different perspectives.

4.3. Reflections on Tampere SUD Program

Here we present an analysis of the SUD program, mapping elements of the degree to the reviewed perspectives. The SUD degree aims to enable students to develop key UNESCO competencies so that they are capable of solving problems with professionals from different backgrounds. In particular, the degree focuses on systems thinking, strategic, collaboration, and integrated problem-solving competencies, as well as intercultural communication. These competencies are strengthened in the common SUD courses, where students from different streams are asked to work together in group projects, workshops, or in reading groups. The project work and teaching methods are designed to strengthen the core competencies, and include problem-based learning, game-based learning, and team-based learning. The methods of implementing core competencies and inter and transdisciplinarity into the degree program is summarized in Table 2.

By including common courses for all SUD students throughout the degree program, it is ensured that students are regularly engaged in interdisciplinary studies, and mixed groupwork ensures that multiple perspectives are included in each group. This helps strengthen and facilitate interdisciplinary collaboration. In addition, the Urban Lab courses provide the opportunity to work on projects outside of the university, promoting transdisciplinary education. By taking a team-based teaching approach, we are able to approach important topics in sustainability from multiple angles within the same lecture. A key means of assessing the common SUD courses are student learning diaries, which prioritize their understanding of concepts across multiple perspectives, and their personal reflections. From this, lecturers are able to gauge students reflective learning and critical thinking. The reflections also entail important feedback, which helps both students and teachers to adjust their behavior and practices to meet the objectives of the program as effectively as possible.

4.3.1. Analysis of the Program from the Student Perspective

The first cohort of 18 SUD students began their studies in Autumn 2020. The students come from diverse backgrounds, including six different nationalities, and with both recent high school graduates and more mature students. Initial feedback collected form the students indicate a mixture of different long-term ambitions, with students hoping for careers in, for example, research and development or governmental organizations, while more than half have no clear career ambitions just yet. A key attraction of the program for the students was the interdisciplinary teaching, providing the students the opportunity to develop their interests in a subject area that they are passionate about.

To tease out students' understanding about transdisciplinarity, a reading group session was organized where the focus was on an article, "Imagining the post-fossil city: why is it so difficult to think of new possible worlds? [43]." One question that was reflected on in breakout rooms (with mixed participants from different streams) and together was "Keys to a post-fossil city: What perspectives or competencies need to be combined to create a walkable city?" in order to understand how students envision transdisciplinary approaches in a practical context. The answers varied from a need for a comprehensive understanding or a vision of sustainable urban development; understanding intended and unintended policy outcomes; behavioral patterns of humans; the art of nudging; understanding urban transportation systems; or the dimensions of urban green as social, ecological, and technical. The benefit of the discussion was that it demonstrated the multiple disciplinary angles needed when considering SUD.

Reflective thinking was encouraged in several ways during our first two courses. Students analyzed issues in learning diaries, where they reflected the contents of each lecture in relation to their own experiences. One of the issues of reflection was a debate exercise facilitated using game-based learning. One student wrote that, "I think that fully listening what another person is has to say can allow questions to arise, like why this person in thinking this way. Are there factors like age, culture or the media, that could be shaping their views? This exercise also showed me that other people's viewpoints are not wrong and perhaps I could examine my own." This quote illustrates the possibilities of transdisciplinarity, as the co-existence of three the disciplines in SUD demand a continuous reflection of disciplinary frames.

Table 2. Analysis of the SUD program.

		Integration into Degree Program
Core Competencies	Systems Thinking	Systems thinking workshops in the common courses are used to draw connections between different elements of urban systems for hypothetical development scenarios and identify unintended consequences.
	Strategic	During the Urban Lab courses, students will be expected to collaborate and develop plans to address sustainability issues of their choosing.
	Collaboration	Students from different discipline streams are tasked with collaborating within groups in workshops and projects in all SUD common courses. In the Introduction to Social Scientific Research and Academic Writing course, the Teams-Based Learning tools are used to teach teams of students about how to read and understand academic articles.
	Critical thinking	Game-based learning is used to encourage students discuss and debate topics from points of view that they personally may not share. In various courses, students are assigned to reading groups, mixed by degree stream, and asked to read and reflect on a selection of assigned newspapers and journal articles and books. In all social science teaching, students are enabled and encouraged to identify and question existing soci(et)al structures, norms, worldviews etc.Students use methodological skills to analyze and interpret urban and societal phenomena.
	Integrated Problem Solving	During the Urban Lab courses, students are divided into groups, with each group containing students from different streams. In collaboration with the university lecturers and stakeholders from city government and local business, they design their own research projects. This provides students with opportunities for contextual learning and problem-based learning.
	Intercultural Communication	All students are required to study a foreign language. In addition, first year students attend an intercultural communication course, during which local and international students work together in groups to learn about cross-cultural communication patterns.
Disciplinarity	Specific	Students attend discipline-specific courses related to their study stream to obtain relevant knowledge and skills.
	Interdisciplinary	A number of interdisciplinary courses are taught throughout the degree program, which all students attend. These courses are taught by each SUD lecturer using team-based teaching and are intended to draw together discipline-specific perspectives on sustainable development.
	Transdisciplinary	Two Urban Lab courses require interdisciplinary collaboration between students from different streams. A key part of the courses is the engagement with organizations outside of the university on real-world problems using the city as a "living lab".
Thinking	Transformative	The Urban Lab courses and some other courses as well are connected to real-life SUD problems, which the students aim at addressing in collaboration with groups external to the university. The purpose of these learning events is to develop understanding on how to make positive changes happen.
	Reflective	We aim to promote reflective thinking in several ways. First, three streams provide different perspectives automatically, second by engaging students into discussion of possible scientific approaches and ways to construct an argument through scientific books, journal articles and systematic writing practice and third, adopting transformative learning.

To understand the views of students on how the degree could support their learning, we asked students, "How can you become an active agent in sustainable urban development?" and "How can university studies support you in this task?" and made an interactive Flinga mind map to record their responses. Some responses focused on skills such as "learning to make an argument in layman's terms" and "integration of technological solutions and understanding discourses and imageries". Other students focused on actions such as "participating in municipal politics", "sharing spaces [such as saunas or maker-spaces) and items and repairing them", "supporting public transportation", "getting to know real life actors in urban development and their relationships", and "participate in events available to youth". The answers to the second question were in line to our thinking about project-based courses, as students mentioned "Urban Labs" as a possibility to meet real life SUD actors and networking and gain "knowledge about resources, where to get information, what to read and follow", which we believe is best tackled through co-operation between the university and real-world development projects. Finally, there was also an understandable wish to get "knowledge about future career options". The answers reflect first, the broad field of SUD a demand for integrative understanding, and second, the fact that many students are coming straight from high school to the SUD program and their professional futures are still somewhat uncertain.

4.3.2. Initial Reflections from SUD Lecturers

While the program is relatively new, there have been a number of valuable lessons learned during its development. Course development and team-based teaching has required that the three lecturers from different faculties—with their own disciplinary backgrounds—have had to spend considerable time learning about sustainability issues of the other two disciplinary fields. This requires a significant time commitment, as well as a need to be open minded to different teaching traditions, perspectives on topics and understand the importance of these perspectives for sustainable development. For this reason, it has been advantageous to recruit new lecturers who are enthusiastic about interdisciplinary collaboration, and are aware of the working requirements prior to beginning, rather than expecting existing staff to take on the extra commitment. It has also required the commitment from the university administration to provide the lecturers with the time and support to engage in this "invisible" work.

The team-based teaching approach and emphasis on student interaction has required flexibility in teaching schedules, as students may show particular interest in an idea or perspective which means teaching takes longer and there is less time to cover other perspectives. The coronavirus situation in Finland in 2020 has meant that teaching began using hybrid in-person and online methods, enabling students who were not able to attend to learn remotely. This has, however, provided interesting opportunities to use the coronavirus as a topic to learn about key sustainability issues globally, such as global urban connectivity, density, behavior changes, emissions, biodiversity, and human encroachment into nature.

According to Keeley and Benton-Short [48], one of the challenges of interdisciplinary team-based teaching is that teachers from different fields do not necessarily agree on what should be included in a course or even what the definitions of key concepts are. While the lecturers have spent time in identifying the core content, team-teaching has nevertheless been an act of balancing between having a shared understanding of the content and goals of teaching and using heterogeneous views to engage students in exploring the complexity of inherently interdisciplinary issues. Heterogeneity of voices may be valuable in its own right, because, as Sennett [63] argues, it encourages ethical practice of communication based on outward-looking orientation, willingness to listen to others, and cooperation rather than confrontation. This kind of communication is needed when "coping with complex realities [of cities] in which finesse and skill replace naked pushiness".

The hybrid teaching—with students both present in class and online—presented a number of challenges. Lecturers used Flinga to do groupwork and create discussion with the whole group with those who were present and who were online. In addition, Mentimeter was used to rank and discuss the ideas and themes within topics with the students, and Kumu was used for mapping systems thinking discussions. Zoom breakout rooms were utilized for having smaller group discussions and game-based learning activities. These online tools worked well, especially in finding out how students were reflecting on SUD issues. However, the interaction with the students was harder to create than in face to face situation, and there was mixed success fostering engagement.

5. Discussion and Conclusions

The review of existing teaching of SUD describes the changing relationship between universities and wider society, and how ESD graduates need to have a set of core competencies that will enable them to drive sustainable change. It also discusses the importance of interdisciplinary and transdisciplinary teaching of SUD, and the critical need to students to experience critical reflection and transformative learning. However, as noted in the review, such teaching is difficult to integrate into existing degree programs, and many programs are focused on integrating sustainability into existing degree programs rather than in pedagogical frameworks developed for ESD. The Tampere SUD program has been developed in response to this need, incorporating interdisciplinary teaching and learning from the ground-up, with a pedagogical framework based on available information on best practice in ESD.

This paper promotes a new line of thinking in educating future professionals on sustainable development issues. It provides a description of the theoretical framework for ESD curricula to address SUD and subsequent development of the Tampere SUD program. It addresses one of the key challenges of sustainable development for HEI teaching at the moment: How to educate professionals able of addressing economic, environmental, and social aspects of sustainability in the living environments and global scales, and capable of developing viable, inclusive, and equitable solutions that promote sustainable development respectively?

Our paper contributes to ESD curricula by synthesizing extant literature on education and learning of SUD. It reviews competencies required to tackle wicked problems in SUD and different pedagogic practices to achieve these competences in an inter- and transdisciplinary teaching program. The synthesis provides a structured overview of relevant perspectives to be considered and approaches to be used in developing SUD education. It applies the synthesis of competences and pedagogical practices in SUD education to describe and analyze a new Tampere SUD Bachelor degree program, contributing a practical understanding of how to apply ESD curricula to educate future experts on sustainable urban development issues. The program aims to provide students with the holistic knowledge necessary to consider development problems from multiple perspectives in addition to providing them with discipline-specific fundamental skills.

Both the literature-based synthesis and the case study provide useful insights for scholars developing more nuanced understanding of SUD education and for HEI actors developing SUD education initiatives in practice. While the literature review revealed that most of the existing SUD education initiatives include individual SUD courses, add-on elements to existing programs and master's level programs, the Tampere SUD is an example of a bachelor level program constructed especially to equip future SUD professionals with the skills and transdisciplinary understanding to enact sustainable change in a global context. The planning of Tampere SUD program and first implementation provide an understanding of how SUD educational approaches can be applied in practice at university degree programs.

The Tampere SUD program is relatively new, and further research and evaluation is necessary to determine how students are achieving the program objectives for core competencies, interdisciplinary thinking, and transformative and critical thought. There will be opportunities to refine the degree program in coming years using student feedback and assessments of the achievements of competence objectives. All this will provide practical knowledge on how to restructure ESD curricula to support learning of sustainable urban development also in other HEIs and how research should be developed to support this process.

Author Contributions: Conceptualization, A.L., J.R., P.J., and L.H.; writing—original draft preparation, J.T., S.J., and M.L.; writing—review and editing, J.T., S.J., M.L., A.L., J.R., and P.J.; funding acquisition, A.L., L.H., and P.J. All authors have read and agreed to the published version of the manuscript.

Funding: This research received no external funding.

Informed Consent Statement: Informed consent was obtained from students to quote their written text and share their reflections on the degree programme.

Data Availability Statement: Data is contained within the article.

Acknowledgments: The Tampere University SUD program has been developed during a lengthy process involving several people in addition to the authors of article. The authors want to acknowledge the valuable contribution of several faculty members, students, and external stakeholders at various stages of program development. The authors would also like to acknowledge the contributions of the SUD students in providing their reflections on SUD and the degree program.

Conflicts of Interest: The authors declare no conflict of interest.

References

1. Steffen, W.; Richardson, K.; Rockstrom, J.; Cornell, S.E.; Fetzer, I.; Bennett, E.M.; Biggs, R.; Carpenter, S.R.; de Vries, W.; de Wit, C.A.; et al. Planetary boundaries: Guiding human development on a changing planet. *Science* **2015**, *347*, 1259855. [CrossRef] [PubMed]
2. Dempsey, N.; Bramley, G.; Power, S.; Brown, C. The social dimension of sustainable development: Defining urban social sustainability. *Sustain. Dev.* **2011**, *19*, 289–300. [CrossRef]
3. Meuleman, L. Metagovernance for Sustainability: A Framework for Implementing the Sustainable Development Goals. *Public Sect. Econ.* **2019**, *43*, 109–113. [CrossRef]
4. Tàbara, J. Social learning to cope with global environmental change and unsustainability. In *Routledge International Handbook of Social and Environmental Change*; Lockie, S., Sonnenfeld, D., Fisher, D.R., Eds.; Routledge: New York, NY, USA, 2013.
5. United Nations. *World Urbanization Prospects: The 2018 Revision*; United Nations: New York, NY, USA, 2018.
6. Swilling, M.; Robinson, B.; Marvin, S.; Hodson, M.; Hajer, M. City-Level Decoupling: Urban Resource Flows and the Governance of Infrastructure Transitions. In *A Report of the Working Group on Cities of the International Resource Panel*; United Nations Environment Programme: Nairobi, Kenya, 2013.
7. Giles-Corti, B.; Vernez-Moudon, A.; Reis, R.; Turrell, G.; Dannenberg, A.L.; Badland, H.; Foster, S.; Lowe, M.; Sallis, J.F.; Stevenson, M.; et al. City planning and population health: A global challenge. *Lancet* **2016**, *388*, 2912–2924. [CrossRef]
8. Kleinert, S.; Horton, R. Urban design: An important future force for health and wellbeing. *Lancet* **2016**, *388*, 2848–2850. [CrossRef]
9. Watts, N.; Amann, M.; Arnell, N.; Ayeb-Karlsson, S.; Belesova, K.; Boykoff, M.; Byass, P.; Cai, W.; Campbell-Lendrum, D.; Capstick, S.; et al. The 2019 report of The Lancet Countdown on health and climate change: Ensuring that the health of a child born today is not defined by a changing climate. *Lancet* **2019**, *394*, 1836–1878. [CrossRef]
10. Grandin, J.; Haarstad, H.; Kjærås, K.; Bouzarovski, S. The politics of rapid urban transformation. *Curr. Opin. Environ. Sustain.* **2018**, *31*, 16–22. [CrossRef]
11. Yigitcanlar, T.; Teriman, S. Rethinking sustainable urban development: Towards an integrated planning and development process. *Int. J. Environ. Sci. Technol.* **2015**, *12*, 341–352. [CrossRef]
12. Mullally, G.; Byrne, E.; Sage, C. Disciplines, Perspectives and Conversations. In *Transdisciplinary Perspectives on Transitions to Sustainability*; Byrne, E., Mullally, G., Sage, C., Eds.; Routledge: London, UK, 2017.
13. Jahn, T. Theory of sustainability? Considerations on a basic understanding of "sustainability science". In *Theories of Sustainable Development*; Enders, J., Remig, M., Eds.; Routledge: London, UK, 2015.
14. Diefenbacher, H. Ten theses on a research agenda for sustainable development. In *Theories of Sustainable Development*; Enders, J., Remig, M., Eds.; Routledge: London, UK, 2015; pp. 188–194.
15. Kester, K. The contribution (or not) of UN higher education to peacebuilding: An ethnographic account. *Glob. Soc. Educ.* **2017**, *15*, 464–481. [CrossRef]
16. Craft, A.; Cremin, T.; Hay, P.; Clack, J. Creative primary schools: Developing and maintaining pedagogy for creativity. *Ethnogr. Educ.* **2014**, *9*, 16–34. [CrossRef]
17. Wiek, A.; Withycombe, L.; Redman, C.L. Key competencies in sustainability: A reference framework for academic program development. *Sustain. Sci.* **2011**, *6*, 203–218. [CrossRef]
18. Wowk, K.; McKinney, L.; Muller-Karger, L.; Moll, R.; Avery, S.; Escobar-Briones, E.; Yoskowitz, D.; McLaughlin, R. Evolving academic culture to meet societal needs. *Palgrave Commun.* **2017**, *3*. [CrossRef]
19. Biberhofer, P.; Rammel, C. Transdisciplinary learning and teaching as answers to urban sustainability challenges. *Int. J. Sustain. High. Educ.* **2017**, *18*, 63–83. [CrossRef]

20. Remington-Doucette, S.M.; Connell, K.Y.H.; Armstrong, C.M.; Musgrove, S.L. Assessing sustainability education in a transdisciplinary undergraduate course focused on real-world problem solving: A case for disciplinary grounding. *Int. J. Sustain. High. Educ.* **2013**, *14*, 404–433. [CrossRef]
21. Rieckmann, M. *Education for Sustainable Development Goals: Learning Objectives*; UNESCO: London, UK, 2017.
22. Wiek, A.; Withycombe, L.; Redman, C.; Mills, S.B.; Banas, S. Moving Forward on Competence in Sustainability Research and Problem Solving. *Environment* **2011**, *53*, 3–13. [CrossRef]
23. Rieckmann, M. Future-oriented higher education: Which key competencies should be fostered through university teaching and learning? *Futures* **2012**, *44*, 127–135. [CrossRef]
24. Steiner, G.; Posch, A. Higher education for sustainability by means of transdisciplinary case studies: An innovative approach for solving complex, real-world problems. *J. Clean. Prod.* **2006**, *14*, 877–890. [CrossRef]
25. Wiek, A.; Bernstein, M.J.; Laubichler, M.; Caniglia, G.; Minteer, B.; Lang, D.J. A Global Classroom for International Sustainability Education. *Creat. Educ.* **2013**, *4*, 19–28. [CrossRef]
26. Karvinen, M.; Vehmaa, A.; Keskinen, M. Muuttuvien työelämätaitojen sisällyttäminen tekniikan alan koulutukseen: Tapaustutkimus Aalto-yliopiston vesi- ja ympäristötekniikan maisteriohjelmasta. *Yliopistopedagogiikka* **2019**, *2019*, 20–41.
27. Lang, D.J.; Wiek, A.; Bergmann, M.; Stauffacher, M.; Martens, P.; Moll, P.; Swilling, M.; Thomas, C.J. Transdisciplinary research in sustainability science: Practice, principles, and challenges. *Sustain. Sci.* **2012**, *7*, 25–43. [CrossRef]
28. Spelt, E.J.H.; Biemans, H.J.A.; Tobi, H.; Luning, P.A.; Mulder, M. Teaching and learning in interdisciplinary higher education: A systematic review. *Educ. Psychol. Rev.* **2009**, *21*, 365–378. [CrossRef]
29. Kates, R.W.; Clark, W.C.; Corell, R.; Hall, J.M.; Jaeger, C.C.; Lowe, I.; McCarthy, J.J.; Schellnhuber, H.J.; Bolin, B.; Dickson, N.M.; et al. Environment and development: Sustainability science. *Science* **2001**, *292*, 641–642. [CrossRef] [PubMed]
30. Brint, S.G.; Turk-Bicakci, L.; Proctor, K.; Murphy, S.P. Expanding the social frame of knowledge: Interdisciplinary, degree-granting fields in american colleges and universities, 1975–2000. *Rev. High. Educ.* **2009**, *32*, 155–183. [CrossRef]
31. Klein, J.T.; Newell, W.H. Advancing Interdisciplinary Studies. In *Handbook of the Undergraduate Curriculum: A Comprehensive Guide to Purposes, Structures, Practices, and Change*; Gaff, J., Ratcliff, J., Eds.; Jossey-Bass: San Francisco, CA, USA, 1997; pp. 393–415.
32. Miller, M.A.; Mccartan, A.-M. At the Crossings: Making the Case for New Interdisciplinary Programs. *Chang. Mag. Higher Learn.* **1990**, *22*, 28–35. [CrossRef]
33. Thomas, I. Critical Thinking, Transformative Learning, Sustainable Education, and Problem-Based Learning in Universities. *J. Transform. Educ.* **2009**, *7*, 245–264. [CrossRef]
34. Greenhalgh-Spencer, H.; Frias, K.; Ertas, A. Transdisciplinary content pedagogy in undergraduate engineering education: Being pulled up short. In *Transdisciplinary Higher Education: A Theoretical Basis Revealed in Practice*; Springer International Publishing: New York, NY, USA, 2017; pp. 73–89, ISBN 9783319561851.
35. Howlett, C.; Ferreira, J.A.; Blomfield, J. Teaching sustainable development in higher education: Building critical, reflective thinkers through an interdisciplinary approach. *Int. J. Sustain. High. Educ.* **2016**, *17*, 305–321. [CrossRef]
36. Paul, R. Critical thinking in every domain and belief. In Proceedings of the 27th Annual International Conference on Critical Thinking, Berkeley, CA, USA, 23–26 July 2007.
37. Lozano, R.; Merrill, M.; Sammalisto, K.; Ceulemans, K.; Lozano, F. Connecting Competences and Pedagogical Approaches for Sustainable Development in Higher Education: A Literature Review and Framework Proposal. *Sustainability* **2017**, *9*, 1889. [CrossRef]
38. Dewey, J. *How We Think*; D. C. Heath and Company: Boston, MA, USA, 1910.
39. Rodgers, C. Defining Reflection: Another Look at John Dewey and Reflective Thinking. *Teach. Coll. Rec.* **2002**, *104*, 842–866. [CrossRef]
40. Wals, A.E.J. Between knowing what is right and knowing that is it wrong to tell others what is right: On relativism, uncertainty and democracy in environmental and sustainability education. *Environ. Educ. Res.* **2010**, *16*, 143–151. [CrossRef]
41. UNESCO. *Teaching and Learning for a Sustainable Future: A Multimedia Teacher Education Programme*; UNESCO: Paris, France, 2002.
42. Cranton, P. *Understanding and Promoting Transformative Learning*, 2nd ed.; Jossey-Bass: San Francisco, CA, USA, 2006.
43. Hajer, M.; Versteeg, W. Imagining the post-fossil city: Why is it so difficult to think of new possible worlds? *Territ. Polit. Gov.* **2019**, *7*, 122–134. [CrossRef]
44. Korobar, V.P.; Siljanoska, J. Challenges of teaching sustainable urbanism. *Energy Build.* **2016**, *115*, 121–130. [CrossRef]
45. Li, N.; Chan, D.; Mao, Q.; Hsu, K.; Fu, Z. Urban sustainability education: Challenges and pedagogical experiments. *Habitat Int.* **2018**, *71*, 70–80. [CrossRef]
46. Wiek, A.; Kay, B. Learning while transforming: Solution-oriented learning for urban sustainability in Phoenix, Arizona. *Curr. Opin. Environ. Sustain.* **2015**, *16*, 29–36. [CrossRef]
47. Molnar, C.; Ritz, T.; Heller, B.; Solecki, W. Using Higher Education-Community Partnerships to Promote Urban Sustainability. *Environ. Sci. Policy Sustain. Dev.* **2010**, *53*, 18–28. [CrossRef]
48. Keeley, M.; Benton-Short, L. Holding complexity: Lessons from team-teaching an interdisciplinary collegiate course on urban sustainability. *Soc. Sci.* **2020**, *9*, 76. [CrossRef]
49. Brewer, G.D. The Challenges of Interdisciplinarity. *Policy Sci.* **1999**, *32*, 327–337. [CrossRef]
50. Bursztyn, M.; Drummond, J. Sustainability science and the university: Pitfalls and bridges to interdisciplinarity. *Environ. Educ. Res.* **2014**, *20*, 313–332. [CrossRef]

51. Brodin, E.M.; Avery, H. Cross-Disciplinary Collaboration and Scholarly Independence in Multidisciplinary Learning Environments at Doctoral Level and Beyond. *Minerva* **2020**, *58*, 409–433. [CrossRef]
52. Foley, G. Reflections on interdisciplinarity and teaching chemical engineering on an interdisciplinary degree programme in biotechnology. *Educ. Chem. Eng.* **2016**, *14*, 35–42. [CrossRef]
53. Lindvig, K.; Lyall, C.; Meagher, L.R. Creating interdisciplinary education within monodisciplinary structures: The art of managing interstitiality. *Stud. High. Educ.* **2019**, *44*, 347–360. [CrossRef]
54. Lambrechts, W.; Mulà, I.; Ceulemans, K.; Molderez, I.; Gaeremynck, V. The integration of competences for sustainable development in higher education: An analysis of bachelor programs in management. *J. Clean. Prod.* **2013**, *48*, 65–73. [CrossRef]
55. Bina, O.; Balula, L.; Varanda, M.; Fokdal, J. Urban studies and the challenge of embedding sustainability: A review of international master programmes. *J. Clean. Prod.* **2016**, *137*, 330–346. [CrossRef]
56. Argento, D.; Einarson, D.; Mårtensson, L.; Persson, C.; Wendin, K.; Westergren, A. Integrating sustainability in higher education: A Swedish case. *Int. J. Sustain. High. Educ.* **2020**, *21*, 1131–1150. [CrossRef]
57. Barth, M. *Implementing Sustainability in Higher Education: Learning in an Age of Transformation*; Routledge: London, UK, 2014.
58. Balsiger, J. Transdisciplinarity in the class room? Simulating the co-production of sustainability knowledge. *Futures* **2015**, *65*, 185–194. [CrossRef]
59. Filho, W.L.; Manolas, E.; Pace, P. The future we want key issues on sustainable development in higher education after Rio and the UN decade of education for sustainable development. *Int. J. Sustain. High. Educ.* **2015**, *16*, 112–129. [CrossRef]
60. Ceulemans, K.; De Prins, M.; Cappuyns, V.; De Coninck, W. Integration of sustainable development in higher education's curricula of applied economics: Large-scale assessments, integration strategies and barriers. *J. Manag. Organ.* **2011**, *17*, 621–640. [CrossRef]
61. Neuhauser, L.; Pohl, C. Integrating transdisciplinarity and translational concepts and methods into graduate education. In *Transdisciplinary Professional Learning and Practice*; Springer: Cham, Switzerland, 2015; pp. 99–120.
62. Christiansen, L.; Fischer, N. Teaching Urban Sociology and Urban Sustainability on Two Feet, Two Wheels, and in Three Cities. *Teach. Sociol.* **2010**, *38*, 301–313. [CrossRef]
63. Sennet, R. *Building and Dwelling: Ethics for the City*; Penguin Books: London, UK, 2019.

sustainability MDPI

Review

Social Inclusion Indicators for Building Citizen-Centric Smart Cities: A Systematic Literature Review

Jalaluddin Abdul Malek [1], Seng Boon Lim [1,*] and Tan Yigitcanlar [2]

1 Social, Environmental and Developmental Sustainability Center (SEEDS), Faculty of Social Sciences and Humanities, University Kebangsaan Malaysia, Bangi 43600, Selangor, Malaysia; jbam@ukm.edu.my

2 School of Built Environment, Queensland University of Technology, 2 George Street, Brisbane 4000, QLD, Australia; tan.yigitcanlar@qut.edu.au

* Correspondence: lims@ukm.edu.my

Abstract: Despite the rhetoric of "citizen-first," which has been tokenized in recent years by the smart city administrations, what it means has long been unclear to many, including the public at large. Put simply, this rhetoric concerns the mindset of the members of a local community and places them at the heart of the smart city initiatives. In order to bring further clarity to this issue under the current neoliberal urbanism, this study aimed to identify the key indicators of citizen-centric smart cities from the perspective of participative governance practices and citizens' responsibilities. To achieve this aim, this study involved a systematic literature review of the social inclusion indicators for building citizen-centric smart cities. The social inclusion indicators that were formed were verified by practitioners to suit the local contexts of an emerging and developing country, in this case, Malaysia. The findings of the review revealed that: (a) the acceptance of social inclusion indicators was mainly limited to the realm of democratic developed countries, leaders' understanding of citizenship, the delegation of decision-making power in governance practices, the participative culture of societies, and individual citizens' self-discipline; (b) the social inclusion indicators may not be welcomed in emerging and developing countries; (c) in the long term, these indicators would shed light on the rise of self-organizing cities that will gain popularity in potential city developments, be it in developed or developing countries.

Keywords: citizen centrism; citizen-centric smart cities; neoliberal urbanism; public participation; participative governance; participatory planning; right to the city; smart city; smart citizenship; social inclusion indicator; sustainable urban development

Citation: Malek, J.A.; Lim, S.B.; Yigitcanlar, T. Social Inclusion Indicators for Building Citizen-Centric Smart Cities: A Systematic Literature Review. *Sustainability* **2021**, *13*, 376. https://doi.org/10.3390/su13010376

Received: 11 October 2020
Accepted: 17 November 2020
Published: 4 January 2021

1. Introduction

To date, considering citizens' perceptions about and perspectives of smart city development is seen as a sound strategy for many political and administrative leaders. Particularly, this has taken the form of promoting eGov (citizen centricity in e-government) that has been upheld in Europe since the mid-2000s [1] and is rooted in the perspective of "citizens as customers" under the new public management [2]. Based on this influence, apart from technological needs or smart cities, in recent years, city administrators have shifted their focus to co-creating smart cities with their citizens [3–6].

The rhetorical smart city visions in emerging and developing countries [7,8], such as the slogans of the federal government of Malaysia and the state government of Selangor's "Peduli Rakyat" (literally care for citizens) [9], have rightly inspired and motivated the general public, who are entirely depending on government resources or actions. Nevertheless, the targeted passive users, beneficiaries, or the public are unaware of their responsibilities, even though "citizen-centric smart city initiatives are rooted in stewardship, civic paternalism, and a neoliberal conception of citizenship" [10]. These neoliberal conceptions "prioritized choice of consumption and individual autonomy within a framework of state and corporate-defined constraints that focused on market-led solutions to urban issues,

Nevertheless, even the citizens' involvement in the consultation process is considered an act of tokenism [52], where the power of decision-making would not be truly delegated to the people [53].

Notwithstanding the aforementioned, the challenges in building participatory governance are mounting. First, only the emphasis on technological corporate factors in smart governance has been criticized for failing to solve the issues concerning smart cities [10,26]. Beyond this, the human factor involved in the smart city program needs serious consideration [38,54] and is seen as a "critical intervention" in a dominant type of corporate smart city [55]. Second, the fundamental role that can be played by people remains vague, as citizens are often regarded as passive users whose opinions are not taken seriously [49,56,57].

Third, global development agendas, such as the New Urban Agenda (NUA) pioneered by the United Nations (UN), has lauded the importance of citizen involvement and inclusion to all parties [58,59]. Nonetheless, the current situation has been criticized by many parties regarding the fact that corporate smart cities are against the current global agenda and have often manipulated the issue of people's involvement and popularized social polarization [60–63]; furthermore, parties have also criticized the NUA framework and emphasized that smart cities would fail if the community refused to get involved [64]. Communities' refusal to participate is most likely due to a lack of understanding in terms of the involvement, type, or process contained in authentic engagement matters [65]. As such, [64] also suggested "dissensus" as a living indicator. This proposal is against the current practice of "building consensus," where the different opinions of the people should be considered, even if the consultation process is "painstaking."

In Malaysia, the planned development of smart cities, such as Cyberjaya, which is located within the Multimedia Super Corridor (MSC), is often criticized by scholars due to a lack of participatory governance [66–68]. On the other hand, existing planned development, such as Petaling Jaya City, is now integrated into the Smart Selangor Blueprint and faces challenges in terms of coordinating the people's role in the new smart city and Local Agenda 21 projects that were launched two decades ago [9]. The main challenge faced by both types of smart city governances in Malaysia is the lack of local context indicators for implementing the element of involvement and the people's role if there is an intention to develop toward citizen-centric city development.

2.3. Theoretical Framework

Two schools of thought influenced the development of smart city scholarship, namely, the technology-driven method and the human-driven method [26]. For the smart city concept, two elements were highlighted by the seminal work of [35], which are related to this study's problems, namely, smart people and smart governance. Furthermore, the three main parties supporting the success of smart cities are the authorities, technological corporations, and citizens, as outlined by [48].

On another note, two citizen-centric ideologies were formulated by [1,36] as follows: (a) "to the people," such as authorities using technology to meet the needs of the people, and (b) "with the people," in terms of the collective thinking of the authorities, technological corporations, and the people when resolving urban issues. By combining both concepts of smart and citizen-centric cities, a basic understanding of the citizen-centric concept in smart cities could be developed.

Furthermore, to develop the concept of "citizen centricity," it was found that the understanding, types, and processes of involvement were essential aspects worthy of being studied. In addition to involvement, the literature on citizenship, other than salvaging citizens' rights as studied by [13,17], were not included in this study, as the people's responsibilities are important in building a smart city [14,56]. This issue of responsibility can be divided into the roles and characters of the people. Based on these literature findings, a theoretical framework (Figure 1) was formed. In the context of this study, a "citizen-centric smart city" is a concept based on humanism with a focus on the participation and responsibilities of the people.

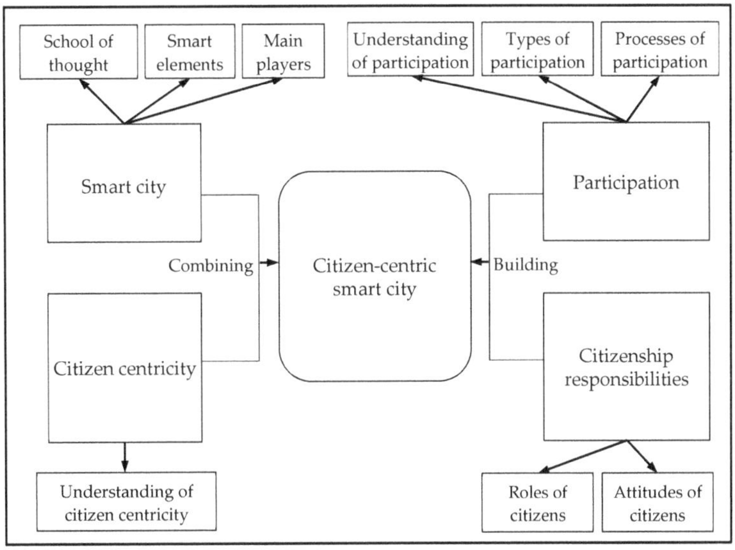

Figure 1. Theoretical framework (source: authors). Smart city [35,48]; Participation [52,69,70]; Citizen-centric smart city [28,71]; Citizen centricity [1,36]; Citizenship responsibilities [14,72].

2.4. Conceptual Framework

Based on the conceptualization of items through literature reviews and the verification of items by practitioners, the study's conceptual framework was formed (Figure 2). This conceptual framework consisted of a construct of a dependent variable—(DV 1) a citizen-centric smart city—and five independent variable constructs—(IV 1) understanding of participation, (IV 2) type of participation, (IV 3) process of participation, (IV 4) role of citizens, and (IV 5) attitude of citizens. This conceptual framework and the questionnaire items were recommended for testing with a five-point Likert scale (refer to [73]), with further quantitative analysis, such as a regression analysis to be conducted.

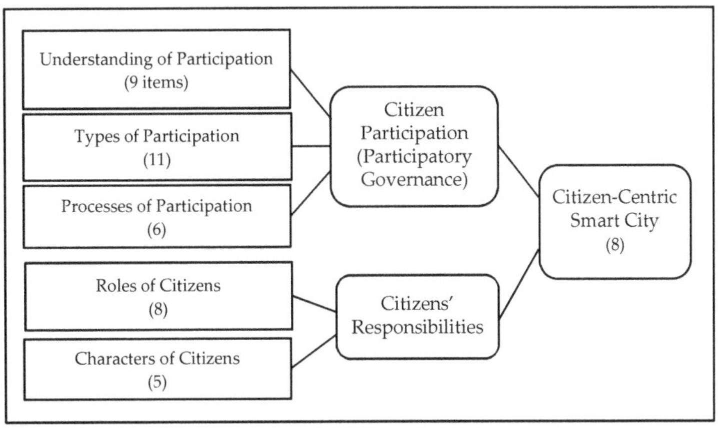

Figure 2. Conceptual framework for a citizen-centric smart city (source: authors).

The authors argued that in constructing a citizen-centric smart city framework, the involvement of urban stakeholders is especially important from the people's perspective, as the citizens require an understanding of the meanings, types, and processes of participation, and the people's role and attitude in accordance to the type of citizen-centric development.

3. Methodology

In expanding the theoretical framework to measurable indicators and answering the research question, this study utilized a systematic literature review method, which practitioners verified at the second stage.

3.1. *Systematic Literature Review*

A complete literature search should not be limited to a single knowledge channel, methodology, journal, or geographical region, but the search should cover all the aspects related to the research topic [74]. Therefore, a systematic literature review method was selected due to the numerous unique procedures in this study. The unique procedures emphasized transparency, a pre-defined research question and protocol, pre-defined search strings, and standard inclusion and exclusion criteria, and justified the data analysis [75]. This study aimed to answer the research question: "What are the key social inclusion indicators supporting the emergence of a citizen-centric smart city?", as elaborated through the research protocols on the subsections below.

3.1.1. Search Strategy

This study used Google Scholar as a database, as it is more open and inclusive compared to other platforms, such as WoS and Scopus. This database was also found to be used by other smart city scholars, such as [76–78]. Through the platform of Google Scholar, the authors were able to refer to articles and other research sources, such as conference proceedings, theses, books, book chapters, reports, and qualified working papers.

The comprehensive search strings developed by the authors were the keywords and their analogous concepts included "smart city(ies)," "citizen-centric," "citizen centricity," "people-centered (centred)," "citizen participation," and "citizenship responsibility(ies)." Unlike [75], the search strings were applied individually without using "OR" as a one-off search. The authors found that too many results were generated under an individual search. Therefore, repeated individual searches were more organized regarding converting the results to a master spreadsheet for data synthesis and analysis.

3.1.2. Inclusion and Exclusion Criteria in the Stage of Identification

According to [79], the preferred reporting items for systematic reviews include the stages of identification, screening, eligibility, and inclusion. In the first stage, namely, identification, five inclusion and exclusion criteria were established to guide the authors when gathering the relevant studies for this review. First, both the boxes of "include patents" and "include citations" were unchecked on the Google Scholar page. Patents were irrelevant to this academic study, while citations were found to be repetitive and challenging regarding finding online sources for further investigation.

Second, the "Advanced Search" setting of "Find Articles" was utilized. Articles with the *exact phrase* were identified and the selected keyword strings were keyed in, as mentioned earlier. The selection of "with *all* of the words," "with *at least one* of the words," and "*without* the words" were ignored, as the search results would be too general to review. Third, only the word occurring *in* the article title but not *anywhere* was identified. This selection also narrowed down the search results. Fourth, both the boxes of "return articles *authored* by" and "return articles *published* in" were ignored, with no limitation to authors and publishers.

Finally, the first box for the search period was left blank, with "2017" written down in the second box. Though the search period included up to December 2017, the authors did not set a specific limit for the beginning of the search period. Unlike [76–78], the initial search periods were set in 1992, 1993, and 1997, in line with the argument that these years appeared in the literature that is relevant to smart city concepts. Nonetheless, this study revealed that the topic of citizen centricity in smart cities was relatively discursive and multidisciplinary, with insufficient and reliable sources for an estimated period of two decades. Furthermore, the authors intended to acquire as many articles as possible to define the indicators. The detailed numbers of exclusion records in the first stage of identification are shown in Table 1.

Table 1. Records of patents and citations that were excluded in the identification stage (source: authors).

Keywords	Search	Exclusion of Patents and Citations	Records after Patents and Citations Were Removed
Smart city	6690	2420	4270
Smart cities	6390	2320	4070
Citizen-centric	378	127	251
Citizen centricity	8	4	4
People-centered	652	339	313
People-centred	632	333	299
Citizenship responsibility	36	16	20
Citizenship responsibilities	30	9	21
Citizen participation	6210	3170	3040
Total	21,026	8738	12,288

3.1.3. Data Extraction in the Screening Stage

All the studies found through the initial search process using the selected keyword strings were incorporated in a master Excel worksheet. In the second stage of screening, first, duplicates were deleted. As the Google Scholar database was algorithmically autogenerated, the duplications needed to be manually deleted by filtering the worksheet. Furthermore, paper exclusions concerning absent or irrelevant title sources (i.e., PowerPoint presentations, white papers, book introductions, calls for papers, competition announcements, and all non-English works) were performed.

3.1.4. Data extraction in the Eligibility Stage

In stage 3 (eligibility), justified full-text article exclusions were performed, where we read the article titles and keywords. For example, the most-cited source document of [35] in smart city literature outlined the divergent roots of smart city development, including the elements of a smart economy, people, governance, mobility, environment, and living, thus indicating the diversity of the various smart city papers in the Google Scholar database.

In answering the research question concerning the identification of the indicators of citizen centricity in smart cities, the scope was further narrowed down to papers involving smart people and governance. Nonetheless, both topics were scarce, with there being more literature on ICT-related urban innovations [80]. The consequent exclusion of papers concerning smart economy, mobility, living, and environment drastically reduced the number of eligible papers to 2350 studies.

Furthermore, the papers involving citizen centricity in smart cities were carefully assessed by reading the abstracts and contents, such as introductions and conclusions. In this regard, the irrelevant abstracts and contents were identified, with the number of papers included in the qualitative synthesis narrowed down to 71 studies.

3.1.5. Data Extraction in the Inclusion Stage

In stage 4 (inclusion), the authors performed an additional backward–forward search [74,81] on each identified article besides the identified studies from the Google Scholar database search.

Table 4. Items for the construct of "types of participation" (source: authors).

No.	Themes	Sources	Types of Participation Item
1	The power of the citizens is strong	[52,69,71,121,132,133]	We (I and the community) have full control over smart city programs.
2	Delegation of authority for mutual benefit	[52,109,112,116,121,129,130,132]	LAs delegate the power to us to make decisions that benefit the community.
3	Consultation enables citizens to achieve dominant decision-making power	[49,52,69,93,109,112,121,122,124, 125,127,130]	Through consultation, we managed to reach the final word (decision) in smart city programs.
4	Citizens are allowed to co-produce	[49,52,69,70,92,105,121,130]	We are allowed by LAs to jointly run smart city programs.
5	Offering grants (financial incentives)	[52,69,112,116,128,129]	LAs offer grants (financial incentives) to run smart city programs.
6	Offering rewards (material incentives)	[52,69,116,128]	LAs offer rewards, but LAs are still in full power.
7	Conducting questionnaires	[52,69]	LAs conduct questionnaires related to our perception of smart city programs.
8	Hold community meetings	[52,116,122,132]	LAs hold meetings with the community.
9	Inform (educate) through information releases	[49,52,69,92,103,104,116,121,123–125,130]	LAs broadcast the correct information to us.
10	Explain misunderstandings (therapy)	[52,69,100,108,114,116,132]	LAs explain misunderstandings of the smart city program to us.
11	Manipulating (emphasizing the purpose of administrative legitimacy, rather than the actual function of the people)	[52,100,114,115,128,129]	LAs manipulate or use us for a reason.

4.1.4. The Items for the Construct of Processes of Participation (IV 3)

The processes of participation refer to public engagement in the value chain of a program or activity from the initial process of drafting the agenda to the final evaluation process. Although the idea of the people involved in the value chain process of a city program was quoted from [70], there were no further explanations of the appropriate items. Thus, most of the relevant items were derived from [69].

In explaining the initial "processes" of participation in a program's value chain, citizens should be involved in formulating the agenda, decision-making, planning, and designing the program's content. Consequently, citizens should be involved in managing and implementing the programs, together with LAs, in the middle process. Finally, citizens should oversee and evaluate the program after the implementation. Table 5 summarizes the themes, sources, and questionnaire items in layperson's terms.

Table 5. Items for the "processes of participation" (source: authors).

No.	Themes	Sources	Processes of Participation Item
1	Formulate agendas, where the power of the people is limited	[56,69,117,118,133–137]	We are involved in formulating the program agenda.
2	Make the right decision	[56,69,111,116,125,133,134, 136–139]	We are involved in decision-making.
3	Plan program content	[52,69,70,98,116,121,125,140]	We are involved in planning the program's activities.
4	Design program details	[37,70,120,141]	We are involved in designing the program's content.
5	Manage program processes	[37,69,70,94,140]	We are involved in managing the program's processes.
6	Implement the program	[52,69,70,124,125,136,139,141]	We implement the program.
7	Monitor (supervise) the program's continuity	[70,135,141]	We oversee the course of the program.
8	Evaluate the program after its implementation	[69,70,100,125,139]	We evaluate the program after its implementation.

did not set a specific limit for the beginning of the search period. Unlike [76–78], the initial search periods were set in 1992, 1993, and 1997, in line with the argument that these years appeared in the literature that is relevant to smart city concepts. Nonetheless, this study revealed that the topic of citizen centricity in smart cities was relatively discursive and multidisciplinary, with insufficient and reliable sources for an estimated period of two decades. Furthermore, the authors intended to acquire as many articles as possible to define the indicators. The detailed numbers of exclusion records in the first stage of identification are shown in Table 1.

Table 1. Records of patents and citations that were excluded in the identification stage (source: authors).

Keywords	Search	Exclusion of Patents and Citations	Records after Patents and Citations Were Removed
Smart city	6690	2420	4270
Smart cities	6390	2320	4070
Citizen-centric	378	127	251
Citizen centricity	8	4	4
People-centered	652	339	313
People-centred	632	333	299
Citizenship responsibility	36	16	20
Citizenship responsibilities	30	9	21
Citizen participation	6210	3170	3040
Total	21,026	8738	12,288

3.1.3. Data Extraction in the Screening Stage

All the studies found through the initial search process using the selected keyword strings were incorporated in a master Excel worksheet. In the second stage of screening, first, duplicates were deleted. As the Google Scholar database was algorithmically autogenerated, the duplications needed to be manually deleted by filtering the worksheet. Furthermore, paper exclusions concerning absent or irrelevant title sources (i.e., PowerPoint presentations, white papers, book introductions, calls for papers, competition announcements, and all non-English works) were performed.

3.1.4. Data extraction in the Eligibility Stage

In stage 3 (eligibility), justified full-text article exclusions were performed, where we read the article titles and keywords. For example, the most-cited source document of [35] in smart city literature outlined the divergent roots of smart city development, including the elements of a smart economy, people, governance, mobility, environment, and living, thus indicating the diversity of the various smart city papers in the Google Scholar database.

In answering the research question concerning the identification of the indicators of citizen centricity in smart cities, the scope was further narrowed down to papers involving smart people and governance. Nonetheless, both topics were scarce, with there being more literature on ICT-related urban innovations [80]. The consequent exclusion of papers concerning smart economy, mobility, living, and environment drastically reduced the number of eligible papers to 2350 studies.

Furthermore, the papers involving citizen centricity in smart cities were carefully assessed by reading the abstracts and contents, such as introductions and conclusions. In this regard, the irrelevant abstracts and contents were identified, with the number of papers included in the qualitative synthesis narrowed down to 71 studies.

3.1.5. Data Extraction in the Inclusion Stage

In stage 4 (inclusion), the authors performed an additional backward–forward search [74,81] on each identified article besides the identified studies from the Google Scholar database search.

The backward–forward search (involving articles dated before or after the identified article) observed the detailed themes or indicators relevant to this study, as the potential article titles were indirectly related to citizen centricity and smart cities.

The authors considered the backward–forward search to be an important step in finding the themes or indicators that were directly related to answering the research question. Hence, five records were found in the backward search, whereas three studies were found in the forward search. Finally, a total of 79 articles were finalized for performing a thematic analysis. The search protocol process is presented in Figure 3 below.

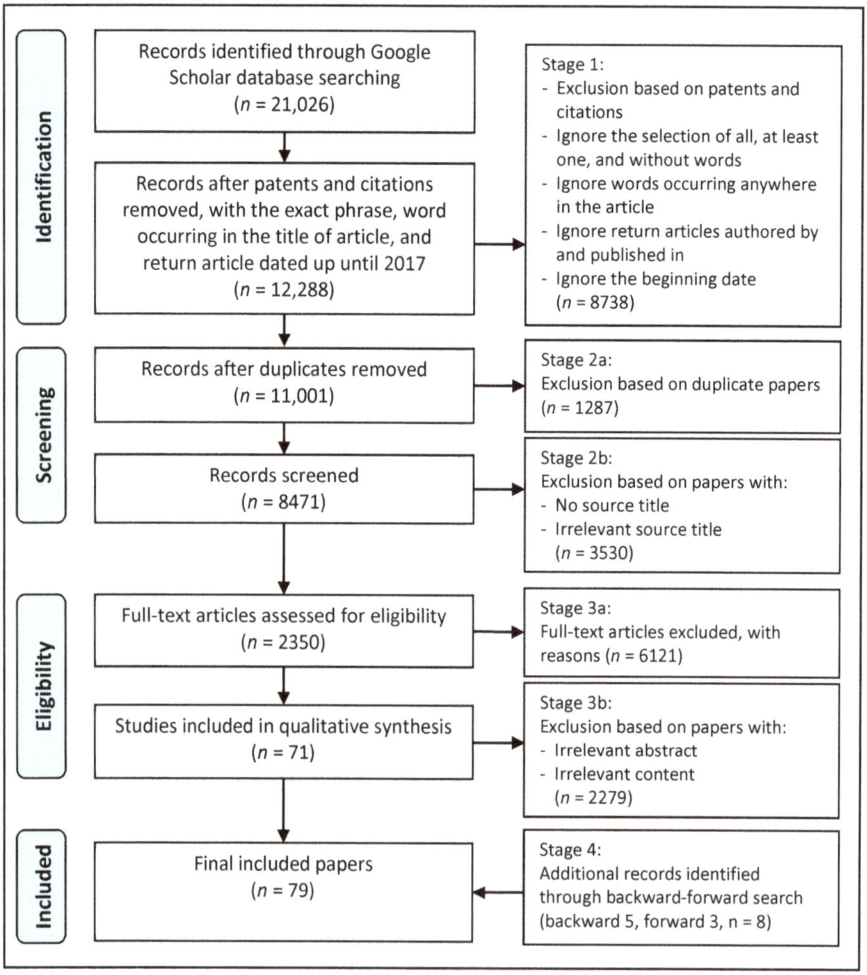

Figure 3. The execution stages of the systematic literature review (source: adapted from [79,82]).

3.1.6. Data Analysis and Risk of Bias

Based on the selected articles, the authors conceptualized the formation of items. A common name was used to describe similar concepts from various authors during the line-by-line coding of the thematic analysis process [74]. For example, "participation" was used as a common name to describe similar concepts of "engagement, collaboration and involvement" [83]. Figure 4 depicts the distribution of the reviewed papers per year, with the earliest article appearing in 1969, which was authored by [52], and the highest number of articles in a year was 14 articles in 2016.

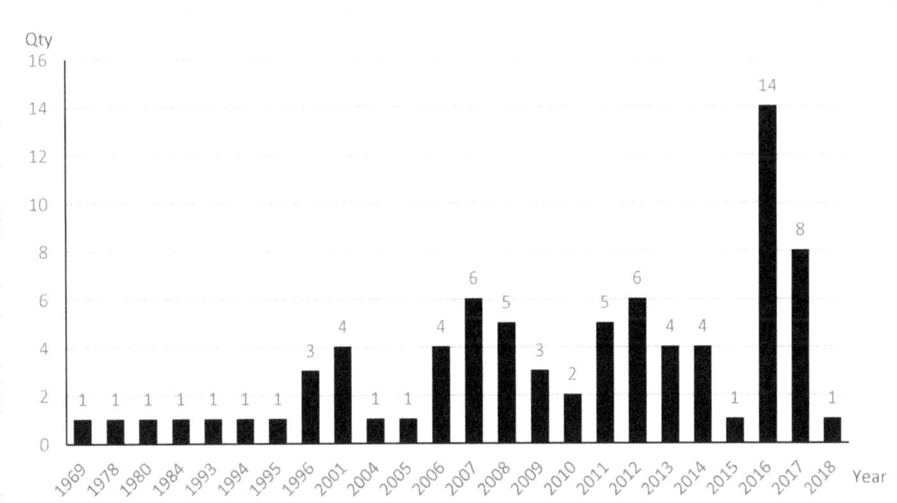

Figure 4. The distribution of the reviewed papers per year (source: authors).

Furthermore, it was discovered that the four most selected publishers were Wiley-Blackwell (nine articles), Springer (eight articles), Taylor & Francis (six articles), and Elsevier (six articles). Organizations such as the OECD, the World Bank, and universities also contributed to this group of articles (Appendix **??**).

According to [79], no guarantees were provided on the assessment or interpretation of the appropriate content by systemic reviewers. Thus, the authors aimed to report the possible risk of bias across selected studies, depending on the qualitative judgment from the authors when interpreting the contents and extracting suitable themes or items to answer the research question.

3.2. *Verification by Practitioners*

With the structuring of indicators through the rigorous process of the systematic literature review, all the indicators were verified again by practitioners in the fields of smart city and participation. The verification process also assisted in reducing the risk of bias from qualitative judgment by the authors. The selected practitioners consisted of two groups, which were the "power holders" with 19 people (11 local authority officers, 5 federal or state officers, and 3 politicians), and the "have-not citizens" as the remaining half (8 representatives of community organizations or residents, 2 non-governmental organizations, 3 academicians, and 6 private sector representatives). Overall, the selected informants provided ideas and suggestions to validate the studied items.

Nonetheless, some practitioners were uneasy about how ordinary citizens could understand the technical terms in the verification process. With such limitations, more examples relevant to common terms were applied in the items, such as grants, as opposed to contracts in the form of finance in running city programs (refer to Appendix **??** for the terms adjusted by practitioners).

4. Findings

4.1. Systematic Literature Review

4.1.1. The Items for the Construct of a Citizen-Centric Smart City (DV 1)

In searching for and suggesting the items for building the study constructs, the authors found sources with few statistical tests. On the other hand, most of the sources were related to conceptual discussions. According to works of literature, the decisions made by local authorities (LAs) should be more focused on citizens' needs and not merely on technology to reach a consensus with citizens regarding realizing the state of citizen centricity in a smart city. LAs also learned to delegate power to citizens, specifically at the initial level of smart city programs. From the citizens' perspective, individuals should be free to participate, play voluntary roles, and continuously contribute information. Both the LAs and citizens should play a role in building a good relationship and understanding and trusting each other.

Table 2 was developed to show the comparisons between themes, collected sources, and items [84]. General terms were applied to explain similar concepts from the articles. The majority of the items were derived from [1,35,36]. The suggested items for measuring the construct of citizen centricity in smart cities were converted into understandable sentences.

Table 2. Items of "citizen centricity in a smart city" that was derived from the literature (source: authors).

No.	Themes	Sources	Citizen-Centric Smart City Item
1	Focus on citizens' needs, not just technology	[1,35,48,54–56,85–97]	The decision by the local authorities (LAs) is more focused on our needs (i.e., both mine and the community's) and not merely on technology.
2	Decision through consensus with citizens	[1,91,93,95,98–104]	The decisions by LAs need to reach a consensus with us.
3	Learn from users/citizens	[1,56,87,94,103–106]	LAs learn from users like us.
4	Power needs to be delegated	[1,35,36,103–105,107–110]	LAs delegate power to us, especially at the initial level of smart city programs.
5	Freedom to participate	[1,35–37,92,100,103,104,107,110–112]	We are free to involve ourselves in any smart city programs.
6	Volunteers needed	[35,36,113]	We play the role of volunteers and contribute information continuously.
7	Build good relationships	[1,100,101,109,114,115]	We understand each other and build a good relationship with LAs.
8	Mutual trust	[1,93,108,110,111,114–118]	We and LAs trust each other.

The design of the items was geared toward the citizens as the respondents. Thus, from the perspective of citizens, the term "we" was chosen and used with caution in representing "I and the community." Respondents were guided to answer the survey according to the "ideal" situation, but not in the "existing" situation that occurs in reality. Highlighting this point is essential to steer the respondents to answer the survey as objectively as possible and think collectively in terms of personal opinions and the community's perceptions of the respondents.

4.1.2. The Items for the Construct of Understanding of Participation (IV 1)

An understanding of citizen participation is considered important and influences the effectiveness of citizens' engagement in city programs [37,119]. This construct attempted to measure the level of citizens' understanding of the participation concept from the perspective of the citizens.

As such, citizens should have a clear understanding of the objective or aim and be aware of the benefits and obstacles of participating in smart city programs. Furthermore, citizens should be confident in playing relevant roles, evaluating the available options, and choosing to reject any programs that are deemed to be inappropriate. Furthermore, citizens should have the desire to influence priorities, attend the programs without going through a representative, and assist in forming the goals and objectives of smart city programs beyond mere participation. Citizens should also understand that the responsibility to make a communal decision and sign an agreement with the LAs, which is beneficial to the community, depends on the citizens.

The themes, primary references, and items in the sentences for the construct of "understanding of participation" are displayed in Table 3.

Table 3. Items for the construct of "understanding of participation" (source: authors).

No.	Themes	Sources	Understanding of the Participation Item
1	Clear understanding of the objective of participation	[98,119–126]	We have a clear understanding of the aim of participating in smart city programs.
2	Be aware of the benefits and obstacles	[52,69,127]	We are aware of the benefits and obstacles to participating in smart city programs.
3	Convinced of your role	[69,102,126,128]	We are confident about the role we can play in smart city programs.
4	Unaware of the minimum rights, responsibilities, and choices	[52,128–130]	We evaluate the options available and choose to reject any programs that are deemed to be inappropriate.
5	Desire to influence the priorities	[52,116,120]	We influence the priorities of the programs to be implemented.
6	Focus on non-superficial involvement	[36,52,100,130]	We are involved in the programs without going through a representative.
7	Help set goals	[70,120]	We assist in the formation of the goals and objectives of the smart city programs.
8	Joint decision-making procedure	[52,105,131]	We evaluate the decision-making procedure together.
9	Power-sharing agreement	[52]	We sign an agreement with the LAs, which is beneficial to the community.

4.1.3. The Items for the Construct of Types of Participation (IV 2)

Type of participation refers to the level or stage of participation. This classification distinguished the approach of participation and the distribution of power, where there may be a co-occurrence without a precise point at the beginning or end. This construct attempted to measure the differences at the level or stage of participation, primarily known as the [52] participation ladder.

The highest level of "citizen-power" participation entailed that citizens should ideally have complete control over smart city programs or delegated power to make decisions benefitting the community. Through consultations, citizens should reach the final word (decision) and allow for the joint management of smart city programs. For the middle level of encouraging "token" participation, LAs offered grants (financial incentives), rewards, and conducted questionnaires related to citizens' perception of smart city programs. For the lowest level of "non-participation," LAs held communal meetings and broadcasted accurate information to citizens.

Relevant themes were published in the literature, as shown in Table 4. It was revealed that various types of "actions" could be classified as involvement and was also a source of confusion for the LAs, citizens, and the writing of articles by the scholars or organizations concerned. Most importantly, the last item in the construct was deliberately designed as a negative item to test respondents who answered the questionnaire unethically.

Table 4. Items for the construct of "types of participation" (source: authors).

No.	Themes	Sources	Types of Participation Item
1	The power of the citizens is strong	[52,69,71,121,132,133]	We (I and the community) have full control over smart city programs.
2	Delegation of authority for mutual benefit	[52,109,112,116,121,129,130,132]	LAs delegate the power to us to make decisions that benefit the community.
3	Consultation enables citizens to achieve dominant decision-making power	[49,52,69,93,109,112,121,122,124, 125,127,130]	Through consultation, we managed to reach the final word (decision) in smart city programs.
4	Citizens are allowed to co-produce	[49,52,69,70,92,105,121,130]	We are allowed by LAs to jointly run smart city programs.
5	Offering grants (financial incentives)	[52,69,112,116,128,129]	LAs offer grants (financial incentives) to run smart city programs.
6	Offering rewards (material incentives)	[52,69,116,128]	LAs offer rewards, but LAs are still in full power.
7	Conducting questionnaires	[52,69]	LAs conduct questionnaires related to our perception of smart city programs.
8	Hold community meetings	[52,116,122,132]	LAs hold meetings with the community.
9	Inform (educate) through information releases	[49,52,69,92,103,104,116,121,123–125,130]	LAs broadcast the correct information to us.
10	Explain misunderstandings (therapy)	[52,69,100,108,114,116,132]	LAs explain misunderstandings of the smart city program to us.
11	Manipulating (emphasizing the purpose of administrative legitimacy, rather than the actual function of the people)	[52,100,114,115,128,129]	LAs manipulate or use us for a reason.

4.1.4. The Items for the Construct of Processes of Participation (IV 3)

The processes of participation refer to public engagement in the value chain of a program or activity from the initial process of drafting the agenda to the final evaluation process. Although the idea of the people involved in the value chain process of a city program was quoted from [70], there were no further explanations of the appropriate items. Thus, most of the relevant items were derived from [69].

In explaining the initial "processes" of participation in a program's value chain, citizens should be involved in formulating the agenda, decision-making, planning, and designing the program's content. Consequently, citizens should be involved in managing and implementing the programs, together with LAs, in the middle process. Finally, citizens should oversee and evaluate the program after the implementation. Table 5 summarizes the themes, sources, and questionnaire items in layperson's terms.

Table 5. Items for the "processes of participation" (source: authors).

No.	Themes	Sources	Processes of Participation Item
1	Formulate agendas, where the power of the people is limited	[56,69,117,118,133–137]	We are involved in formulating the program agenda.
2	Make the right decision	[56,69,111,116,125,133,134, 136–139]	We are involved in decision-making.
3	Plan program content	[52,69,70,98,116,121,125,140]	We are involved in planning the program's activities.
4	Design program details	[37,70,120,141]	We are involved in designing the program's content.
5	Manage program processes	[37,69,70,94,140]	We are involved in managing the program's processes.
6	Implement the program	[52,69,70,124,125,136,139,141]	We implement the program.
7	Monitor (supervise) the program's continuity	[70,135,141]	We oversee the course of the program.
8	Evaluate the program after its implementation	[69,70,100,125,139]	We evaluate the program after its implementation.

4.1.5. The Items for the Construct of Roles of Citizens (IV 4)

The role of the people was discussed as one of the important factors in building citizens' responsibilities in a citizen-centric smart city. Eight roles were identified in the construct items using language that is easily understood by the average respondents, as shown in Table 6.

Table 6. Items for the "roles of citizens" (source: authors).

No.	Theme	Source	Roles of Citizens Item
1	Leaders—lead local authorities to make decisions	[52,100]	As leaders, we lead LAs to make decisions.
2	Local champion—takes the initiative	[112,113,116,136]	As local champions, we take the initiative to start the program and connect to the relevant parties.
3	Co-producers—work together	[39,70,112,134,138,142]	As co-producers, we work together with LAs.
4	Entrepreneurs—bring economic innovation	[113,143]	As entrepreneurs, we bring economic and financial innovation to the community.
5	Solution proposers—advise and propose	[71,103,105,109,138,141,144,145]	As solution proposers, we suggest alternatives and advise LAs.
6	Human sensors—supply data, reports, or complaints	[38,56,57,71,72,100,141]	As human sensors, we contribute data.
7	Volunteers—contribute time and energy	[72,102,106,112,113,124,125,136,142,145]	As volunteers, we contribute time and help regardless of returns.
8	Experts—share competencies or experience	[52,72,100–102,106,108,116,132,136,146]	As experts, we contribute our expertise to help smart city programs.

The first role that ideal or radical citizens played in smart city programs was in leading authorities to make decisions [147]. The second role was as local champions, where citizens took the initiative to initiate the program and connect to relevant parties. The third role was as co-producers in collaboration with LAs. Furthermore, citizens could play the role of entrepreneurs in producing economic and financial innovations for the community. Citizens could also be solution proposers by suggesting alternatives and advising LAs. For the sixth role, citizens contributed data, consciously or subtly, as human sensors. Citizens could also volunteer to contribute time and help intrinsically. Finally, citizens contributed knowledge and expertise as experts. All the proposed citizen roles would eventually assist LAs with the shortage of human and financial resources involving smart city management.

4.1.6. The Items for the Construct on Characters of Citizens (IV 5)

Along with the people's responsibility for developing a citizen-centric city, people's attitudes were equally crucial to the study. Here, the researchers summarized the relevant themes and resources and synthesized the items that fit this study in Table 7.

Table 7. Items for the "characters of citizens" (source: authors).

No.	Theme	Source	Characters of Citizens Item
1	Active involvement is known to be important	[102,103,123,128,136,137,141,148]	We are active and care about each other.
2	Aware and sensitive to what is happening	[35,100,103,116,132,136,146,149]	We are aware and sensitive to the programs that take place in the city and the surroundings.
3	Independence in problem-solving	[35,38,117,118,129,131,136,137,149]	We choose not to rely on governmental resources.
4	Higher education allows for meaningful involvement	[100,103,108,109,136,149,150]	Being educated is important for us to get involved in city programs.
5	Interest in public life and public values	[5,35,37,38,48,101,151–153]	We are interested in public life, public values, and act quickly against things that disrupt community life.

The first attitude that citizens should possess is the active care of each other, such as neighbors and communities. Furthermore, citizens should pay attention and be sensitive to the programs occurring in the city and the surrounding environment. Citizens should be independent when problem-solving instead of merely relying on governmental resources. Being educated in colleges or universities would also allow for meaningful involvement. Finally, ideal citizens in smart cities should have an attitude that reflects an interest in public life, public values, and acting quickly against the disruptions in community life.

4.2. Verification by Practitioners

Once the six constructs and 49 items were derived from the literature, further verification of the feasibility was conducted by practitioners. The authors confirmed that the six constructs were deemed appropriate to explain a citizen-centric smart city model. However, the exact items needed some modifications, such as including the appropriate usage of words, sentences, and examples to guide respondents in answering the questionnaire with more accuracy [94]. In total, two of the original items were removed, and a total of 47 items were applied in the conceptual framework for future use in the instrument (Appendix **??**). The details of the items adjusted through the practitioners' verifications are presented in Appendix **??**.

5. Discussion and Reflection

The notion of citizen centricity in (smart) city development is not a novel one and is perceived as a continuous trend in the development of e-government, which started in the mid-2000s. It is in line with the concept derived from new public management, where citizens should be viewed as customers to improve public service delivery. As customers, the demands of services have implications that are likely to turn citizens into passive users or beneficiaries who receive and demand from public administrators. It was argued that in constructing a citizen-centric smart city (CCSC), there should be no generalizations in viewing citizens as customers. In fact, there is a need to seriously research and construct a CCSC from the perspectives of both citizenships (to explain the notion of citizen centricity) and literature on the conception of the smart city conception.

The theoretical framework of a CCSC formed in this study is the first structured framework of its kind in the literature on smart cities. This CCSC theoretical framework thoroughly explains the original combination of the source of references for the notion of citizen centricity and the smart city concept. From these combinations, it was revealed that the construction of such detailed indicators and citizenship conception should include three major components, namely, citizens' rights (not included in this study, as it was already detailed by [13,17]), citizens' responsibilities, and the practices of citizen participation in governance. This framework is unique, as the framework is viewed from a fundamental perspective of what citizens can contribute to the formation of a CCSC. The focus of the indicators originated from the perspective of the citizens rather than the government's point of view. Such a perspective would make the role of citizens proactive and similar to the conception of self-organizing cities [49,154–156], which developed beyond the current neoliberal smart cities.

From such a theoretical framework, this study has attempted to construct a conceptual framework consisting of six constructs and 47 items. All the constructs were verified carefully by 38 practitioners in the fields of smart city development and community participation. The first stage of the formation of indicators was derived from the majority of literature reviews from scholars in developed Western countries, hence outlining the holistic scope of citizenship notions. Moreover, practitioners from emerging and developing countries with comparatively lower democracy and citizenship conceptions, such as Malaysia, tended to agree with all the indicators from the Western literature and included additional examples from the local context to gain a better understanding of the residents. Thus, the design of such indicators could be said to suit both developed and developing countries. Concerning future implementations, the examples of particular items should be altered with caution to suit other local contexts.

The authors predict that further studies on empirical survey results could yield a less significant result on some indicators, as the mindset and acceptance of democracy and the rights in emerging and developing countries, such as Malaysia, could be lower compared to developed countries. Hence, the full acceptance of the indicators is not possible, as the democratic innovations of developing countries are faced with challenges and restrictions [157]. The proposed democratic innovation includes expanding the role of citizens as co-producers [158], which is rarely practiced in developing countries. Furthermore, with regard to the usage of political slogans in building a CCSC, leaders may appear insincere if people are tokenized as customers with needs to fulfill but not cultivated and given the opportunities to participate in governance practices. Thus, the authors are fully aware of the challenges and costs of participatory and deliberative governance [136,159]. However, to intervene in such neoliberal smart urbanism [160] and realize the possibility of "self-organizing" smart cities, these CCSC indicators are worthy of reference and can be modified in different contexts.

The potential self-organizing cities led by local stakeholders could emerge as responses to unsatisfactory government-driven processes, market failures [161], the intention to legitimize a government's retreat from sectors that have traditionally played a vital role [162], or the intervention of e-participation through digital technologies [163,164]. This self-organizing and more democratic realm of a CCSC has led to three levels of discussion. The first level is the democratic culture of a country, the leaders' understanding, and the delegation of decision-making power in governance. The second level is society's perception of citizenship, the participative culture of societies, and the lack of links to decision-making [165]. The third level is the individual citizens' discipline and contributions to the country or city.

For an emerging and developing country like Malaysia, the dual forms of Islamic and secular administrations and constitutions are often criticized by scholars in the context of democracy [166–168]. Such a context hints that the highest constitution is not as open to democracy as practiced by Western countries. Canada, for example, has the Citizenship Act, the Charter of Rights and Freedoms, and the Multiculturalism Act [169–171], as opposed to Malaysia. Another example is the Nordic welfare society in the context of a democratic culture, welfarism, and redistributive policies that provide support to the development of participatory and innovative platforms by strengthening social inclusion, regulating the growth mechanisms, and easing the tensions between pro-growth and anti-growth coalitions [143].

In Malaysia, "participatory governance practice" is a tokenized term under the current top-down policy governance practices. The majority of government administrators "listen" and act according to political masters, but less focus is given to grassroots suggestions [172,173]. The lack of participatory governance practices was also similar to other developing countries, such as India, China, and Egypt [23,117,174,175]. In developing India's 100 smart cities, [174] questioned the liberal electoral democracy in India on the extent to which a smart city can deliver de facto inclusion and participation. [176] added that, instead of testing Indian smart cities as the grounds for democratic participation, smart citizens were nudged as subaltern citizens in urban governance. In China's smart-eco city projects, [117,177] reported that citizen input in the decision-making phase was quite limited, hence suggesting legislative reforms and the professionalization of Chinese officials in dealing with bottom-up input. In Egypt, [175] recommended that the Egyptian government focus specifically on smart people, such as giving citizens equal opportunity to participate in public decision-making.

Suppose the future survey results of the proposed indicators in this study receive high acceptance. In that case, participatory governance may become a new norm in local governance and mark a transition from party politics, expert dominance, and siloed bureaucracy to citizen participation, consequently supporting citizens' efforts to co-produce public services and build potential self-organizing smart cities.

At the level of society, the culture of participation in government programs is considered to be low [9,178]. From studies based on the Petaling Jaya and Cyberjaya smart city cases [9], the low level of participation was not interpreted based on the moderate quantity of participative programs involving citizens in the implementation stages, but the interpretation was based on the particularly low (even none) quantity of programs that empowered citizens at the initial stage of decision-making. The situation in Malaysia resembles making "decision by decision," where the community has no liberty to decide, is constrained by decisions from authorities, and is at the mercy of the authorities [9]. Furthermore, as described by [162], in the context of Amsterdam and Amersfoort, The Netherlands, "self-organization seems to take place in the shadow of a government hierarchy: either a fear-based one or a benevolent one," particularly in the context of meta-governance. In the context of Helsinki, Finland, self-organization also lacked links to decision-making, thus constraining new solutions and creative actions [165].

Such contexts indicated that society's mindset is still conservative, with a vague understanding of the citizenship's regime, leading to a possibly high dependency of people on the government. The evidence in Malaysia, such as the withdrawal of participation from the Rome Statute of the International Criminal Court [179] and the human rights issues evoked by racially and religiously motivated political parties, had correctly signaled the relatively low appreciation of equality in human rights when race- and religion-based interests are challenged.

At the level of individual Malaysian citizens, people's self-discipline would increase with the realization of a CCSC. For example, the role of volunteers, local champions, and co-producers with characteristics of proactiveness and awareness of CCSC development are all important responsibilities that a citizen has to contribute to building a CCSC. In a potential majority of highly responsible citizens, this contributes to the building of sustainable and inclusive societies, cities, and a wider scope of progressiveness in Malaysia.

6. Conclusions

The construction of a CCSC is much like developing a democratic society, which requires a higher appreciation of a society with a wider citizenship regime and a self-disciplined and responsible culture for individual citizens. As technology has been identified as a mere catalyst in solving societal issues [180], the new main focus is now on the people (along with good policy) in rightly building a CCSC based on public values and upholding the realms of democracy and citizenship.

Nonetheless, [12] pointed out that hegemonic neoliberal urban growth is mostly incompatible with citizen-centered ideas. Thus, this study's limitation concerned the non-appreciation of neoliberal thinking parties (specifically ruthless capitalists and far-right and antidemocratic national and local authorities), who have become too used to and trapped under the mainstream of neoliberal smart urbanism. These parties may criticize and strongly oppose the viability of the proposed indicators and provide negative empirical results. Still, the authors encourage further investigations to apply the proposed indicators in the contexts of both developed and developing countries.

More information and refinements on the indicators will eventually yield the acceptance of the CCSC model. As such, the conception of a CCSC is the ideal state of building cities resembling "self-organizing" types, but with possibilities to be rejected by neoliberal-thinking leaders and administrators in practice. However, the authors strongly believe that in an estimated period of 30 to 50 years, the CCSC will receive wider acceptance with the emergence of drawbacks in neoliberal smart cities (see [57,181,182]). In the meantime, the smart cities' trend of branding, be it the alternative new brandings of cities or the conception of the citizen centricity perspective, will remain valid and strive for realization in a sustainable, just, and humane form of city development [183].

Author Contributions: Conceptualization and methodology, J.A.M. and S.B.L.; writing—original draft preparation, software, formal analysis, investigation, and data curation, S.B.L.; supervision, validation, resources, funding acquisition, and project administration, J.A.M.; review and editing, T.Y. All authors have read and agreed to the published version of the manuscript.

Funding: The article processing charges (APCs) were funded by the Malaysian Ministry of Higher Education (grant number FRGS/1/2019/SS06/UKM/02/2). The funder was not involved in the planning, execution, write-up of the systematic review, or other contents of this article.

Institutional Review Board Statement: Not applicable.

Informed Consent Statement: Informed consent was obtained from all subjects involved in the study.

Data Availability Statement: Data is contained within the article and appendixes.

Conflicts of Interest: The authors declare no conflict of interest.

Appendix A

Table A1. Literature discussing the direction of citizen centricity in smart cities' development (Source: Authors).

No.	Perspective	Methodology	Scholars and Descriptions
1	Technology-driven method: Stresses the importance of digital technology and big data in helping to attain citizen (centricity) participation in the smart city (services).	1(a) Conceptual discussions	[184]—To achieve the Dubai Happiness Agenda through digital technology, which is tokenized on the essence of participation. [185]—Using (customer relationship management) data analysis to measure the dimensions of smart city services that match the needs of the citizens (i.e., the needs for hedonic, contact, value co-creation, or a real-time response).
		1(b) Empirical proofs of the design of web/mobile urban service applications toward achieving a citizen-centeredness state.	[32]—Through the design of the IES (Internet Enabled Services) Cities platform to enable citizens to act as prosumers (double consumer and producer). [186]—Viewed citizens as primary beneficiaries and empowering citizens (deep participation * and data-literate citizenry) as challenges, thus proposed a solution of an Open City Toolkit (OCT) ** through GIScience and open data. * Deep participation is about raising awareness and enabling communities to have their say in matters related to city life. ** Examples of OCT can be accessed through http://geo-c.eu/opencitytoolkit (European Union's funding project). Related papers: [187,188]. [189]—Proposed an open model (open government, open innovation, open data, and open services paradigms) oriented toward the design, production, and deployment of public services and mobile apps ***, where citizens were viewed as co-creators of ideas. *** Examples of the WeLive platform can be accessed through https://www.welive.eu/ (European Union's funding project). Related papers: [190,191]. [192]—Proposed a citizen-centric typology for smart city services from marketing and service science, where citizens were viewed as users and customers. Others: [31,193–198].

Table A1. *Cont.*

No.	Perspective	Methodology	Scholars and Descriptions
2	Human-driven method: Stresses the importance of citizenship/social participation perspective in attaining the direction of citizen centricity in smart city development.	2(a) Conceptual discussions	Component of governance practice: [33]—Achieving Smart Nation Singapore through citizen-oriented smart city (governance) policies. Component of a general mention of citizenship: [34]—Stressed genuine citizen centricity will either happen when citizens were engaged in a partnership or delegated power and involved in the decision-making process in smart city programs. Others: [36–39].
		2 (b) Propose empirical measurements of citizen participation and citizens' rights in the smart city (standard).	[10]—Proposed a "Scaffold of Smart Citizen Participation" to assess the citizen-centric nature of smart city initiatives in Dublin. They concluded that these "citizen-centric" smart city initiatives were rooted in stewardship, civic paternalism, and a neoliberal conception of citizenship, rather than being grounded in civil, social, and political rights, and the common good. [13,17]—Proposed indicators for safeguarding citizens' rights. [35]—Proposed indicators for smart people and smart governance.

Note: * Deep participation is about raising awareness and enabling communities to have their say in matters related to city life; ** Examples of OCT can be accessed through http://geo-c.eu/opencitytoolkit (European Union's funding project); *** Examples of the WeLive platform can be accessed through https://www.welive.eu/ (European Union's funding project).

Appendix B

Table A2. List of selected reviewed articles (source: authors).

No.	Source's Title	Publisher	Publication	Author
1	Citizen-centric approaches to e-government and the back-office transformation	Association for Computing Machinery (ACM)	(Proceedings)	[85]
2	Conceptualizing smart city with dimensions of technology, people & institutions	ACM	(Proceedings)	[48]
3	Citizen-centered e-government services: benefits, costs, and research needs	ACM	(Proceedings)	[86]
4	Assessment methodology in smart cities based on public value	ACM	(Proceedings)	[153]
5	Models of e-democracy	Association for Information Systems (AIS)	*Communications of the Association for Information Systems*	[140]
6	Citizen-centric demand model for transformational government systems	AIS	(Proceedings)	[95]
7	A ladder of citizen participation	American Institute of Planners	*Journal of the American Institute of Planners*	[52]
8	The smart city from a public value perspective	Atlantis Press	(Proceedings)	[152]
9	Public value from co-production by clients	Australia and New Zealand School of Government	(Working paper)	[151]
10	Using the transformational government framework to deliver public sector services	Brunel University	(Working paper)	[91]
11	A review on public participation in Environmental Impact Assessment in Malaysia	Academy of Economic Studies, Bucharest, Romania	*TERUM-Theoretical and Empirical Research in Urban Management*	[146]

Table A2. *Cont.*

No.	Source's Title	Publisher	Publication	Author
12	"Citizens as analysts" redux: revisiting Aaron Wildavsky on public participation	University of Westminster Press	*Journal of Public Deliberation*	[145]
13	An institutional analysis of Environmental Impact Assessment (EIA) in Malaysia: social conflict and credibility	Delft University of Technology	(Thesis)	[114]
14	The guide to effective participation	Delta Press, Brighton	(Book)	[69]
15	Participatory learning for sustainable agriculture	Elsevier	*World Development*	[128]
16	From e-government to we-government: defining a typology for citizen coproduction in the age of social media	Elsevier	*Government Information Quarterly*	[141]
17	Transformational change and business process reengineering (BPR): lessons from the British and Dutch public sector	Elsevier	*Government Information Quarterly*	[127]
18	Public participation in waste management decision making: analysis and management of conflicts	Elsevier	*Journal of Hazardous Materials*	[137]
19	Citizen participation in China's eco-city development. Will 'new-type urbanization' generate a breakthrough in realizing it?	Elsevier	*Journal of Cleaner Production*	[117]
20	Participation's place in rural development: seeking clarity through specificity	Elsevier	*World Development*	[139]
21	Citizens' attitudes towards e-government and e-governance: a UK study	Emerald	*International Journal of Public Sector Management*	[87]
22	The smart city and its citizens: governance and citizen participation in Amsterdam Smart City	Erasmus University	(Thesis)	[49]
23	E-participation—a key factor in developing smart cities	European Citizen and Public Administration	(Proceedings)	[101]
24	A handbook for citizen-centric eGovernment	European Commission	(Book)	[1]
25	A citizen-centric public sector: why citizen centricity matters and how to obtain it	International Academy, Research and Industry Association (IARIA)	(Proceedings)	[36]
26	Citizen-centric eGovernment services: use of indicators to measure degree of user involvement in eGovernment service development	IARIA	(Proceedings)	[105]
27	Citizen participation in smart cities: evaluation framework proposal	Institute of Electrical and Electronics Engineers (IEEE)	(Proceedings)	[120]
28	A web 2.0 citizen-centric model for t-government services	IEEE	*IEEE Intelligent Systems*	[97]
29	Understanding smart cities: an integrative framework	IEEE	(Proceedings)	[109]
30	Involving citizens in smart city projects: systems engineering meets participation	IEEE	(Proceedings)	[106]
31	E-governance and development: service delivery to empower the poor	Idea Group (IGI) Publishing	*International Journal of Electronic Government Research*	[92]
32	Citizens as sensors/ information providers in the co-production of smart city services.	Luiss University Press	(Proceedings)	[38]
33	Where's wally? In search of citizen perspectives on the smart city	Multidisciplinary Digital Publishing Institute (MDPI)	*Sustainability*	[126]

Table A2. *Cont.*

No.	Source's Title	Publisher	Publication	Author
34	City-as-a-platform: the rise of participatory innovation platforms in Finnish cities	MDPI	*Sustainability*	[143]
35	Making local democracy work: municipal officials' views of public participation	National League of Cities	(Book)	[142]
36	New politics: towards a mature Malaysian democracy	National Translation Institute of Malaysia	(Book)	[99]
37	Being a 'citizen' in the smart city: up and down the scaffold of smart citizen participation.	National University of Ireland Maynooth	(Working paper)	[71] *
38	Citizens as partners: OECD handbook on information, consultation and public participation in policy-making	Organisation for Economic Co-operation and Development (OECD)	(Report)	[123]
39	Engaging citizens in policy-makings: information, consultation and public participation	OECD	(Report)	[122]
40	Models of democracy: from representation to participation?	Oxford University Press	(Book chapter from) *The Changing Constitution*	[131]
41	Critical interventions into the corporate smart city	Oxford University Press (Cambridge Political Economy Society)	*Cambridge Journal of Regions, Economy and Society*	[55]
42	Assessing public participation in U.S. cities	Sage	*Public Performance & Management Review*	[135]
43	Conflicting perceptions on participation between citizens and members of local government	Springer	*Quality & Quantity*	[116] *
44	The role of citizen participation in municipal smart city projects: lessons learned from Norway	Springer	(Book chapter from) *Smarter as the New Urban Agenda*	[56]
45	Technology helps, people make: a smart city governance framework grounded in deliberative democracy	Springer	(Book chapter from) *Smarter as the New Urban Agenda*	[54]
46	'Mind the gap': e-government and e-democracy	Springer	(Book chapter from) *International Conference on Electronic Government*	[103]
47	'Mind the gap II': e-government and e-governance	Springer	(Book chapter from) *International Conference on Electronic Government*	[104]
48	Co-production makes cities smarter: citizens' participation in smart city initiatives	Springer	(Book chapter from) *Co-production in the Public Sector*	[37]
49	Smart city projects and citizen participation: the case of London	Springer	(Book chapter from) *Public Sector Management in a Globalized World*	[144]
50	The citizens in e-participation	Springer	*(Book chapter from) International Conference on Electronic Government*	[132]
51	Depoliticising development: the uses and abuses of participation	Taylor & Francis	*Development in Practice*	[129]
52	Assessing public participation initiatives in local government decision-making in Malaysia	Taylor & Francis	*International Journal of Public Administration*	[118]

Table A2. *Cont.*

No.	Source's Title	Publisher	Publication	Author
53	Citizen participation: models and methods	Taylor & Francis	*International Journal of Public Administration*	[134]
54	Will the real smart city please stand up?	Taylor & Francis	*City*	[62]
55	Contemporary public involvement: toward a strategic approach	Taylor & Francis	*Local Environment*	[98]
56	Caught in the middle: Community Development Corporations (CDCs) and the conflict between grassroots and instrumental forms of citizen participation	Taylor & Francis	*Journal of the Community Development Society*	[130]
57	Smart cities in Europe	Technical University of Košice	(Proceedings)	[107] *
58	A comparatives study on public participation in Environmental Impact Assessment (EIA) in Malaysia and European Union	Tilburg University, Netherlands	(Thesis)	[115]
59	E-government survey 2012: e-government for the people	United Nations	(Report)	[96]
60	A critical review of citizen participation in smart cities: the citizens at the core of Smart Namur	Universite De Namur	(Thesis)	[102]
61	Influence of citizen-centric perspective on the effectiveness of e-governance systems in Malaysia	Universiti Putra Malaysia	(Thesis)	[94]
62	Factors influencing participation of rural women in Padzey Project in Taiz Governorate, Yemen	Universiti Putra Malaysia	(Thesis)	[149]
63	Citizen-centric demand model for transformational government	Universiti Teknologi Mara	(Thesis)	[93]
64	Strong democracy: participatory politics for a new age	University of California Press	(Book)	[111]
65	The role of citizens in "smart cities"	University of Presov, Slovakia	(Proceedings)	[72]
66	Engaging democracy: an institutional theory of participatory budgeting	University of Washington	(Thesis)	[112]
67	Smart cities: ranking of European medium-sized cities	Vienna University of Technology	(Report)	[35]
68	Engaging citizens in democratic governance and the decision-making process with congressional committees	Walden University	(Thesis)	[100]
69	Models of democracy	Wiley-Blackwell	(Book)	[110]
70	Why are smart cities growing? Who moves and who stays	Wiley-Blackwell	*Journal of Regional Science*	[150]
71	Citizen participation: can we measure its effectiveness?	Wiley-Blackwell (American Society for Public Administration)	*Public Administration Review*	[119]
72	Beyond engagement and participation: user and community coproduction of public services	Wiley-Blackwell	*Public Administration Review*	[70]
73	Varieties of participation in complex governance	Wiley-Blackwell	*Public Administration Review*	[121]
74	Citizen participation in decision making: is it worth the effort?	Wiley-Blackwell	*Public Administration Review*	[136]
75	Citizen, customer, partner: rethinking the place of the public in public management	Wiley-Blackwell	*Public Administration Review*	[138]
76	Putting the "public" back in public values research: designing participation to identify and respond to values	Wiley-Blackwell	*Public Administration Review*	[133]

Table A2. *Cont.*

No.	Source's Title	Publisher	Publication	Author
77	Further dissecting the black box of citizen participation: when does citizen involvement lead to good outcomes?	Wiley-Blackwell	*Public Administration Review*	[108]
78	The World Bank participation sourcebook	World Bank	(Report)	[124]
79	Strategic framework for mainstreaming citizen engagement in World Bank Group operations	World Bank	(Report)	[125]

Note: * Ref. [71]—this working paper was published in GeoJournal, 2019; Ref. [116]—this journal article was online first in 2017; Ref. [107]—this proceeding paper was published in Journal of Urban Technology, 2011.

Appendix C

Table A3. Selected and adjusted items after the verification by practitioners (source: authors).

No.	Item from Literature	Adjusted Item with Verification by Practitioners
	The Construct of "Citizen-Centric Smart City"	
2	The decision by LAs needs to reach a consensus with us.	Decisions made by LAs are through an agreement, *such as dialogues with us.*
4	LAs delegate power to us, especially at the initial level of smart city programs.	LAs delegate authority to us at the initial stages of urban programs, *such as meeting with us for decision-making.*
6	We play the role of volunteers and contribute information continuously.	We are *responsible and together* build a smart city.
	The Construct of "Understanding of Participation"	
6	We are involved in the programs without going through a representative.	We *attend (without a representative) and join along* in the program organization.
8	We value the decision-making process together.	We value the decision-making process together, *and not just comply with LAs.*
	The Construct of "Types of Participation"	
5	LAs offer grants (financial incentives) to run smart city programs.	LAs offer *grants/contracts in the form of finance* to run city programs.
6	LAs offer rewards, but LAs still have full power.	LAs offer *rewards/gifts, such as shirts, bins, and others.*
8	LAs hold meetings with the community.	LAs hold meetings *with us.*
9	LAs broadcast the correct information to us.	LAs *publish accurate information.*
	The Construct of "Processes of Participation"	
	(A combination of items 1 and 2)	We are involved in *setting the program agenda.*
	(A combination of items 3 and 4)	We are involved in *planning program activities.*
	(Original item was 6)	We *present and join along* in the program.
	The Construct of "Roles of Citizens"	
3	As co-producers, we work together with the LAs.	As co-producers, we work together with LAs *and contribute relevant resources.*
6	As human sensors, we contribute data.	As human sensors, we *report issues to the LAs.*
	The Construct of "Characters of Citizens"	
3	We choose not to rely on government resources.	We choose *a less dependent approach* to government resources.
4	Being educated is important for us to get involved in city programs.	*Efforts to obtain higher education (BSc and above)* are important *to prepare us* to be involved in the program.
5	We are interested in public life, public values, and acting quickly on things that disrupt community life.	We are interested in *public affairs* and act quickly on things that disrupt community life.

Appendix D

Table A4. Instrument construct and items (source: authors).

Construct	Num. of Original Items (Based on Literature Review)	Num. of New Items (Improved after the Interview)	Detail Num. of New Items after Adjustment	Detail Num. of New Items Remaining
(1) Citizen-centric smart city	8	8	3	5
(2) Understanding of participation	9	9	2	7
(3) Types of participation	11	11	4	7
(4) Processes of participation	8	6	5	3
(5) Roles of citizens	8	8	3	5
(6) Characters of citizens	5	5	3	2
Total	49	47		

References

1. Undheim, T.A.; Blakemore, M. *A Handbook for Citizen-Centric eGovernment*; Version 2.1; European Commission: Brussels, Belgium, 2007.
2. Hood, C. The "New Public Management" in the 1980s: Variations on a theme. *Account. Organ. Soc.* **1995**, *20*, 93–109. [CrossRef]
3. Van der Hoogen, A.; Scholtz, B.; Calitz, A.P. Using theories to design a value alignment model for smart city initiatives. In *International Federation for Information Processing*; Hattingh, M., Ed.; Springer: Berlin/Heidelberg, Germany, 2020; pp. 55–66. [CrossRef]
4. United Cities and Local Governments (UCLG). *Executive Summary of Co-creating the Urban Future: The Agenda of Metropolises, Cities and Territories*; UCLG: Barcelona, Spain, 2016.
5. Nabatchi, T.; Sancino, A.; Sicilia, M. Varieties of participation in public services: The who, when, and what of Coproduction. *Public Adm. Rev.* **2017**, *77*, 766–776. [CrossRef]
6. Allwinkle, S.; Cruickshank, P. Creating smart-er cities: An overview. *J. Urban Technol.* **2011**, *18*, 1–16. [CrossRef]
7. Watson, V. The allure of 'smart city' rhetoric: India and Africa. *Dialogues Hum. Geogr.* **2015**, 36–39. [CrossRef]
8. Datta, A. A 100 smart cities, a 100 utopias. *Dialogues Hum. Geogr.* **2015**, *5*, 49–53. [CrossRef]
9. Lim, S.B. Membina Model Bandar Pintar Berpusatkan Rakyat di Malaysia. Ph.D. Thesis, Universiti Kebangsaan Malaysia, Bangi, Malaysia, 2020.
10. Cardullo, P.; Kitchin, R. Being a 'citizen' in the smart city: Up and down the scaffold of smart citizen participation in Dublin, Ireland. *GeoJournal* **2019**, *84*, doi. [CrossRef]
11. Kummitha, R.K.R. Entrepreneurial urbanism and technological panacea: Why Smart City planning needs to go beyond corporate visioning? *Technol. Forecast. Soc. Chang.* **2018**, *137*, 330–339. [CrossRef]
12. Fainstein, S. *The Just City*; Cornell University Press: Ithaca, NY, USA, 2010.
13. Marsal-Llacuna, M.-L. City indicators on social sustainability as standardization technologies for smarter (citizen-centered) governance of cities. *Soc. Indic. Res.* **2016**, *128*, 1193–1216. [CrossRef]
14. Jenson, J. Redesigning citizenship regimes after Neoliberalism. Moving towards Social Investment. In *What Future for Social Investment?* Morel, N., Palier, B., Palme, J., Eds.; Institute for Future Studies: Stockholm, Sweden, 2009; pp. 27–44.
15. Joss, S.; Cook, M.; Dayot, Y. Smart cities: Towards a new citizenship regime? A discourse analysis of the British Smart City Standard. *J. Urban Technol.* **2017**, *24*, 29–49. [CrossRef]
16. Lim, S.B.; Jalaluddin, A.M.; Mohd Yusof, H.; Zurinah, T. Malaysia Smart City Framework: A trusted framework for shaping smart Malaysian citizenship? In *Handbook of Smart Cities*; Augusto, J.C., Ed.; Springer International Publishing: Cham, Switzerland, 2020; pp. 1–24. [CrossRef]
17. Marsal-Llacuna, M.-L. Building universal socio-cultural indicators for standardizing the safeguarding of citizens' rights in smart cities. *Soc. Indic. Res.* **2017**, *130*, 563–579. [CrossRef]
18. Martin, C.; Evans, J.; Karvonen, A.; Paskaleva, K.; Yang, D.; Linjordet, T. Smart-sustainability: A new urban fix? *Sustain. Cities Soc.* **2019**, *45*, 640–648. [CrossRef]
19. Nesti, G. Defining and assessing the transformational nature of smart city governance: Insights from four European cases. *Int. Rev. Adm. Sci.* **2018**, *86*, 20–37. [CrossRef]
20. Glasmeier, A.K.; Nebiolo, M. Thinking about smart cities: The travels of a policy idea that promises a great deal, but so far has delivered modest results. *Sustainability* **2016**, *8*, 1122. [CrossRef]
21. Castelnovo, W.; Misuraca, G.; Savoldelli, A. Smart cities governance: The need for a holistic approach to assessing urban participatory policy making. *Soc. Sci. Comput. Rev.* **2015**, *34*, 724–739. [CrossRef]
22. Meijer, A.; Bolívar, M.P.R. Governing the smart city: A review of the literature on smart urban governance. *Int. Rev. Adm. Sci.* **2016**, *82*, 392–408. [CrossRef]

23. Tan, S.Y.; Taeihagh, A. Smart city governance in developing countries: A systematic literature review. *Sustainability* **2020**, *12*, 2020. [CrossRef]
24. Oxford University Press. Oxford Learner's Dictionaries. Available online: https://www.oxfordlearnersdictionaries.com/definition/english/citizenship?q=citizenship (accessed on 2 August 2019).
25. Jenson, J.; Phillips, S.D. Regime shift: New citizenship practices in Canada. *Int. J. Can. Stud.* **1996**, *14*, 111–136.
26. Kummitha, R.K.R.; Crutzen, N. How do we understand smart cities? An evolutionary perspective. *Cities*. **2017**, *67*, 43–52. [CrossRef]
27. Cardullo, P.; Kitchin, R. Smart urbanism and smart citizenship: The neoliberal logic of 'citizen-focused' smart cities in Europe. *Environ. Plan. C Polit. Sp.* **2019**, *37*, 813–830. [CrossRef]
28. De Waal, M.; Dignum, M. The citizen in the smart city. How the smart city could transform citizenship. *IT Inf. Technol.* **2017**, *59*, 263–273. [CrossRef]
29. Cowley, R.; Joss, S.; Dayot, Y. The smart city and its publics: Insights from across six UK cities. *Urban Res. Pract.* **2018**, *11*, 53–77. [CrossRef]
30. Dirks, S.; Keeling, M.; Dencik, J. *IBM Global Business Services Executive Report: How Smart is your City? Helping Cities Measure Progress*; International Business Machines Corporation: New York, NY, USA, 2009. [CrossRef]
31. Yonezawa, T.; Matranga, I.; Galache, J.A.; Maeomichi, H.; Gurgen, L.; Shibuya, T. A citizen-centric approach towards global-scale smart city platform. In Proceedings of the International Conference on Recent Advances in Internet of Things, Singapore, 7–9 April 2015; pp. 1–6.
32. Aguilera, U.; Peña, O.; Belmonte, O.; López-de-Ipiña, D. Citizen-centric data services for smarter cities. *Futur. Gener. Comput. Syst.* **2017**, *76*, 234–247. [CrossRef]
33. Chang, F.; Das, D. Smart nation Singapore: Developing policies for a citizen-oriented smart city initiative. In *Developing National Urban Policies: Ways Forward to Green and Smart Cities*; Kundu, D., Sietchiping, R., Kinyanjui, M., Eds.; Springer: Singapore, 2020; pp. 425–440.
34. Lim, S.B.; Jalaluddin, A.M.; Mohd, Y.H.; Zurinah, T. Citizen participation in building citizen-centric smart cities. *Geogr. Malays. J. Soc. Space* **2018**, *14*, 42–53. [CrossRef]
35. Giffinger, R.; Fertner, C.; Kramar, H.; Kalasek, R.; Pichler, N.; Meijers, E. *Smart Cities: Ranking of European Medium-Sized Cities*; TU Vienna: Wien, Astria, 2007.
36. Berntzen, L.; Johannesen, M.R.; Ødegård, A. A citizen-centric public sector: Why citizen centricity matters and how to obtain it. In Proceedings of the 9th International Conference on Advances in Human-oriented and Personalized Mechanisms, Technologies, and Services (AHPMTS), Rome, Italy, 21–25 August 2016; pp. 14–20.
37. Castelnovo, W. Co-production makes cities smarter: Citizens' participation in smart city initiatives. In *Co-Production in the Public Sector*; Fugini, M., Ed.; Springer: Cham, Switzerland, 2016; pp. 97–117.
38. Castelnovo, W. Citizens as sensors/information providers in the co-production of smart city services. In Proceedings of the 12th Italian Chapter of the Association for Information Systems (IT AIS), Carisolo, TN, Italy, 10–13 February 2016; pp. 51–62.
39. Mainka, A.; Bech-Petersen, S.; Castelnovo, W.; Hartmann, S.; Miettinen, V.; Stock, W.G. Open innovation in smart cities: Civic participation and co-creation of public services. *Proc. Assoc. Inf. Sci. Technol.* **2016**, *53*, 1–5. [CrossRef]
40. Yigitcanlar, T.; Kamruzzaman, M. Planning, development and management of sustainable cities: A commentary from the guest editors. *Sustainability* **2015**, *7*, 14677–14688. [CrossRef]
41. Kundu, D.; Sietchiping, R.; Kinyanjui, M. *Developing National Urban Policies: Ways Forward to Green and Smart Cities*; Springer: Singapore, 2020. [CrossRef]
42. Angelidou, M. Smart cities: A conjuncture of four forces. *Cities* **2015**, *47*, 95–106. [CrossRef]
43. Yigitcanlar, T.; Han, H.; Kamruzzaman, M.; Ioppolo, G.; Sabatini-Marques, J. The making of smart cities: Are Songdo, Masdar, Amsterdam, San Francisco and Brisbane the best we could build? *Land Use Policy* **2019**, *88*, 104187. [CrossRef]
44. Melissen, J.; Caesar-Gordon, M. "Digital diplomacy" and the securing of nationals in a citizen-centric world. *Glob. Aff.* **2016**, *2*, 321–330. [CrossRef]
45. Alawadhi, S.; Scholl, H.J. Smart governance: A cross-case analysis of smart city initiatives. In Proceedings of the 49th Hawaii International Conference on System Sciences, Koloa, HI, USA, 5–8 January 2016; pp. 2953–2963. [CrossRef]
46. Babu, C.M.; Sasankar, A.B.; Prasuna, K. Smart city framework strategies for citizen centric governance. *Int. J. Adv. Res. Comput. Sci. Manag. Stud.* **2016**, *4*, 113–121.
47. Dameri, R.P.; Benevolo, C. Governing smart cities: An empirical analysis. *Soc. Sci. Comput. Rev.* **2016**, *34*, 693–707. [CrossRef]
48. Nam, T.; Pardo, T. Conceptualizing smart city with dimensions of technology, people & institutions. In Proceedings of the 12th Annual International Conference on Digital Government Research, College Park, MD, USA, 12–15 June 2011; pp. 282–291.
49. Capra, C.F. The Smart City and its Citizens: Governance and Citizen Participation in Amsterdam Smart City. Master Thesis, Erasmus University, Rotterdam, The Netherlands, 2014.
50. Sengupta, A.K. Conceptualizing the right to development for the twenty-first century. In *Realizing the Right to Development: Essays in Commemoration of 25 Years of the United Nations Declaration on the Right to Development*; United Nations: New York, NY, USA, 2013; pp. 67–87.
51. Cornwall, A.; Nyamu-Musembi, C. Putting the "rights-based approach" to development into perspective. *Third World Q.* **2004**, *25*, 1415–1437. [CrossRef]
52. Arnstein, S.R. A ladder of citizen participation. *J. Am. Inst. Plan.* **1969**, *35*, 216–224. [CrossRef]

53. Mohseni, H. Public engagement and smart city definitions: A classifying model for the evaluation of citizen power in 2025 Tehran. *GeoJournal* **2020**. [CrossRef]
54. Alonso, R.G.; Castro, S.L. Technology helps, people make: A smart city governance framework grounded in deliberative democracy. In *Smarter as the New Urban Agenda; A Comprehensive View of the 21st Century City*; Gil-Garcia, J.R., Pardo, T.A., Nam, T., Eds.; Springer: Cham, Switzerland, 2016; pp. 333–347.
55. Hollands, R.G. Critical interventions into the corporate smart city. *Camb. J. Reg. Econ. Soc.* **2015**, *8*, 61–77. [CrossRef]
56. Berntzen, L.; Johannessen, M.R. The role of citizen participation in municipal smart city projects: Lessons learned from Norway. In *Smarter as the New Urban Agenda; A Comprehensive View of the 21st Century City*; Gil-Garcia, J.R., Pardo, T.A., Nam, T., Eds.; Springer: Cham, Switzerland, 2016; pp. 299–314.
57. Vanolo, A. Is there anybody out there? The place and role of citizens in tomorrow's smart cities. *Futures* **2016**, *82*, 26–36. [CrossRef]
58. Caprotti, F.; Cowley, R.; Datta, A.; Broto, V.C.; Gao, E.; Georgeson, L.; Herrick, C.; Odendaal, N.; Joss, S. The New Urban Agenda: Key opportunities and challenges for policy and practice. *Urban Res. Pract.* **2017**, *10*, 367–378. [CrossRef]
59. Gil-Garcia, J.R.; Pardo, A.T.; Nam, T. *Smarter as the New Urban Agenda: A Comprehensive View of the 21st Century City*; Springer: Cham, Switzerland, 2016.
60. Benedikt, O. The value citizens of smart cities: The case of Songdo City. Grad. *J. Soc. Sci.* **2016**, *12*, 17–36.
61. Graham, S. Bridging urban digital divides? Polarisation and Information and Communications Technologies (ICTs). *Urban Stud.* **2002**, *39*, 33–56. [CrossRef]
62. Hollands, R.G. Will the real smart city please stand up? *City* **2008**, *12*, 303–320. [CrossRef]
63. Vanolo, A. Smartmentality: The smart city as disciplinary strategy. *Urban Stud.* **2014**, *51*, 883–898. [CrossRef]
64. Kaika, M. "Don't call me resilient again!": The New Urban Agenda as immunology ... or ... What happens when communities refuse to be vaccinated with 'smart cities' and indicators. *Environ. Urban.* **2017**, *29*, 89–102. [CrossRef]
65. King, C.S.; Feltey, K.M.; Susel, B.O. The question of participation: Toward authentic public participation in public administration. *Pub. Admin. Rev.* **1998**, *58*, 317–326. [CrossRef]
66. Kobie, N. Inside Cyberjaya, Malaysia's failed Silicon Valley. Available online: http://www.wired.co.uk/article/malaysia-cyberjaya-silicon-valley-smart-cities (accessed on 3 July 2017).
67. Salman, A. Cyberjaya: Malaysia's promised Silicon Valley a central plan, which failed. Available online: http://www.ideas.org.my/cyberjaya-malaysias-promised-silicon-valley-a-central-plan-which-failed/ (accessed on 19 November 2018).
68. Yusof, N.; van Loon, J. Engineering a global city: The case of Cyberjaya. *Sp. Cult.* **2012**, *15*, 298–316. [CrossRef]
69. Wilcox, D. *The Guide to Effective Participation*; Delta Press: Brighton, UK, 1994.
70. Bovaird, T. Beyond engagement and participation: User and community coproduction of public services. *Public Adm. Rev.* **2007**, *67*, 846–860. [CrossRef]
71. Cardullo, P.; Kitchin, R. *Being a 'Citizen' in the Smart City: Up and down the Scaffold of Smart Citizen Participation*; Programmable City Working Paper No. 30; National University of Ireland Maynooth: County Kildare, Ireland, 15 May 2017.
72. Berntzen, L.; Johannessen, M.R. The role of citizens in "smart cities". In Proceedings of the Management International Conference, University of Presov; 2016.
73. Bosch, P.; Jongeneel, S.; Rovers, V.; Neumann, H.-M.; Airaksinen, M.; Huovila, A. *CITYkeys Indicators for Smart City Projects and Smart Cities*; European Commission: Brussels, Belgium, 2017.
74. Shaffril, H.A.M.; Samsuddin, S.F.; Samah, A.A. The ABC of systematic literature review: The basic methodological guidance for beginners. *Qual. Quant.* **2020**, 1–28. [CrossRef]
75. Greyson, D.; Rafferty, E.; Slater, L.; MacDonald, N.; Bettinger, J.A.; Dubé, È.; MacDonald, S.E. Systematic review searches must be systematic, comprehensive, and transparent: A critique of Perman et al. *BMC Public Health.* **2019**, *19*, 1–6. [CrossRef]
76. Cocchia, A. Smart and digital city: A systematic literature review. In *Smart City*; Dameri, R.P., Rosenthal-Sabroux, C., Eds.; Springer: Cham, Switzerland, 2014; pp. 13–43.
77. Anthopoulos, L.; Janssen, M.; Weerakkody, V. A Unified Smart City Model (USCM) for smart city conceptualization and benchmarking. *Int. J. Electron. Gov. Res.* **2016**, *12*, 77–93. [CrossRef]
78. Mora, L.; Deakin, M.; Reid, A. Combining co-citation clustering and text-based analysis to reveal the main development paths of smart cities. *Technol. Forecast. Soc. Chang.* **2019**, *142*, 56–69. [CrossRef]
79. Moher, D.; Liberati, A.; Tetzlaff, J.; Altman, D.G. The PRISMA Group. Preferred reporting items for systematic reviews and meta-analyses: The PRISMA statement. *PLoS Med.* **2009**, *6*, e1000097. [CrossRef]
80. Mora, L.; Bolici, R.; Deakin, M. The first two decades of smart-city research: A bibliometric analysis. *J. Urban Technol.* **2017**, *24*, 3–27. [CrossRef]
81. Xiao, Y.; Watson, M. Guidance on conducting a systematic literature review. *J. Plan. Educ. Res.* **2019**, *39*, 93–112. [CrossRef]
82. Purwanto, A.; Zuiderwijk, A.; Janssen, M. Citizen engagement with open government data: A systematic literature review of drivers and inhibitors. *Int. J. Electron. Gov. Res.* **2020**, *16*, 1–25. [CrossRef]
83. Gudes, O.; Kendall, E.; Yigitcanlar, T.; Pathak, V.; Baum, S. Rethinking health planning: A framework for organising information to underpin collaborative health planning. *Heal. Inf. Manag. J.* **2010**, *39*, 18–29. [CrossRef] [PubMed]
84. Webster, J.; Watson, R.T. Analyzing the past to prepare for the future. *MIS Q.* **2002**, *26*, 13–23.

85. Luna-Reyes, L.F.; Gil-Garcia, J.R.; Celorio Mansi, J.A. Citizen-centric approaches to e-government and the back-office transformation. In Proceedings of the 12th Annual International Conference on Digital Government Research (ICDGR), College Park, MD, USA, 12–15 June 2011; pp. 213–218.
86. Bertot, J.C.; Jaeger, P.T.; Mcclure, C.R. Citizen-centered e-government services: Benefits, costs, and research needs. In Proceedings of the 9th Annual International Digital Government Research Conference (ICDGR), Montreal, QC, Canada, 18–21 May 2008; pp. 137–142.
87. Kolsaker, A.; Lee-kelley, L. Citizens' attitudes towards e-government and e-governance: A UK study. *Int. J. Public Sect. Manag.* **2008**, *21*, 723–738. [CrossRef]
88. Sigwejo, A.; Pather, S. A citizen-centric framework for assessing e-government effectiveness. *Electron. J. Inf. Syst. Dev. Ctries.* **2016**, *74*, 1–27. [CrossRef]
89. Wang, J.; Tao, Z. Citizen-centered e-government strategy governance framework: Case of China. In Proceedings of the International Conference on Web Information Systems and Mining, Shanghai, China, 7–8 November 2009; pp. 589–593.
90. Wang, L.; Bretschneider, S.; Gant, J. Evaluating web-based e-government services with a citizen-centric approach. In Proceedings of the 38th Hawaii Annual International Conference on System Sciences, Waikoloa, HI, USA, 3–6 July 2005; pp. 129–137.
91. Borras, J. Using the transformational government framework to deliver public sector services. In Proceedings of the Transforming Government Workshop, Brunel University, London, UK, 8–9 May 2012.
92. Zambrano, R. E-governance and development: Service delivery to empower the poor. *Int. J. Electron. Gov. Res.* **2008**, *4*, 1–13. [CrossRef]
93. Kamalia Azma, K. Citizen-Centric Demand Model for Transformational Government. Ph.D. Thesis, Universiti Teknologi Mara, Shah Alam, Malaysia, 2018.
94. Jafari, S.M. Influence of Citizen-Centric Perspective on the Effectiveness of E-Governance Systems in Malaysia. Ph.D. Thesis, Universiti Putra Malaysia, Seri Kembangan, Malaysia, 2012.
95. Kamalia Azma, K.; Nor Laila, M.N. Citizen-centric demand model for transformational government systems. In Proceedings of the 21st Pacific Asia Conference on Information Systems (PACIS), Langkawi, Malaysia, 16–20 July 2017; AISeL: Milan, Italy, 2017. Paper No. 139.
96. United Nations (UN). *E-Government Survey 2012: E-Government for the People*; United Nations: New York, NY, USA, 2012.
97. Dais, A.; Nikolaidou, M.; Anagnostopoulos, D. A web 2.0 citizen-centric model for t-government services. *IEEE Intell. Syst.* **2013**, *1*, 10–18. [CrossRef]
98. Jackson, L.S. Contemporary public involvement: Toward a strategic approach. *Local Environ.* **2001**, *6*, 135–147. [CrossRef]
99. Abdullah, S. *New Politics: Towards a Mature Malaysian Democracy*; National Translation Institute of Malaysia: Kuala Lumpur, Malaysia, 2008.
100. Williams, S.I. Engaging Citizens in Democratic Governance and the Decision-Making Process with Congressional Committees. Ph.D. Thesis, Walden University, Minneapolis, MN, USA, 2014.
101. Vrabie, C.I.; Tîrziu, A.-M. E-participation—A key factor in developing smart cities. In Proceedings of the 11th International Conference of European Integration Realities and Prospectives (EIRP), Danubius University, Galati, Romania, 20–21 May 2016; pp. 123–128.
102. Simonofski, A. A Critical Review of Citizen Participation in Smart Cities: The Citizens at the Core of Smart Namur. Master's Thesis, Universite De Namur, Namur, Beigium, 2016.
103. Kolsaker, A.; Lee-kelley, L. 'Mind the gap': E-government and e-democracy. In Proceedings of the International Conference on Electronic Government, Kraków, Poland, 4–8 September 2006; Wimmer, M.A., Scholl, H.J., Grönlund, Å., Andersen, K.V., Eds.; Springer: Berlin/Heidelberg, Germany, 2006; pp. 96–106. [CrossRef]
104. Kolsaker, A.; Lee-kelley, L. 'Mind the gap II'E-government and e-governance. In Proceedings of the International Conference on Electronic Government, Regensburg, Germany, 3–7 September 2007; Wimmer, M.A., Scholl, J., Grönlund, Å., Eds.; Springer: Berlin/Heidelberg, Germany, 2007; pp. 35–43. [CrossRef]
105. Berntzen, L. Citizen-centric eGovernment services: Use of indicators to measure degree of user involvement in eGovernment service development. In Proceedings of the 6th International Conference on Advances in Human-oriented and Personalized Mechanisms, Technologies, and Services (CENTRIC), Venice, Italy, 27 October–1 November 2013; IARIA: Barcelona, Spain, 2013; pp. 132–136.
106. Vacha, T.; Přibyl, O.; Lom, M.; Bacúrová, M. Involving citizens in smart city projects: Systems engineering meets participation. In Proceedings of the Smart Cities Symposium Prague (SCSP), Prague, Czech Republic, 26–27 May 2016; IEEE: Piscataway, NJ, USA, 2016; pp. 1–6.
107. Caragliu, A.; Bo, C.D.; Nijkamp, P. Smart cities in Europe. In Proceedings of the 3rd Central European Conference in Regional Science (CERS), Technical University of Košice, Košice, Slovak Republic, 7–9 October 2009; pp. 45–59.
108. Yang, K.; Pandey, S.K. Further dissecting the black box of citizen participation: When does citizen involvement lead to good outcomes? *Public Adm. Rev.* **2011**, *71*, 880–892. [CrossRef]
109. Chourabi, H.; Nam, T.; Walker, S.; Gil-Garcia, J.R.; Mellouli, S.; Nahon, K.; Pardo, T.A.; Scholl, H.J. Understanding smart cities: An integrative framework. In Proceedings of the 45th Hawaii International Conference on System Sciences Understanding (ICSS), Maui, HI, USA, 4–7 January 2012; IEEE: Piscataway, NJ, USA, 2012; pp. 2289–2297.
110. Held, D. *Models of Democracy*; Wiley-Blackwell: Oxford, UK, 1996.
111. Barber, B.R. *Strong democracy: Participatory Politics for a New Age*; University of California Press: Berkeley, CA, USA, 1984.

112. Johnson, C. Engaging Democracy: An Institutional Theory of Participatory Budgeting. Ph.D. Thesis, University of Washington, Washington, DC, USA, 2017.
113. Harrington, K. Smart city leaders, champions, and entrepreneurs. In *Smart Economy in Smart Cities*; Kumar, V., Ed.; Springer: Singapore, 2017; pp. 1005–1012.
114. Nor Hisham, M.S. An Institutional Analysis of Environmental Impact Assessment (EIA) in Malaysia: Social Conflict and Credibility. Ph.D. Thesis, Delft University of Technology, Delft, The Netherlands, 2016.
115. Mokhtar, N. A Comparatives Study on Public Participation in Environmental Impact Assessment (EIA) in Malaysia and European Union. Master's Thesis, Tilburg University, Tilburg, The Netherlands, 2016.
116. Mohammadi, S.H.; Norazizan, S.; Nikkhah, H.A. Conflicting perceptions on participation between citizens and members of local government. *Qual. Quant.* **2018**, *52*, 1761–1778. [CrossRef] [PubMed]
117. Li, H.; de Jong, M. Citizen participation in China's eco-city development. Will "new-type urbanization" generate a breakthrough in realizing it? *J. Clean. Prod.* **2017**, *162*, 1085–1094. [CrossRef]
118. Halimah, A.M.; Ahmad Martadha, M.; Lawton, A. Assessing public participation initiatives in local government decision-making in Malaysia. *Int. J. Public Adm.* **2016**, *39*, 812–820.
119. Rosener, J.B. Citizen participation: Can we measure its effectiveness? *Public Adm. Rev.* **1978**, *38*, 457–463. [CrossRef]
120. Simonofski, A.; Asensio, E.S.; De Smedt, J.; Snoeck, M. Citizen participation in smart cities: Evaluation framework proposal. In Proceedings of the 19th Conference on Business Informatics (CBI), Thessaloniki, Greece, 24–27 July 2017; IEEE: Piscataway, NJ, USA, 2017; pp. 227–236. [CrossRef]
121. Fung, A. Varieties of participation in complex governance. *Public Adm. Rev.* **2006**, *66*, 66–75. [CrossRef]
122. Organization for Economic Co-operation and Development (OECD). *Engaging Citizens in Policy-Makings: Information, Consultation and Public Participation*; OECD Public Management Policy Brief; OECD: Paris, France, 2001.
123. Organization for Economic Co-operation and Development (OECD). *Citizens as Partners: OECD Handbook on Information, Consultation and Public Participation in Policy-Making*; OECD: Paris, France, 2001.
124. World Bank. *The World Bank Participation Sourcebook*; The World Bank: Washing DC, USA, 1996.
125. World Bank. *Strategic Framework for Mainstreaming Citizen Engagement in World Bank Group Operations*; The World Bank: Washing DC, USA, 2014.
126. Thomas, V.; Wang, D.; Mullagh, L.; Dunn, N. Where's wally? In search of citizen perspectives on the smart city. *Sustainability* **2016**, *8*, 207. [CrossRef]
127. Weerakkody, V.; Janssen, M.; Dwivedi, Y.K. Transformational change and business process reengineering (BPR): Lessons from the British and Dutch public sector. *Gov. Inf. Q.* **2011**, *28*, 320–328. [CrossRef]
128. Pretty, J.N. Participatory learning for sustainable agriculture. *World Dev.* **1995**, *23*, 1247–1263. [CrossRef]
129. White, S.C. Depoliticising development: The uses and abuses of participation. *Dev. Pract.* **1996**, *6*, 6–15. [CrossRef]
130. Silverman, R.M. Caught in the middle: Community Development Corporations (CDCs) and the conflict between grassroots and instrumental forms of citizen participation. *J. Community Dev. Soc.* **2005**, *36*, 35–51. [CrossRef]
131. Morison, J. Models of democracy: From representation to participation? In *The Changing Constitution*; Jowell, J., Oliver, D., Eds.; Oxford University Press: Oxford, UK, 2007; pp. 134–156.
132. Hansen, H.S.; Reinau, K.H. The citizens in e-participation. In Proceedings of the International Conference on Electronic Government, Kraków, Poland, 4–8 September 2006; Wimmer, M.A., Scholl, H.J., Grönlund, Å., Andersen, K.V., Eds.; Springer: Berlin/Heidelberg, Germany, 2006; pp. 70–82.
133. Nabatchi, T. Putting the "public" back in public values research: Designing participation to identify and respond to values. *Public Adm. Rev.* **2012**, *72*, 699–708. [CrossRef]
134. Callahan, K. Citizen participation: Models and methods. *Int. J. Public Adm.* **2007**, *30*, 1179–1196. [CrossRef]
135. Wang, X. Assessing public participation in U.S. Cities. *Public Perform. Manag. Rev.* **2001**, *24*, 322–336. [CrossRef]
136. Irvin, R.A.; Stansbury, J. Citizen participation in decision making: Is it worth the effort? *Pub. Admin. Rev.* **2004**, *64*, 55–65. [CrossRef]
137. Wiedemann, P.M.; Femers, S. Public participation in waste management decision making: Analysis and management of conflicts. *J. Hazard. Mater.* **1993**, *33*, 355–368. [CrossRef]
138. Thomas, J.C. Citizen, customer, partner: Rethinking the place of the public in public management. *Public Adm. Rev.* **2013**, *73*, 786–796. [CrossRef]
139. Cohen, J.; Uphoff, N. Participation's place in rural development: Seeking clarity through specificity. *World Dev.* **1980**, *8*, 213–235. [CrossRef]
140. Paivarinta, T.; Saebo, O. Models of e-democracy. *Commun. Assoc. Inf. Syst.* **2006**, *17*, 818–840. [CrossRef]
141. Linders, D. From e-government to we-government: Defining a typology for citizen coproduction in the age of social media. *Gov. Inf. Q.* **2012**, *29*, 446–454. [CrossRef]
142. Barnes, W.; Mann, B. *Making Local Democracy Work: Municipal Officials' Views of Public Participation*; National League of Cities: Washington, DC, USA, 2010.
143. Anttiroiko, A. City-as-a-platform: The rise of participatory innovation platforms in Finnish cities. *Sustainability* **2016**, *8*, 922. [CrossRef]
144. Willems, J.; Van den Bergh, J.; Viaene, S. Smart city projects and citizen participation: The case of London. In *Public Sector Management in a Globalized World*; Andeßner, R., Ed.; Springer: Wiesbaden, Germany, 2017; pp. 249–266.
145. Forest, P.G. "Citizens as analysts" redux: Revisiting Aaron Wildavsky on public participation. *J. Public Delib.* **2013**, *9*, 1–13.

146. Marzuki, A. A review on public participation in Environmental Impact Assessment in Malaysia. *Theor. Empir. Res. Urban Manag.* **2009**, *3*, 126–136.
147. Yigitcanlar, T. Australian local governments' practice and prospects with online planning. *URISA J.* **2006**, *18*, 7–17.
148. Ekelin, A. To be or not to be active: Exploring practices of e-participation. In Proceedings of the International Conference on Electronic Government, Kraków, Poland, 4–8 September 2006; Wimmer, M.A., Scholl, H.J., Grönlund, Å., Andersen, K.V., Eds.; Springer: Berlin/Heidelberg, Germany, 2006; pp. 107–118.
149. Al-Areqi, B.M.A. Factors influencing participation of rural women in Padzey Project in Taiz Governorate, Yemen. Master's Thesis, Universiti Putra Malaysia, Seri Kembangan, Malaysia, 2010.
150. Winters, J.V. Why are smart cities growing? Who moves and who stays. *J. Reg. Sci.* **2011**, *51*, 253–270. [CrossRef]
151. Alford, J. *Public Value from Co-Production by Clients*; Australia and New Zealand School of Government: Carlton, Australia, 2009.
152. Cosgrave, E.; Tryfonas, T.; Crick, T. The smart city from a public value perspective. In Proceedings of the 2nd International Conference on ICT for Sustainability (ICT4S), Stockholm, Sweden, 24–27 August 2014; Atlantis Press: Paris, France, 2014; pp. 369–377.
153. Porto, J.; Macadar, M. Assessment methodology in smart cities based on public value. In Proceedings of the 18th Annual International Conference on Digital Government Research, Staten Island, NY, USA, 7–9 June 2017; ACM: New York, NY, USA, 2017; pp. 461–470.
154. Savini, F. Self-organization and urban development: Disaggregating the city-region, deconstructing urbanity in Amsterdam. *Int. J. Urban Reg. Res.* **2016**, *40*, 1152–1169. [CrossRef]
155. Oosterlynck, S.; González, S. "Don't waste a crisis": Opening up the city yet again for neoliberal experimentation. *Int. J. Urban Reg. Res.* **2013**, *37*, 1075–1082. [CrossRef]
156. Cabannes, Y.; Douglass, M.; Padawangi, R. *Cities by and for the People*; Amsterdam University Press: Amsterdam, The Netherlands, 2018. [CrossRef]
157. Mikami, N. Trends in democratic innovation in Asia. In *Handbook of Democratic Innovation and Governance*; Elstub, S., Escobar, O., Eds.; Edward Elgar Publishing Limited: Cheltenham, UK, 2020; pp. 421–434.
158. Rotta, M.J.R.; Sell, D.; dos Santos Pacheco, R.C.; Yigitcanlar, T. Digital commons and citizen coproduction in smart cities: Assessment of Brazilian municipal e-government platforms. *Energies* **2019**, *12*, 2813. [CrossRef]
159. Jalaluddin, A.M.; Lim, S.B.; Zurinah, T. Understanding the issues of citizen participation. *J. Nusant. Stud.* **2019**, *4*, 1–22. [CrossRef]
160. Cardullo, P.; Di Feliciantonio, C.; Kitchin, R. *The Right to the Smart City*; Emerald: Bingley, UK, 2019. [CrossRef]
161. Boonstra, B.; Boelens, L. Self-organization in urban development: Towards a new perspective on spatial planning. *Urban Res. Pract.* **2011**, *4*, 99–122. [CrossRef]
162. Nederhand, J.; Bekkers, V.; Voorberg, W. Self-organization and the role of government: How and why does self-organization evolve in the shadow of hierarchy? *Public Manag. Rev.* **2016**, *18*, 1063–1084. [CrossRef]
163. Allen, P.M. *Cities and Regions as Self-Organizing Systems: Models of Complexity*; Routledge: London, UK, 1997. [CrossRef]
164. Portugali, J. *Self-Organization and the City*; Springer: Heidelberg, Germany, 2000. [CrossRef]
165. Horelli, L.; Saad-Sulonen, J.; Wallin, S.; Botero, A. When self-organization intersects with urban planning: Two cases from Helsinki. *Plan. Pract. Res.* **2015**, *30*, 286–302. [CrossRef]
166. Shamsul, A.B. Nations-of-Intent in Malaysia. In *Asian Forms of Nations*; Tonnesson, S., Antlov, H., Eds.; Curzon: London, UK, 1996; pp. 323–347.
167. Abdul Mutalib, M.F.M.; Wan Zakaria, W.F.A. Pasca-Islamisme dalam PAS: Analisis terhadap kesan Tahalluf Siyasi. *Int. J. Islam. Thought.* **2015**, *8*, 52–60. [CrossRef]
168. Al-Attas, S.M.N. *Islam and Secularism*; Muslim Youth Movement of Malaysia: Kuala Lumpur, Malaysia, 1978.
169. Canada Government. Canadian Citizenship Act. 1947. Available online: https://pier21.ca/research/immigration-history/canadian-citizenship-act-1947#footnote-6 (accessed on 10 December 2019).
170. Goodman, N.; Zwick, A.; Spicer, Z.; Carlsen, N. Public engagement in smart city development: Lessons from communities in Canada's Smart City Challenge. *Can. Geogr./Le Géographe. Can.* **2020**, 1–17. [CrossRef]
171. Jenson, J. Fated to live in interesting times: Canada's changing citizenship regimes. *Can. J. Polit. Sci.* **1997**, *30*, 627–644. [CrossRef]
172. Abdullah, H.S.; Kalianan, M. From customer satisfaction to citizen satisfaction: Rethinking local government service delivery in Malaysia. *Asian Soc. Sci.* **2008**, *4*, 87–92. [CrossRef]
173. Chin, J. History and context of public administration in Malaysia. In *Public Administration in Southeast Asia: Thailand, Philippines, Malaysia, Hong Kong and Macao*; Berman, E.M., Ed.; Taylor & Francis: Boca Raton, FL, USA, 2011; pp. 141–153. [CrossRef]
174. Hoelscher, K. The evolution of the smart cities agenda in India. *Int. Area Stud. Rev.* **2016**, *19*, 28–44. [CrossRef]
175. Hamza, K. Smart city implementation framework for developing countries: The case of Egypt. In *Smarter as the New Urban Agenda; A Comprehensive View of the 21st Century City*; Gil-Garcia, J.R., Pardo, T.A., Nam, T., Eds.; Springer: Cham, Switzerland; pp. 171–187. [CrossRef]
176. Datta, A. The digital turn in postcolonial urbanism: Smart citizenship in the making of India's 100 smart cities. *Trans Inst Br Geogr.* **2018**, *43*, 405–419. [CrossRef]
177. Cheng, H.; Hu, Y. Planning for sustainability in China's urban development: Status and challenges for Dongtan eco-city project. *J. Environ. Monit.* **2010**, *12*, 119–126. [CrossRef]

178. Mariana, M.O. Stakeholder Participation in the Implementation of Local Agenda 21 in Malaysia. Ph.D. Thesis, Universiti Putra Malaysia, Seri Kembangan, Malaysia, 2008.
179. Noor, E. Foreign and Security Policy in the New Malaysia. Available online: https://www.lowyinstitute.org/publications/foreign-and-security-policy-new-malaysia#sec41256 (accessed on 8 September 2020).
180. Mata, A.M. Is smart city an utopia? Lessons learned and final reflections. In *Smart and Sustainable Cities for Innovative Urban Planning in Mexico*; Editorial Academica Espanola: Beau Bassin, Mauritius, 2018; pp. 198–207. [CrossRef]
181. Hollands, R.G. Beyod the corporate smart city? Glimpses of other possibilities of smartness. In *Smart Urbanism: Utopian Vision or False Dawn? Marvin, S., Luque-Ayala, A., McFarlane, C., Eds.*; Routledge: London, UK, 2016; pp. 168–184.
182. Yigitcanlar, T.; Foth, M.; Kamruzzaman, M. Towards post-anthropocentric cities: Reconceptualizing smart cities to evade urban ecocide. *J. Urban Technol.* **2019**, *26*, 147–152. [CrossRef]
183. Chang, D.L.; Sabatini-Marques, J.; da Costa, E.M.; Selig, P.M.; Yigitcanlar, T. Knowledge-based, smart and sustainable cities: A provocation for a conceptual framework. *J. Open Innov. Technol. Mark. Complex.* **2018**, *4*, 1–17. [CrossRef]
184. Zakzak, L. Citizen-centric smart city development: The case of Smart Dubai's "Happiness Agenda". In Proceedings of the 20th Annual International Conference on Digital Government Research (dg.o 2019), Dubai, UAE, 18–20 June 2019; ACM: New York, NY, USA, 2019; pp. 141–147. [CrossRef]
185. Kulkarni, U.; Robles-Flores, J.A. Data analytics to improve citizen-centric smart city services. In Proceedings of the 25th Americas Conference on Information Systems (AMCIS), Cancún, Mexico, 15–17 August 2019; pp. 1–10.
186. Degbelo, A.; Granell, C.; Trilles, S.; Bhattacharya, D.; Casteleyn, S.; Kray, C. Opening up smart cities: Citizen-centric challenges and opportunities from GIScience. *ISPRS Int. J. Geo-Inform.* **2016**, *5*, 16. [CrossRef]
187. Granell, C.; Bhattacharya, D.; Casteleyn, S.; Degbelo, A.; Gould, M.; Kray, C.; Painho, M.; Trilles, S. GEO-C: Enabling open cities and the open city toolkit. In Proceedings of the International Archives of the Photogrammetry, Remote Sensing and Spatial Information Sciences (ISPRS), Dar es Salaam, Tanzania, 29–31 August 2018; pp. 61–68. [CrossRef]
188. Degbelo, A.; Bhattacharya, D.; Granell, C.; Trilles, S. Toolkits for smarter cities: A brief assessment. In *Ubiquitous Computing and Ambient Intelligence*; García, C.R., Caballero-Gil, P., Burmester, M., Quesada-Arencibia, A., Eds.; Springer: Heidelberg, Germany, 2016; pp. 431–436. [CrossRef]
189. Emaldi, M.; Aguilera, U.; López-de-Ipiña, D.; Pérez-Velasco, J. Towards citizen co-created public service apps. *Sensors* **2017**, *17*, 1265. [CrossRef] [PubMed]
190. Lopes-de-Ipina, D.; Vanhecke, S.; Pena, O.; De Nies, T.; Mannens, E. Citizen-centric linked data apps for smart cities. In *Ubiquitous Computing and Ambient Intelligence. Context-Awareness and Context-Driven Interaction*; Urzaiz, G., Ochoa, S.F., Bravo, J., Chen, L.L., Oliveira, J., Eds.; Springer: Cham, Switzerland, 2013; pp. 70–77.
191. Kauppinen, S.; Luojus, S.; Lahti, J. Involving citizens in open innovation process by means of gamification: The case of WeLive. In Proceedings of the 9th Nordic Conference on Human-Computer Interaction (NordiCHI'16), Gothenburg, Sweden, 23–27 October 2016; ACM: New York, NY, USA, 2016. Article No. 23. [CrossRef]
192. Lee, J.; Lee, H. Developing and validating a citizen-centric typology for smart city services. *Gov. Inf. Q.* **2014**, *31*, S93–S105. [CrossRef]
193. Alshibly, H.; Chiong, R. Customer empowerment: Does it influence electronic government success? A citizen-centric perspective. *Electron. Commer. Res. Appl.* **2015**, *14*, 393–404. [CrossRef]
194. Delmastro, F.; Arnaboldi, V.; Conti, M. People-centric computing and communications in smart cities. *IEEE Commun. Mag.* **2016**, *54*, 122–128. [CrossRef]
195. Giesbrecht, T.; School, H.J.; Schwabe, G. Smart advisors in the front office: Designing employee-empowering and citizen-centric services. *Gov. Inf. Q.* **2016**, *33*, 669–684. [CrossRef]
196. Ilhan, A.; Möhlmann, R.; Stock, W.G. Citizens' acceptance of u-life services in the ubiquitous city Songdo. In *Citizen's Right to the Digital City: Urban Interfaces, Activism, and Placemaking*; Foth, M., Ed.; Springer: Singapore, 2015; pp. 215–229.
197. Mukhtyar, K. Frost & Sullivan's Citizen Centric Smart City Development Model. Available online: https://www.linkedin.com/pulse/frost-sullivans-citizen-centric-smart-city-model-kavan-mukhtyar%0AFrost (accessed on 15 November 2017).
198. Purao, S.; Seng, T.C.; Wu, A. Modeling citizen-centric services in smart cities. In *Conceptual Modeling*; Ng, W., Storey, V.C., Trujillo, J.C., Eds.; Springer: Berlin/Heidelberg, Germany, 2013; pp. 438–445.

sustainability

Article

Alternative Governance Model for Historical Building Conservation in China: From Property Rights Perspective

Nan Guo , Edwin Hon Wan Chan * and Esther Hiu Kwan Yung

Department of Building and Real Estate, and Research Institute for Sustainable Urban Development, The Hong Kong Polytechnic University, Hong Kong 999077, China; nannan.guo@connect.polyu.hk (N.G.); esther.yung@polyu.edu.hk (E.H.K.Y.)

* Correspondence: edwin.chan@polyu.edu.hk; Tel.: +852-2766-5800

Abstract: With the rapid advancement of urbanisation, the adaptive reuse of heritage plays a key role in achieving sustainable development, which is widely recognised by UNESCO and International Council on Monuments and Sites (ICOMOS). In the process of urban renewal, unclear property rights have seriously hindered the relocation of old houses, compensation and the adaptive reuse of historical buildings, even causing a series of social contradictions, such as violence. Moreover, forced evictions and controversy in dealing with the rights of residents, particularly the so-called 'nail households' have attracted public attention. However, few studies have analysed the problems and countermeasures from the perspective of unclear property rights. This study focuses on analysing the unclear property rights of historical buildings to propose an Alternative Governance Model for Historical Building Conservation in China. Founded on the Coase Theorem of externalities and property rights to examine the existing complex property ownership and rights patterns of 63 historical buildings in the famous Pingjiang Historic Block in Suzhou, China, the model provides reasonable and feasible reconstruction schemes for each situation. The operation model can also provide a symbiosis of new and old building solutions for urban renewal in developing countries, which may encounter a similar challenge of urbanisation.

Keywords: sustainable urban development; governance; urban renewal; historical buildings protection; property rights

Citation: Guo, N.; Chan, E.H.W.; Yung, E.H.K. Alternative Governance Model for Historical Building Conservation in China: From Property Rights Perspective. *Sustainability* **2021**, *13*, 203. https://doi.org/10.3390/su13010203

Received: 6 November 2020
Accepted: 24 December 2020
Published: 28 December 2020

1. Introduction

To meet the sustainability agenda of urbanisation in the process of smart and sustainable urban development, there has been a certain consensus that, compared with new buildings, the protection and transformation of old buildings have been generally accepted, due to either social effects or environmental benefits [1]. In terms of the environment, renovating building materials of old buildings can reduce the pressure on energy consumption. By recycling existing materials, which can reduce carbon emissions, the impact on climate change can be reduced [1,2]. On the other hand, from a social perspective, old building protection projects can maintain social continuity and protect buildings that have been in disrepair for a long time [3].

However, given the insufficient awareness about the protection of historical buildings and facing a shortage of urban land, in numerous densely populated countries and regions, particularly in the fast-growing East Asia regions, historical buildings are demolished to meet various development needs. This practice is pervasive in many cities that are in the initial stage of economic development. For example, after the Second World War, Japan's large-scale urbanisation development resulted in the arbitrary destruction and demolition of many ancient buildings [4]. In the 1950s, the ancient walls of Beijing were demolished on a large scale [5]. Moreover, at the beginning of this century, Lee Tung Street in Hong Kong was demolished and rebuilt by the government as a pedestrian shopping street without any distinctive local style [6]. The continuous growth of modern cities is gradually compressing the space of historical buildings. As such buildings are demolished, a city's features and its people's sense of identity in terms of culture are disappearing. Historical buildings are under many threats in many historic cities, including rapid urbanisation, an increasing housing demand, and socio-cultural and climate changes [7,8]. From the perspective of sustainable development, the adaptive reuse of historical buildings can play a decisive role not only in terms of heritage conservation, but also as an urban strategy capable of contributing to environmental protection and generating new economic, cultural, and social values [9,10].

In recent years, an increasing number of successful adaptive reuse projects have given the government confidence, and policy makers are also trying to apply this method more to protect historical buildings and blocks [11]. However, there are numerous cases of obstacles caused by unknown ownership and property rights, which seriously hinder the construction of smart and sustainable urban development. Specifically, unclear property rights can neither deter the destruction of important historical buildings nor effectively protect existing ones [12]. On the one hand, the government experiences difficulties paying high amounts of money for repairs. On the other hand, the excessive use of such historically valuable buildings with existing conditions may devastate numerous households. The government has encouraged the public to participate in adaptive reuse projects, which is an important measure for achieving sustainable urban development [13], but chaotic property rights, particularly as a legacy of transition from state-ownership property in China, seriously hinder the transformation of these buildings by developers. Incidents such as demolitions and violence against demolitions have occurred owing to mistrust among the government, residents, and developers [14,15]. Meanwhile, residents who do not cooperate in demolitions waste time and increase the cost of reconstruction projects [16,17]. Property rights issues hinder the renovation of old buildings, making sustainable urban development difficult to pursue. Such problems are particularly serious in developing countries such as China, which has transformed from a centrally planned to a market economy. Ancient buildings with rich historical and artistic value have become a sweet burden in the urbanisation of a city.

This study selected the most representative case, namely, the Pingjiang Historic Block in China, as the research test bed. An on-site field investigation was conducted in this district. A total of 63 historical buildings have been, are being, or are about to be transformed in this area. By examining these historical buildings, this study analyses the influence of the complex property rights structure on the protection and redevelopment of historical buildings. The problems in this test bed case are typical of numerous other historic cities in China and worldwide. Based on the Coase Theorem, the authors combined the reality and difficulties of historical building protection to establish an operational governance model founded on property rights. Therefore, the findings can provide a feasible operational model that can solve the primary problem in sustainable urban development and be applied to similar historical buildings in China and other countries during the smart and sustainable urban development process. The result may provide policy makers with measures to solve urbanisation challenges, such as the inadequacy of urban conservation management and sustainable development policies, dilemma in balancing public and private rights, the lack of resources amongst decision makers, and persistent conflict and competition between heritage conservation needs and developers' interests [7].

2. Theoretical Framework

2.1. Property Rights

Property rights refer to the behavioural relationships of people resulting from the existence of objects and their use. Property rights are rights unified by government enforcement and market forces. The property rights system refers to the systemised property rights relationships and system support for dividing, determining, defining, protecting, and exercising one's property rights [18]. The property rights system can be further divided into formal rules, including laws, regulations, social contracts, the construction and determination of organisations, and informal rules, including cultural traditions, habits, conventions, moral rules, and so on [19,20].

Property rights include ownership, power of possession, right of control, and right of use. Coase holds a view in *The Problem of Social Cost*, which won the Nobel Prize, that the initial delimitation of legal rights affects the efficiency of an economic system [21]. If property rights are not explicitly defined, then individuals cannot have reasonable expectations whilst making transactions with others, and society will lose benefits generated by labour division and cooperation. Additionally, opportunistic behaviours, such as encroachment on the rights and interests of others, will emerge, and social disruption will be inevitable. Similarly, when individual property rights cannot be protected by society, individuals cannot make long-term economic plans and arrangements and thus lose incentive to accumulate and protect resources. Furthermore, negative drawbacks include the act of wasting and destroying resources [10]. If we want to optimally allocate limited resources, we need to consider and avoid the damaging impact caused by two parties whilst determining their actions [21].

The Coase Theorem shows the decisive effect of property rights on transaction costs [21]. The key points are summarised by Chappelow [22] as:

- *Under the right conditions parties to a dispute over property rights will be able to negotiate an economically optimal solution, regardless of the initial distribution of the property rights.*
- *The Coase Theorem offers a potentially useful way to think about how to best resolve conflicts between competing businesses or other economic uses of limited resources.*
- *In order for the Coase Theorem to apply fully, the conditions of efficient, competitive markets and, most importantly, zero transactions costs must occur.*

The clear delineation of private property rights is an essential prelude to market transactions. The clearer the property rights, the lower the transaction costs and the higher the efficiency. This theorem can be reflected in the protection of historical buildings; that is, when the property rights of historical buildings are clear, the operation of the market mechanism and adaptive reuse of historical buildings are efficient and can be used appropriately. However, when property rights are unclear, the protection of historical buildings is hindered and their destruction is exacerbated.

Coase holds the view that 'The delimitation of rights is an essential prelude to market transactions' [19]. In the process of urban renewal and adaptive reuse of historical buildings, clarifying property rights is a prime concern [17]. Coase further states that a government is a super firm because it can influence the use of production factors through administrative decisions. For example, a given government can establish a set of legal systems about rights that can be adjusted through market transactions. It can also compulsorily stipulate laws and require people to obey them. The adjustment of one kind of right produces more output values than other arrangements [21].

2.2. Hypothesis

On the premise of conforming to the law of Chinese Cultural Relics, according to the Coase Theorem, an operational model can be established to properly maintain the historical buildings as far as possible. In this governance model (Figure 1), the government is responsible for the management and maintenance of the historical buildings with public property rights, while the private property owners are responsible for the maintenance of their buildings. In this research, Pingjiang Historic Block is selected to test the feasibility of this model.

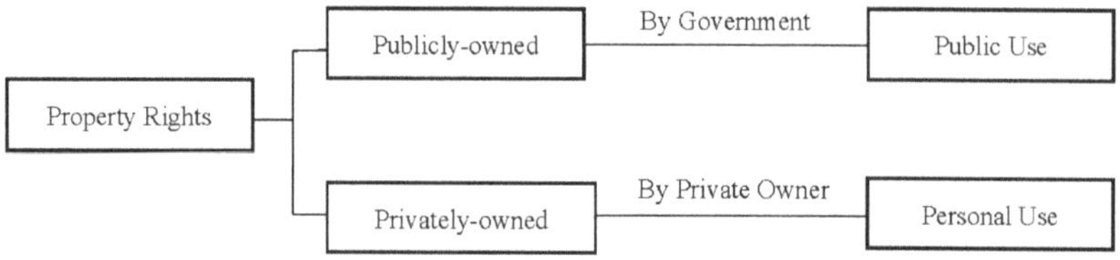

Figure 1. Governance Model for Historic Buildings Conservation.

3. Methodology

This research selected Pingjiang Historic Block as the test-bed object because the historical buildings in this district cover nearly all modes of property rights existing in China, which are highly concentrated in the current reconstruction of old cities, as well as in the protection and renewal of historical blocks in China [23]. The government has taken numerous effective measures to protect the historic block, such as the Protection and Restoration of Historical Buildings Project. It was highly praised for the revitalisation project. In 2005, UNESCO awarded the Asia-Pacific Heritage Award to the Pingjiang Historic Block. According to the organising committee, *'The revitalisation project is a commendable example of integrated urban rehabilitation, which has restored the physical, social and commercial fabric of one of China's most well-known waterway historic towns'* [24]. However, certain renovation projects still have been delayed owing to property rights issues. Pingjiang Historic Block is highly representative and is of prohibitive research value. The present study focuses on the discussion of historical buildings with complex property rights structures.

There are a total of 63 historic sites arranged along both sides of the street (Figure 2), including one material cultural World Heritage Site, one nonmaterial cultural World Heritage exhibit place, three national-level cultural relic protection units, and 15 provincial and municipal cultural relic protection units and controlled and protected buildings (as of 2015). The property rights analysis conducted in this study was based on real and accurate data. These sites provided a considerable number of research subjects [25,26]. Official data obtained from functional government departments would have been the best choice. However, the government claimed that it lacks accurate property rights data. Based on the literature, the property rights structure of several historical buildings in Suzhou are only recorded in the 'Protection and Utilisation of Controlled and Protected Buildings in the Ancient City of Suzhou'. However, obtaining property data that genuinely and thoroughly reflect the current state was difficult. Therefore, to fill this gap, the authors selected the Pingjiang Historic Block, which is a representative historic quarter, as a pilot for a field investigation for two years and discovered and sorted out information that contradicted certain literature descriptions.

The survey was based on the latest version of the Suzhou Pingjiang Historical and Cultural Street Protection Plan (2014) issued by the Suzhou Municipal Planning Bureau. In conjunction with the 'List of Suzhou Controlled and Protected Buildings' of the same year provided by the Suzhou Bureau of Cultural Relics, a field investigation was conducted in the 63 existing national, provincial, and municipal cultural relic protection units and controlled and protected buildings, which were set up by the Suzhou municipal government. Registered property right structures, households, resident compositions, the current protection situation, existing functions, and other aspects were also investigated.

Face-to-face interviews were conducted with experts, administrators, and residents (Table 1). The authors also visited competent government departments in Suzhou, including the Department of Housing Management; Department of Planning, Land and Resources Bureau; and Urban Construction Archives Bureau. A massive amount of first-hand material was obtained, documents with approximately 80,000 characters transcribed from audio recordings were compiled, and nearly 3000 photos and videos were sorted. The investigation lasted for two years of 2017–2019, and the data were updated in August 2019.

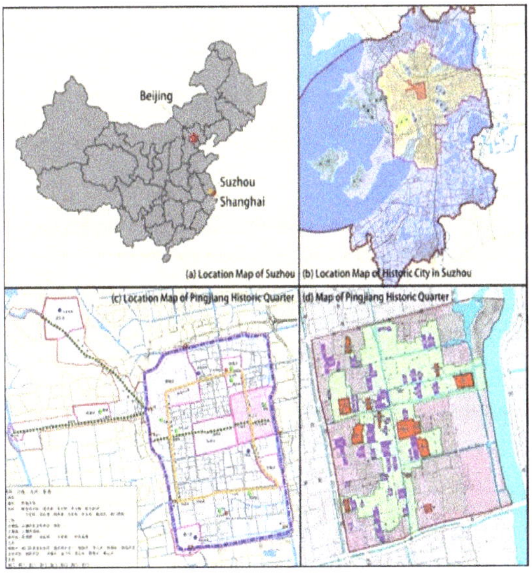

Figure 2. (**a**) Map showing the location of Suzhou; (**b**) map showing the location of the historic city of Suzhou; (**c**) map showing the location of Pingjiang Historic Block; (**d**) map of Pingjiang Historic Block (source: Suzhou Planning Bureau website).

Table 1. Interviewees in face-to-face interviews.

Category	Name of Interviewees or Institutions
Experts	AA (expert in ancient city protection) BB (director of China's Famous Historic and Cultural City Protection Research Academy) CC (expert in ancient city protection in Suzhou) DD (general manager of Suzhou Ancient City Investment and Construction Co., Ltd.) EE (director of Suzhou Pingjiang Historic District Protection and Maintenance Co., Ltd.) FF (director of Department of Planning of Suzhou University of Science and Technology) GG (founder of Cat's Castle in the Sky Concept Book Store)
Departments	Department of Housing Management Urban Construction Archives Bureau Department of Planning, Land and Resources Bureau Planning Bureau
Enterprises	Suzhou Pingjiang Historic District Protection and Renovation Co., Ltd. Suzhou Culture and Tourism Development Group (SCTDG)
NGOs	Suzhou Institute for the Conservation of National Historic Cities
Residents	Residents and users in the 63 historical relics in Pingjiang Historic Block

4. Results

4.1. Survey Results

The research results presented certain views that verified and corrected published literature. In the actual situation, records of the 63 historical relics in Pingjiang showed that 35 buildings were publicly owned, accounting for 55.6% of the total number of buildings. A total of 22 publicly and privately owned buildings had property rights, accounting for approximately 35%. However, six buildings were privately owned, accounting for 9.5% of the total (Table 2). Among the controlled and protected buildings, 40 public houses were under the direct administration of the Housing Management Department, accounting for 63.5% of all historical relics. This portion was mainly used as low-rent houses provided for low-income families in Suzhou.

of Private Rental Housing that 'private homeowners could not recover the houses which already rented by the state'. At the time, the property rights structure appeared in three forms, namely, public property, private property, and rental houses. Rental houses were products of the socialist transformation. Former private homeowners could no longer retrieve their property, but the state had not clearly defined the ownership of property. During the Cultural Revolution in 1966, the Red Guards forcibly seized the property rights of private homeowners through violent means. All houses had one property rights attribute, that is, state ownership [27].

4. From 1983 to 2004: Implementation of private housing policy and resolution of the problem of 'standard houses' After the Cultural Revolution, the government implemented the private housing policy and returned the property rights of private houses to private homeowners. However, the government claimed that the property rights of 'rental houses' belonged to the state and implementing the private housing policy was not feasible [27,28]. From the 1980s, historical houses became private properties and rental houses. However, the most dramatic difference during the 1950s was that the property rights of 'rental houses' changed from being private to public property, thereby becoming part of the directly managed public housing, which was the most complex existing property rights structure in China.

In summary, the structure of real estate property rights was extremely complex in 1949 and the new government then issued a series of measures to clarify it. However, the original problem of property rights during the Cultural Revolution was not clarified and was worsened by the political struggle.

After the reform and opening up in 1980s, the government attempted to readdress the problem, but failed to achieve the desired effect owing to insufficient implementation. Since the 1980s, China has entered the stage of large-scale civil construction. In the process of the demolition and reconstruction of old cities, removal and relocation have become significant problems for the government, and the drawbacks of the unclear property rights structure have become apparent. However, no one has been willing to tackle such sensitive topics, and the government and academia have prevaricated with 'problems left over by history', claiming that 'property rights are too complex' to be resolved. Although the awareness of the protection of traditional culture has increased since 2000 and the government has invested large amounts of capital to preserve and renovate historical blocks, property rights problems remain between the ideal and reality, thereby severely delaying the pace of urban renewal. The property rights issue affects the entire process of urban development and cannot be avoided. Therefore, the issue, that is, the existing complex property rights structure in China, should be addressed and resolved with sound theoretical basis.

4.2.2. Legal Factor

To protect outstanding traditional architecture, the government implemented the Law of the People's Republic of China on the Protection of Cultural Relics in 1982. Historical buildings were classified as national, provincial, and municipal cultural relic protection units according to their historical, cultural, and artistic value. In addition, local governments established protection lists for outstanding historical buildings excluded from cultural relic protection units, such as the Excellent Historical buildings in Shanghai, the Excellent Modern Buildings in Beijing, the Historic Feature Buildings in Xiamen, and the Controlled and Protected Buildings in Suzhou [3,29]. However, the original complex property rights problem worsened when the cultural relic labels were added to the buildings [28].

Article 5 of the 1982 Law stipulates that for all memorial buildings, historical buildings, and cultural relics publicly or privately owned, ownership is under the protection of national laws, and the owners of these cultural relics must abide by the regulations of the country on cultural relic protection and management. In addition, the law stipulates that buildings owned by the state and protected as cultural relics cannot be sold on the market [30,31]. The property rights of these buildings are redefined by national laws; that is, residents and users only have usage rights. The buildings are publicly owned and the local government represents the state in terms of supervision and control. Given the feature described above, an intersection between these two types of property rights emerges when historical buildings are listed as cultural relics, thereby making it difficult to determine and obtain rights to yields derived from ownership [32]. The value form cannot be separated from the value entity. Therefore, defining the boundary of property rights is difficult, as it is obscure in an actual operation.

In addition to unclear property rights, inappropriate modes of property rights can damage historical buildings. Certain historical buildings are public houses with property rights under the direct administration of the government. Most of such buildings are overused, endure exposure to wind and rain, and incur degrees of damage that are extremely high. Therefore, maintenance expenses are considerably large, and residents have no incentive to contribute to the maintenance of such buildings. Article 6, Chapter 1 of the 1982 Law stipulates that expenses for cultural relic protection and management shall be included in the financial budget of the central and local government; that is, the government should arrange the protection and management of historical resident houses and fund their repair and protection. Although residents are not required to make contributions, they can enjoy the benefits. In such cases, historical houses repaired with government funds can be regarded as public articles and thus enjoyed by residents free of charge. Therefore, people ignore the protection of historical buildings, as they can enjoy the benefits without assuming responsibility. To maximise such benefits, residents may overuse or damage buildings under controlled protection. Conflicts between the use and maintenance of historical buildings will result in a conflict between the government's invalid investment in such buildings and residents' overconsumption.

Therefore, currently, historical buildings that suffer from the most severe damages are historical buildings whose property rights belong to the state. This problem is common in historical cities in China. As the maintenance and upgrading of historical buildings require large amounts of financial and technical resources, the government's budget cannot satisfy this demand. Therefore, for such buildings, the government has maintained an attitude of 'not aspiring for the ownership but the existence', in order to transfer usage or property rights, activate and reuse historical buildings by cooperating with social forces, and extend the longevity of such buildings. In the case of complicated property rights, the clarification of property rights and adaptive reuse were problems that needed to be resolved.

4.3. Causes of Changes in Property Rights in Pingjiang Historic Block

4.3.1. External Causes

Most external reasons were influenced by policies from different eras. The timeline shows a series of changes. Figure 3 illustrates the timeline of the evolution of property ownership from 1949. In the early 1950s, every parcel of land in the country was registered, government authorities issued the Property Ownership Certificate, and the 1954 Constitution protected lands and houses.

Figure 4 shows the timeline for changes in property rights in Pingjiang Historic Block. In 1956, the government initiated the socialist transformation of industry and commerce. In the same year, the central government approved the 'Basic Conditions of Private Property in the Cities' and 'Opinions on Initiating Socialist Transformation'. Moreover, private houses in these cities were subject to 'transformation'. In 1958, the compulsory transformation of private property campaign emerged. The primary form was 'lease by the country', which meant that 'the country will be responsible for the overall leasing, distribution and maintenance', targeting certain privately leased houses and drawing a starting line for the transformation; that is, if a property leased by its owner ventured beyond the starting line, then it would be subject to transformation. The intended purpose of this policy was to 'use the methods like purchase, which pays fixed rent during a certain period, to gradually change their ownership'. However, in practice, the government only collected compulsory rent from owners without purchase or similar activities, and ownership remained in the hands of owners with a Property Ownership Certificate. When the Cultural Revolution began in 1966, the Red Guards seized private properties through violence and claimed ownership. According to Article X of the 1982 Constitution, urban lands shall be the property of the state. Between 1982 and 1988, the government cleared private properties on a massive scale, owners were granted Property Ownership Certificates, and the concept of land ownership ceased to exist. However, according to residents, they lost ownership not only of their land, but also of their houses.

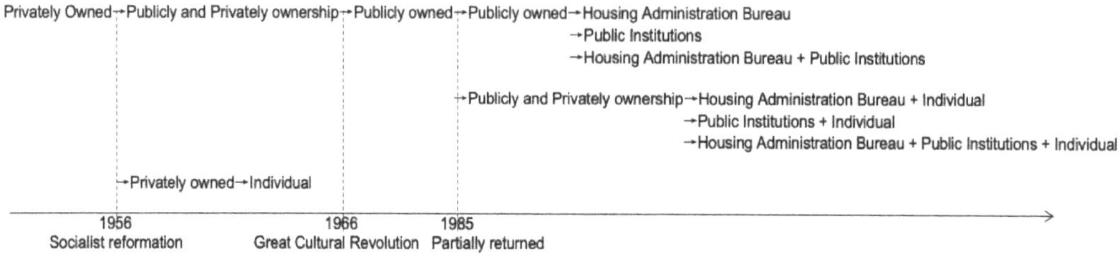

Figure 4. Timeline for changes in property rights in Pingjiang Historic Block (summarised by the authors).

4.3.2. Internal Causes

Internal causes included the residential function of houses entering the market as a commodity since the 1980s, thereby resulting in the circulation and transaction of property or usage rights, which could lead to a change of property rights, as shown in Figure 5. It presents a summary of the circulation pattern of properties in history according to market conditions. After the Cultural Revolution, numerous private houses transformed into directly controlled public houses under the ownership of the housing authority. In recent years, the number of low-income individuals gradually decreased. Therefore, residents of affordable housing have disregarded the value of houses and overused them. To protect historical buildings, the government intends to recall the right to use such houses. In terms of ownership changes, properties are transferred to the government or private enterprises after purchase and primarily consist of public-interest and commercial houses. The circulation of property rights mainly depends on the government's redemption and is relocated through the transfer of property rights or currency.

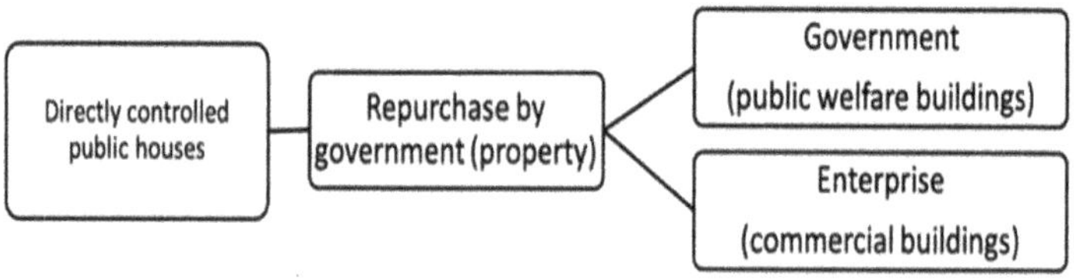

Figure 5. Circulation of property rights of historical buildings.

5. Discussions

Coase's view involves a key word that is a problem to be solved, i.e., resource redistribution. In the field of urban renewal, historical buildings have multiple attributes. As cultural relics, historical buildings have very high historical and artistic values, so knowledgeable people are needed to maintain these buildings through professional means. Houses have residential functions, but houses are dilapidated and entail high maintenance costs, which are hard for residents to bear. Whether to protect them as cultural relics or to continue to allow them to serve as houses is the redistribution of production factors in urban renewal. When the ownership is fragmented it will bring many constraints. Taking back the usage right and unified planning by the state is more conducive to the preservation of historical buildings [33,34].

Indeed, historical and artistic values cannot be copied and reconstructed, as opposed to the dispensable residential function. The government is a super enterprise, which can control production factors via administrative means [35]. The government also plays such a role in promoting urban renewal [36]. For the protection of cultural relics, the government should promulgate compulsory regulations to restrict the behaviour of users, whether property rights are publicly or privately owned, under the protection of the law, whilst the use function should be considered. When property rights are public, the government has the right to take back the house and optimally allocate resources. While retaining the artistic and historical value and the use function of traditional architecture, the government can protect the historical buildings according to the public and market demand. Through policy adjustment, the government can expand the market of historical buildings [35]. Experiential reconstruction will become the trend in future historical building protection. This type of reconstruction resembles 'haematopoiesis', which can also solve the problem of a shortage of funds for the protection of historical buildings. For houses with use functions belonging to residents, when the property right is clear and the transaction cost is relatively low, the government hopes to maintain the decision-making power of the housing function. Residents who live here are willing to move out after they receive satisfactory compensation, given that they do not have the property rights of historical buildings and cannot afford high maintenance costs. Generally, the compensation they get is much higher than the value of the house because the high-value properties tend to be overcompensated [37]. Therefore, those who have no property rights but only the right to use them are willing to cooperate with the government. Even the nail households just want to haggle over the compensation rather than move out. Once an agreement is reached, the government can take back the property rights.

However, when property rights are private, property owners know that permanent private property rights are scarce resources. The property owners have the exclusive right to alienate or to not alienate the buildings [38]. In addition, they have lived in their own houses for generations and are not willing to hand them over to the state easily. In their minds, the transaction price is very expensive. The government only plays a supervisory role and does not pay high relocation fees, thereby forcing property owners to protect their houses.

With due reference to key points of the Coase Theorem highlighted in the above Section 2: Theoretical Framework, when these rights belong to the same subject, the owner has initiative and motivation to protect and maintain a historical building effectively. The research results substantiate the authors' view as proposed in the beginning. For example, during the survey, privately owned buildings, such as the Fang residence on Xuan Qiao Alley, were in a satisfactory condition. Private owners had lived here for generations. However, when the right to use is transferred, ensuring that the user will seriously protect the historical building is difficult. As shown in Figure 3, the property rights system has continuously changed since 1958. Therefore, defining the rights and responsibilities of parties is challenging. Residents only possess the right to use a historical building and thus lack a cultural identity and sense of belongingness. They have no enthusiasm to maintain the 'public house', but hope to obtain tangible benefits from its demolition. The rent of such houses is too low to afford maintenance costs. These housings are dilapidated, suffering from the daylighting and dampness for years. There are various safety hazards in the house with a danger of collapse at any time. Meanwhile, the Housing Management Department, which has right of ownership, also lacks sufficient funds to protect historical buildings. 'Rental houses' and 'public houses' suffer from damages or are destroyed, similar to the Xu residence.

Moreover, when the government attempts to recall right to use, large numbers of residents become unwilling to move out. The term 'nail households' refers to people who refuse to move out of buildings for demolition, often owing to disagreements about compensation, which severely delays the progress of the redevelopment process. The DeLin Hall Wu residence is an example of such a case, which was delayed for five years. In the above case, the non-zero-sum game between residents was a typical prisoner's dilemma model. Some people are willing to move out early and get rewards and resettlement houses. However, some nail households are willing to get more compensation. The personal best choice is not the group's best choice. In the face of interests, the relationship between people becomes subtle. While competing for interests, they also have selfish desires to get more. In a group, individuals making rational choices often lead to collective irrationality.

Through interviews with developers and contractors of many projects, it was understood that property rights and funds are common problems in all current historical building reconstruction projects. The biggest uncertainty in funds is the demolition compensation, and the contractor must overcome the difficulties that cannot be overcome by traditional crafts.

For the 10 pilot projects of the first batch of ancient building protection and repair projects in Suzhou, some ancient buildings abandoned the reconstruction, and some projects were promoted. The whole project progressed very smoothly when all tenants accept the reconstruction, such as with the former Fang residence, the current Xiao Hui Wang Art Centre. Regarding the projects that all the residents refused to relocate, such as private property owners or projects with proper building maintenance and living environment, the government gave up the relocation and allocated funds for the optimisation of living conditions, from the reconstruction project to the people's livelihood project. For most projects, some residents coordinated with the relocation and some residents refused. Demolition was repeated in each process along with the betrayal and cooperation between households, and the Nash equilibrium gradually became Pareto Optimality. That is, in this situation, no property right owners or residents could be better off without making at least one individual worse off [21]. It was considered to be a minimal notion of efficiency that did not necessarily result in a socially desirable distribution of resources [39,40]. Although this was not the optimal solution, it was the only way to solve the property rights problem of historical building adaptive reuse.

Such cases can be seen everywhere in China where cities are developed constantly, such as in this red-headed document, *Announcement of the Suzhou Municipal People's Government on the Determination of Houses on State-Owned Lands* (Su Fu Gong [2013] No. 3), which is the housing compensation plan of Pan Zuyin's former residence (a historical building in Pingjiang Historic Block) project:

> *The expropriated person may choose monetary compensation or exchange property rights. The expropriated households who choose the monetary compensation can purchase the fixed-quotas commercial houses provided by the government. If the property rights exchange is chosen, the value of the property exchange houses and the expropriated houses will be calculated after the price is calculated at the same evaluation point, and the difference is settled.*

Among them, (1) monetary compensation means that in the house expropriation compensation, the owner of the expropriated house is compensated in monetary form based on the market evaluation price. (2) Property rights exchange means the exchange of houses provided by the house expropriation department for the exchange of property rights and the houses to be expropriated. After the price is calculated, the difference is settled. For the residents who cooperate with the relocation, different relocation incentive fees can be obtained depending on the degree of cooperation.

Through the above analysis, the model can be improved as shown in Figure 6. It is consistent with the hypothesis model that privately owned historical buildings are still protected and repaired by private property owners. In order to protect the interests of the private property owners, the private parts of the public–private ownership buildings will still belong to the private property owner, and the buildings will be maintained by the private property owner. Similarly, the Public Institutions shall be responsible for the management of the parts in public–private historical buildings and the public–private historical buildings that belong to them.

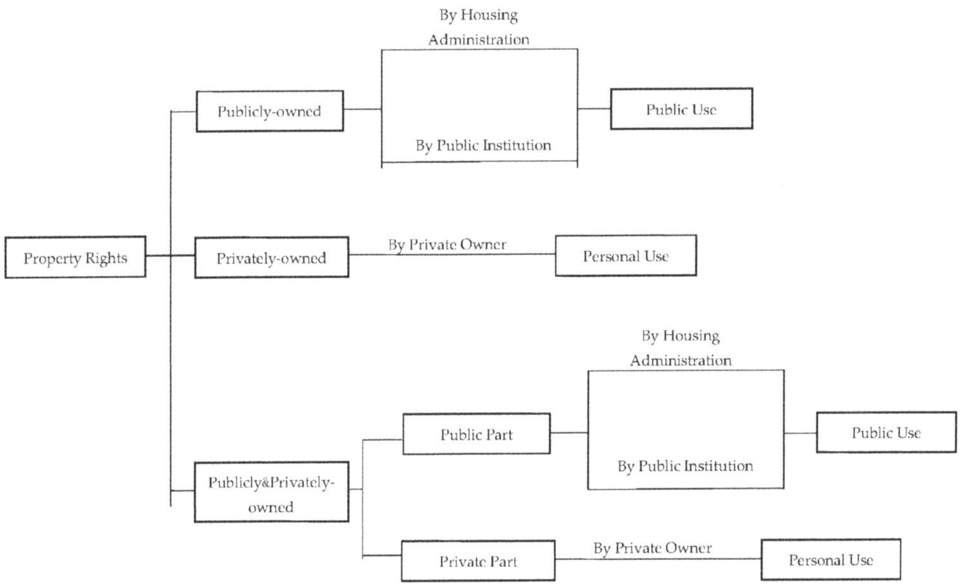

Figure 6. Governance Model for Historic Buildings Conservation in China (proposed by the authors).

In this governance model, the most complicated part belongs to the Housing Administration Bureau, which is called direct management of public housing. Due to the historical reasons mentioned before, this type of housing is rented to citizens with poor economic conditions. To avoid excessive use, local governments should take back the usage rights of these buildings gradually through two ways, that is, property rights replacement and monetary compensation. For a successful adaptive reuse project, two principles should be followed, that is, to obey the government's unified organisation and management, and to follow the government's planning scheme and functional orientation.

The analysis shows that whenever a whole building is publicly owned, the transaction cost to the government is close to zero, and the government has the autonomy to renovate the building. Among the 35 public historical buildings in Pingjiang District, 20 of them have been renovated and are managed by the government, and seven are under renovation, accounting for 77% of effectively protected public historical buildings. Residents in the remaining eight buildings have expressed their desire to move away from the dilapidated buildings. They are willing to move to new homes if the government will give a satisfactory compensation. Meanwhile, six buildings are owned by private individuals, accounting for 100% of private buildings. These private residents refuse to make deals with the government. Yet, the government is not willing to pay high fees to obtain the property rights of such houses. These types of property rights are distinctive, and thus, renovation or maintenance work should be carried out smoothly. However, the public–private ownership of property rights is considerably complicated. A total of 22 historical buildings in Pingjiang Historic Block are public–private owned, and 14 of them (accounting for 64%) cannot be promoted for renovation. Five buildings have been transformed, and two are under renovation. However, from the perspective of renovation projects, several owners of private properties, including such buildings, are willing to trade with the government, whereas the rest of them refuse. For private property owners who refuse to trade, the government will transform the public areas of their buildings. The private property owners can still reside their own houses, but they have to take up the responsibility of routine maintenance of the historical buildings in the areas that they own. The government will help them with the exterior transformation. Without affecting the daily life of private property owners, the appearance of buildings with private property rights should be made consistent with that of the public parts, resulting in consensus between landscape and art value. Such complex cases can explored be in detail in future studies. The 'bargain' mentioned by Coase also exists in the protection of historical buildings. In the process of repeated communication, the government and private owners can achieve the optimal allocation of resources. When both sides cannot agree on the price, they should arrive at a symbiosis to work together.

6. Conclusions

Taking Pingjiang road as a test bed for analysis of property rights in historical buildings, this study highlights two key findings through the application of the Coase Theorem in the conservation of historical buildings. First, the impact of property rights, that is, whether the property rights are clear or not, will affect the progress of protection work. Second, the governance model for heritage conservation, that is, different property rights of historical buildings lead to different transformation methods.

Firstly, in the case of Pingjiang Historic Block, the result of the survey showed that unclear property rights seriously hinder urban renewal. Unclear property rights will lead to unclear responsibilities and rights, and lack of responsible persons to maintain historical buildings. Unclear property rights can hinder transactions and incur higher costs. Given the unclear property rights, historical renovation projects experience difficulties in proceeding according to their plans. Gradually eliminating the unclear parts of the chaotic property rights and implementing transformation in batches are necessary. The reconstruction project can be implemented gradually according to the unique layout of historical buildings.

Secondly, urban renewal is a process of resource optimisation and reuse. The historical and artistic values of historical buildings are valuable resources to society as a whole. In the process of sustainable urban development, it is necessary to retain the artistic and historical value and the use function of traditional architecture. When property rights are clear, public buildings are protected by the government. Similarly, when property rights are private, private property owners should be responsible for the maintenance. With a regulatory role to play, the government has the obligation to protect historical buildings through designing good policy. Therefore, historical buildings with public property rights should be gradually taken over and transformed by the government. Historical buildings with private property rights should still be maintained and managed by private property owners, which is the low transaction costs option, and the government should assume the supervisory and regulatory role.

This study has its limitations. Firstly, urban renewal is a dynamic process, and the role and capacity of government or individuals are constantly evolving or even changing functions. The data listed by the author can only cover a certain period of time, and subsequent changes are excluded. Secondly, the authors attempted to obtain official data from the government, but the property rights issue is sensitive, and the government is reluctant to provide such data. In addition, we find that the data obtained from some literature differ from the actual data. Thus, the authors spent several years collecting the data on-site themselves. However, it is admitted that some deviations were observed in cases that involved oral history, but they are verified as far as possible. This condition can be regarded as both a limitation and a unique contribution of this paper. Firsthand data are closer to the truth, and they can be used to verify the governance model in this paper. Lastly, the proposed model only provides a method, but each historical building has a unique story and value in practice. As such, the protection work must be analysed on a case-by-case basis, which can be elaborated in subsequent studies.

This model, based on sound theories, successfully solved the primary problem of adaptive reuse projects and provided solutions for other old building renovation projects with disordered property rights, so that the property rights problem was no longer an obstacle to sustainable urban development but could be explored to provide the drive force. An increasing number of vacant and dilapidated old buildings have been given new life, which is conducive to sustainable urban development from social and environmental perspectives. While reducing the impact of buildings on climate change, at the same time, the social and artistic value of buildings has also been extended.

Author Contributions: All authors contributed to this paper with N.G. taking the leading role. E.H.W.C. and N.G. helped design the framework. N.G. conducted the investigation, analyzed the data, and drafted the article. E.H.W.C. and E.H.K.Y. revised the article and approved the final manuscript. All authors have read and agreed to the published version of the manuscript.

Funding: This research was partially funded by the Research Grants Council of the Hong Kong SAR Government Project No.15200614 (grant B-Q42P), and supported by a PhD study grant of The Hong Kong Polytechnic University.

Institutional Review Board Statement: Not applicable.

Informed Consent Statement: The data presented in this study are available on request from the corresponding author. The data are not publicly available due to the subsequent studies.

Acknowledgments: The authors would like to thank all the experts who shared their valuable opinions on heritage conservation and contributed their experiences and views on the heritage conservation in Suzhou. In addition, the authors would like to express their gratitude to Cao Xiuan and Zhou Sijia, who helped with data collection.

Conflicts of Interest: The authors declare no conflict of interest.

References

1. Langston, C.; Chan, E.H.W.; Yung, E.H.K. Hybrid input-output analysis of embodied carbon and construction cost differences between new-build and refurbished projects. *Sustainability* **2018**, *10*, 3229. [CrossRef]
2. Pomponi, F.; de Wolf, C.; Moncaster, A. (Eds.) *Embodied Carbon in Buildings: Measurement, Management and Mitigation*; Springer: Berlin, Germany, 2018.
3. Yung, E.H.K.; Chan, E.H.W.; Xu, Y. Sustainable development and the rehabilitation of a historic urban district—Social sustainability in the case of Tianzifang in Shanghai. *Sustain. Dev.* **2014**, *22*, 95–112. [CrossRef]
4. Kobayashi, M. *Rekishi Teki Machinami Saisei No Design Shuho*, 1st ed.; Zhang, G., Translator; Tsinghua University Press: Beijing, China, 2015; p. 105.
5. People's Daily Online. Available online: http://en.people.cn/90780/8486419.html (accessed on 8 June 2019).
6. Qian, X.; Yin, C. From Redevelopment to Gentrification in Hong Kong: A Case Study of Kwun Tong Town Center Project. *Open House Int.* **2018**, *43*, 83–93.
7. Udeaja, C.; Trillo, C.; Awuah, K.G.B.; Patel, D.A.; Mansuri, L.E.; Jha, K.N. Urban heritage conservation and rapid urbanization: Insights from Surat, India. *Sustainability* **2020**, *12*, 2172. [CrossRef]
8. Ren, S.B. The Conflicts and Plan Coordination in Urban Renewal. *Modern Urban Res.* **2011**, *1*, 12–18.
9. Della Spina, L. Adaptive sustainable reuse for cultural heritage: A multiple criteria decision aiding approach supporting urban development processes. *Sustainability* **2020**, *12*, 1363. [CrossRef]
10. De Medici, S.; De Toro, P.; Nocca, F. Cultural Heritage and Sustainable Development: Impact Assessment of Two Adaptive Reuse Projects in Siracusa, Sicily. *Sustainability* **2020**, *12*, 311. [CrossRef]
11. Hsing, Y.-T. *The Great Urban Transformation: Politics of Land and Property in China*; Oxford University Press: Oxford, UK, 2010.
12. Bao, M.J. Issues on the Ownership of Unmovable Cultural Relics. *China Cult. Herit. Sci. Res.* **2013**, *23*, 34–37.
13. Ruan, Y.S.; Yuan, F.; Xiao, J. Analysis and Suggestions on the Existing Wave of Old City Reconstruction. *Urban Plan. Forum* **2014**, *1*, 14–17.
14. The Guardian. Available online: https://www.theguardian.com/world/2007/apr/03/china.jonathanwatts (accessed on 6 November 2019).
15. China Daily Website. Available online: http://www.chinadaily.com.cn/imqq/china/2010-09/13/content_11294773.htm (accessed on 6 November 2019).
16. Fighting Rages over Tokyo Airport, Nashua Telegraph. 20 May 1978. Available online: https://news.google.com/newspapers?nid=2209&dat=19780520&id=Xl1YAAAAIBAJ&sjid=-jQNAAAAIBAJ&pg=7085,4794702 (accessed on 6 November 2019).
17. Bowen, R.W. The Narita Conflict. *Asian Sur.* **1975**, *15*, 598–615. [CrossRef]
18. Durlauf, S.N.; Blume, L. *The New Palgrave Dictionary of Economics*; Palgrave Macmillan: Basingstoke, UK, 2008.
19. Coase, R.H. *The Firm, the Market and the Law*; University of Chicago Press: Chicago, IL, USA, 1988.
20. Klein, D.B.; Robinson, J. Property: A Bundle of Rights? Prologue to the Symposium. *Econ J. Watch* **2011**, *8*, 193–204.
21. Coase, R.H. The Problem of Social Cost. *J. Law Econ.* **1960**, *3*, 1–44. [CrossRef]
22. Chappellow Coase Theorem. Available online: https://www.investopedia.com/terms/c/coase-theorem.asp (accessed on 22 October 2020).

23. Peng, J.H.; Hu, X.M. Reflection on the "From Bottom to Top" City Renewal Model from the perspective of Property Right. *Archit. Cult.* **2014**, *2*, 117–119.
24. Media Release: Asia-Pacific-Heritage-Awards. Available online: https://bangkok.unesco.org/sites/default/files/assets/article/Asia-Pacific%20Heritage%20Awards/files/2005-winners.pdf (accessed on 10 September 2020).
25. Liu, Q.Y. The Issue of the Property Rights of the Protection of Historical and Cultural Heritage of Suzhou. *Shanghai Urban Plan. Rev.* **2008**, *2*, 48–51.
26. Lin, L.; Ruan, Y.S. Conservation Planning and Practice for Pingjiang Historic District of Suzhou City. *Urban Plan. Forum* **2006**, *3*, 45–51.
27. Liu, L.Z. Analysis on the Evolution of Property Rights in Beijing Old City Courtyard. *Beijing Plan. Rev.* **2011**, *4*, 21–24.
28. Hua, X.M. *For the Hometown That Cannot Be Lost*; New Century Publishing Company: Taipei, Taiwan, 2009.
29. Zhu, J.G.; Li, N.; Wu, K.J. Research on the List of Beijing's Excellent Modern and Contemporary Building. *City Plan. Rev.* **2008**, *10*, 38–41, 49.
30. Cai, Y. Civil resistance and rule of law in China: The defense of homeowners' rights. In *Grassroots Political Reform in Contemporary China*; Perry, E.J., Goldman, M., Eds.; Harvard University Press: Cambridge, UK, 2007; pp. 174–195.
31. The State Council the People's Republic of China Website. Available online: http://www.gov.cn/banshi/2005-07/12/content_13749.htm (accessed on 16 October 2020).
32. Shi, J.H.; Zhou, Y. *Protection and Utilization of Controlled & Protected Buildings in the Ancient City of Suzhou*; Southeast University Press: Nanjing, China, 2010.
33. Adams, D.; Hastings, E.M. Urban renewal in Hong Kong: Transition from Development Corporation to Renewal Authority. *Land Use Policy* **2001**, *18*, 245–258. [CrossRef]
34. Hastings, E.M.; Adams, D. Facilitating urban renewal: Changing institutional arrangements and land assembly in Hong Kong. *Prope. Manag.* **2005**, *23*, 110–121. [CrossRef]
35. Lai, L.W.; Lorne, F.T. The Fourth Coase Theorem: State planning rules and spontaneity in action. *Plan. Theory* **2015**, *14*, 44–69. [CrossRef]
36. Lai, L.W.; Chau, K.W.; Cheung, P.A.C. Urban renewal and redevelopment: Social justice and property rights with reference to Hong Kong's constitutional capitalism. *Cities* **2018**, *74*, 240–248. [CrossRef]
37. Munch, P. An economic analysis of eminent domain. *J. Political Econ.* **1976**, *84*, 473–498. [CrossRef]
38. Lai, L.W. Private property rights not to use, earn from or trade land in urban planning and development: A meeting between Coase and Buchanan. *Habitat Int.* **2014**, *44*, 555–560. [CrossRef]
39. Coase, R.H. *The Problem of Social Cost. Classic Papers in Natural Resource Economics*; Palgrave Macmillan: London, UK, 1960; pp. 87–137.
40. Barr, N. *The Relevance of Efficiency to Different Theories of Society, Economics of the Welfare State*; Oxford University Press: Oxford, UK, 2012.

sustainability

MDPI

Review

A Framework for Stormwater Quality Modelling under the Effects of Climate Change to Enhance Reuse

Buddhi Wijesiri [1], Erick Bandala [2], An Liu [3] and Ashantha Goonetilleke [1,*]

1 School of Civil and Environmental Engineering, Queensland University of Technology, Brisbane, QLD 4000, Australia; b.mahappukankanamalage@qut.edu.au
2 Division of Hydrologic Sciences, Desert Research Institute, Las Vegas, NV 89119, USA; erick.bandala@dri.edu
3 College of Chemistry and Environmental Engineering, Shenzhen University, Shenzhen 518060, China; liuan@szu.edu.cn
* Correspondence: a.goonetilleke@qut.edu.au; Tel.: +61-7-3138-1539

Received: 15 October 2020; Accepted: 10 December 2020; Published: 14 December 2020

Abstract: Water scarcity, which is exacerbated by climate change, is a major challenge to ensure human well-being. Therefore, it is equally important to protect conventional water resources from degradation and at the same time to identify cost-effective alternatives with a low carbon footprint. In this regard, stormwater plays a key role as it is a largely under-utilised resource for both, potable and non-potable use. However, stormwater carries substantial loads of pollutants to receiving waters such as rivers. Unfortunately, the lack of comprehensive stormwater quality modelling strategies, which account for the effects of climate change, constrains the formulation of effective measures to improve the quality of stormwater. Currently, there is a significant knowledge gap in the merging of stormwater quality modelling and climate modelling. This paper critically reviews current stormwater quality modelling approaches (quantity and quality) and the role of climate modelling outputs in stormwater quality modelling. This is followed by the presentation of a robust framework to integrate the impacts of climate change with stormwater quality models.

Keywords: climate change; stormwater modelling; stormwater quality; stormwater reuse; water scarcity

1. Introduction

Urban stormwater is considered a nuisance because it causes flooding and has an impact on aquatic ecosystem health [1–4]. This mindset overlooks the potential to transform stormwater into a safe-to-use resource. While population increase and the impact of climate change exacerbate water scarcity, authorities commonly continue to fail to see urban stormwater as the last available uncommitted water resource for our cities. For example, in 2015, the Australian Senate recommended formulating strategies to optimise stormwater management [5]. Nevertheless, the National Water Account for 2018 (Australia) disregarded treated stormwater as a source of recycled water [6]. The significance of stormwater reuse is related not only to its contribution to meeting the water demand in urban areas, but also to safeguarding conventional water resources [7,8]. Given the inevitability of climate change effects on the amount of water resources available to urban settlements, Goonetilleke et al. [9] have urged taking advantage of the opportunities offered by stormwater.

Turning this potentially valuable source of water into a safe-to-use resource requires the removal of pollutants entrained in stormwater. To remove pollutants, understanding the processes that determine their loads/concentrations, including build-up on urban surfaces during dry weather periods and wash-off during wet weather, is a fundamental need. Even though these processes have been investigated under the influence of urbanisation [10–12], pollutant behaviour is subject to variations due to the impacts of climate change [13]. Global warming (or more critically, regional warming) results in changes to typical patterns of dry and wet weather periods [14–17], increasing the complexity of changes in stormwater pollutant loads and their characteristics [17,18].

Most studies on the impact of climate change on stormwater only address the changes in runoff quantity in response to changes to rainfall patterns [19–21]. Further, the studies that assess or develop stormwater management measures such as Low-Impact Development (LID) mainly address the control of runoff volume in response to climate change [22–27]. Only a limited number of studies have highlighted the adverse impacts of climate change on stormwater quality [28,29]. However, these studies fail to draw attention to the changes in the patterns of pollutant build-up and wash-off during dry and wet weather periods. The paucity of information and guidance makes it difficult to accurately predict future changes to stormwater quality essential for planning and management decision making in the context of safeguarding stormwater quality and thereby enhancing its reuse.

Predicting stormwater quality is undertaken using mathematical models. The two primary modelling approaches currently used are: (1) physically-based modelling, which replicates temporal and spatial variations in stormwater quality using established physical theory; and (2) statistical modelling for simulating approximations of scenarios subject to a set of observed (field) data. For accurate stormwater quality predictions, the models are required to: (a) encompass robust physical relationships between the changes in stormwater quality and influential factors; and (b) be able to quantify uncertainty in model predictions [30].

Currently, neither (a) nor (b) as articulated above can be measured accurately. Regarding requirement (a) above, the current mathematical formulations of stormwater pollution processes consider dry and wet weather as a static system and do not allow for their dynamic nature resulting from changes driven by global warming [31,32]. Consequently, the accuracy of such mathematical formulations is questionable as weather altered by global warming can change the behaviour of pollutants. For example, large amounts of particulate solids, which carry toxic pollutants, can accumulate on impervious surfaces over longer dry periods, while heavy rainfall can wash-off increasingly large shock-loads of pollutants into receiving waters, exceeding their assimilation capacity. Regarding requirement (b) above, it is inevitable, given the limitations in requirement (a) noted above, that current stormwater quality models do not account for uncertainties that can arise due to the effects of global warming.

In short, stormwater treatment measures lack resilience given that global warming continues to alter dry and wet weather patterns [17]. Hence, the receiving waters remain vulnerable to degradation, and the availability of safe-to-use stormwater will be limited to fulfil the water demands of cities.

This paper critically evaluates the current practices in stormwater quality modelling to identify the changes needed to enhance stormwater quality prediction accuracy. Accordingly, current climate modelling approaches are critically reviewed to identify key aspects of dry and wet weather conditions that need to be accounted for in stormwater quality modelling. This review establishes the platform for climate impact assessment within the context of urban stormwater quality modelling.

2. Current Practice in Stormwater Quality Modelling and Its Deficiencies

At a generic level, current stormwater models consist of two modules that operate simultaneously (see Figure 1). The rainfall-runoff module generates information about runoff during a rainfall event. The runoff quality module generates information about pollutant accumulation (build-up) on catchment surfaces during the dry weather period and subsequent wash-off via stormwater runoff [33].

The current stormwater models have been built based on either physically-based (mechanistic) or statistical (or hybrid) modelling approaches. Both approaches have strengths and deficiencies and the resulting models are likely to generate information on stormwater quantity and quality that may not be sufficiently reliable. Hence, there is no stand-alone model that can accurately replicate the entire process of stormwater pollution. For example, commonly used Stormwater Management Model (SWMM) [34,35] and Model for Urban Stormwater Improvement Conceptualisation (MUSIC) [36,37] are appropriate for planning and management decision making, but lack mathematical formulations of complex interactions between pollutants that are accounted for in Hydrologic Simulation Program-Fortran (HSPF) [38] model. Further, only a few models such as Mike URBAN [39,40] and HSPF are capable of accounting for the processes of both, dissolved and particulate pollutants. The capabilities and limitations of different modeling tools are detailed in Table 1.

Table 1. Capabilities and limitations of commonly used stormwater quality modelling tools (adapted from Wijesiri [41]).

Type	Modelling Tool	Capabilities	Limitations
Physically-Based	Stormwater Management Model (SWMM) developed by the US Environmental Protection Agency [34,35]	- Ten different pollutants. - Four build-up models: linear, power, exponential and saturation. - Three wash-off models: First order decay, rating curve and event mean concentration (EMC). - Used as a planning and design tool.	- Does not account for pollutant interactions. - Lack of accuracy in process replication. - Does not facilitate uncertainty analysis.
	Mike URBAN developed by the Danish Hydraulics Institute [39,40]	- Number of different pollutants in both, dissolved and particulate phases. - Linear and exponential build-up models. - Wash-off model is based on pollutant detachment due to the impact of raindrops. - Integrated with the capabilities of SWMM. - Used as a planning and design tool.	- Greater complexity in the model structure. - Lack of accuracy in process replication. - Does not facilitate uncertainty analysis.
	Storage, Treatment, Overflow and Runoff Model (STORM) developed by the US Corps of Engineers [42]	- Six different pollutants. - Two approaches for build-up modelling: proportionality between pollutants and accumulated solids, and linear build-up model as a function of time. - Wash-off modelling: first order exponential decay (proportionality between washed-off and remaining pollutants). - Used as a planning tool.	- Does not account for pollutant interactions. - Lack of accuracy in process replications. - Does not facilitate uncertainty analysis.
	Hydrologic Simulation Program-Fortran (HSPF) developed by the US Environmental Protection Agency [38]	- Ten different pollutants in dissolved and particulate phases. - Accounts for pollutant interactions. - Linear build-up models. - Wash-off models consider direct proportionality between wash-off rate of different pollutants and runoff. - Used as a planning and design tool.	- Lack of accuracy in process replications. - Inefficient calibration. - Does not facilitate uncertainty analysis.
	Distributed Routing Rainfall Runoff Model-Quality (DR3M-QUAL) developed by the US Geological Survey [43,44]	- Four different pollutants. - Exponential build-up and wash-off models. - Pollutant association with solids. - Used as a planning and design tool.	- Does not account for pollutant interactions. - Lack of accuracy in process replication. - Does not facilitate uncertainty analysis.
	Model for Urban Stormwater Improvement Conceptualisation (MUSIC) developed by Cooperative Research Centre for Catchment Hydrology (CRCCH), Australia [36,37]	- Three different pollutants. - Primarily a planning and management decision support system. - Appropriate for conceptual modelling of stormwater pollution mitigation strategies, and to evaluate their performance.	- Not a detailed modelling tool. - Does not account for pollutant interactions.
Statistical	Derived probability distribution approach [45,46]	- Generates probability distributions of runoff and pollutant loads using the probability distributions of rainfall characteristics and build-up and wash-off models. - Facilitates to improve the conceptualization of pollutant processes. - Facilitates to develop stormwater quality control measures.	
	Bayesian approach with Metropolis algorithm [47]	- Enhances model calibration. - Generates model parameters that optimise the model functions and probability distributions of those parameters. - Facilitates to improve the conceptualization of pollutant processes. - Lack of sufficient data for model calibration may limit application.	

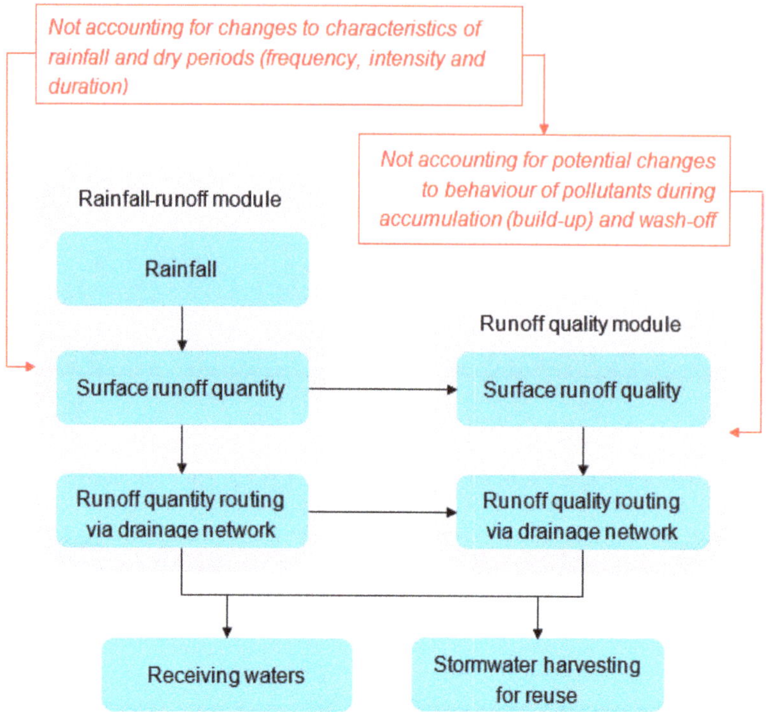

Figure 1. Schematic of the generic structure of current stormwater models (major deficiencies identified are in red text).

In particular, stormwater quality predictions involve various types of uncertainty (due to model structure, data, and parameters) because considerably fewer improvements have been made to the stormwater quality module, compared to the improvements made to the rainfall-runoff (quantity) module. This lack of advancement in stormwater quality modelling is because greater attention has been paid to the mitigation of stormwater quantity-related adverse impacts (e.g., flooding) rather than to improving stormwater quality for reuse [48].

The mathematical formulations included in the runoff quality module are adversely affected by the effects of climate change because the patterns of pollutant build-up and wash-off could change in response to changes in dry and wet weather periods. Therefore, the current mathematical formulations of pollutant build-up and wash-off need enhancements in order to take into consideration the effects of climate change.

Improvements to current mathematical formulations should first consider temporal and spatial patterns of pollutant build-up and wash-off under dry and wet weather conditions influenced by the regional effects of climate change. For example, several regions around the world are projected to experience longer dry periods and more frequent short duration intense rainfall, which have already become evident over the past decade [17]. Under these weather conditions, typical patterns of pollutant build-up (e.g., asymptotic increase towards an equilibrium at around 7–9 antecedent dry days) and wash-off (exponential decay) [49,50] could change as depicted in Figure 2. The pollutant load accumulated is expected to remain at the equilibrium level over a longer period of time (blue line in Figure 2) than typical dry weather events (red line in Figure 2). Consequently, the expected changes to wet weather could result in wash-off of a large amount of accumulated pollutants over a shorter period of time (blue line in Figure 2) than relatively slower wash-off that could occur during typical rainfall events (red line in Figure 2). As such, while typical rainfall events would wash-off only a fraction of accumulated pollutants, the rainfall events influenced by climate change could wash-off almost all the accumulated pollutants from catchment surfaces. Further, lesser known effects such as first-flush phenomenon (wash-off of shock loads of pollutants at the initial portion of a rainfall event, see Figure 3) would play a greater role in influencing the quality of stormwater runoff at the catchment outlet.

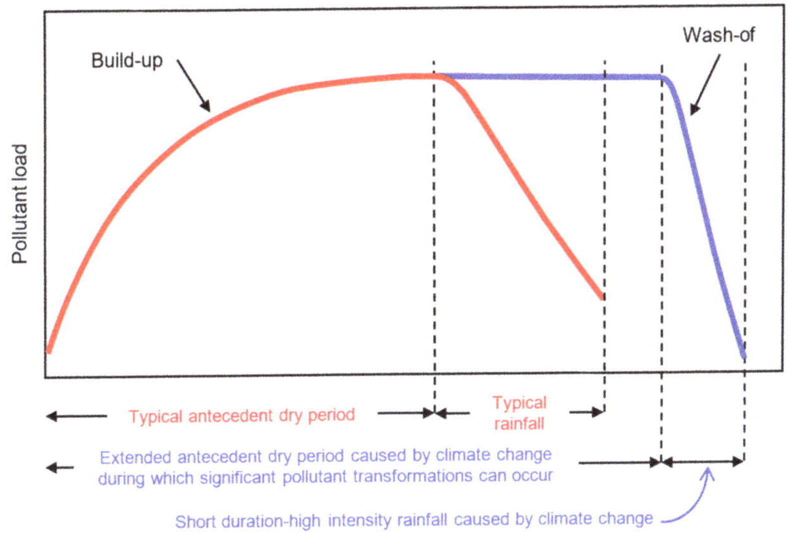

Figure 2. Pollutant behaviour during build-up and wash-off in response to impacts of climate change (X-axis not to scale).

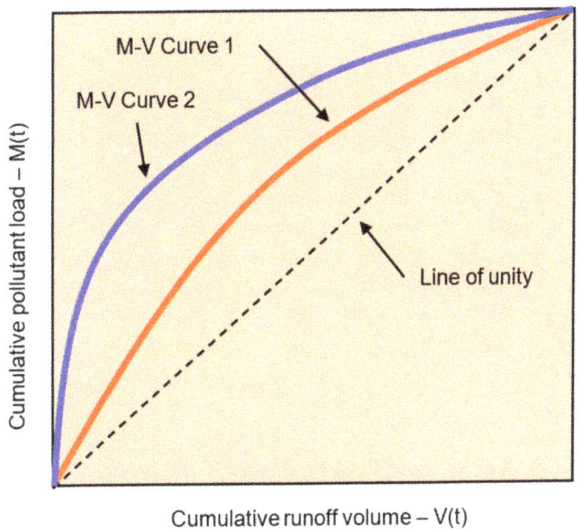

Figure 3. Hypothetical depiction of first-flush. Note: The Mass-Volume (M-V) curves show the relationship between dimensionless M and V at a given point in time (t). As such, first flush occurs when the M-V curve lies above the line of unity. M-V curve 1 shows first-flush during a typical rainfall event, while M-V curve 2 shows first flush during a rainfall event affected by climate change (adapted from Perera et al. [51]).

3. Role of Climate Modelling Outputs in Stormwater Quality Modelling

Climate modelling entails the mathematical formulation and integration of energy and matter transfer on the land, atmosphere and ocean. The outcomes of climate models (general circulation models (GCMs)) indicate the probability of a given area having relatively warmer/cooler or wetter/drier climate, which is different from typical weather reports of daily wet/dry conditions.

With the objective of broadening the understanding of climate change in the historical, current, and future contexts, the Coupled Model Intercomparison Project (CMIP) brings together multiple climate modelling groups from across the world. The CMIP makes standardised output of climate models available to other researchers who are involved in climate change impact assessments. Detailed information about CMIP phases can be accessed via World Climate Research Program (WCRP) and the Program for Climate Model Diagnosis and Intercomparison (PCMDI) of the US Department of Energy. Meanwhile, the most recent phase, CMIP6, provides information that may be useful for the mathematical formulation of stormwater pollution processes under the effects of climate change.

CMIP6 addresses three issues broadly related to the Earth's climate system under 12 scientific themes (clouds/circulation, regional phenomena, ocean/land/ice, impacts, scenarios, decadal prediction, geo-engineering, land use, carbon cycle, chemistry/aerosols, characterising forcing, and paleo) [52]:

Issue 1: Response of Earth's climate system to radiative forcing: The radiative forcing (difference between energy absorbed by and radiated back from the Earth) changes due to natural/anthropogenic emissions. Consequently, land, ocean and atmospheric temperatures could change, and in turn, variables such as precipitation could also change.

Stormwater quality perspective: Changes to dry and wet weather patterns are the two key responses that need to be incorporated into the mathematical formulations of stormwater pollution processes such as build-up and wash-off. Fundamental changes to the structure of current mathematical functions are necessary due to changes to pollution processes (such as lengthy equilibrium periods during build-up and intense first flush events that lasts for shorter periods during wash-off, see Figures 2 and 3).

Issue 2: Systematic model biases: Climate models are expected to replicate complex climate processes, which can exhibit substantial inherent variability, particularly at the regional scale. As such, model biases are likely to arise, leading to over/under estimations of future climate.

Stormwater quality perspective: Current stormwater models already have deficiencies relating to model structure, input and calibration data, and model parameters specific to underlying processes such as coefficients of pollutant build-up and wash-off [53,54]. Understanding systematic biases in climate models and minimising their effects are necessary when using climate projections for stormwater quality modelling.

Issue 3: Assessment of future changes in climate under climate variability, predictability and uncertainty in future scenarios: The future climate scenarios are developed based on the expected changes to socio-economic systems that contribute to emissions. These scenarios are associated with uncertainty, which are accounted for when undertaking climate impact assessments.

Stormwater quality perspective: Stormwater pollution processes can be influenced by the variability in climate processes (precipitation and dry period), which leads to creating uncertainty in stormwater quality projections. Therefore, it is also necessary to address the effects of such uncertainty when projecting stormwater quality into the future.

Accordingly, five themes can be proposed (out of the 12 listed above) that have potential implications for stormwater quality modelling. These include cloud/circulation, regional phenomena, land use, scenarios and impacts. The outputs of the model intercomparison projects (MIPs) under clouds/circulation theme [55,56] are a guide to identifying the appropriate climate models that produce precipitation/dry period projections, which can contribute to stormwater quality modelling (see Figure 4).

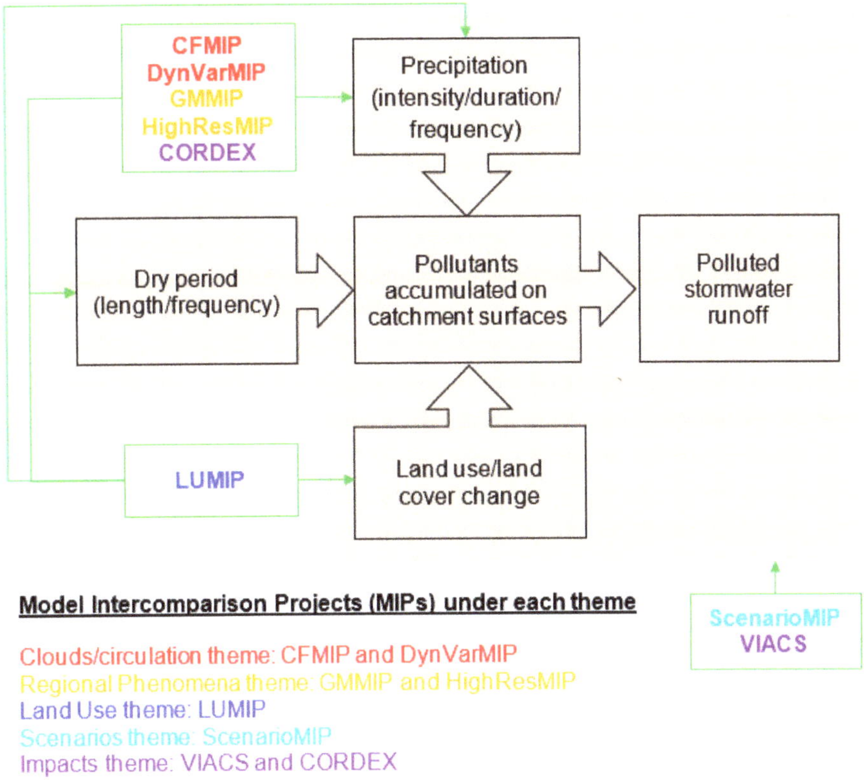

Figure 4. Framework for incorporating various climate modelling outputs of Model Intercomparison Projects (MIPs) under the proposed five themes into specific elements of stormwater quality modelling. Note: CFMIP—Cloud Feedback Model Intercomparison Project [55]; DynVarMIP—Dynamics and Variability Model Intercomparison Project [56]; GMMIP—Global Monsoons Model Inter-comparison Project [57]; HighResMIP—High Resolution Model Intercomparison Project [58]; LUMIP—Land Use Model Intercomparison Project [59]; ScenarioMIP—Scenario Model Intercomparison Project [60]; VIACS—Vulnerability, Impacts, Adaptation and Climate Services [61]; CORDEX—Coordinated Regional Downscaling Experiment [62].

The outputs of the MIPs under Regional Phenomena theme [57,58] provide key information on the variability in monsoon system, which could substantially alter the length of dry periods (mega droughts) and result in heavy storms and large flood events. These significantly exacerbate the quality of stormwater, while flooding contributes to disperse pollutants across multiple areas.

In a recent study conducted in the North Australia region [13], the likely impacts of the climate change-driven variations in dry and wet weather events have been demonstrated. Due to an increase in atmospheric temperature by 1.5 °C (around year 2030–2052) above pre-industrial levels, North Australia region is projected to experience (compared to year 2007) dry periods that are extended by 4.72 days (range of variation at RCP 8.5 warming scenario: 3.67–11.93 days) and increase in rainfall by 0.31 mm (range of variation at RCP 8.5 warming scenario: 0–4.50 mm). These projections are set to worsen due to an additional 0.5 °C increase in temperature (i.e., 2 °C above pre-industrial levels), such that dry periods in North Australia can be 7.34 days longer (range of variation at RCP 8.5 warming scenario: 5.11–17.96 days) and the region can expect to receive 1.40 mm more rainfall (range of variation at RCP 8.5 warming scenario: 0–5.58 mm). As a result of these projected changes in dry and wet weather, Wijesiri et al. [13] estimated (compared to year 2007) a more than 90% increase in the build-up of particle-bound toxicants such as heavy metals and nearly a 50% increase in those pollutants in stormwater runoff.

Given that extreme weather events mostly occur at smaller spatial scales, reliable climate projections under this Regional Phenomena theme could be crucial for stormwater quality modelling (see Figure 4), as stormwater quality exhibits substantial variability at catchment scale as well as between geographic regions [12].

Furthermore, land use is a key determinant of the quantity and type of pollutants released into stormwater runoff, although the effects of land use change have not been adequately accounted for in current stormwater models [63]. On the other hand, anthropogenic activities specific to different types of land use contribute to the emission of aerosols and greenhouse gases. This has implications to the climate such as extreme droughts and heavy precipitation [64]. The outputs of the MIPs under Land Use theme [59] play a key role in influencing the relationships between spatial and temporal changes in land use/land cover that contribute to both, climate change (and its effects on dry and wet weather) and pollutant accumulation on catchment surfaces (see Figure 4).

Various scenarios of how physical and human systems are expected to change in the future and their impacts on the climate system are important for projecting future climate change. The MIPs outputs under scenarios theme [60] provides climate projections of future scenarios of emissions and land use change. These scenarios are developed based on future pathways of inherently uncertain socio-economic developments, which also contribute to stormwater pollution, including pollutant generation, deposition and subsequent re-distribution during dry weather periods and wash-off during rainfall events. Therefore, the future climate scenarios could be the basis for projecting stormwater quality into the future (see Figure 4) and in turn for designing robust measures to enhance stormwater reuse.

Regarding climate impact assessment, usually a two-way dialogue is established between the climate modelling community and those who expect to utilise model outcomes for assessing the impacts on various human-environmental systems [61]. Further, one of the crucial elements of impact assessment is downscaling of regional climate projections. This plays a key role in stormwater quality modelling due to the regional variability in the factors that influence pollutant generation and subsequent distribution during dry and wet weather events. Accordingly, the outputs of MIPs under the impacts theme [62] provide a common framework to produce downscaled regional climate projections and to assess associated uncertainties, which can be considered for stormwater quality modelling to ensure the robustness in the projection of regional scale stormwater quality (see Figure 4).

4. Future Research Directions

Despite the call for actions towards the mitigation of the impacts of climate change [65], urban areas are still far from ensuring human well-being and the safety of ecosystems. In relation to stormwater, there are two major areas of research being undertaken, namely, flood mitigation and stormwater treatment. Efforts to predict flood events and identifying flood-prone areas have at least started to address the effects of changing weather patterns [66–68]. However, research on stormwater treatment is still largely focused on resizing existing treatment measures such as Water Sensitive Urban Design (WSUD) systems. For example, Zhang et al. [69] report that the treatment performance of WSUD systems would not significantly change in future climate change scenarios, and recommend slightly larger systems for reliable performance. However, they have only accounted for the impact of simulated future weather patterns on the volume of stormwater runoff. There is a knowledge gap in the accounting of how these simulated rainfall and dry periods influence pollutant behaviour on urban impervious surfaces, and in particular, phenomena such as first flush. As such, the current state of research on stormwater treatment does not affirm the resilience of these systems, without which they would be ineffective in the long-term in response to climate change. Therefore, a significant step change is necessary in terms of merging stormwater and climate research outcomes, rather the application of resources for simply improving available technologies.

5. Conclusions

This review has identified deficiencies in existing stormwater quality models in relation to addressing the effects of climate change. As such, it is recommended that: (1) the rainfall-runoff (quantity) module should be incorporated with mathematical formulations of the changes to the characteristics of rainfall and dry periods (frequency, intensity and duration) caused by global/regional warming; and (2) the runoff quality module should be incorporated with mathematical formulations of the potential changes to the patterns of pollutant accumulation (build-up) and wash-off. Additionally, the runoff quality module needs significant improvements due to the complex behaviour of pollutants in response to climate-change-influenced changes to dry and wet weather events. Therefore, five themes within the latest phase of CMIP (CMIP6) are proposed as a basis to account for the effects of climate change on stormwater quality. Collectively, the five themes (cloud/circulation, regional phenomena, land use, scenarios and impact assessment) are expected to provide guidance for accounting for the changes to dry and wet weather patterns by radiative forcing under various future socio-economic developments (future climate scenarios), effects of systematic climate model biases on climate projections, natural variability in climate processes, and uncertainties in climate projections, in the formulation of technically robust stormwater quality models.

Author Contributions: Conceptualization, B.W., E.B., A.L. and A.G.; writing—original draft preparation, B.W.; writing—review and editing, E.B., A.L. and A.G. All authors have read and agreed to the published version of the manuscript.

Funding: This research was funded by Guangdong Basic and Applied Basic Research Foundation, China, grant number 2019A1515110353 and 2019A1515010843, and Key-Area Research and Development Program of Guangdong Province, grant number 2019B110205003.

Conflicts of Interest: The authors declare no conflict of interest. The funders had no role in the design of the study; in the collection, analyses, or interpretation of data; in the writing of the manuscript; or in the decision to publish the results.

References

1. Mignot, E.; Li, X.; Dewals, B. Experimental modelling of urban flooding: A review. *J. Hydrol.* **2019**, *568*, 334–342. [CrossRef]
2. Petit-Boix, A.; Sevigné-Itoiz, E.; Rojas-Gutierrez, L.A.; Barbassa, A.P.; Josa, A.; Rieradevall, J.; Gabarrell, X. Floods and consequential life cycle assessment: Integrating flood damage into the environmental assessment of stormwater Best Management Practices. *J. Clean. Prod.* **2017**, *162*, 601–608. [CrossRef]
3. Burns, M.J.; Schubert, J.E.; Fletcher, T.D.; Sanders, B.F. Testing the impact of at-source stormwater management on urban flooding through a coupling of network and overland flow models. *Wires Water* **2015**, *2*, 291–300. [CrossRef]
4. Beaudry, M. *From Nuisance to Resource: Understanding Microbial Sources of Contamination in Urban Stormwater-Impacted Bodies of Water Intended for Water Reuse Activities*; University of Alberta: Edmonton, AB, Canada, 2019.
5. ECRC. *Stormwater Management in Australia*; Commonwealth of Australia: Canberra, Australia, 2015.
6. BoM. *Water in Australia 2017–2018*; Commonwealth of Australia, Bureau of Meteorology: Melbourne, Australia, 2019.
7. Walsh, C.J.; Fletcher, T.D.; Bos, D.G.; Imberger, S.J. Restoring a stream through retention of urban stormwater runoff: A catchment-scale experiment in a social–ecological system. *Freshw. Sci.* **2015**, *34*, 1161–1168. [CrossRef]
8. Goonetilleke, A.; Liu, A.; Gardner, T. *Briefs for GSDR-Urban Stormwater Reuse: An Agenda for Sustainable Development*; UN Global Sustainable Development Report; UN Department of Economic and Social Affairs: New York, NY, USA, 2016.
9. Goonetilleke, A.; Liu, A.; Managi, S.; Wilson, C.; Gardner, T.; Bandala, E.R.; Walker, L.; Holden, J.; Wibowo, M.A.; Suripin, S. Stormwater reuse, a viable option: Fact or fiction? *Econ. Anal. Policy* **2017**, *56*, 14–17. [CrossRef]

10. Hamel, P.; Daly, E.; Fletcher, T.D. Source-control stormwater management for mitigating the impacts of urbanisation on baseflow: A review. *J. Hydrol.* **2013**, *485*, 201–211. [CrossRef]
11. Wei, T.; Wijesiri, B.; Jia, Z.; Li, Y.; Goonetilleke, A. Re-thinking classical mechanistic model for pollutant build-up on urban impervious surfaces. *Sci. Total Environ.* **2019**, *651*, 114–121. [CrossRef]
12. Wijesiri, B.; Liu, A.; Gunawardana, C.; Hong, N.; Zhu, P.; Guan, Y.; Goonetilleke, A. Influence of urbanisation characteristics on the variability of particle-bound heavy metals build-up: A comparative study between China and Australia. *Environ. Pollut.* **2018**, *242*, 1067–1077. [CrossRef]
13. Wijesiri, B.; Liu, A.; Goonetilleke, A. Impact of global warming on urban stormwater quality: From the perspective of an alternative water resource. *J. Clean. Prod.* **2020**, 121330. [CrossRef]
14. King, A.D.; Karoly, D.J.; Henley, B.J. Australian climate extremes at 1.5 C and 2 C of global warming. *Nat. Clim. Chang.* **2017**, *7*, 412. [CrossRef]
15. Chevuturi, A.; Klingaman, N.P.; Turner, A.G.; Hannah, S. Projected Changes in the Asian-Australian Monsoon Region in 1.5 °C and 2.0 °C Global-Warming Scenarios. *Earth's Future* **2018**, *6*, 339–358. [CrossRef]
16. IPCC. *Managing the Risks of Extreme Events and Disasters to Advance Climate Change Adaptation. A Special Report of Working Groups I and II of the Intergovernmental Panel on Climate Change*; Field, C.B., Barros, V., Stocker, T.F., Qin, D., Dokken, D.J., Ebi, K.L., Mastrandrea, M.D., Mach, K.J., Plattner, G.-K., Allen, S.K., et al., Eds.; Cambridge University Press: Cambridge, UK; New York, NY, USA, 2012.
17. IPCC. *Global Warming of 1.5 °C. An IPCC Special Report on the Impacts of Global Warming of 1.5 °C above Pre-Industrial Levels and Related Global Greenhouse Gas Emission Pathways, in the Context of Strengthening the Global Response to the Threat of Climate Change, Sustainable Development, and Efforts to Eradicate Poverty*; Masson-Delmotte, V., Zhai, P., Pörtner, H.O., Roberts, D., Skea, J., Shukla, P.R., Pirani, A., Moufouma-Okia, W., Péan, C., Pidcock, R., et al., Eds.; World Meteorological Organization: Geneva, Switzerland, 2018.
18. IPCC. *Climate Change 2014: Mitigation of Climate Change. Contribution of Working Group III to the Fifth Assessment Report of the Intergovernmental Panel on Climate Change*; Edenhofer, O., Pichs-Madruga, R.Y., Sokona, E.F., Kadner, S., Seyboth, K., Adler, A., Baum, I., Brunner, S., Eickemeier, P., Kriemann, B., et al., Eds.; Cambridge University Press: Cambridge, UK; New York, NY, USA, 2014.
19. Donnelly, C.; Greuell, W.; Andersson, J.; Gerten, D.; Pisacane, G.; Roudier, P.; Ludwig, F.J.C.C. Impacts of climate change on European hydrology at 1.5, 2 and 3 degrees mean global warming above preindustrial level. *Clim. Chang.* **2017**, *143*, 13–26. [CrossRef]
20. Zhai, R.; Tao, F.; Xu, Z. Spatial–temporal changes in runoff and terrestrial ecosystem water retention under 1.5 and 2 °C warming scenarios across China. *Earth Syst. Dyn.* **2018**, *9*, 717–738. [CrossRef]
21. Semadeni-Davies, A.; Hernebring, C.; Svensson, G.; Gustafsson, L.-G. The impacts of climate change and urbanisation on drainage in Helsingborg, Sweden: Suburban stormwater. *J. Hydrol.* **2008**, *350*, 114–125. [CrossRef]
22. Cook, L.M.; McGinnis, S.; Samaras, C. The effect of modeling choices on updating intensity-duration-frequency curves and stormwater infrastructure designs for climate change. *Clim. Chang.* **2020**, *159*, 289–308. [CrossRef]
23. Brudler, S.; Arnbjerg-Nielsen, K.; Hauschild, M.Z.; Rygaard, M. Life cycle assessment of stormwater management in the context of climate change adaptation. *Water Res.* **2016**, *106*, 394–404. [CrossRef]
24. Thakali, R.; Kalra, A.; Ahmad, S. Understanding the effects of climate change on urban stormwater infrastructures in the Las Vegas Valley. *Hydrology* **2016**, *3*, 34. [CrossRef]
25. Zahmatkesh, Z.; Burian, S.J.; Karamouz, M.; Tavakol-Davani, H.; Goharian, E. Low-Impact Development Practices to Mitigate Climate Change Effects on Urban Stormwater Runoff: Case Study of New York City. *J. Irrig. Drain. Eng.* **2015**, *141*, 04014043. [CrossRef]
26. Pyke, C.; Warren, M.P.; Johnson, T.; LaGro, J.; Scharfenberg, J.; Groth, P.; Freed, R.; Schroeer, W.; Main, E. Assessment of low impact development for managing stormwater with changing precipitation due to climate change. *Landsc. Urban Plan.* **2011**, *103*, 166–173. [CrossRef]
27. Giese, E.; Rockler, A.; Shirmohammadi, A.; Pavao-Zuckerman, M.A. Assessing Watershed-Scale Stormwater Green Infrastructure Response to Climate Change in Clarksburg, Maryland. *J. Water Resour. Plan. Manag.* **2019**, *145*, 05019015. [CrossRef]
28. Saraswat, C.; Kumar, P.; Mishra, B.K. Assessment of stormwater runoff management practices and governance under climate change and urbanization: An analysis of Bangkok, Hanoi and Tokyo. *Environ. Sci. Policy* **2016**, *64*, 101–117. [CrossRef]

29. Wang, M.; Zhang, D.; Adhityan, A.; Ng, W.J.; Dong, J.; Tan, S.K. Assessing cost-effectiveness of bioretention on stormwater in response to climate change and urbanization for future scenarios. *J. Hydrol.* **2016**, *543*, 423–432. [CrossRef]
30. Bell, C.D.; McMillan, S.K.; Clinton, S.M.; Jefferson, A.J. Hydrologic response to stormwater control measures in urban watersheds. *J. Hydrol.* **2016**, *541 Pt B*, 1488–1500. [CrossRef]
31. Wijesiri, B.; Egodawatta, P.; McGree, J.; Goonetilleke, A. Assessing Uncertainty in Pollutant Build-up and Wash-off Processes. *Environ. Pollut.* **2016**, *212*, 48–56. [CrossRef] [PubMed]
32. WWAP. *The United Nations World Water Development Report 4: Managing Water under Uncertainty and Risk (Volume 1)*; UNESCO: Paris, France, 2012.
33. Liu, A.; Goonetilleke, A.; Egodawatta, P. *Role of Rainfall and Catchment Characteristics on Urban Stormwater Quality*; Springer: Singapore, 2015.
34. Rossman, L.A. *Storm Water Management Model User's Manual Version 5.1*; US EPA Office of Research and Development: Washington, DC, USA, 2015.
35. Huber, W.C.; Dickinson, R.E. *Storm Water Management Model, Version 4: User's Manual*; US Enviornmental Protection Agency: Athens, GA, USA, 1988.
36. Wong, T.H.; Fletcher, T.D.; Duncan, H.P.; Coleman, J.R.; Jenkins, G.A. A model for urban stormwater improvement: Conceptualization. In *Global Solutions for Urban Drainage*; Srecker, E.W., Huber, W.C., Eds.; American Society of Civil Engineers: Reston, VA, USA, 2002; pp. 1–14.
37. Wong, T.H.F.; Fletcher, T.D.; Duncan, H.P.; Jenkins, G.A. Modelling Urban Stormwater Treatment—A Unified Approach. *Ecol. Eng.* **2006**, *27*, 58–70. [CrossRef]
38. Bicknell, B.R.; Imhoff, J.C.; Kittle, J.L., Jr.; Donigian, A.S., Jr.; Johanson, R.C. *Hydrological Simulation Program-Fortran: User's Manual Version 11*; U.S. Environmental Protection Agency, National Exposure Research Laboratory: Athens, GA, USA, 1997.
39. MikeUrban. *Mouse Pollution Transport: Reference Manual—Surface Runoff Quality*; Danish Hydraulic Institue: Hørsholm, Denmark, 2017.
40. MikeUrban. *Mike Urban Collection System—User Guide*; Danish Hydraulic Institute: Hørsholm, Denmark, 2014.
41. Wijesiri, B. *Assessing Uncertainty in Relation to Urban Stormwater Pollutant Processes*; Queensland University of Technology (QUT): Brisbane, Australia, 2016.
42. HEC. *Storage, Treatment, Overflow, Runoff Model, STORM, Generalized Computer Program 723-58-L7520*; Hydrologic Engineering Center, U.S. Corps of Engineers: Davis, CA, USA, 1977.
43. Alley, W.M.; Smith, P.E. *Multi-Event Urban Runoff Quality Model*; U.S. Geological Survey: Reston, VA, USA, 1982.
44. Alley, W.M.; Smith, P.E. *Distributed Routing Rainfall-Runoff Model: Version 2*; U.S. Geological Survey: Denver, CO, USA, 1982.
45. Chen, J.; Adams, B.J. A Derived Probability Distribution Approach to Stormwater Quality Modeling. *Adv. Water Resour.* **2007**, *30*, 80–100. [CrossRef]
46. Chen, J.; Adams, B.J. A Framework for Urban Storm Water Modeling and Control Analysis with Analytical Models. *Water Resour. Res.* **2006**, *42*, 1–13. [CrossRef]
47. Kanso, A.; Gromaire, M.-C.; Gaume, E.; Tassin, B.; Chebbo, G. Bayesian Approach for the Calibration of Models: Application to an Urban Stormwater Pollution Model. *Water Sci. Technol.* **2003**, *47*, 77–84. [CrossRef]
48. Wijesiri, B.; Egodawatta, P.; McGree, J.; Goonetilleke, A. Understanding the Uncertainty Associated with Particle-bound Pollutant Build-up and Wash-off: A Critical Review. *Water Res.* **2016**, *101*, 582–596. [CrossRef]
49. Egodawatta, P.; Thomas, E.; Goonetilleke, A. Mathematical Interpretation of Pollutant Wash-off from Urban Road Surfaces Using Simulated Rainfall. *Water Res.* **2007**, *41*, 3025–3031. [CrossRef]
50. Wijesiri, B.; Egodawatta, P.; McGree, J.; Goonetilleke, A. Influence of Pollutant Build-up on Variability in Wash-off from Urban Road Surfaces. *Sci. Total Environ.* **2015**, *527–528*, 344–350. [CrossRef] [PubMed]
51. Perera, T.; McGree, J.; Egodawatta, P.; Jinadasa, K.B.S.N.; Goonetilleke, A. Taxonomy of influential factors for predicting pollutant first flush in urban stormwater runoff. *Water Res.* **2019**, *166*, 115075. [CrossRef] [PubMed]
52. Eyring, V.; Bony, S.; Meehl, G.A.; Senior, C.A.; Stevens, B.; Stouffer, R.J.; Taylor, K.E. Overview of the Coupled Model Intercomparison Project Phase 6 (CMIP6) experimental design and organization. *Geosci. Model Dev.* **2016**, *9*, 1937–1958. [CrossRef]

53. Bonhomme, C.; Petrucci, G. Should we trust build-up/wash-off water quality models at the scale of urban catchments? *Water Res.* **2017**, *108*, 422–431. [CrossRef] [PubMed]
54. Hong, Y.; Bonhomme, C.; Le, M.-H.; Chebbo, G. A new approach of monitoring and physically-based modelling to investigate urban wash-off process on a road catchment near Paris. *Water Res.* **2016**, *102*, 96–108. [CrossRef]
55. Webb, M.J.; Andrews, T.; Bodas-Salcedo, A.; Bony, S.; Bretherton, C.S.; Chadwick, R.; Chepfer, H.; Douville, H.; Good, P.; Kay, J.E.J.G.M.D. The cloud feedback model intercomparison project (CFMIP) contribution to CMIP6. *Geosci. Model Dev.* **2017**, *2017*, 359–384. [CrossRef]
56. Gerber, E.P.; Manzini, E.J.G.M.D. The Dynamics and Variability Model Intercomparison Project (DynVarMIP) for CMIP6: Assessing the stratosphere-troposphere system. *Geosci. Model Dev.* **2016**, *9*, 3413–3425. [CrossRef]
57. Zhou, T.; Turner, A.G.; Kinter, J.L.; Wang, B.; Qian, Y.; Chen, X.; Wu, B.; Liu, B.; Zou, L.; Bian, H.J.G.M.D. GMMIP (v1. 0) contribution to CMIP6: Global monsoons model inter-comparison project. *Geosci. Model Dev.* **2016**, *9*, 3589–3604. [CrossRef]
58. Haarsma, R.J.; Roberts, M.J.; Vidale, P.L.; Senior, C.A.; Bellucci, A.; Bao, Q.; Chang, P.; Corti, S.; Fučkar, N.S.; Guemas, V.J.G.M.D. High resolution model intercomparison project (HighResMIP v1. 0) for CMIP6. *Geosci. Model Dev.* **2016**, *9*, 4185–4208. [CrossRef]
59. Lawrence, D.M.; Hurtt, G.C.; Arneth, A.; Brovkin, V.; Calvin, K.V.; Jones, A.D.; Jones, C.D.; Lawrence, P.J.; de Noblet-Ducoudré, N.; Pongratz, J.J.G.M.D. The Land Use Model Intercomparison Project (LUMIP) contribution to CMIP6: Rationale and experimental design. *Geosci. Model Dev.* **2016**, *9*, 2973–2998. [CrossRef]
60. O'Neill, B.C.; Tebaldi, C.; van Vuuren, D.P.; Eyring, V.; Friedlingstein, P.; Hurtt, G.; Knutti, R.; Kriegler, E.; Lamarque, J.F.; Lowe, J.; et al. The Scenario Model Intercomparison Project (ScenarioMIP) for CMIP6. *Geosci. Model Dev.* **2016**, *9*, 3461–3482. [CrossRef]
61. Ruane, A.C.; Teichmann, C.; Arnell, N.W.; Carter, T.R.; Ebi, K.L.; Frieler, K.; Goodess, C.M.; Hewitson, B.; Horton, R.; Kovats, R.S. The vulnerability, impacts, adaptation and climate services advisory board (VIACS AB v1. 0) contribution to CMIP6. *Geosci. Model Dev.* **2016**, *9*, 3493–3515. [CrossRef]
62. Gutowski, W.J., Jr.; Giorgi, F.; Timbal, B.; Frigon, A.; Jacob, D.; Kang, H.S.; Raghavan, K.; Lee, B.; Lennard, C.; Nikulin, G.; et al. WCRP COordinated Regional Downscaling EXperiment (CORDEX): A diagnostic MIP for CMIP6. *Geosci. Model Dev.* **2016**, *9*, 4087–4095. [CrossRef]
63. Gunawardena, J.M.; Liu, A.; Egodawatta, P.; Ayoko, G.A.; Goonetilleke, A. *Influence of Traffic and Land Use on Urban Stormwater Quality: Implications for Urban Stormwater Treatment Design*; Springer: Singapore, 2017.
64. Greve, P.; Orlowsky, B.; Mueller, B.; Sheffield, J.; Reichstein, M.; Seneviratne, S.I.J.N.g. Global assessment of trends in wetting and drying over land. *Nat. Geosci.* **2014**, *7*, 716–721. [CrossRef]
65. WEF. *The Global Risks Report 2019*, 14th ed.; World Economic Forum: Geneva, Switzerland, 2019.
66. Pasquier, U.; Few, R.; Goulden, M.C.; Hooton, S.; He, Y.; Hiscock, K.M. "We can't do it on our own!"—Integrating stakeholder and scientific knowledge of future flood risk to inform climate change adaptation planning in a coastal region. *Environ. Sci. Policy* **2020**, *103*, 50–57. [CrossRef]
67. Cloke, H.L.; Wetterhall, F.; He, Y.; Freer, J.E.; Pappenberger, F. Modelling climate impact on floods with ensemble climate projections. *Q. J. R. Meteorol. Soc.* **2013**, *139*, 282–297. [CrossRef]
68. Västilä, K.; Kummu, M.; Sangmanee, C.; Chinvanno, S. Modelling climate change impacts on the flood pulse in the Lower Mekong floodplains. *J. Water Clim. Chang.* **2010**, *1*, 67–86. [CrossRef]
69. Zhang, K.; Manuelpillai, D.; Raut, B.; Deletic, A.; Bach, P.M. Evaluating the reliability of stormwater treatment systems under various future climate conditions. *J. Hydrol.* **2019**, *568*, 57–66. [CrossRef]

sustainability

MDPI

Article

Analysis of Alternatives for Sustainable Stormwater Management in Small Developments of Polish Urban Catchments

Joanna Boguniewicz-Zabłocka [1] and Andrea G. Capodaglio [2,*]

[1] Department of Thermal Engineering and Industrial Facilities, Faculty of Mechanical Engineering, Opole University of Technology, 45040 Opole, Poland; j.boguniewicz-zablocka@po.edu.pl

[2] Department of Civil Engineering and Architecture, University of Pavia, 27100 Pavia, Italy

* Correspondence: andrea.capodaglio@unipv.it

Received: 20 November 2020; Accepted: 4 December 2020; Published: 6 December 2020

Abstract: Sustainable stormwater management approaches in accordance with the EU Water Framework Directive (WFD) allow a source control to handle the quality and quantity of the runoff at local level or near the source. The most popular technologies applied in Europe are green roofs, porous pavements, retention basins and bioswales/raingardens. In this article, two of these solutions (retention tank with reuse, and rain garden, respectively), applied to single dwelling case studies in a suburban area in the Silesia Region (Poland), are illustrated and analyzed. The selected cases consider technical and economic aspects as the most important factors for decision on the selection of onsite stormwater management approach. Both systems have been operational for approximately two years. The retention tank proved a good solution, reducing stormwater overflows and allowing local water reuse for lawn irrigation; however, investment and maintenance costs in this case are relatively higher. The raingarden proved to work efficiently in this small scale implementation and implied much lower initial investment and costs. The economic sustainability of these interventions at single dwelling scale was analyzed, showing interesting returns, with outcome depending on the degree of possible water reuse (lower water bills) and availability of fiscal or fee incentives. Introduction of financial incentive schemes will encourage homeowners and developers to implement stormwater control solutions, allowing rapid amortization of investment costs with additional benefits to the community, such as reduced environmental impact of stormwater overflows and possible economies in the construction and management of stormwater systems.

Keywords: stormwater management; retention basin; rain garden; low impact development (LID); green infrastructure; cost analysis

1. Introduction

Sustainable stormwater management has been and still is a long-time issue in urban drainage systems. In addition to potentially causing adverse impacts on wastewater treatment operations in traditional combined systems [1], flooding during storm events has always been a common problem in urban areas: in fact, almost every city, regardless of the type of sewer network, is potentially vulnerable to this phenomenon, whose frequency has exacerbated due to the increased intensity and recurrence of extreme hydro-meteorological events linked to long term climate variability. Increases in impervious urban surfaces, poor resiliency of urban drainage system design and increased frequency of downpours in urban areas can increase peak storm runoff with corresponding impact on human life and health, property and water security. The Intergovernmental Panel on Climate Change (IPCC) reported that the number of heavy precipitation events has significantly increased in inland areas worldwide [2].

Climate indicators for the last decades show generalized statistical increase of event-specific maximum precipitation in many cities [3,4]. In addition, extreme hydro-meteorological events impact on the physical, chemical and biological parameters of water in urban water bodies, both through direct runoff, and separate or combined sewer overflows (CSOs) [5]. Pollutants conveyed by storm flows in addition to organic matter include pathogens (Fecal Indicator Bacteria, FIB), nutrients, metals and emerging contaminants [6,7], the latter often at low level concentrations, which are difficult to monitor by traditional means [8].

Safe urban stormwater management is becoming a major concern: the conventional approach based on piped drainage is currently criticized as poorly efficient as, only partly effective during meteorological extremes, it does not eliminate environmental problems [9]. In the 1980s, structurally intensive approaches were proposed such as the "deep tunnel" concept: large, underground collectors designed to relieve urban sewer systems from excess stormwater flow and curtail overflow frequency. These systems were built in large cities in the U.S. (e.g., Milwaukee, Chicago, Boston, Atlanta) and around the world (e.g., Hong Kong, Guangzhou, Singapore) [10,11]. In addition to the high cost involved (the deep tunnel project in Milwaukee, one of the first of this kind, required 14 years, at a cost in excess of US$2.3 billion, to complete [12]), the features of these systems may raise unexpected management challenges [13]. Furthermore, their operation requires high energy inputs for pumping and subsequent treatment of dilute sewage, increasing the already high greenhouse gases (GHG) emission footprint of water systems [14].

While these may have proven use in large urban areas, this approach may not be fully resolutive for the targeted impacts. Low-tech, more sustainable methods in accordance with the EU Water Framework Directive (WFD) [15] may be effective in many cases, especially in smaller urbanizations [16]. Sustainable storm water management should promote source control methods at, or nearby, the source. The most effective approach, which could successfully complement technologically-intensive approaches, consists of trapping stormwater and storing it into temporary impoundments for evaporation or ground infiltration. Rain and roof gardens, grassy swales or ditches (bioswales and bioretention basins) and permeable pavements could be highly beneficial [17]. Many mitigation measures are also being proposed to increase urban systems' resilience against floods [5,18]. These include discharge separation at source [19], local water reuse [20] and implementation of decentralized water management [21]. Modern stormwater management requires separation of rainwater from sewage, providing a higher level of service and benefits such as: elimination of CSOs, pollution prevention and possible use of stormwater as an alternative resource.

Sustainable storm water management is connected with so-called blue-green infrastructure (BGI). BGI foresees the implementation of either natural or man-provided solutions to enhance management of water resources and water infrastructure and services risk resilience. Innovative fiscal and non-fiscal tools, which may include payment for ecosystem services schemes [22], may be introduced to encourage their implementation on public and private property [23]. These practices are described in the literature under various labels, such as Low Impact Development (LID) [24], nature-based solutions (NBS) [25] or Sustainable Urban Drainage Systems (SUDS) [26]. All reduce the impact of constructed impervious surface areas (ISA) and of their hydrological and ecological disturbances. A generally accepted scale to assess ISA impact on urban watersheds indicates stress conditions, with ISA between 1% and 10%, impacted if between 10–25% and degraded if greater than 25% [27]. The global ISA average is estimated at 93 m^2 per person. As a comparison, ISA in Poland is about 110 m^2/person, similar to other European countries (Germany, Italy, the United Kingdom, the Czech Republic, Hungary, Austria, Switzerland, Bulgaria, Slovakia, and Denmark) with ISAs around 100-150 m^2/person. In France, Portugal, Belgium, Ireland, Sweden and Spain, ISA is between 150–220 m^2/person, in the USA, close to 300 m^2/person, and in China around 68 m^2/person [28].

This article presents two case studies of sustainable stormwater management practices in small urban developments in Polish catchments, respectively concerning: (1) on-site retention and reuse, and (2) rainwater infiltration gardens. The aim of the analysis of the two approaches is to highlight their application's sustainability, considering also related legal and economic aspects. With this objective, a simple cost-benefit analysis and a general discussion on urban stormwater management are also presented.

2. An Overview of Sustainable Stormwater Management Practices

Sustainable urban stormwater management solutions face increasing complexity from conflicting demands resulting from increasing urbanization, influence of expected climate variability and financial and budgetary constraints of cities. To address these, the current approach has evolved to accommodate increasing use of LID techniques: these aim to restore urban watersheds functions to pre-development stage hydrology, increasing resilience to external stresses, without compromising the requirements of modern urbanization. Figure 1 summarizes the effects of urban development on flow volumes and frequency. LID can be optimally applied in new urbanization planning, but also as retrofit of existing infrastructure. In addition to purely hydrologic issues, recent stormwater quality regulations are a major factor in LID adoption. While traditional urban stormwater management relies on fast conveyance of excess water away from affected areas, LID relies mainly on infiltration, evapotranspiration and the incorporation of natural hydrologic features extending water retention and reducing runoff, peak flows and pollutant loads. A review of implementation and performance of low impact development approaches was recently published [29].

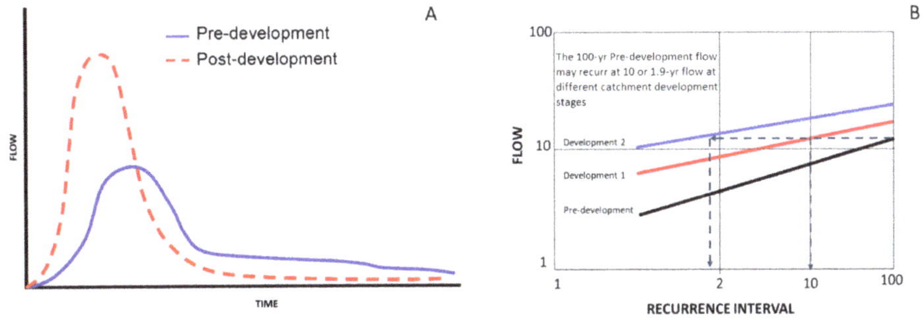

Figure 1. Effect of urban development on stormwater flow (**A**) and return frequency (**B**). With increasing urbanization, the recurrence interval of a 100-year flood can be reduced by one or more orders of magnitude, increasing the probability of risk to life and property.

LID stormwater management occurs mainly by means of two approaches: infiltration-based and retention-based. Both reduce an urban basin's effective impervious surface [30]; however, neither of the two individual approaches is generally sufficient to successfully restore a natural flow regime. Management of rainwater may also include installations for collection and reuse of precipitation [21], which can help decrease consumption of water, treated to drinking quality. Several combined solutions to limit runoff and promote use of collected rainwater were recommended by the European Commission [31]. Possible uses that do not require potable water are car washing, garden and lawn watering, laundry making, or toilet flushing. These have been already implemented around the world. Depending on climatic conditions, type of building and use, the reduction of demand for mains water may be as high as 60% [32].

Infiltration-based approaches assist in baseflow restoration by recharging subsurface and groundwater flow [31]. They include swales, infiltration trenches and basins, unlined bioretention systems (e.g., rain-gardens), and porous pavements. Their effectiveness is highly affected by site conditions, hence the wide reported range of performances.

Swale systems (open channels filled with vegetation) can be used to replace traditional curbs and for erosion control in peri-urban areas. They are designed to induce infiltration, sedimentation, and filtration during flow conveyance, resulting in some degree of water quality improvement. Infiltration trenches usually consist of gravel-filled channels covered with soil and vegetation. Bioretention ponds (also called rain gardens) are landscaped low areas where onsite reduction and treatment of runoff occurs. They are generally vegetated with shrubs, perennials, or trees, and covered with bark mulch. Finally, permeable pavements (including paving blocks, plastic grids, porous asphalts and concretes) allow slow runoff infiltration, promoting pollutant removal by entrapment, adsorption or biological degradation.

Retention-based approaches are designed to hold collected stormflow and reduce outflow from a catchment. They have substantial influences on local flow regime, reducing peak flow, but may result in increased outflow persistence. They include wetlands, ponds, green roofs, and decentralized rainwater harvesting; some have been used extensively for years. Although they can be quite effective for pollutant removal, they have limited effect in reducing overall runoff volumes, since this occurs mainly by evapotranspiration.

Green roofs have proven beneficial for stormwater control in many studies. It has been claimed that green roofs, in addition to runoff control, may induce other additional environmental benefits, such as air quality improvement, urban heat-island effect mitigation and urban aesthetics amelioration. According to existing experiences, there are few disadvantages to this solution, the cost of installation being the main one. In addition to the cost of the vegetated component, which may vary according to execution and aesthetic requirements, the added expense to install a green roof, compared to a traditional flat roof, consists mainly in higher structural costs, as the underlying structure may have to be strengthened to cope with the extra structural load. Green roofs show effectiveness in stormwater retention, with ability to attenuate runoff peaks from events with 2–100 years recurrence intervals, reducing the need for detention basins, with beneficial environmental impact. Green roofs can effectively retain 100% of rainfall in events with precipitation less than 12 mm, and significantly delay hydrologic responses, slowing onset of runoff by an average of 5.7 h, and peak runoff response by an average of 2 h. Annual runoff reduction between 38% to 54% and peak flow reductions up to 90%, were reported [33]. Green roofs have an effect on buildings' energy requirements, raising winter roof temperatures by up to 6 °C, and lowering it by up to 19 °C in the summer, with much narrower ranges of diurnal fluctuations [34]. Several municipalities in Europe provide economic incentives for this practice. Over the last 25 years, many such projects have been completed in the USA and Northern European countries, including Germany and Switzerland.

Decentralized stormwater harvesting may significantly improve water retention within a catchment, reducing annual runoff volumes. Stormwater harvesting is more efficient in terms of runoff reduction if designed to supply water on a daily (short-term), rather than seasonal, basis. Harvested water may be readily used onsite, e.g., for irrigation, or can be further treated for high-quality uses, becoming a significant component of urban hydrology. It is increasingly seen as a valuable resource, especially in water-scarce areas. The potential range of rainwater use as public water substitute is limited by quality and cost of any necessary treatment [20], but it can also imply significant energy requirements and emissions reductions for supply systems [14]. In some areas, due to particularly favorable environmental quality, rainwater could be directly used as a drinking water supply source [35].

In Poland, although the University of Warsaw Library is considered one of the most beautiful and the largest roof garden in Europe, with surface of 1 hectare [36], issues concerning sustainable, low impact stormwater management have largely remained outside the mainstream of research and application interest, with few significant examples.

Legal and Economic Aspects of Stormwater Management

The adoption of alternative approaches to storm-water management bears an implicit economic impact on land development scenarios and water resource protection. Studies have been carried out, showing that the impact of LID and similar practices on property value is quite complex and variable.

While this impact may be compensated by appropriate taxation policies, it reduces externalities due to avoided pollution [37].

Through the implementation of the Water Framework Directive into Polish law (Polish 'Water Law'), an obligation to manage stormwater runoff according to sustainable development rules came into existence [38]. Rainwater should be retained as much as possible at or near the location where precipitation occurred, through the use of surface or underground retention, and in-ground infiltration. Pursuant to the Polish Act, the "discharge of rainwater or meltwater into waters or into water facilities, contained in open or closed rainwater drainage systems for the discharge of atmospheric precipitation or into collective sewage systems within the administrative boundaries of cities" is part of "water services". Regardless of legal obligations related to discharge, financial obligations also exist. As provided for in Art. 389 IP, for this discharge it is necessary to obtain a suitable permit and abide by payment of fees to the service operator.

In the USA, a similar situation exists: The Water Quality Act of 1987 mandated the implementation of a comprehensive program to address stormwater runoff. Since no specific funding provision was established, municipalities introduced various fees in order to fund stormwater projects, effectively implementing a separate local taxation system. Other countries, like Canada, the United Kingdom, Germany, and Australia, have similarly introduced stormwater charges [39,40]. Communities currently use a combination of instruments: some introduced user fees to provide dedicated funding to "stormwater utilities" (separated from wastewater utilities), notwithstanding the complexity of identifying who pays and benefits for what, since drainage systems are often interconnected. Alternate funding methods include local bonds for infrastructural improvements, developer extension fees (capital costs shared among new developers), impact fees (based on mitigation costs of new developments' impact), special assessment fees, property taxes, and stormwater user (i.e., service) fees. User fee schemes could provide an equitable, dedicated source of funding, with charges commensurate with the cost of service, but they are not always applicable. Stormwater utility fees are considered more efficient and environmentally sustainable, allowing long-term planning and solutions, at the political cost of high visibility. Unlike other water cycle related fees, these could be reduced by stormwater credits for introducing best management practices (BMPs), such as those described.

In European countries, stormwater fees have been introduced for many years: in most German Lands, fees are calculated based on the impervious surface area. In Hamburg, for example, it is calculated according to the total costs attributed to the impervious area connected to the public sewer, and currently amounts to 0.73 €/m^2 impervious area [41]. In Italy there are no specific stormwater management fees so far. The cost of water services is covered under the formula of an "Integrated water tariff" paid to the local sewer operator, which takes care of all the aspects of water services from supply to collection and treatment, based on metered water consumption.

In Poland, the issue of stormwater fees foresees a water service fee, due for discharge of rainwater or snowmelt. The purpose is to encourage users to rationally manage water and limit pollution, as well as cover costs associated with drainage and facilities for its treatment. Fees consist of a fixed and a variable component: the former, sometimes referred to as "subscription" fee, depends on the maximum allowable rainwater discharge specified in the water permit. The base fee is PLN 0.75/m^3 (about 0.17 Euro) yearly, but if water retention devices with capacity >30% of annual runoff from the area are installed, it is reduced tenfold. In areas covered by combined sewer systems, the fee represents the cost of sewage collection. In areas with separate sewers, the fee is based on the volume of collected runoff or on the impervious area. The latter could range from 0.31–7.06 PLN/m^2 (about 0.07–1.63 Euro/m^2) [42].

In addition, Polish Water Law sets conditions for runoff drainage from industrial areas. In the case of areas greater than 3500 m^2, a variable fee is applied if more than 70% of the surface is excluded from biologically-active areas (Article 268.1) as shown in Table 1. The fee is also assessed on all real estate located in areas not served by sewers. This applies to all real estate meeting the cited criteria, including large surface commercial sites (e.g., supermarkets, warehouses), residential estate, office buildings, and housing communes.

Table 1. Variable stormwater discharge fees for industrial areas in Poland.

Site Characteristics	Fee Amount [PLN (€)/m²/yr)] *
Site without retention devices permanently connected	1.00 (0.23 €)
Site with retention devices with capacity of up to 10% of the annual runoff, permanently connected	0.60 (0.14 €)
Site with retention devices with capacity of 10–30% of the annual runoff, permanently connected	0.30 (0.07 €)
Site with retention devices with capacity of more than 30% of the annual runoff, permanently connected	0.10 (0.02 €)

* 1 PLN ≅ 0.2278 € (average between October 2019 and October 2020). This exchange rate will be used in all subsequent cost figures exposed in the paper.

In 2003, the city of Pila pioneered the introduction of a stormwater fee, quickly followed by other towns (Ostrow Wielkopolski, Nysa, Bielsko-Biala, Poznan, Biala Podlaska and Boleslawiec). Today, almost 95% of Polish town have implemented stormwater fee structures.

3. Case Studies of Stormwater Management Upgrade in Two Urban Developments in Silesia

Notwithstanding a clear global trend towards the increasing development and implementation of sustainable urban drainage systems, this issue is not addressed on a widespread scale at the moment in Poland, where a conventional stormwater management approach still remains the most common in urban management. In Poland, which has an unfavorable water balance, rainwater still constitutes an unappreciated contribution to the urban water cycle and is still mostly treated as a nuisance to be disposed of, and discharged as quickly as possible to a receiving water body. Only recently, following the momentum of predicted effects of climate change, and the problem of a lowered groundwater table across the country [43], is stormwater starting to be considered as a possible alternative resource. An analysis of national domestic water consumption trends showed that approximately 50% of public drinking-quality water consumption could be substituted by reused rainwater, with peak of about 65% in public buildings [44].

Two case studies of sustainable stormwater solutions implemented in small buildings in the Kobierzyce commune in the Silesia Region of south Poland are presented and analyzed herein. These concern on-site retention and subsequent water reuse, and a rainwater infiltration garden installation, respectively.

3.1. Case Study 1: Onsite Rainwater Retention and Subsequent Reuse

The first case examined concerns a community center building with playground and parking, built on the site of a demolished establishment, where a local rainwater retention system was implemented. The total plot area of 3300 m^2 consists of directly connected impervious areas (roof and parking) of about 700 m^2 (21% of the lot surface). Built area (including terrace) is 380 m^2, total paved surfaces 1500 m^2, playground 150 m^2, and biologically active area (lawns and trees) 1250 m^2 (37.88% of total) (Figure 2). The organic soil layer consists of low-permeability compacted sand, clay and sandy loams. Groundwater occurs at a depth of 1.5–1.8 m below surface, and therefore is poorly suitable for stormwater infiltration.

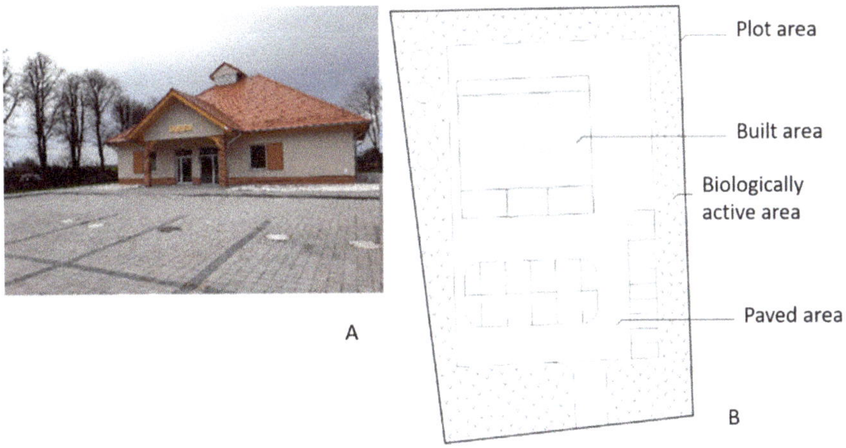

Figure 2. (**A**) New rural community center building, (**B**) plot area.

In the original site conditions, stormwater was discharged directly to a sewer network; with the increased impervious area (larger building and paved area), higher runoff and overflow events were expected. Aside from pure ecological considerations, the main factor suggesting the adoption of an alternative stormwater management solution was related to the increased fee for its discharge. During redevelopment, the site drainage was therefore re-designed with the implementation of a retention basin to reduce stormwater release into the sewer from the property area.

The design of the retention system is based on the estimate of the amount of rainwater and snowmelt on site: for small-scale solutions (e.g., single-family housing, service construction, single public facility building) as in the described case this does not require complex dynamic flow calculations, unlike the case of large catchments [45]. The site's 20-year average annual precipitation (rainwater plus snowmelt), obtained from records of a nearby meteorological station, was estimated at about 600 mm (Figure 3). In the last two years, values of 812 mm and 544 mm were observed, respectively.

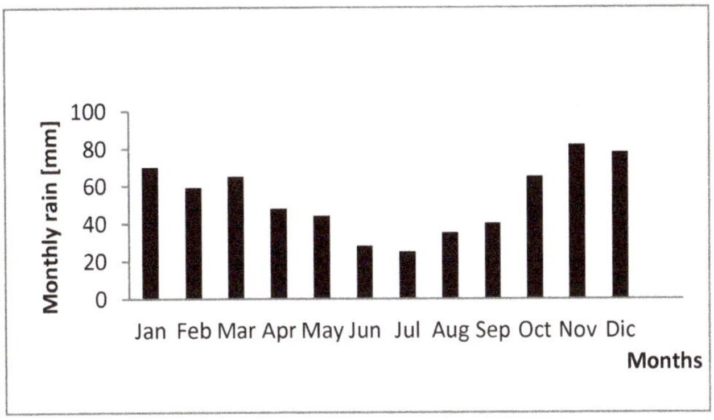

Figure 3. 20-year, monthly average rainfall for the case study site.

The widely used Polish standard (conforming to European standards) PN-EN 75 indicates the following formula for calculating the runoff rate for surfaces <10,000 m^2:

$$Q = \Psi \times I \times A \tag{1}$$

where Q is the maximum flow (L/s), Ψ the permeability coefficient, I the rainfall intensity (per ha), and A the area (ha) considered.

The maximum rainfall rate can be calculated using the Błaszczyk method, for 15-min events (Figure 4) and areas with annual rainfall H < 800mm:

$$q = \left(6631\ \sqrt[3]{H^2C}\right)/t_3^2\Big) \tag{2}$$

where C is the return frequency during which rain occurs with duration t and intensity q, and H the average annual rainfall, in mm.

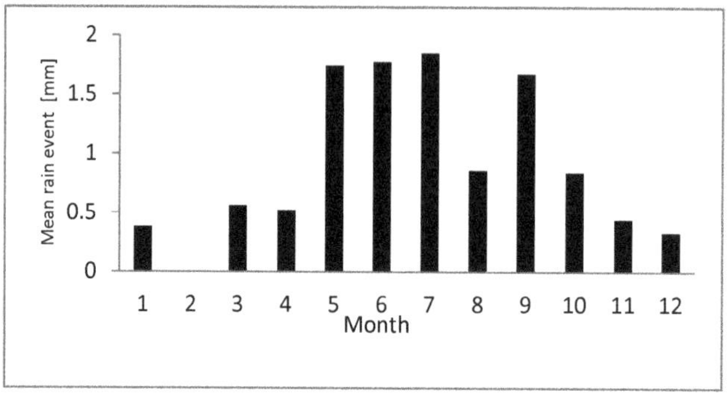

Figure 4. 20-year average values of rainfall on the Kobierzyce site for 15-min events.

The observed behavior of monthly and burst precipitation time series has significant seasonal variability, with relatively dry summers of intense, short rainfall events. This makes local storage an issue of primary relevance in runoff management. Total average runoff of 1134 and 642 m^3/year were thus calculated for parking area and roof. Based on the maximum value of 1.85 mm intensity for the 20-year 15-min event, runoff volumes of 27.75 m^3 and 11 m^3 were calculated for the parking and roof areas, respectively, and targeted for temporary storage. Excess runoff during the more intense events would be diverted to the sewer. Storage tank volume was determined by the practical formula [46]:

$$V_u = V_j \times F_{zr} = 0.06\ [q_{max}(t) - q_{dl}] \times t_d \times f_a \times f_z \times F_{zr} \tag{3}$$

where: V_j indicates the unit retention volume (m^3), F_{zr} the reduced area (ha) of the contributing surface, $q_{max}(t)$ the maximum unit rainfall intensity (L/ha) with duration t_d [min], q_{dt} maximum specific outflow from storage (L/ha), f_a a reduction factor (≤1), depending on corrivation time in the network, t_p (min) and the frequency of rainfall C [years], and f_z a safety factor for volume exceedance (1.1–1.2).

Runoff management was thus reconfigured as follows: roof runoff directed immediately to underground storage (11 m^3 capacity); runoff from car parking, processed in a class I oil separator according to PN-EN 858:2005 standard, to reduce residual concentration of petroleum substances below 5 mg/L. Parking runoff is conveyed by the site's drainage network (137 m of DN400, 49 m of DN315, 54 m of DN200 and 30 m of DN160 pipes, for a total available free volume of 23.3 m^3, enough to hold approximately 85% of the maximum parking runoff volume during and after the design event), and ends in the underground tank from which the excess overflows to the storm sewer. The maximum 15-min runoff calculated from (2) amounts to 132 L/s; the maximum downstream conveyance capacity of the drainage network is 14.3 L/s. A prefabricated oil separator, with integrated settling tank, type OKSYDAN-P 15 (OKSYDAN Sp.z.o.o, Gliwice, Poland)) with nominal capacity of 15 L/s and integrated settling tank of 1.5 m^3 was installed upstream of the underground tank. Local prescriptions on maximum overflow into storm sewers prescribe a limit of 10 L/s, hence a flow regulator (AQUANTIS 330598, diam. 160 mm) was installed at the outlet of the tank.

Stored runoff is targeted for local non-potable reuse: green area watering, surface washing or car washing. According to locally adopted design criteria, the retention tank volume could be suitable to irrigate a green area close to 1000 m^2, as shown in Table 2.

Table 2. Storage tank sizing requirements.

Roof Area (m^2)	350	400
Max. green watered area (m^2)	830	950
Minimum retention tank volume (m^3)	10	11

The underground tank, the pivot element of the system, is fitted with a replaceable cartridge filter (sieve size 25 μm) to retain suspended solids prior to overflow into the municipal storm system. The filter operates with a limited head loss (1–3 cm, depending on fouling), and does not require additional energy inputs, but it must be periodically replaced. A recirculation pump is provided to feed lawn irrigation and other reuse options. This design is able:

1. to reduce and delay runoff drainage into the sewer system;
2. to retain rainwater at source;
3. to infiltrate irrigation water, enhancing evapotranspiration from biologically-active surfaces;
4. to reuse retained water for local uses and reduce water bills;
5. to optimize storm sewer network operation, reducing flood risk in the neighboring area and pollution of receiving waters.

In addition to the reduction of costs associated with lower volumes of stormwater discharged into the drainage system, collected runoff can be used, according to local regulations, for non-potable purposes, thereby reducing water bills at the site. Drawbacks include the need for periodic cleaning of filters and gutters from debris. From a cost-balance standpoint, if runoff from impervious surfaces were to be discharged in full to the sewage system, as in a conventional system, costs of discharge fess would be applicable. This amount can be determined based on existing regulation as 0.34 € per square meter-year of impervious area, at the amount of about 230 € per year. An additional fixed fee for discharge is assessed at about 46 € per year. With the designed system, the annual discharge fee amount is reduced to less than 90 €. Additional costs for the installation of the retention system (storage tank and flow regulator only as local drainage network, the oil separator being required by regulations in either case) amount to about 1600 €, system maintenance (periodic cleaning) and operation cost was assessed at 0.2 €/m^3 storage volume. These figures are summarized in Table 3.

Table 3. Economic balance of stormwater management in Case study 1.

	Without Runoff Retention System	**With Runoff Retention System**
Water reuse from collected runoff (est. average) [m^3/year]	0	640
Water consumption for irrigation (average) [m^3/year]	400	0
Construction costs of additional storage [€]	0	1600
Maintenance & operating costs [€/year]	20	75
Water tariff (1.25 €) /m^3 × 400 m^3) [€/year]	500	0
Stormwater fee [€/year]	276	137
Fee for discharge to sewer network [€/year]	256	<90
Total annual costs [€/year]	1052	302

Figures presented in Table 3 are based on "design" data and actual billing for municipal and water and stormwater services (2017, prior to retention system installation, and 2019, after). Figures include actual fees paid, including changes introduced in 2018 due to new local regulations. Considering an average initial interest rate of 1.5% per year (average historic Polish discount rate till February 2020; it is now 0.1%), and considering the annual cost difference, the additional investment for the retention system was recovered in the first 2.3 years. Considering a design lifespan of 20 years, water bills' saving at year 20 would total about 13,200 €, assuming no tariff variations, over eight times the amount of the initial investment.

3.2. *Case Study 2: Rainwater Infiltration Garden*

The second case study concerns a rural community center building constructed near a residential area in the Kobierzyce commune. The community center plays an important role for the local community: it is the place for town meetings, public participation and other organized events. The roof area of the building is about 650 m^2, a small underpass and parking cover 500 m^2 of a total plot area of 2100 m^2, in which the biologically active area amounts to 950 m^2 (Figure 5). Rainwater was originally discharged into the sewer network, but due to the high discharge fees it was decided to seek alternative stormwater management practices. A solution contemplating the implementation of a rain garden was selected after consultation with the residents, as a perceived adequate approach for the local community, since is relatively simple to implement, may constitute an occasion for enhanced public involvement and participation, which is an important factor in all matters of sustainable management, and required minimal disruption to the existing site. Participation in the construction, planting and ongoing maintenance of the garden can in fact be treated as a form of integration within the local community.

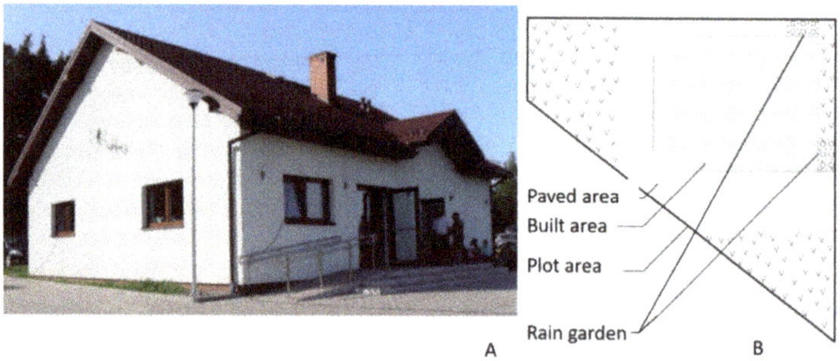

A B

Figure 5. (**A**) Rural community center building, (**B**) surrounding area.

The rain garden system provides runoff infiltration, temporary retention and pre-treatment. A rain garden is built in a shallow depression of the terrain, receiving rainwater from the roof with a gutter and downpipes. The water may flood temporarily on the garden surface (immediately after precipitation), but for the most part of the year it functions as a dry (unirrigated) garden. Its construction implies a proper layering of the subsoil with substrates of good permeability and porosity, which ensure water penetration into buried drainage pipes connected to a storm sewer network or into the underlying aquifer. Coarse sand, limestone and volcanic rock are used for substrate layering.

A rain garden surface includes increased permeability soil (gravel) and vegetated areas with specially selected plants (usually species original to wetlands) that can play an important role in water purification from nutrients and heavy metals. An appropriate soil and vegetation choice can not only fulfill functions of water storage, but also those of biological pollutants removal. A rain garden is specifically designed to collect roof and paved surfaces runoff, store it temporarily, and infiltrate it to underground drainage pipes. Guidelines indicate that a suitable required garden area should be at least 2% of the effective drained area (total area multiplied by a runoff coefficient, depending on the type of surface). Based on the site's characteristics, the total calculated area, with runoff coefficient equal to 1, should be at least 13 m^2, therefore two raingarden plots, each of 8 m^2, located on opposite building corners, were planned, each receiving runoff from an opposite roof pitch.

The installation of the garden started with the excavation of a trench with a depth of about 1 m, with a bottom filled with two 10 cm layer of gravel aggregates (8–16 mm and 2–8 mm size) over which a perforated drainage pipe (diameter 90 mm), enveloped in coconut braid cloth, was laid. A vertical overflow pipe (also 90 mm in diameter) installed into the drainage pipe and protruding about 10 cm over the decorative surface gravel of the garden, allows rapid infiltration in case of high intensity events. The horizontal drainage pipe is covered by a 30 cm layer of fine gravel aggregate. The space between the drainage layer and the surface (45 cm) is filled with a mix of coarse sand, brick ore and dolomite aggregate. (Figure 6). The drainage pipe is connected to the public storm water drainage system to avoid overflow and local flooding in case of extreme events. The rain garden top layer consists of a 15–20 cm thick decorative gravel and stonecrop (*Sedum* spp.) vegetation, with a 2% slope from sides to center. Stonecrop is a succulent perennial plant ideal for dry areas, with easy maintenance and low culture requirements. The chosen plants are Jade and Echeveria.

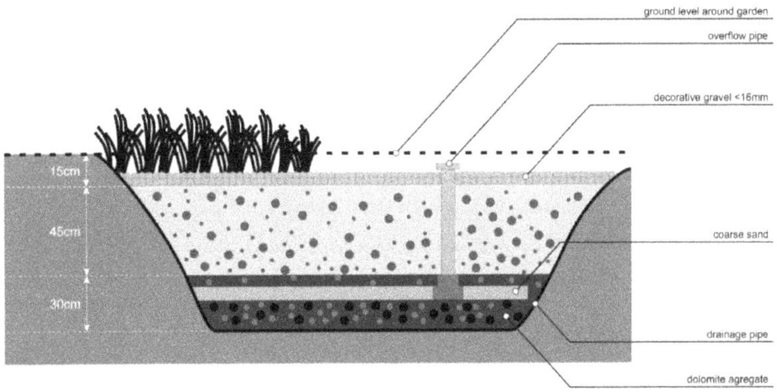

Figure 6. Cross-section of the rain garden layers.

Runoff volumes were calculated (using the same rainfall data of case 1) as: runoff from parking, 226 m^3/year, and from roof, 1047 m^3/year; the average volume of rainwater overflow to the sewage system after rain garden implementation was estimated at less than 10 m^3/yr. In this case, parking area runoff, representing a small fraction of overall runoff, was discharged directly to the storm sewer, at a cost estimated at 90 €/year.

Before construction of the raingarden, the cost of stormwater discharge into the sewer involved payment of a fee of 220 €/year. The rain garden construction cost was quite low (12.3 €/m^2, excluding vegetation), and on the grounds of analyses carried out during the first year of operation, annual maintenance and operating costs were estimated at about 0.5 €/m^2/year. The economic summary of the solution is presented in Table 4.

Figures in Table 4 are based on "design" rainfall volumes and actual billings received. With the same assumptions adopted in case 1, it can be seen that the entire investment for raingarden installation was recovered in less than one year. Assuming a useful project life of 8–10 years (after which some intervention to restore soil permeability would probably be necessary) accrued tariff savings between 2900 and 4100 € would have accumulated, or between 13 and 18 times the initial cost for raingarden operation.

Table 4. Economic balance of stormwater management in Case study 2.

	Without Raingarden	With Raingarden
Construction costs [€]	0	227 (199 + 28 for plants)
Maintenance & operating costs [€/year]	0	7.40
Water services fee [€/year]	60	0
Stormwater fee [€/year]	250	0
Fee for discharge to sewer network [€/year]	220	90 (parking area only)
Total annual costs	530	97.40

4. Discussion

In both case studies presented, benefits of an alternative stormwater management approach result from two factors, runoff delivery reduction, and reduction of discharge fees due to implementation of mitigation measures. Introduction of such solutions brings environmental benefits in addition to economic ones: the possibility of using water for irrigation (case 1) allows improved maintenance of green areas, even during long intervals between rainfall events; in addition, lower water consumption from the public supply network reduces not only water bills but also energy consumption and related emissions. Improved maintenance of green areas will reduce surface soil erosion during intense storms, decreasing the load of directly mobilized sediments (and associated pollutants). The use of storage to relieve the flow load on rainwater drainage during heavy rainfall can prevent network overload and local flooding. Both systems have been operated now for two years. The retention tank proved a good solution, notwithstanding its relatively higher cost, and the rain garden also proved to work efficiently. In the first two years of operation (one of which with 35% higher than average cumulative precipitation), no direct local overflows to the Ślęza river were observed from either site's outfall, even during the most intense events. Prior to the introduction of these systems, overflows of various intensity occurred approximately 10 times per year, as per qualitative records of the sewer operator. While it is early to assess the long-term performance of these systems, they have shown a positive effect in their operation so far. In case 1, the implemented solution showed direct positive effects on receiving water visual quality near the outfall, preventing pollutants from impervious areas being discharged into the stream, and reducing hydrocarbon residues and suspended solids compared to previous conditions (control analyses carried out twice per year for compliance purposes showed no detectable traces of these pollutants).

The cost analysis carried out for the two rainwater management solutions allows the following conclusions:

- the lowest investment costs were obtained for the rain garden solution. In that case, construction cost amounted to 0.22 €/m^3 of estimated captured annual runoff. In case 1, the investment cost amounted to 0.9 €/m^3.
- operating and maintenance costs (excluding investment amortization) of the two solutions are respectively 0.17 €/m^3/year (case 1) and 0.09 €/m^3/year (case 2). This includes systems' maintenance and municipal fees. Without any intervention, cost of the operation and maintenance would be respectively 0.60 €/m^3 and 0.50 €/m^3, for the most part due to municipal fees.

According to the analyzed figures, the rain garden solution appears to be the most appropriate from the cost point-of view; however, site subsoil conditions must be conductive to rapid infiltration. Furthermore, this solution excludes any subsequent local water reuse. Application of a rain garden solution in Case 1 (although subsoil conditions were not ideal) would yield a scenario summarized in Table 5.

Table 5. Comparison of rain garden vs. local storage solution in Case study 1 site.

	With Rain Garden (Estimated)	With Existing Runoff Retention System
Water reuse from collected runoff (est. average) [m^3/year]	0	640
Water consumption for irrigation (average) [m^3/year]	400	0
Construction costs [€]	570	1600
Maintenance & operating costs [€/year]	18.20	77.50
Water tariff (1.25 €/m^3 × 400 m^3) [€/year]	500	0
Stormwater fee [€/year]	137	137
Fee for discharge to sewer network [€/year]	<90	<90
Total annual costs [€/year]	745	304.50

As illustrated, the lower cost of rain garden installation would be quickly offset by the remaining cost of water supply (as in the case of no intervention), since this solution does not allow water reuse. On the other hand, it can be seen that in the absence of discharge fees (and related incentives) the economic analysis of the two cases would be completely different, as shown in Table 6.

Table 6. Economic analysis in the case of no discharge fees.

	Case 1 (Retention and Reuse)	Case 2 (Rain Garden)
Construction costs [€]	1600	227
Maintenance & operating costs [€/year]	304.50	7.40
Water tariff saved (1.25 €/m^3 × 400 m^3) [€/year]	500	0
Annual net cost [€/year]	−195.50	7.40

In case 1, in fact, investment costs would be recovered in less than four years, with a subsequent accrued gain of 16,950 € due to water bill savings until project year 20, i.e., over 10 times the amount of the initial investment. This contradicts the conclusions of a previous review of rainwater collection and usage systems for single-family houses in Poland, conducted in 2009 and based on the offer from manufacturers' catalogues, with costs estimated at 265–1225 €/m^3_{stored} (1050–5000 PLN/m^3), much higher than in the case study presented. It is clear that under these assumptions, investments would show recovery after a period of 59–100 years, depending on water demand and capacity, and thus hardly be justifiable [44].

In case 2, no economic advantage would exist, and initial cost would never be recovered. This analysis confirms that a financial incentives policy are paramount to the achievement of sustainable stormwater management at the local level.

Reducing, on average, by 89% and 81.5% the runoff volumes from local site discharges in the two cases, considerable savings on sewer network construction could be achieved by the city, offsetting the missed income from discharge fees. An estimate of the overall cost/benefit balance of such citywide policies requires, however, a complex approach that goes beyond the purpose of this paper.

While non-potable reuse of rainwater may provide significant conservation of potable water supplies, the possible relationship between reuse and microbiological risks should be carefully considered. Very few studies are available to date on pathogen risk related to such onsite reuse, and clearly the contamination potential is highly dependent on the specific site. Generally, studies showed that a large percentage of roof runoff samples were non-detects with reference to pathogens (95%–90%, depending on microorganism), while stormwater samples from residential and commercial/light industrial areas showed most probable number (MPN) lognormal organisms distributions in the range of 1.3 ± (1.3–2.5) MPN 10 L^{-1} [47]. While some studies indicated that ingestion of untreated, onsite-collected roof rainwater and stormwater may result in gastrointestinal infection risks occasionally greater than that traditionally acceptable (10^{-3} ppy), they also determined that conventionally collected and treated wastewater pathogen log-reductions may be too restrictive when applied to stormwater, with conflicting evidence about the level of treatment (if any) required for health protection [48]. Decentralized treatment for stormwater may eventually be necessary in specific cases, in the direction of what has already been proposed for greywater reuse: applied technologies may include membranes and microbial fuel cell applications, both showing a compatible degree of pathogenic organism reduction [49,50]. Pathogen cells commonly range from about 1 to 10 microns in length, hence higher degrees of filtration than the one already used in Case 1 (25 μm) may be required.

4.1. General Considerations on Stormwater Control Practices

Although the effectiveness of LIDs on storm flow control has been demonstrated in a number of cases, barriers still exist to their broader implementation in new urban developments, due in part to the additional upfront costs for their implementation and long-term maintenance. Costing tools have been developed to allow designers to assess life-cycle costing of different LID practices and evaluate their efficiency [51,52]. These provide a framework to facilitate for capital, operation and maintenance costs estimation, and assess present life-cycle value. Their use, however, is limited by the availability of actual system components costs for specific areas, which sometimes cannot be easily estimated due to lack of previous installations.

The effect of LID practices should not be underestimated even in areas traditionally subject to high volume storms, since these practices successfully trap and filter a considerable portion of runoff, alleviating pressure on existing conveyance systems and reducing runoff side-effects such as downstream erosion, pollutant loadings, and damage to stream and riparian area habitats. Even in high-density urbanization areas, such as the center of the city of Athens (Greece), simulated introduction of LID practices showed potential peak flow reduction in the range of 13.4%–28.2%, and total runoff volume reduction in the range of 24.5%–29% [53]. A U.S. EPA review of 17 LID application case studies in the country showed that capital cost savings in infrastructure development following LID methods application ranged from 15% to 80% [54]. A model-based study concerning the selection of cost-effective LID strategies in Graz (Austria) considering the entire water balance and life-cycle-cost (including land costs) issues showed that there is not one specific optimal LID strategy, but that application of LID treatment trains, consisting of multiple interventions, shows high potential for cost-effective runoff reduction and control [55]. Cost-benefit analysis of LID for stormwater management in an urban catchment in Norway showed that these methods reduce combined sewer overflow (CSO) and that basin-wide optimized solutions in terms of maximum effects and minimum cost can be identified through the use of hydrological modelling [56]. Although no published studies have so far quantified the generalized impact of basin-wide LID practices in urban settings on storm sewers sizing requirements, it can be assumed that their wide-scale adoption could provide long-term benefits in terms of infrastructure design and investment costs.

A key factor in selecting the appropriate LID practice for a specific site lies in the understanding of the site specifics. For example, vegetated filter strips or rain gardens may be an ideal solution for small developments as in case 2 presented herein, but not for sites with large drainage areas. Some other limitations on potential LID installations include requirement of local codes' approval, possible increased pavement failures at LID/curb interfaces, liability and safety concerns, and reduced performance over time.

As interest in rainwater harvesting increases even in humid regions with well-developed water supply infrastructures, it is important to understand the functions and quantify the impacts of these systems. The most popular rainwater harvesting option for homeowners, the so-called "rain barrel" (or small buried storage, less than 1 m^3), often provides inadequate storage even for small irrigation demands in dry periods, and overflows frequently in response to intense storm events. Rain barrels, while providing a valuable demonstration and awareness function, do little to limit runoff, except in particular cases. Studies indicate that only larger rainwater harvesting systems, such as that described in case 1, may have substantial impact on both runoff volume capture and replacement of typical household irrigation demands [32].

In urban settings, regulations in most countries do not allow the use of harvested rainwater for domestic applications other than toilet flushing at the moment, but water utilities are increasingly confronted with customers aiming to decrease their household water footprint by treating rainwater onsite for drinking uses. According to literature, rainwater quality may be better than some surface waters, especially when these are mixed with treatment plant effluents that could still contain pharmaceutical residues or microbial contamination [57]. However, depending on local conditions, rainwater might also be susceptible to microbiological contamination from local pests or wildlife (e.g., avian or rodent species, and even large animals, such as dogs, boars or deer), hence precautionary methods or treatments should be adopted in such cases. A study of a new development district in the Amsterdam area, with total impervious area of about 93,600 m^2, estimated that 64,000 m^3 of water could be harvested adopting current practices, covering about 51% of the drinking water demand of future residents [58]. A combined supply scheme (rainwater harvesting plus central drinking water production) was proposed; however, in order to maintain sufficient supply capacity to deliver drinking water at any time (including dry periods), treatment process and network design would be identical to a traditional system. Site specific economic and energy simulation would be needed in these cases to ascertain any advantage of such solutions. Several studies concluded that cost-efficiency of rainwater harvesting strategies for drinking water provision is strictly linked to local water prices, and that such systems should be preferably installed at the neighborhood level in new construction areas to be cost-effective [59–61].

In addition to cost factors, green infrastructure projects should include early community involvement and communication, and clear evaluation based on project motivation and outcomes. Public perception may be one of the greatest hurdles to overcome, since studies suggest that industrial and commercial users often choose to use municipal water over harvested rainwater, despite its availability [62]. Case study 2, where residents were actually involved in the planning and in the management of the rain garden, is a good example of such practices.

4.2. Potential for Rainwater Reuse in Poland

It was estimated that, at current costs, under Polish conditions rainwater provision for non-potable uses is nearly twenty times cheaper than purchase from public supply, considering only energy consumption for the harvesting system, but excluding initial investment [44]. Climatic conditions in Poland would generally enable effective functioning of such systems, as confirmed by the results of simulation studies [63]. Although it was shown that under the current conditions investment costs are much lower than they had been assessed a few years ago, the financial sustainability of such choices should be evaluated on a case-by-case basis. The introduction of local incentives and fee reduction is without doubt a strong boost to the adoption of these systems in new developments, improving the economic aspects.

In Poland, approximately 85% of potable water comes from surface waters, and requires subsequent purification processes [64]; water supply infrastructure was designed in the 1970s and 1980s to meet the increasing water demand of intensively developing, water-absorbing industries, as well as the high water consumption in the residential sector, and still has considerable reserve capacity. Limiting flow in these networks could require measures to detect and counteract possible secondary contamination during longer in-pipe residence, if this and other water savings solutions were to be adopted on a large scale. Improved systems for in-line detection of waterborne pollutants, such as the application of drinking water contaminant warning systems that are currently being developed may be of use [65].

From this point of view, drastically reducing consumption of water from existing municipal systems may not always be desirable, however, as the price of tap water will become higher; due to the need to modernize the existing treatment and distribution systems, the introduction of mixed supply schemes may become appealing.

5. Conclusions

At the moment, the scale of rainwater reuse is still small in Poland, therefore the issue of sustainable rainwater management is highly relevant and an object of active technical debate. The most popular, relatively simple, solution is to adopt small retention/infiltration devices for single development units, allowing increased use/infiltration of rainwater. As shown, this could bring significant environmental and financial advantages to the community and property owners. The economic sustainability of these measures is strictly correlated to the existence of fiscal/fee incentives, although those solutions that contemplate local reuse also benefit from lower public water purchase costs.

Alternative rainwater management solutions aim primarily at relief and possible replacement of traditional sewage systems: as shown, raingarden is a cost-effective and resilient approach, that provide a number of advantages for facility managers interested in sustainability, but its financial viability is limited by the lack of reuse options. Onsite rainwater storage (harvesting) for reuse implies direct financial savings on water supply costs that could make the solution appealing for individual users even in the short term.

Although the generalized use of these systems would add resiliency to stressed water supply networks, additional aspects such as longer flow residence times in distribution systems and resulting quality issues that may arise should be evaluated and addressed.

Author Contributions: J.B.-Z. followed field applications and gathered related data. A.G.C. designed the initial manuscript framework and organization. Both authors equally contributed to data interpretation and manuscript preparation. Both authors have read and agreed to the published version of the manuscript.

Funding: This research received no external funding.

Conflicts of Interest: The authors declare no conflict of interest.

References

1. Harremoes, P.; Capodaglio, A.G.; Hellstrom, B.G.; Henze, M.; Jensen, K.N.; Lynggaard-Jensen, A.; Otterpohl, R.; Soeberg, H. Wastewater treatment plants under transient loading- Performance, modelling and control. *Water Sci. Technol.* **1993**, *27*, 71–115. [CrossRef]
2. IPCC. *Climate Change 2014: Synthesis Report. Contribution of Working Groups I, II and III to the Fifth Assessment Report of the Intergovernmental Panel on Climate Change*; Pachauri, R.K., Meyer, L.A., Eds.; IPCC: Geneva, Switzerland, 2014; p. 151.
3. Fu, C.C.; Liu, J.H.; Wang, H.; Xiang, C.Y.; Fu, X.R.; Luan, Q.H. Urban storm flooding: Characteristics and management in Beijing. *MATEC Web Conf.* **2018**, *246*, 01042. [CrossRef]
4. Puczko, K.; Jekatierynczuk-Rudczyk, E. Extreme hydro-meteorological events influence to water quality of small rivers in urban area: A case study in Northeast Poland. *Sci. Rep.* **2020**, *10*, 10255. [CrossRef] [PubMed]
5. Novotny, V. *Integrated Sustainable Urban Water, Energy, and Solids Management*, 1st ed.; John Wiley & Sons: Hoboken, NJ, USA, 2020.
6. Viviano, G.; Valsecchi, S.; Polesello, S.; Capodaglio, A.G.; Tartari, G.; Salerno, F. Combined use of caffeine and turbidity to evaluate the impact of CSOs on river water quality. *Water Air Soil Pollut.* **2017**, *228*, 330. [CrossRef]
7. Copetti, D.; Marziali, L.; Viviano, G.; Valsecchi, L.; Guzzella, L.; Capodaglio, A.G.; Tartari, G.; Polesello, S.; Valsecchi, S.; Mezzanotte, V.; et al. Intensive monitoring of conventional and surrogate quality parameters in a highly urbanized river affected by multiple combined sewer overflows. *Water Supply* **2019**, *19*, 953–966. [CrossRef]

8. Capodaglio, A.G.; Callegari, A.; Molognoni, D. Online monitoring of priority and dangerous pollutants in natural and urban waters: A state-of-the-art review. *Manag. Environ. Qual. Int. J.* **2016**, *27*, 507–536. [CrossRef]
9. Qiao, X.J.; Kristoffersson, A.; Randrup, T.B. Challenges to implementing urban sustainable stormwater management from a governance perspective: A literature review. *J. Clean. Prod.* **2018**, *196*, 943–952. [CrossRef]
10. Palmitessa, R.; Borup, M.; Mikkelsen, P.S. Urban tunnel systems for conveyance and storage of storm- and wastewater: Features, classification, and modelling. In Proceedings of the 11th International Conference on Urban Drainage Modelling, Palermo, Italy, 23–26 September 2018.
11. Wu, H.; Huang, G.; Meng, Q.; Zhang, M.; Li, L. Deep tunnel for regulating combined sewer overflow pollution and flood disaster: A case study in Guangzhou city, China. *Water* **2016**, *8*, 329. [CrossRef]
12. Beech, G.D. Lessons Learned—Milwaukee water pollution abatement program. *J. Manag. Eng.* **1992**, *8*, 186–191. [CrossRef]
13. Liang, Z.S.; Sun, J.; Chau, H.K.; Leong, E.I.M.; Chen, D.W.G.; Jiang, F. Experimental and modelling evaluations of sulfide formation in a mega-sized deep tunnel sewer system and implications for sewer management. *Environ. Int.* **2019**, *131*, 105011. [CrossRef]
14. Capodaglio, A.G.; Olsson, G. Energy issues in sustainable urban wastewater management: Use, demand reduction and recovery in the urban water cycle. *Sustainability* **2020**, *12*, 266. [CrossRef]
15. EU Water Framework Directive. Directive 2000/60/EC of the European Parliament and of the Council of 23 October 2000 establishing a framework for Community action in the field of water policy. *Off. J. Eur. Communities* **2000**, *22*, 1–73.
16. Capodaglio, A.G.; Ghilardi, P.; Boguniewicz-Zablocka, J. New paradigms in urban water management for conservation and sustainability. *Water Pract. Technol.* **2016**, *11*, 176–186. [CrossRef]
17. Trincheria, J.; Yemaneh, A. *New Knowledge on Urban Stormwater Management*; Final Report, Project No. 319923 BalticFlows—"Monitoring and Management of Flowing Rain Water in Baltic Sea Catchment Areas". Available online: http://www.lvif.gov.lv/uploaded_files/sadarbiba/balticflows/Rezult%C4%81ti/D3%206%20Final%20Report%20on%20New%20Knowledge%20on%20USWM_160526.pdf. (accessed on 10 September 2020).
18. Andimuthu, R.; Kandasamy, P.; Mudgal, B.V.; Jeganathan, A.; Balu, A.; Sankar, G. Performance of urban storm drainage network under changing climate scenarios: Flood mitigation in Indian coastal city. *Sci. Rep.* **2019**, *9*, 7783. [CrossRef] [PubMed]
19. Capodaglio, A.G. Taking the water out of "wastewater": An ineluctable oxymoron for urban water cycle sustainability. *Water Environ. Res.* **2020**, 1–11. [CrossRef] [PubMed]
20. Capodaglio, A.G. Fit-for-purpose urban wastewater reuse: Analysis of issues and available technologies for sustainable multiple barrier approaches. *Crit. Rev. Environ. Sci. Technol.* **2020**. [CrossRef]
21. Capodaglio, A.G.; Callegari, A.; Cecconet, D.; Molognoni, D. Sustainability of decentralized wastewater treatment technologies. *Water Pract. Technol.* **2017**, *12*, 463–477. [CrossRef]
22. Capodaglio, A.G.; Callegari, A. Can payment for ecosystem services schemes be an alternative solution to achieve sustainable environmental development? A critical comparison of implementation between Europe and China. *Resources* **2018**, *7*, 40. [CrossRef]
23. Brears, R.C. *Blue and Green Cities. The Role of Blue-Green Infrastructure in Managing Urban Water Resources*; Palgrave Macmillan: London, UK, 2018.
24. EPA. Urban Runoff: Low Impact Development. Available online: https://www.epa.gov/nps/urban-runoff-low-impact-development (accessed on 29 June 2020).
25. Oral, H.V.; Carvalho, P.; Gajewska, M.; Ursino, N.; Masi, F.; van Hullebusch, E.D.; Kazak, J.K.; Exposito, A.; Cipolletta, G.; Andersen, T.R.; et al. A review of nature-based solutions for urban water management in European circular cities: A critical assessment based on case studies and literature. *Blue Green Syst.* **2020**, *2*, 112–136. [CrossRef]
26. Ellis, J.B.; D'Arcy, B.J.; Chatfield, P.R. Sustainable urban-drainage systems and catchment planning. *Water Environ.* **2002**, *16*, 286–291. [CrossRef]
27. Arnold, C.L.; Gibbons, C.J. Impervious surface coverage. *Am. Plan. Assoc. J.* **1996**, *62*, 243–258. [CrossRef]

28. Elvidge, C.D.; Tuttle, B.T.; Sutton, P.C.; Baugh, K.E.; Howard, A.T.; Milesi, C.; Bhaduri, B.; Nemani, R. Global distribution and density of constructed impervious surfaces. *Sensors* **2007**, *7*, 1962–1979. [CrossRef] [PubMed]
29. Eckart, K.; McPhee, Z.; Bolisetti, T. Performance and implementation of low impact development—A review. *Sci. Total Environ.* **2017**, *607–608*, 413–432. [CrossRef] [PubMed]
30. Fletcher, T.D.; Andrieu, H.; Hamel, P. Understanding, management and modelling of urban hydrology and its consequences for receiving waters: A state of the art. *Adv. Water Resour.* **2013**, *51*, 261–279. [CrossRef]
31. BIO Intelligence Service. *Water Performance of Buildings*; Final Report for the European Commission; BIO Intelligence Service: Paris, France, 2012.
32. Jones, M.; Hunt, W. Performance of rainwater harvesting systems in the southeastern United States. *Resour. Conserv. Recycl.* **2010**, *54*, 623–629. [CrossRef]
33. Moran, A.; Hunt, W.F.; Jennings, G.D. A North Carolina field study to evaluate greenroof runoff quantity, runoff quality, and plant growth. In Proceedings of the ASCE World Water and Environmental Resources Congress 2003, Philadelphia, PA, USA, 23–26 June 2003.
34. DeNardo, J.C.; Jarrett, A.R.; Manbeck, H.; Beattie, D.J.; Berghage, R. Stormwater mitigation and surface temperature reduction by green roofs. *Trans. ASAE* **2005**, *48*, 1491–1496. [CrossRef]
35. Sazakli, E.; Alexopoulos, A.; Leotsinidis, M. Rainwater harvesting, quality assessment and utilisation in Kefalonia island, Greece. *Water Res.* **2007**, *41*, 2039–2047. [CrossRef]
36. Kowalczyk, A. Green roofs as an opportunity for sustainable development in urban areas. In *Sustainable Development Applications*; Bergier, T., Kronenberg, J., Eds.; Sendzimir Foundation: Zielonki, Poland, 2011; p. 102.
37. Zhao, J.Z.; Fonseca, C.; Zeerak, R. Stormwater utility fees and credits: A funding strategy for sustainability. *Sustainability* **2019**, *11*, 1913. [CrossRef]
38. Act 'Water Law'. *Polish Journal of Laws*, 23 August 2017.
39. Williams, E.S.; Wise, W.R. Economic impacts of alternative approaches to storm-water management and land development. *J. Water Resour. Plan. Manag.* **2009**, *135*, 537–546. [CrossRef]
40. Tasca, F.A.; Assunção, L.B.; Finotti, A.R. International experiences in stormwater fee. *Water Sci. Technol.* **2017**, *1*, 287–299. [CrossRef]
41. Bertram, N.P.; Waldhoff, A.; Bischoff, G.; Ziegler, J.; Meinzinger, F.; Skambraks, A.K. Synergistic benefits between stormwater management measures and a new pricing system for stormwater in the city of Hamburg. *Water Sci. Technol.* **2017**, *76*, 1523–1534. [CrossRef] [PubMed]
42. Burszta-Adamiak, E. The financial mechanisms of urban stormwater management. In *Water in the City Sustainable Development Applications Series 5*; Sendzimir Foundation: Krakow, Poland, 2004; pp. 57–69. Available online: https://sendzimir.org.pl/wp-content/uploads/2019/11/ZRZ5_web.pdf (accessed on 3 September 2020).
43. Szalinska, E. Water quality and management changes over the history of Poland. *Bull. Environ. Contam. Toxicol.* **2018**, *100*, 26–31. [CrossRef] [PubMed]
44. Słys, D. Potential of rainwater utilization in residential housing in Poland. *Water Environ. J.* **2009**, *23*, 318–325. [CrossRef]
45. Liu, W.; Feng, Q.; Deo, R.C.; Yao, L.; Wei, W. Experimental study on the rainfall-runoff responses of typical urban surfaces and two green infrastructures using scale-based models. *Environ. Manag.* **2020**, *66*, 683–693. [CrossRef] [PubMed]
46. Kotowski, A. *Podstawy Bezpiecznego Wymiarowania Odwodnień Terenów (Wydanie II), Tom I—Sieci Kanalizacyjne, Tom II—Obiekty Specjalne (Basics of Safe Dimensioning of Drainage of Areas, Volume I—Sewer Networks, Volume II—Special Objects)*, 2nd ed.; Seidel-Przywecki: Warsaw, Poland, 2015.
47. Schoen, M.; Ashbolt, N.; Jahne, M.; Garland, J. Risk-based enteric pathogen reduction targets for non-potable and direct potable use of roof runoff, stormwater, and greywater. *Microb. Risk Anal.* **2017**, *5*, 32–43. [CrossRef] [PubMed]
48. Schoen, M.; Garland, J. Review of pathogen treatment reductions for onsite non-potable reuse of alternative source waters. *Microb. Risk Anal.* **2017**, *5*, 25–31. [CrossRef]
49. Cecconet, D.; Callegari, A.; Hlavínek, P.; Capodaglio, A.G. Membrane bioreactors for sustainable, fit-for-purpose greywater treatment: A critical review. *Clean Technol. Environ. Policy* **2019**, *21*, 745–762. [CrossRef]

50. Cecconet, D.; Bolognesi, S.; Piacentini, L.; Callegari, A.; Capodaglio, A.G. Bioelectrochemical greywater treatment for non-potable reuse and energy recovery. *Water* **2020**, in press.
51. Houdeshel, C.D.; Pomeroy, C.A.; Hair, L.; Moeller, J. Cost-estimating tools for low-impact development best management practices: Challenges, limitations, and implications. *J. Irrig. Drain. Eng.* **2011**, *137*. [CrossRef]
52. Uda, M.; Van Seters, T.; Graham, C.; Rocha, L. *Evaluation of Life Cycle Costs for Low Impact Development Stormwater Management Practices*; Sustainable Technologies Evaluation Program, Toronto and Region Conservation Authority: Toronto, Canada, 2013.
53. Kourtis, I.M.; Tsihrintzis, V.A.; Baltas, E. Simulation of Low Impact Development (LID) Practices and Comparison with Conventional Drainage Solutions. *Proceedings* **2018**, *2*, 640. [CrossRef]
54. US EPA. *Reducing Stormwater Costs through Low Impact Development (LID) Strategies and Practices EPA 841-F-07-006*; United States Environmental Protection Agency, Nonpoint Source Control Branch: Washington, DC, USA, 2007.
55. Leimgruber, J.; Krebs, G.; Camhy, D.; Muschalla, D. Model-based selection of cost-effective low impact development strategies to control water balance. *Sustainability* **2019**, *11*, 2440. [CrossRef]
56. Xu, C.-Y.; Li, H. Cost and benefit analysis of Low Impact Development (LID) for stormwater management in an urban catchment in Norway. In Proceedings of the EGU General Assembly 2020, held online. 4–8 May 2020. EGU2020-21881. [CrossRef]
57. Angrill, S.; Petit-Boix, A.; Morales-Pinzón, T.; Josa, A.; Rieradevall, J.; Gabarrell, X. Urban rainwater runoff quantity and quality—A potential endogenous resource in cities? *J. Environ. Manag.* **2017**, *189*, 14–21. [CrossRef] [PubMed]
58. Hofman-Caris, R.; Bertelkamp, C.; de Waal, L.; van den Brand, T.; Hofman, J.; van der Aa, R.; van der Hoek, J.P. Rainwater harvesting for drinking water production: A sustainable and cost-effective solution in The Netherlands? *Water* **2019**, *11*, 511. [CrossRef]
59. Roebuck, R.M.; Oltean-Dumbrava, C.; Tait, S. Whole life cost performance of domestic rainwater harvesting systems in the United Kingdom. *Water Environ. J.* **2011**, *25*, 355–365. [CrossRef]
60. Domènech, L.; Saurí, D. A comparative appraisal of the use of rainwater harvesting in single and multi-family buildings of the Metropolitan Area of Barcelona (Spain): Social experience, drinking water savings and economic costs. *J. Clean. Prod.* **2011**, *19*, 598–608. [CrossRef]
61. Morales-Pinzón, T.; Lurueña, R.; Rieradevall, J.; Gasol, C.M.; Gabarrell, X. Financial feasibility and environmental analysis of potential rainwater harvesting systems: A case study in Spain. *Resour. Conserv. Recycl.* **2012**, *69*, 130–140. [CrossRef]
62. Chini, C.M.; Canning, J.F.; Schreiber, K.L.; Peschel, J.M.; Stillwell, A.S. The green experiment: Cities, green stormwater infrastructure, and sustainability. *Sustainability* **2017**, *9*, 105. [CrossRef]
63. Słys, D. The economic result of system application for utilizing precipitation waters in family housing construction. *Instal* **2006**, *6*, 66–69.
64. Statistical Yearbook of the Republic of Poland. 2018. Available online: https://stat.gov.pl/en/topics/statistical-yearbooks/ (accessed on 3 September 2020).
65. Capodaglio, A.G. In-stream detection of waterborne priority pollutants, and applications in drinking water contaminant warning systems. *Water Sci. Technol. Water Supply* **2017**, *17*, 707–725. [CrossRef]

Publisher's Note: MDPI stays neutral with regard to jurisdictional claims in published maps and institutional affiliations.

sustainability MDPI

Case Report

Strategizing Smart, Sustainable, and Knowledge-Based Development of Cities: Insights from Florianópolis, Brazil

Jamile Sabatini-Marques [1], Tan Yigitcanlar [2], Tatiana Schreiner [1], Tatiana Wittmann [1], Debora Sotto [3] and Tommi Inkinen [4,*]

1 Department of Engineering and Knowledge Management, Federal University of Santa Catarina, Campus Universitario, Trindade, Florianópolis, SC 88040-900, Brazil; jamile.sabatini@abes.org.br (J.S.-M.); tati@labchis.com (T.S.); tatianaw@labchis.com (T.W.)
2 School of Built Environment, Queensland University of Technology, 2 George Street, Brisbane, QLD 4000, Australia; tan.yigitcanlar@qut.edu.au
3 Institute of Advanced Studies, University of São Paulo, R. do Anfiteatro, 513, Butantã, São Paulo, SP 05508-060, Brazil; dsotto@usp.br
4 Faculty of Science and Engineering, University of Turku, 20014 Turku, Finland
* Correspondence: tommi.inkinen@utu.fi; Tel.: +358-400-882-818

Received: 6 October 2020; Accepted: 24 October 2020; Published: 25 October 2020

Abstract: Unarguably, smart, sustainable, and knowledge-based development is critical for securing a livable future for our rapidly urbanizing world. The aim of this study is to generate insights into determining effective and efficient strategies to increase sustainability and innovation capabilities of cities to achieve long-term desired urban outcomes. This paper places the city of Florianópolis (Brazil) under the smart, sustainable, and knowledge-based urban development microscope. The methodological approach of the study involves a qualitative analysis through surveys (100 submitted forms, 55 responses received) and interviews (12) with key experts and stakeholders from Florianópolis. The findings of the study reveal that Florianópolis' innovation ecosystem has high potential to thrive, but the city still has structural issues to deal with first, related to the gap between the potential to grow, and acknowledgement from key actors of the city to support the overall territory development considering the complex dimensions. This issue suggests amplifying the ecosystem's vision, including different sectors and, especially, addressing innovation for the common good. The insights generated from the investigation of Florianópolis' case are also invaluable to other cities' planning for strategizing their transformation, and seeking smart, sustainable, and knowledge-based development pathways.

Keywords: smart city; smart urbanism; smart and sustainable urban development; sustainable development; knowledge-based urban development; urban governance; urban transformation; innovation; Florianópolis; Brazil

1. Introduction

Smart, sustainable, and knowledge-based urban development (KBUD) is the catchphrase of our time, as the whole world struggles with climate, pandemic, and financial emergencies [1,2]. Particularly, in the case of emerging economies, this has had a severe impact for Brazil, among others [3–5]. In this case report, we placed an emerging innovation capital of Brazil, namely Florianópolis, under the microscope to generate insights for determining effective and efficient strategies to increase sustainability and innovation capabilities of cities to achieve long-term desired urban outcomes in the age of climate, pandemic, and financial catastrophes.

Florianópolis is the capital of the Brazilian southern state of Santa Catarina. Mostly located in the Santa Catarina Island, Florianópolis has a population of approximately half a million inhabitants [6], a per capita GDP of R 40,162 (around USD 7452) [7] and a Human Development Index of 0.847, the third best city in Brazil [8]. Thanks to its coastal and isolated nature and subtropical climate, the city is a popular tourist destination for both Brazilians and South American travelers.

Around the 1980s, the city started looking for alternatives to tackle its evident economic growth limitations. As its territory is mostly insular, Florianópolis could only offer employment in the public, retail, fishing, and tourism sectors, as industrial developments were, and still are, strictly forbidden by environmental regulations. Through federal, state, and municipal incentives, and taking advantage of the presence of high-quality universities, mostly focused on engineering, the technology sector was established as a viable form of development that coupled with environmental concerns while bringing high added value and generating employment and income. In these circumstances, in the 1980s, two important institutions were founded in Florianópolis to foster an ecosystem of technology and innovation, i.e., Centers of Reference in Innovative Technologies Foundation (CERTI) and the Technology Association of Santa Catarina (ACATE).

During the 1990s, a significant number of software companies, a technological park, ParqTec Alfa, and two incubators, CELTA and MIDI, were established in Florianópolis. A few years later, in 2006, an innovation center, named Sapiens Park, was established in the city as a joint initiative from the state government and the CERTI Foundation. As a result of the joint efforts conducted by the state and city government, universities and entrepreneurs, the city successfully established itself as an innovation center in the late 2000s, thus, recognized both in Brazil and abroad. Currently, the technology sector in Florianópolis generates over R 4.3 billion (around USD 798 million) in revenue per year and employs more than 17,000 people [9]. According to the 2020 edition of the Connected Smart Cities Ranking, which maps the cities with the greatest development potential in Brazil, Florianópolis was ranked as the second most intelligent and connected city in the country [10]. The city won third place in the economy category, in addition to fourth place in the technology and innovation category, fifth place in mobility and accessibility, seventh place in entrepreneurship and health, and 10th in security in the same ranking [10]. Florianópolis was also ranked as the second best entrepreneurial ecosystem amongst the major cities in Brazil, according to the Entrepreneur Cities Index [11].

Local tax exemptions and incentives and public policy programs set up by the Municipal Law of Innovation, issued in 2012, have played, and still play, an important role in Florianópolis economic transition to technology and innovation. A Municipal Innovation Council (CMI) was established as the main managing body of innovation policy and, since 2007, the city mobilized the local innovation ecosystem through the so-called "Innovation Promotion Arrangements (API)", aimed to fund selected projects with both public and private resources, including the Municipal Innovation Fund and the Tax Incentive Program for Innovation. Correspondingly, the local government mapped out a "Route of Innovation", in order to identify and interconnect the city's innovation and technology strategic points as a means to present and promote Florianópolis' innovation ecosystem through a roadmap, extending from the city center to the North of the island. The "Route of Innovation" also aims to connect government, academia, companies, and the community in general, to leverage the entrepreneurial and innovative potential of all actors involved [12].

Recently, two major events have marked Florianópolis, also known as Floripa, development towards a more sustainable city, the new operation of the Florianópolis International Airport and the reopening of the Hercílio Luz Bridge. The Floripa Airport has been managed by a company that also manages Zurich Airport, assuming the city's international airport operations for a period of 30 years and investing in the construction of a new passenger's terminal, inaugurated in 2019. Additionally, the iconic Hercílio Luz Bridge, the longest suspension bridge in Brazil and the first link from Santa Catarina island to the mainland, was restored and reopened to the public on 30 December 2019 after 28 years of closure.

In this context, a research project named the "Smart Floripa Project" was developed between 2018 and 2019 in order to determine whether or not, and to what extent, the KBUD framework could contribute to transform Florianópolis into a smart city of innovation by 2030. The Smart Floripa Project was promoted as a joint effort of a set of institutions, including the Federal University of Santa Catarina, the Institute of Advanced Studies of the University of São Paulo, the Queensland University of Technology, Fecomércio SC, Senac SC, the Brazilian Association of Software Companies, the Lixo Zero Brasil Institute, the Government of the State of Santa Catarina, and the Municipality of Florianópolis.

The investigation was developed as a qualitative study, with surveys and in-depth interviews with multiple stakeholders engaged in the city's innovation ecosystem. The results, as indicated below, revealed that Florianópolis' potential for innovation stemmed from its natural, human, social, and intellectual characteristics, and the main obstacles opposing the city's ambition to become a smart city of innovation stemmed from significant political and governance gaps [13–15].

These political and governance gaps consist, first, of an alarming lack of experience in evidence-based public policymaking amongst all the city's stakeholders. As a result, decision-making processes do not rely on concrete data or scientific evidence, which, bottom line, leads to poor quality and unsustainable public policies, as both public and private managers lack the knowledge and consistent data on sustainable urban development opportunities and challenges [16]. In other words, the city needs to establish a support model for public policy decision making, which could be provided by the KBUD framework.

Finally, Florianópolis needs to balance its innovative ambition with sustainable urban development standards, especially in regards to social equity and environmental protection. In this case, it must be explored whether or not KBUD can articulate these elements and what the existing benchmarks and references are in the matter. In this setting, this paper aims to collect inputs and reflections for the drafting of evidence-based public policies in favor of Florianópolis' ambition to become a smart city of innovation within sustainable development standards, that is, balancing economic prosperity, social equity, and environmental protection through technology and innovation.

2. Knowledge-Based Urban Development of Florianópolis

As mentioned above, KBUD is one of those contemporary terms that has been successfully applied for explaining the overall development of urban locations [17]. The model has gained popularity as it combines four main elements that are considered to be crucial (positive) to development trajectories and pathways [18]. The KBUD framework has become increasingly popular during the last two decades and was conceived in the mid-1990s [19] and has gone through a significant number of updates and detailing in contemporary literature. Because the advance of economies is being radically altered by dynamic processes of economic and territorial restructuring, supported by knowledge and innovation [20–22], the KBUD framework offers a potentially beneficial set of instruments (Figure 1), which can improve the well-being and competitiveness of cities [23,24].

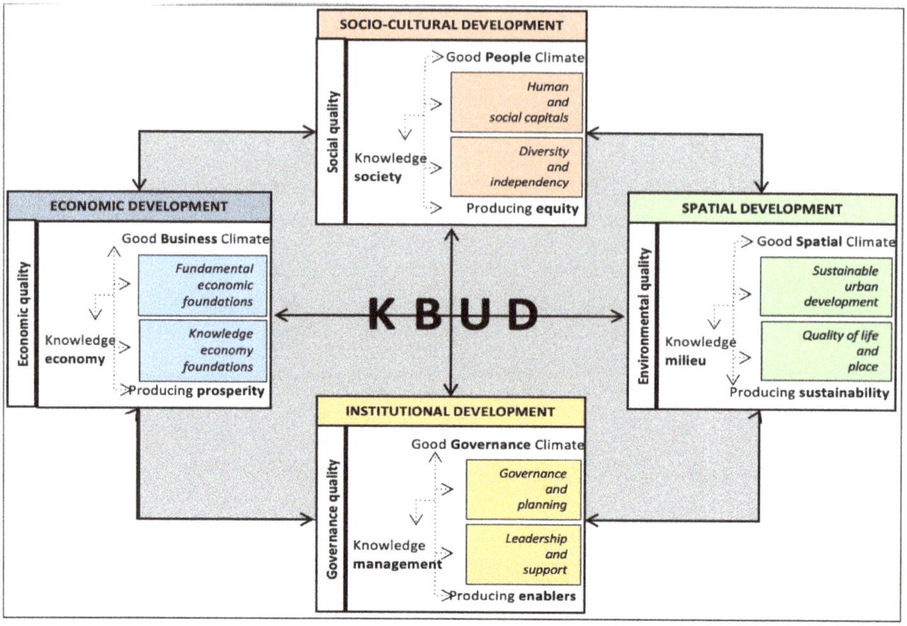

Figure 1. The knowledge-based urban development framework, derived from [18].

Cities (as cases) are interesting study subjects as they are always unique and different in their contextual properties. Locational awareness is fundamental in KBUD [25]. For example, people and organizations tend to follow their local and national cultures in the ways they organize, conduct, and perceive work, leisure, and, particularly, the use of (smart) services as a part of daily living rhythm. Industrial heritage and path-dependency matters as they are intertwined with the social condition of the particular urban contexts under study, Florianópolis being the case here. The growing concentration of technology companies and higher education institutions, associated with KBUD practices, has offered an opportunity to transform Florianópolis into a hub for innovation and knowledge in Brazil [26,27].

Figure 1 outlines the overall structure of the framework and mindset that is important for KBUD. Urban development is considered through four main elements, namely economic, sociocultural, spatial (or environmental), and institutional domains. These four elements provide a tested and well-functioning categorization system for empirical analyses [28]. Under this perspective, the KBUD process in Florianópolis comprises the development of innovation hubs with national relevance and impact, especially in the information technology sector. The development of these innovation districts in the city offers strong potential to produce codified and tacit knowledge, supporting a shift to the knowledge and innovation economy and boosting economic, social, and human gains [29].

The spatial development category, in Figure 1, is the most important one in relation to sustainable and environmentally sound urban development. Whereas the economic approach is considered to be a straightforward assessment of economic performance in time and place, the spatial layer instead stresses the importance of quality of life and clean urban environment as a part of it [30]. There are close connections to the sociocultural dimension of KBUD as an integrative platform for understanding the diversity and complexity of theoretical challenges related to social and environmental structuring of cities [31]. This is supported by an easily verifiable notion that "sustainability" is, nowadays, often connected with urban "smartness", as indicated by Figure 1. One of the early contributions to digitalization and technology-based (smart) urban development was done by Graham and Marvin [32], roughly around the same time when Knight [19] wrote about knowledge-based development concept. Their work pointed out the importance of communication technologies in the evolution of cities.

After these initial considerations, technological advancements have expanded to include green and environmentally friendly new solutions aiding cities to improve their surroundings, such as in the new ways of improving their surroundings, working locations, and leisure hotspots [33,34].

Finally, institutional aspects are needed in KBUD as the development of smartness and environmental conditions are founded on governance quality. It is commonly acknowledged that governmental operability, well established forms of citizen participation, and transparency in public decision making are fundamental components of KBUD (as well as democracy) [35]. The developments of inclusive smart services are often realized in public-private partnerships and in other forms of collaborative arrangements. Advancements in planning theory (e.g., actor networks) provide a decisive intellectual capacity for KBUD when applied in case studies. Overall, the KBUD framework provides insight into the highly complex phenomena of urban evolution that requires an understanding of the time-space trajectories in relation to progression in the field of smart technologies and decline of environmental conditions [16].

3. Methodology

The methodological approach consisted of a set of qualitative analysis methods. Firstly, an online survey was conducted to capture the perspectives of experts from the State of Santa Catarina on KBUD and innovation performances of Florianópolis. Secondly, in-depth face-to-face interviews were undertaken with key policymakers of the State of Santa Catarina to evaluate the survey findings and gather additional insights into Florianópolis' innovation ecosystem and asset base.

3.1. Online Survey

In order to structure the online survey questionnaire, some indicators of the 2018 Global Innovation Index were selected in order to measure Florianópolis' performance. The Global Innovation Index aims to capture the multidimensional facets of innovation by providing a rich database of detailed metrics for 126 national economies, which represent 90.8% of the world's population and 96.3% of the global gross domestic product (GDP). A wide range of high-, medium-, and low-income countries use the index as a tool for understanding, benchmarking, and improving their innovation performances [36]. Taking advantage of the Global Innovation Index framework, a questionnaire was set up for experts from the State of Santa Catarina to evaluate the performance of Florianópolis in relation to Brazil's ranking, according to main indicators clustered under three categories, as presented in Table 1.

Table 1. Global Innovation Index indicators used in the design of the questionnaire (source, authors).

Innovation Input Performance	Innovation Output Performance	Innovation Overall Performance
Institutions	Outputs of knowledge and technology	Strengths
Human capital and research	Creative outputs	Weaknesses
Infrastructure		Opportunities
Market sophistication		Threats
Business sophistication		Overall performance

The questionnaire also used Brazil's assessment in comparison with the Organization for Economic Co-operation and Development (OECD) and Latin American countries as guiding benchmarks. The respondents were asked to rate Florianópolis' innovation performance considering the national performance, defining it using a 7-scale Likert system. For each category, respondents were also asked to state the main reasons for their rating, as well as what could be done to improve the performance in that specific area.

The first version of the questionnaire included a total of 55 questions, 22 questions sought quantitative responses and the remaining 33 questions sought qualitative responses. A pilot version of the questionnaire was placed into a Google Form and sent to 10 people. After considering the feedback received, the questionnaire was shortened to 35 questions, 22 of them quantitative and 13 qualitative.

The final version of the questionnaire was sent by e-mail to 100 selected experts, i.e., innovation and knowledge-based development specialists belonging to the following four sectors: public, private, academic, and not-for-profit. The survey was left open for 15 days and a strong follow-up was made by phone and WhatsApp to receive as many responses as possible. In total, 55 valid responses were received (55% response rate). Among these respondents, 17 respondents were from the not-for-profit sector, 16 respondents were from the private sector, 10 respondents were from public sector, and 12 respondents were from the academic sector.

Quantitative responses were analyzed with the google forms' graphic generation tools, whereas the analysis of the open-ended questions was made by a coding system, based on KBUD dimensions. A total of 16 codes were identified, 4 for each dimension (economic development, sociocultural development, spatial development, and institutional development), and highlighted with a different color (Table 2). Each open-ended response was also coded considering the subcodes for each code category. In total, 660 responses were read, coded, and subcoded, accordingly. Additionally, the main issues from the responses were manually identified and added to a strengths, weaknesses, opportunities, and threats (SWOT) table for further analysis.

Table 2. The coding system (source, authors).

No	Code	Subcode
		Economic development dimension
1	Business context	ICT companies; international; companies; large companies start-ups; business model; business culture; industry of services
2	Innovation activity	United ecosystem; engaged actors; innovative businesses; entrepreneurship environment; creative companies; business organization; entrepreneurial culture; innovation events; market size; programs for start-ups; competitive market; innovation cluster; innovation culture
3	Business infrastructure	Laboratories; investment in infrastructure; technological park; innovation centers; living labs; research centers; infrastructure planning; technology; education centers
4	Investment	Funds; credit; support for economic development; foreign direct investment support for the entrepreneur; investment in R&D; companies' attraction; finance mechanism; venture capital
		Sociocultural development dimension
5	Labor force	Talented people; knowledge workers; human capital; talents attraction
6	Education institutes	university prestige; university ranking; university role; innovative academy; academy corporatism; ideological bias of university
7	Skill sets	Education investment; education quality; technical courses; English language knowledge; training programs; specialization; knowledge absorption; knowledge transformation; knowledge exchange; knowledge application; academia curriculum; intellectual property; incentives in schools and universities to discover talents; applied research; knowledge sharing
8	Cultural assets	Local culture; tolerance; city branding; provincialism; city profile; receptivity
		Spatial development dimension
9	Locational characteristics	Geographically isolated; island limitations (no industries); space for companies; non-industrial area
10	Quality of life and place	Living cost; security
11	Spatial infrastructure	Urban mobility; airport; housing; distance from airport to innovation clusters; sanitation
12	Sustainability	Investments in sustainable infrastructure; environmental impact; nature; preservation areas; energy (replace use of fossil fuel); conscious development; environmental and sustainable technologies

Table 2. *Cont.*

No	Code	Subcode
		Institutional development dimension
13	Governance characteristics	Government effectiveness; political continuity; political stability Public agencies; organs effectiveness; ICTs/online government; accountability; political environment (provincial); political articulation
14	Institutions, partnerships, and international relations	Quadruple-helix partnerships; public-private partnerships; university-company partnership; integration between different sectors; networking; cooperation; integration with the international environment; international exchange; international publications; exchange programs; foreign students and researchers; partnership with international entities; invest in high tech international events; internationalized market; international positioning
15	Support mechanisms	Institutional arrangement; strong entities, e.g., ACATE, Business hub, CERTI; Associationism; active leadership; service to entrepreneurship; governmental incentives, Fapesc; public; investment/incentives; tax incentive programs
16	Regulatory environment	Municipal Law of Innovation; legislation; bureaucracy; public policies; high cost of production; patent applications; legal security

3.2. Face-to-Face Interviews

To gain an in-depth understanding of Florianópolis' innovation performance and potentials, the survey results were further explored and validated through in-depth interviews with 12 key policymakers and policy influencers of the State of Santa Catarina. Due to the importance of the quadruple helix for innovation ecosystems, the interviewees were selected at the ratio of 3 representatives of each sector: public, private, academic, and non-for-profit, as shown in Table 3. Seniority and high-level innovation knowledge were also factors considered for the interviewee's selection.

Table 3. Sectoral and institutional representation of the interviewees (source, authors).

Public Sector	Academic Sector	Not-For-Profit Sector	Private Sector
Federal Ministry of Science, Technology, Innovation, and Communications	Federal Institute of Santa Catarina, Department of Innovation	National Sustainable Waste Organization Institute	Tech Solution Company with Applications on Smart City Communications
State Government Florianópolis Metropolitan Area Authority	Federal University of Santa Catarina, Research & Innovation Hospital	National Technology & Innovation Center	Tech Solution Company with Applications on Financial and Commercial Transactions
Municipal Technology and Innovation Authority	Federal University of Santa Catarina, Administration	Santa Catarina Sate Industry Federation	Real Estate Development Group

The semi-structured interviews gathered an average of nine recording hours. Interviewees were asked to sign a consent and confidentiality agreement to participate in the research. After giving interviewees a brief introduction on the investigation and the online survey results, the researchers directed the following six key conversation starter questions, adding complementary questions whenever necessary:

- What are the main challenges and opportunities in the development of Florianópolis as a smart innovation island?
- What roles have the academic, public, not-for-profit, and private sectors been playing in Florianópolis development as a smart innovation island?
- How can Florianópolis perform well in promoting, attracting and talent retention, and investments to achieve its goal of becoming a smart innovation island?
- What is your vision of Florianópolis to become a smart innovation island?
- What actions can be taken to support Florianópolis' vision to become a thriving smart innovation island?
- Do you have any other comments or perspectives to add on?

Each transcribed interview text was carefully read and analyzed to identify each interviewee's perspectives on the opportunities and challenges for Florianópolis's transition towards becoming a smart city of innovation considering all four KBUD dimensions.

4. Results

4.1. Survey Results

As previously mentioned, the online expert survey targeted to evaluate the performance of Florianópolis in relation to Brazil's ranking in the 2018 Global Innovation Index, in which Brazil ranked 64th (out of a total of 126 countries). In comparison with the OECD and Latin American countries, Brazil was respectively ranked in the 37th and sixth position.

The respondents rated Florianópolis' innovation performance considering the national performance, defining it on a seven-scale Likert system as follows: 1 (far below), 2 (moderately below), 3 (slightly below), 4 (at national average), 5 (slightly above), 6 (moderately above), and 7 (far above). For each category, respondents were also asked to state the main reasons for their rating, as well as what could be done to improve the performance in that specific area.

The questionnaire was sent by e-mail to 100 eminent innovation specialists, of which 55 provided responses. Overall, 83% of the respondents rated Florianópolis' innovation performance as better than Brazil's performance, among which, 26% considered it slightly above, 35% moderately above, and 22% far above. Respondents also evaluated Florianópolis performance as superior to the national performance in all innovation input and output areas, as shown in Tables 4 and 5 below.

Table 4. Respondent views on the specific innovation input performance areas of Florianópolis (source, authors).

Innovation Input Performance							
Institutions (innovation input)	Far below	Moderately below	Slightly below	At the national average	Slightly above	Moderately above	Far above
Political environment (political stability and safety, and government effectiveness)	2%	7%	7%	15%	25%	31%	13%
Regulatory environment (regulatory quality, rule of law, and cost of redundancy dismissal)	4%	7%	13%	32%	27%	15%	2%
Business environment (ease of starting a business, and ease of resolving insolvency)	2%	13%	11%	18%	29%	18%	9%
Human capital and research (innovation input)	Far below	Moderately below	Slightly below	At the national average	Slightly above	Moderately above	Far above
Education (expenditure on education, government funding/pupil, school life expectancy, pupil/teacher ratio, and PISA scales in reading, math, and science)	1.5%	4%	1.5%	15%	40%	29%	9%
Tertiary education (tertiary enrolment, graduates in science and engineering, and tertiary inbound mobility)	2%	2%	5%	13%	27%	40%	11%
Research and development (R&D) (researchers, gross expenditure on R&D, global R&D companies, and international university ranking)	0%	5.5%	7.5%	22%	27%	29%	9%
Infrastructure (innovation input)	Far below	Moderately below	Slightly below	At the national average	Slightly above	Moderately above	Far above
Information and communication technologies (ICTs) (ICT access, ICT use, government's online service, and e-participation)	0%	7%	4%	20%	31%	27%	11%
General infrastructure (electricity output, logistics performance, and gross capital formation)	0%	0%	13%	29%	35%	18%	5%
Ecological sustainability (gross domestic product (GDP)/unit of energy use, environmental performance, and ISO 14001 environmental certificates)	1.5%	3.5%	7%	29%	35%	22%	2%

Table 4. *Cont.*

Innovation Input Performance							
Market sophistication (innovation input)	Far below	Moderately below	Slightly below	At the national average	Slightly above	Moderately above	Far above
Credit (ease of getting credit, domestic credit to private sector, and microfinance gross loans)	2%	7%	9%	29%	37%	9%	7%
Investment (ease of protecting minority investors, market capitalization, and venture capital deals)	4%	11%	5%	25%	33%	20%	2%
Trade, competition and market scale (applied tariff rate, intensity of local competition, and domestic market scale)	3.5%	3.5%	20%	24%	36%	11%	2%
Business sophistication (innovation input)	Far below	Moderately below	Slightly below	At the national average	Slightly above	Moderately above	Far above
Knowledge workers (knowledge-intensive employment, firms offering formal training, gross domestic expenditure on R&D (GERD) performed by business, GERD financed by business, and females employed with advanced degrees)	0%	1.5%	3.5%	9%	29%	42%	15%
Innovation linkages (university/industry research collaboration, state of cluster development, GERD financed by abroad, joint venture (JV)–strategic alliance deals, and patent families 2+ offices)	0%	1.5%	11%	12.5%	33%	33%	9%
Knowledge absorption (intellectual property payments, high-tech net imports, ICT services imports, foreign direct investment (FDI) net inflows, and research talent in business enterprise)	2%	4%	7%	20%	45%	18%	4%

Table 5. Respondent views on the specific innovation output performance areas of Florianópolis (source, authors).

Innovation Output Performance							
Knowledge and technology outputs (innovation output)	Far below	Moderately below	Slightly below	At the national average	Slightly above	Moderately above	Far above
Knowledge creation (patents by origin, patent cooperation treaty (PCT) patents by origin, utility models by origin, scientific and technical articles, and citable documents' H-index)	2%	7%	11%	18%	31%	25%	6%
Knowledge impact (growth rate of purchasing power parity (PPP), new businesses, computer software spending, ISO 9001 quality certificates, and high- and medium-high-tech manufactures)	2%	4%	9%	9%	45%	22%	9%
Knowledge diffusion (intellectual property receipts, high-tech net exports, ICT services exports, and FDI net outflows)	2%	5%	13%	18%	33%	22%	7%
Creative outputs (innovation output)	Far below	Moderately below	Slightly below	At the national average	Slightly above	Moderately above	Far above
Intangible assets (trademarks by origin, industrial designs by origin, ICTs and business model creation, and ICTs and organizational model creation)	2%	0%	12.0%	14.5%	33%	27%	11%
Creative goods and services (cultural and creative services exports, national feature films, entertainment and media market, printing and other media, and creative goods exports)	4%	5%	18%	22%	29%	11%	11%
Online creativity (generic top-level domains (TLDs), country-code TLDs, Wikipedia edits, and mobile app creation)	3.5%	3.5%	10.5%	24%	33%	15%	10.5%

In order to further deepen the assessments and better understand the provided scores, respondents were asked to justify their answers, by indicating what could be done to improve the city's performance with respect to each indicator and by naming three main strengths, weaknesses, opportunities, and threat areas related to Florianópolis' overall innovation performance. The answers to these open-ended questions were analyzed according to a manually created code list based on the four main dimensions of the KBUD framework, as shown in Table 6 below.

Table 6. Content analysis codes by knowledge-based urban development dimensions (source, authors).

Sociocultural Development	Spatial Development	Institutional Development	Economic Development
Labor force	Locational characteristics	Governance characteristics	Business context
Educational institutes	Quality of life and place	Institutions, partnerships, and international relations	Innovation activities
Skillsets	Spatial infrastructure	Support mechanisms	Business infrastructure
Cultural assets	Sustainability	Regulatory environment	Investment

In the economic development dimension of KBUD, the following four main areas were identified: business context, innovation activities, business infrastructure, and investment. Innovation activity was the subdimension most cited by respondents, with 36.5% of total mentions, 70.12% of them being positive comments. This can be explained by Florianópolis actors' engagement and the ability to act with a cluster approach (named locally as associationism), which has put together a well-structured innovation ecosystem. According to one of the respondents, "Florianópolis is a point off the curve in relation to Brazil. I see a very great leadership, especially from ACATE, in the organization and dynamization of technology companies. The opportunities that are presented by ACATE to business owners, partners or not, are varied and current". Respondents also stated that the possible improvements in this area need to focus on monitoring, evaluating, and consolidating present actions in order to foster and improve prospective innovation development.

As per the business context, it became evident that despite the recognition of expressive information and communication technology (ICT) companies' activities in the city, there was room for improvement. More than 19% of citations were about improvement and 30.23% referred negatively to this area. Respondents considered the market to be "very small", mostly represented by small to medium size enterprises (SMEs) with little intention or expression on exports and a reduced call to increase productivity.

The innovation business infrastructure has been quite well developed in Florianópolis. It comprises innovation districts, innovation centers, recognized incubators, and accredited laboratories (e.g., Embrapii). However, there is a demand for better and modern technologies, such as a comprehensive optical fiber network, and low carbon infrastructure, as well as social innovation experiment environments like the living labs. About 30% of the comments on the business infrastructure were positive, 35% were negative, and 35% were neutral or progressive, but still pointed out the need for improvements.

Public or private sector investment was perceived as a major demand, i.e., 47.61% of citations highlighted the necessity of immediate improvements, and 37.3% referred negatively to investment availability and opportunities. "Sinapse da inovação" or "The Synapse of Innovation", an incentive program for innovative entrepreneurship that offers financial resources, training, and support to transform innovative ideas into successful enterprises, was cited as one of the leading instruments developed by the market. Despite a few business angels and investment banking, the general perception was that the investment capacity was limited and precarious. One of the respondents highlighted that "it is necessary to create a culture of private investment in areas such as culture and tourism, which still rely heavily on public investment". Respondents also pointed out a lack of investment in R&D and little support for entrepreneurs and venture capitalists.

The sociocultural development dimension encompasses the following four subdimensions: labor force, educational institutes, skill sets, and cultural assets. Labor force, with 46% of positive references, is a major asset of the city, which attracts talented and creative people [37]. Nevertheless, as the innovation sector is growing, it is also a concern about 31% of the mentions on that topic were negative, 23% of them highlighted specifically the need for improvements. One respondent stated that "the city must attract even more knowledge workers, including foreign researchers, and maintain the ones that are already working in it". Another respondent also pointed out that "educational institutions must contribute to the creation and diffusion of knowledge, as well as in the training of knowledge workers in new areas of digital transformation". Additionally, three other commonly mentioned concerns regarding the labor force were the following: the lack of entrepreneur women, innovators not being high-tech businesses, and the need for continuous qualification of the existing workforce.

Because Florianópolis has strong educational institutes, 75% of the expert comments on the education system were positive. Federal University of Santa Catarina (UFSC) is recognized as one of the best in Brazil, especially its engineering courses; also, the State University (UDESC), and other metropolitan universities, are committed to innovation and have contributed to the development of the innovation ecosystem by supplying talent. Appointed possibilities of improvement included a greater diffusion of the knowledge generated in universities and better collaboration between the academic and private sectors, to transform scientific knowledge into innovation.

Half of the respondents highlighted education and skill sets as key pillars for improvement in all aspects related to innovation. In this sense, despite the good universities, respondents drew special attention to the need to invest in basic education, technical training programs, knowledge diffusion and application. One of the experts remarked that "innovation should be treated and stimulated in basic education, and academia research should meet the demands and needs of the society and industry", which would require, according to another respondent, "the revision of the education policies, educational methodologies, and lecturer' qualifications".

On the one hand, the main cultural assets highlighted by the respondents were the culture of innovation, city branding, local culture, and receptivity of the city. On the other hand, provincialism, the fact that creativity is not seriously taken as a business, and egocentrism were indicated as prejudicial elements to the innovation ecosystem. According to the respondents, Florianópolis needed to open itself to the world, show what has been done, and timely consolidate itself as a technology and innovation pole. Comments on this subdimension were balanced, i.e., 30% of them were positive, 35% were negative, and about 35% were pointing to the need for consolidation.

Regarding the spatial development dimension, the most cited themes were locational characteristics, quality of life and place, spatial infrastructure, and sustainability.

In relation to the locational characteristics, although the island's natural beauty was acknowledged as an important magnet to the creative class, the concurrent land use limitations, and high prices, as well as the impediment of industrial activities were appointed as negative factors. The experts highlighted concerns with sanitary and energy issues, as well as the physical distance between Florianópolis and other important national innovation hubs such as São Paulo. As a result, most of the comments, around 60%, on locational characteristics were negative.

Quality of life and place was cited mainly as a positive contributor to the city's success (45.45%), but respondents expressed some concern about its decline in recent years. High cost of living, lower wages, and public safety issues were pointed out as sensitive areas to boost the permanence of the creative class, entrepreneurs, and skilled labor force. Most of the respondents, however, believed that Florianópolis should be positioned as "by far the best city to live in Brazil".

Spatial infrastructure was the most cited subject in the spatial dimension, with 68.51% of negative comments, mostly due to the city's grave mobility issues. Among the suggested improvements, the most critical ones were the following: investments in a better road system, multimodal public transportation, and the creation of new urban centralities with housing options accessible to various levels of income (to reduce the need to move from home to work). The distance between the airport and the innovation clusters, as well as the lack of technological structure to support new technologies, were also indicated as important concerns. As one of the respondents highlighted, "Florianópolis' transportation and sanitation are far behind from some Brazilian states, which ultimately undermines corporate productivity and ecological sustainability in a more general way".

Sustainability was also highlighted as a significant concern by 48.27% of the respondents. Citations indicated a desire for more public investments in sustainable infrastructure, renewable energy sources, environmental and sustainable technologies, environment preservation areas, as well as the need for greater compliance with housing, land use, and environmental regulations. Some respondents also suggested that "specific programs related to the UN Sustainable Development Goals should be developed".

Lastly, the institutional dimension was pointed out as the most challenging area in the city. The lack of ideal governance characteristics; institutions, partnerships, and international relations; support mechanisms; and regulatory environment were cited by almost all respondents. Most of the citations (54.56%) mentioned the necessity to improve governance standards and practice, and 36.75% of them were extremely critical of the current governance system.

The main concerns regarding governance characteristics revolved around the non-continuity of political actions, the ineffectiveness of public agencies, the lack of accountability, and inadequate management of innovation. "The state and municipal government has adopted much more rhetoric of valuing innovation than concrete actions coordinated with private agents", argued one of the respondents. Possible improvements could be obtained, according to the respondents, through greater agility and dynamic management of public entities, policy transparency and open data, long-term and community-driven policies, governmental will, interest and openness to innovation, well-conducted and communicated policy evaluations, and e-participation.

Institutions, partnerships, and international relations were pointed out as the governance subdimension of greater importance to innovation development in the city. Respondents emphasized the importance of quadruple-helix partnerships, even though criticizing them for their usual inertia and impracticality. Respondents highlighted the necessity to strengthen public-private partnerships and technical-scientific cooperation agreements between the university and productive sectors. Other cited issues were the need for greater city internationalization to build an adequate environment for international business, through international exchange programs, networking, international missions, and international cooperation programs.

Experts praised the city's associationism and active leadership, consistent with the several innovation support mechanisms developed through the years, such as incubators, institutional arrangements, innovation entities (ACATE, CERTI, and business hubs) and some tax incentives. Nonetheless, respondents pointed out that Florianópolis needs to offer significant tax incentives, benefits, and facilities to attract knowledge-intensive companies and to support even more the already existing creative and innovative industry. As one of the respondents observed, "there is no strong micro-credit financing policy. This year the Municipality of Florianópolis launched zero interest financing for small business owners, but there is a lack of stronger government action".

Finally, the regulatory environment was heavily criticized and pointed out as the sector that, if refined, could boost the development of Florianópolis as a smart innovation island. Even though the Florianópolis Municipal Law of Innovation was referred to as a good starting point, the need for de-bureaucracy was mentioned by many experts, as well as the need to improve legal certainty, to facilitate business implementation, to ease patent registration, to have more flexible and agile regulatory bodies, and to define specific public policies to stimulate the innovative sector.

4.2. Interview Results

This section reports the results of the interviews with key policymakers and policy influencers, selected amongst actors belonging to each of the four sectors, i.e., public, not-for-profit, private, and academic (see Table 7 below).

Table 7. Interviewee profiles (source, authors).

Category	No	Expertise and Relevance
Public sector	Interviewee #1	Federal Ministry of Science, Technology, Innovation, and Communications
	Interviewee #2	State Government Florianópolis Metropolitan Area Authority
	Interviewee #3	Municipal Technology and Innovation Authority
Private sector	Interviewee #4	Technology Solution Company with Applications on Smart City Communications
	Interviewee #5	Technology Solution Company with Applications on Financial and Commercial Transactions
	Interviewee #6	Real Estate development group
Not-for-profit sector	Interviewee #7	National Sustainable Waste Organization Institute
	Interviewee #8	National Technology and Innovation Centre
	Interviewee #9	Santa Catarina State Industry Federation
Academic sector	Interviewee #10	Federal Institute of Santa Catarina, Department of Innovation
	Interviewee #11	Federal University of Santa Catarina, Research & Innovation Hospital
	Interviewee #12	Federal University of Santa Catarina, Administration

In total, 12 key policymakers and policy influencers of Florianópolis participated in the interviews. As an evidence of lack of diversity in gender and race in leadership positions in Florianópolis, all interviewees were white male in every sector.

Each interview was carefully transcribed and analyzed to identify each interviewee perspective on the opportunities and challenges for Florianópolis' transition towards becoming a smart city of innovation. The relevant citations were grouped per KBUD dimension, i.e., institutional, environmental, sociocultural, and economic, as well as per helix, i.e., public (GOV), private (COM), academic (EDU), and non-for-profit (ORG), allowing for the identification of the main challenges and opportunities appointed by each one of the four helixes, in each KBUD dimension and respective codes, as explained below.

Firstly, it is important to highlight that the most cited KBUD dimension was the sociocultural dimension, with 46 references (31.5%); economic development came in second place, with 39 citations (26%), closely followed by the institutional dimension, with 37 citations (26.7%). The spatial dimension came in fourth place, with 23 citations (15.8%).

Regarding the institutional dimension, the most cited code was "institutions, partnerships, and international relations", with 14 citations, followed by governance characteristics and regulatory environment, with nine citations each, and support mechanisms, with five citations. The distribution per helix (KBUD domain) is indicated in Figure 2 below.

As per the code, institutions, partnerships, and international relations, the private and academic sectors perceived, respectively, opportunities for associations and an innovation environment for promoting quadruple helix interactions. The public sector is aligned in the sense of developing public and private partnerships and looks forward to replicating the actual network model in other areas, as well as attracting foreign investors to boost this aspect.

Despite private sector's optimism towards private initiative to collaborate, it also senses the quadruple-helix trust development process as a challenge, which has an echo in the academy's understanding of "islands of innovation within the federal university". The private sector sees the barrier to big companies as a challenge too and the academic sector recognizes "politics" as the reason that prevents innovation from taking off and reaching country and global levels. The not-for-profit sector poses that, despite the lack of direction and market access, the ecosystem allows things to happen.

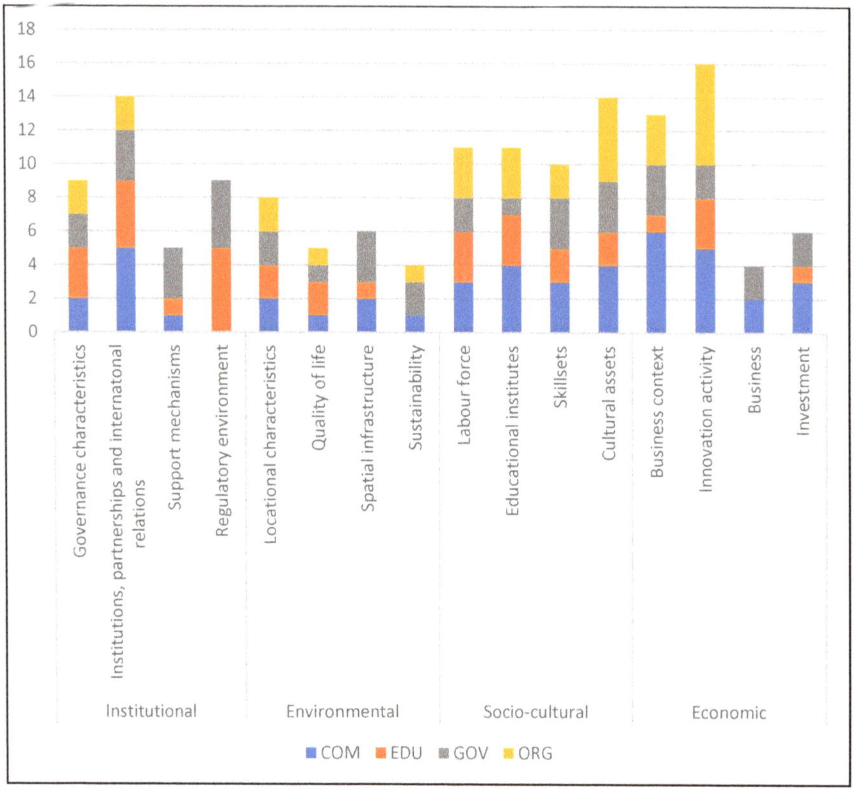

Figure 2. Key topic reference frequency per helix (source, authors).

Governance characteristics are basically seen as a challenge and mostly appointed to the public sector, which is the only helix that perceives opportunities for major players' interactions. The private sector sees governance with a lack of entrepreneurial mindset and is not innovative. One of the interviewees points out that "there is a lack of understanding and perception of the public servers about this new world, the new demands of people and cities. They must know more about Item Response Theory and how to use a database. This way they will not be so immediate and would choose better ways of transport for the city, for example". The academic sector understands the state as a burden, i.e., claims for fewer hindrances and stresses on policy and idea discontinuity. The not-for-profit sector identifies lack of investment in technology and city management.

Regulatory environment was mainly brought up by the academic and public sectors due to the innovation municipal fund, which has raised over R 1 million (approximately USD 185,500) in three projects, evidencing some government involvement towards innovation. As a challenge, both sectors also mentioned the need for de-bureaucratization. Support mechanisms are cited as challenges for both private and public sectors. The academic sector states the lack of local support mechanisms that push companies to develop projects in the northeast region of Brazil, where financial contribution is cut by half. The public sector claims the lack of national and state level development policies in this matter.

The academic sector perceives associations (ACATE) and the innovation park (Sapiens park) as support mechanisms, and the public sector believes that university and private institutes' partnerships should thrive as an opportunity to develop mechanisms for innovation development.

The spatial dimension was the least KBUD dimension cited by the interviewees, with a total of 23 references. Locational characteristics had eight references; quality of life had five references; spatial infrastructure had six references, and sustainability only had four references.

All sectors highlighted the city's locational characteristics, i.e., its natural beauty, attractiveness, pleasantness, and singularity, as an opportunity for development. Nevertheless, all of them also identified challenges linked to the same locational characteristics, due to its geographical limitations and national and local issues pertaining to those limitations. "Floripa has a movement that is contrary to building things. It is a city with a very preservationist concern, also for being an incredibly beautiful environment, and it's difficult and delicate to be able to make any construction in this context", mentions one of the interviewees.

To the public and not-for-profit sectors, a good quality of life and a high-quality police force are opportunities to develop, whereas the private sector highlighted growing concerns with public security. To the academic sector, the city tends to its disorderly growth and must invest heavily in infrastructure.

Spatial infrastructure, particularly, was appointed by all sectors as a challenging issue, with deficiencies in basic infrastructure, green leisure areas, and urban planning. "Even though there are many beaches here, I almost do not go to the beaches because it is a complete chaos, there are no bars, there is nothing, no innovation. And when it rains, everyone goes to the mall, there is no other thing to do. The few theaters and museums lack equipment. We don't have many parks, green leisure areas", points out one of the interviewees. The private sector highlighted that the city, despite being a tourist destination, lacked enough marinas. The academic and the public sectors both remarked that the city must solve its pressing mobility issues, investing in public transport and urban mobility improvements to better organize the flow of people, goods, and services in the city. An interviewee highlights that "we need actions to improve the day to day of the population. It's hard to keep creative people here with no mobility, taking more than one hour to go to the continent area and come back. Besides being good professionals, they want to have a good quality of life".

Sustainability was the least mentioned aspect. The private sector envisages the city's many environmentally protected areas as an asset to tourism, whereas the not-for-profit sector, with a broader point of view, draws special attention to the city's unique balance between social and economic development and environmental preservation as an opportunity for growth. In terms of challenges, the public sector highlighted the difficult coexistence between urban development and environmental protection, as well as the need to expand the use of geoprocessing tools in the city. One of the interviewees mentions that "Florianópolis has a series of associated difficulties to the space that can be used on the island. Many of our rules, regulations, laws and requirements, have difficulty in coexisting with the environment that needs to be preserved, the progress in a sustainable way and initiatives to restructure the city".

The sociocultural dimension was the most cited KBUD dimension, with 46 references in total. Within its codes, cultural assets were the most mentioned (n = 14), followed by labor force (n = 11), educational institutes (n = 11), and skillsets (n = 10). The distribution of references per code and helix is demonstrated in Figure 2.

Cultural assets are foreseen as opportunities by the not-for-profit sector due to a sense of harmony and mutual understanding. The academic sector perceives having a European entrepreneurial influence as an opportunity, and that local culture adaptation to technology is a challenge. Overall, cultural assets are marked by challenges despite the ecosystem's "contagious innovation culture". The private sector points to a lack of leadership, motivation, and willingness; adding to a lack of alignment and diversity by the public sector; and a lack of education and prioritization for the common good by the not-for-profit sector.

Labor force analysis is quite intertwined with the code skill sets. It is generally perceived both as a key and bottleneck aspect by the quadruple helix. Private and academic sectors highlight capacity building and the need for continuous training as the main challenges, whereas the not-for-profit sector observes the aspects that directly challenge labor force such as housing, mobility, and safety.

"We shouldn't want to bring labor force to Florianopolis because it will also overcrowd the city. We should qualify the many talented people that are already here. We should work on two points: qualifying out talents and bringing strategic people", says one of the interviewees. The public sector points to family allocation related to labor force import, a need raised by both public and private sectors due to a labor shortage and lack of qualified workforce, which is a matter related to skill sets. The academic sector perceives a lack of entrepreneurial training and the public sector points to ideological conflict in entrepreneurship programs.

Overall, educational institutes are highly regarded by the private, public, and academic sectors. In particular, the academic sector perceives opportunities at the Federal University because of the "entrepreneurship spirit", the presence of tech leaders, and the private funding for engineering labs. Despite "resistance to change" from the Federal University, according to the public sector, there is an "engineering strength" as an asset to look at.

The private sector points to challenges due to universities' "walls" and the lack of investment and scholarships. This point of view is aligned with the not-for-profit sector's concern in relation to low investment in science and the lack of relationship between companies and universities. The not-for-profit sector also adds the challenge of turning intangible knowledge into tangible capital, outdated teaching, and lack of living labs. "We are not managing to turn science into innovation", highlights one of the interviewees.

Last, but not least, economic development was the second most cited KBUD dimension among interviewees, with 39 mentions. Overall, the business environment, meaning the context of economic activities developed in the city, as well as its innovation practice, presented the major highlights (n = 29), whereas business infrastructure and investment had fewer mentions (n = 10), basically pointed out by the public and private sectors.

The public sector understands the city as an asset. The private sector describes Florianópolis as having four main vectors, tourism, civil construction, maritime economy, and ICT, of which all other three helixes, except for the academic sector, affirm that tourism is underdeveloped. On the one hand, the private sector perceives the lack of integration within business context as a challenge, on the other hand, the public sector believes there are too many participatory councils and a shortage of strong brands.

In terms of innovation activity, the private sector helix pointed out that companies' mix minimizes risk, facilitated networking, and provided a high startup density in the city, whereas the public sector helix acknowledged the important role played by the Federal University of Santa Catarina, UFSC in the triple helix origins in the 70s and further development [38].

The consolidation of the innovation environment is perceived by the private sector, and also by the academic and public sectors. One of the representatives of the not-for-profit sector helix understands that "building a more sophisticated science-based cluster, in areas where the issue of science is important as energy and life science is a medium-term challenge".

Nonetheless, there are critics from both the academy and not-for-profit sector helixes. One academy representative says: "It's a city that wants to do everything (...) It wants to innovate, to be a city of events, to be a tourist city … so it shoots everywhere without a direction". From the not-for-profit sector helix, an interviewee questioned, where the innovation really is by pointing out that "companies here stand out by discipline, technique, growth and applicability, but I can't remember anything that's revolutionary". This respondent realized that "we don't see innovation in the streets" due to the lack of a creative economy: "Great things are content and everyone sells it in this world. We sell tools (...) we do not sell creativeness".

As per business infrastructure, the public sector recognizes institutions CERTI Foundation and ACATE association by playing an important role along with Santa Catarina State's research foundation (FAPESC) for supporting and providing an environment for business. Nevertheless, the private sector identifies there is a lack of capillarity in terms of a "product-led structure not to take big companies, but the startups that today are the basis of the island's innovation, to the world". This leads to another aspect for economic development, i.e., investment. Despite the public sector stating there is a lack of private and international investment, the private sector believes it is not about money but collaborative work, "making the quadruple helix rotate". In this sense, "the role of the state is to generate purchase for the national technology-based industry".

5. Discussion and Conclusions

This paper applied the KBUD framework in order to shed light on wicked problems that are complex, unclear, interdependent, multi-causal, unpredictable, nonlinear, and dynamic [39]. Research poses that instead of finding the correct answer to a wicked problem, the aim is to achieve a shared understanding of possible solutions [40], which was precisely the reason for studying representatives from the local quadruple helix and their different perspectives.

This study assessed labor force, training/education/institutes, and cultural assets under the KBUD framework's sociocultural dimension. This was supported by the data, as Florianópolis' most cited aspect in the interviews and mostly highly evaluated in the survey. In the latter, half of respondents are aware that the labor force is a major asset and that "creative people attract creative people". Nevertheless, expressions used in the interviews to drive Florianópolis into becoming smarter and more sustainable were the "engineering strength" and the "entrepreneurship spirit" of the Federal University of Santa Catarina. Regardless of the balance between positive and negative remarks on cultural assets in the survey, there is a "contagious innovation culture" in the air, present in both survey and interview. Again, this leads to the observation that not much was referred to on behalf of diversity. For that matter, the city as a whole and the innovation ecosystem will need to think beyond the actual visible trait of gender and race. Creative and responsive solutions are longing for diversity and balancing from different perspectives to face wicked problems.

Labor force imports, due to a shortage of qualified workforce, is also a short-term response to a major demand. A harder and longer-term action is to invest in basic education and training programs. Because it is important to retain the city's talents, it is also important not to lose potential ones to organized crime for lack of investment and opportunities. In this sense, it is relevant to stress that social inequality was mentioned by only two interviewees. In addition, a concern with the increase in safety issues, which is a main consequence of social unbalance, is pointed out as a crucial challenge to creatives' permanence in the city, along with housing and mobility issues. In turn, these aspects belong to the spatial discussion, which was the least cited dimension.

The spatial dimension of the KBUD framework is a major issue in the data, because of Florianópolis' basic infrastructure and territorial planning deficiencies and also for not placing the climate change reality on the agenda. Despite presenting itself as a "green industry", the innovation ecosystem has a lot to adapt to and work towards facing this "glocal" matter, with consequences that deeply affect every sector. It starts by shifting from the utilitarian perspective of nature, which has been granting Florianópolis the title of being a popular tourist destination worldwide for its beaches and seafood, to that of having a responsible engagement with the environment, respecting its geographical limitations. An unfolding to this call can be foreseen in the economic and institutional development dimensions.

This study comprehended governance, partnerships, and support mechanisms as main aspects of the institutional dimension of the KBUD framework. Perceptions, which are considered to be the most challenging area for both survey and interview, are mainly targeted to the public sector's lack of efficiency, accountability, innovation, and continuity of actions. In addition, there is room for perspective change towards the public sector whose aim is to serve the common good and to support the quadruple helix working together in this direction. These liabilities need to be sorted out, as well as social participation beyond the "too many city councils", which are not necessarily addressing the city's aspirations.

Commitment and trust development are key ingredients for this dimension to develop well, which were pointed out as a challenge in the interview, and related to municipal and state collaborative planning and programs' developments. Florianópolis' innovation ecosystem has the potential to thrive, despite that the city still has structural issues to deal with, related to the gap between its desire to grow and the acknowledgement from its major actors to support the overall territory to thrive in its complex dimensions. It suggests amplifying the ecosystem's vision, including different sectors and, especially, tackling innovation for the common good [41].

The pressure caused by the growing urbanization generates complex and multifaceted challenges that can only be faced through processes that involve innovation in a shared way. Smart, sustainable, and knowledge-based urbanism approaches can help localities thrive in not only economy and society terms but also spatial and governance wise [42–44].

As a concluding remark, this study was a snapshot of Florianópolis with a focus on the innovation ecosystem reality, an important tool for a transition plan aiming at a more balanced socioecological scenario. It generated useful insights that should help the city retain and inspire more talented and creative people to come and stay, which directly responds to the need for diversity. To keep this on an ongoing basis, in other words sustainability, investment and opportunities must be made in overall education, for all. Here, it goes without saying that as important as it is to develop a strategy plan to boost innovation, basic demands on the socio-environmental front will also need to be acknowledged, analyzed, and addressed in an integrative manner. Therefore, the case study findings reported in this paper are not only useful for the case city Florianópolis, but also invaluable for other cities planning to strategize their transformation, and seeking smart, sustainable, and knowledge-based development pathways in the age of climate, pandemic, and financial catastrophes.

Author Contributions: J.S.-M. and T.Y. designed and supervised the study; T.S., T.W., and D.S. conducted the analysis, and prepared the first full draft of the manuscript; T.I. expanded the manuscript and helped to finalize it by improving its rigor, relevance, and caliber. All authors have read and agreed to the published version of the manuscript.

Funding: The research reported in this paper received funding from industry partners of the Smart Floripa Project, namely the State Government of Santa Catarina, Municipality of Florianópolis, National Confederation of Trade in Goods, Services and Tourism, Senac Santa Catarina.

Acknowledgments: The authors thank anonymous referees for their invaluable comments on an earlier version of the manuscript.

Conflicts of Interest: The authors declare no conflict of interest.

References

1. Chang, D.L.; Sabatini-Marques, J.; Da Costa, E.M.; Selig, P.M.; Yigitcanlar, T. Knowledge-based, smart and sustainable cities: A provocation for a conceptual framework. *J. Open Innov. Technol. Mark. Complex.* **2018**, *4*. [CrossRef]
2. Yigitcanlar, T. *Rethinking Sustainable Development: Urban Management, Engineering, and Design*; IGI Global: Hersey, PA, USA, 2010.
3. Schmalz, S.; Ebenau, M. After Neoliberalism? Brazil, India, and China in the Global Economic Crisis. *Globalizations* **2012**, *9*, 487–501. [CrossRef]
4. Barberia, L.G.; Gómez, E.J. Political and institutional perils of Brazil's COVID-19 crisis. *Lancet* **2020**, *396*, 367–368. [CrossRef]
5. Diele-Viegas, L.M.; Pereira, E.J.D.A.L.; Rocha, C.F.D. The new Brazilian gold rush: Is Amazonia at risk? *For. Policy Econ.* **2020**, *119*, 102270. [CrossRef]
6. Instituto Brasileiro de Geografia e Estatistica. Estimativas da população residente para os municípios e para as unidades da federação com data de referência em 1° de julho de 2019: [notas metodológicas]. 2019. Available online: https://biblioteca.ibge.gov.br/index.php/biblioteca-catalogo?view=detalhes&id=2101662 (accessed on 18 April 2020).
7. Instituto Brasileiro de Geografia e Estatistica. PIB per capita: IBGE, em parceria com os Órgãos Estaduais de Estatística, Secretarias Estaduais de Governo e Superintendência da Zona Franca de Manaus—SUFRAMA. 2017. Available online: https://cidades.ibge.gov.br/brasil/sc/florianopolis/panorama (accessed on 18 April 2020).

8. Programa de las Naciones Unidas para el Desarrollo; Instituto de Pesquisa Econômica Aplicada; Fair and Just Prosecution. Atlas do Desenvolvimento Humano no Brasil. 2010. Available online: http://atlasbrasil.org.br/2013/pt/perfil_m/florianopolis_sc (accessed on 18 April 2020).
9. Ribeiro, J.M.; Bocasanta, S.L.; Ávila, B.O.; Magtoto, M.; Jonck, A.V.; Gabriel, G.M.; de Andrade, J.B. The adoption of strategies for sustainable cities: A comparative study between Seattle and Florianopolis legislation for energy and water efficiency in buildings. *J. Clean. Prod.* **2018**, *197*, 366–378. [CrossRef]
10. Urban Systems. Ranking Connected Smart Cities. 2020. Available online: https://d335luupugsy2.cloudfront.net/cms/files/48668/1599564431Ranking_CSC_2020.pdf (accessed on 14 September 2020).
11. Endeavor Brasil. Índice de Cidades Empreendedoras Brasil. 2017. Available online: https://d335luupugsy2.cloudfront.net/cms/files/6588/1512651268AF-REAL-ICE-2017-web.pdf (accessed on 19 September 2020).
12. Baptista, S.R. Metropolitanization and forest recovery in southern Brazil: A multiscale analysis of the Florianópolis city-region, Santa Catarina State, 1970 to 2005. *Ecol. Soc.* **2008**, *13*, 5. [CrossRef]
13. Inkinen, T. Best practices of the Finnish Government Information Society Policy Programme. *Transform. Gov. People Process. Policy* **2012**, *6*, 167–187. [CrossRef]
14. Zygiaris, S. Smart City Reference Model: Assisting Planners to Conceptualize the Building of Smart City Innovation Ecosystems. *J. Knowl. Econ.* **2012**, *4*, 217–231. [CrossRef]
15. Confortin, S.C.; Antes, D.L.; Pessini, J.; Schneider, I.J.; d'Orsi, E.; Barbosa, A.R. Comparison of sociodemographic profile and health conditions of elderly residents in predominantly rural and urban areas of the Greater Florianópolis, Southern Brazil. *Cad. Saúde Coletiva.* **2016**, *24*, 330–338. [CrossRef]
16. Millar, C.C.J.M.; Choi, C.J. Development and knowledge resources: A conceptual analysis. *J. Knowl. Manag.* **2010**, *14*, 759–776. [CrossRef]
17. Benneworth, P.; Ratinho, T. Reframing the role of knowledge parks and science cities in knowledge-based urban development. *Environ. Plan. C* **2014**, *32*, 784–808. [CrossRef]
18. Yigitcanlar, T.; Dur, F. Making space and place for knowledge communities: Lessons for Australian practice. *Australas. J. Reg. Stud.* **2013**, *19*, 36–63.
19. Knight, R.V. Knowledge-based Development: Policy and Planning Implications for Cities. *Urban Stud.* **1995**, *32*, 225–260. [CrossRef]
20. Chawla, D.; Joshi, H. Knowledge management practices in Indian industries—A comparative study. *J. Knowl. Manag.* **2010**, *14*, 708–725. [CrossRef]
21. Hall, T.; Barrett, H. *Urban Geography*; Routledge: New York, NY, USA, 2018.
22. Evers, H.D.; Gerke, S.; Menkhoff, T. Knowledge clusters and knowledge hubs: Designing epistemic landscapes for development. *J. Knowl. Manag.* **2010**, *14*, 678–689. [CrossRef]
23. Yigitcanlar, T. Australian local governments' practice and prospects with online planning. *URISA J.* **2006**, *18*, 7–17.
24. Inkinen, T.; Kaakinen, I. Economic Geography of Knowledge-Intensive Technology Clusters: Lessons from the Helsinki Metropolitan Area. *J. Urban Technol.* **2016**, *23*, 95–114. [CrossRef]
25. Hortz, T. The Smart State test: A critical review of the Smart State Strategy 2005–2015's Knowledge-Based Urban Development. *Int. J. Knowl. Based Dev.* **2016**, *7*, 75. [CrossRef]
26. Andion, C.; Alperstedt, G.D.; Graeff, J.F. Social innovation ecosystems, sustainability, and democratic experimentation: A study in Florianopolis, Brazil. *Rev. Adm. Pública.* **2020**, *54*, 181–200. [CrossRef]
27. De Andrade, J.B.; Ribeiro, J.M.; Fernandez, F.; Bailey, C.; Barbosa, S.B.; da Silva Neiva, S. The adoption of strategies for sustainable cities: A comparative study between Newcastle and Florianópolis focused on urban mobility. *J. Clean. Prod.* **2016**, *113*, 681–694. [CrossRef]
28. Kiuru, J.; Inkinen, T. Predicting innovative growth and demand with proximate human capital: A case study of the Helsinki metropolitan area. *Cities* **2017**, *64*, 9–17. [CrossRef]
29. Lopez-Saez, P.; Navas-Lopez, J.E.; Martín-de-Castro, G.; Cruz-Gonzalez, J. External knowledge acquisition processes in knowledge-intensive clusters. *J. Knowl. Manag.* **2010**, *14*, 690–707. [CrossRef]
30. Mohamed, M.; Murray, A.; Mohamed, M. The role of information and communication technology (ICT) in mobilization of sustainable development knowledge: A quantitative evaluation. *J. Knowl. Manag.* **2010**, *14*, 744–758. [CrossRef]
31. Petruzzelli, A.M.; Albino, V.; Carbonara, N.; Rotolo, D. Leveraging learning behavior and network structure to improve knowledge gatekeepers' performance. *J. Knowl. Manag.* **2010**, *14*, 635–658. [CrossRef]
32. Graham, S. *Telecommunications and the City*; Routledge: London, UK, 2002.

33. Merisalo, M.; Makkonen, T.; Inkinen, T. Creative and knowledge-intensive teleworkers' relation to e-capital in the Helsinki metropolitan area. *Int. J. Knowl. Based Dev.* **2013**, *4*, 204. [CrossRef]
34. Inkinen, T.; Merisalo, M.; Makkonen, T. Variations in the adoption and willingness to use e-services in three differentiated urban areas. *Eur. Plan. Stud.* **2018**, *26*, 950–968. [CrossRef]
35. Zhao, P. Building knowledge city in transformation era: Knowledge-based urban development in Beijing in the context of globalisation and decentralisation. *Asia Pac. Viewp.* **2010**, *51*, 73–90. [CrossRef]
36. Dutta, S.; Reynoso, R.E.; Garanasvili, A.; Saxena, K.; Lanvin, B.; Wunsch-Vincent, S.; Guadagno, F. *The Global Innovation Index 2018: Energizing the World with Innovation*; Cornell University: Ithaca, NY, USA, 2018.
37. Cabrita, M.D.R.M.F.; Machado, V.; Cabrita, C. Managing creative industries in the context of knowledge-based urban development. *Int. J. Knowl. Based Dev.* **2013**, *4*, 318. [CrossRef]
38. Gohari, S.; Medalen, T.; Aranya, R. Exploring the Impact of Complex Multi-Level Governance Structures on the Societal Contribution of Universities to Knowledge-Based Urban Development. *Soc. Sci.* **2019**, *8*, 279. [CrossRef]
39. Levin, K.; Cashore, B.; Bernstein, S.; Auld, G. Overcoming the tragedy of super wicked problems: Constraining our future selves to ameliorate global climate change. *Policy Sci.* **2012**, *45*, 123–152. [CrossRef]
40. Conklin, J. Wicked Problems and Social Complexity. In *Dialogue Mapping: Building Shared Understanding of Wicked Problems*; Wiley: New York, NY, USA, 2006.
41. Florida, R.; Adler, P.; Mellander, C. The city as innovation machine. *Reg. Stud.* **2016**, *51*, 86–96. [CrossRef]
42. Yigitcanlar, T.; Kamruzzaman, M. Planning, development and management of sustainable cities: A commentary from the guest editors. *Sustainability* **2015**, *7*, 14677–14688. [CrossRef]
43. Alvarez, M.D. Creative cities and cultural spaces: New perspectives for city tourism. International. *J. Cult. Tour. Hosp. Res.* **2010**, *4*, 171–175. [CrossRef]
44. Martin, C.J.; Evans, J.; Karvonen, A. Smart and sustainable? Five tensions in the visions and practices of the smart-sustainable city in Europe and North America. *Technol. Forecast. Soc. Chang.* **2018**, *133*, 269–278. [CrossRef]

sustainability

MDPI

Article

Healthy Power: Reimagining Hospitals as Sustainable Energy Hubs

Nicholas Gurieff [1,2,*], Donna Green [2,3], Ilpo Koskinen [2,4], Mathew Lipson [2,5], Mark Baldry [6], Andrew Maddocks [1], Chris Menictas [7], Jens Noack [8], Behdad Moghtaderi [1] and Elham Doroodchi [9]

1. Priority Research Centre for Frontier Energy Technologies & Utilisation, University of Newcastle, Callaghan, NSW 2308, Australia; andrew.maddocks@newcastle.edu.au (A.M.); behdad.moghtaderi@newcastle.edu.au (B.M.)
2. Digital Grid Futures Institute, University of New South Wales, Sydney, NSW 2052, Australia; donna.green@unsw.edu.au (D.G.); ilpo.koskinen@unsw.edu.au (I.K.); m.lipson@unsw.edu.au (M.L.)
3. Climate Change Research Centre, University of New South Wales, Sydney, NSW 2052, Australia
4. Design Next, University of New South Wales, Sydney, NSW 2052, Australia
5. ARC Centre of Excellence for Climate System Science, University of New South Wales, Sydney, NSW 2052, Australia
6. School of Biomedical Engineering and School of Physics, University of Sydney, Sydney, NSW 2006, Australia; mark.baldry@sydney.edu.au
7. School of Mechanical and Manufacturing Engineering, University of New South Wales, Sydney, NSW 2052, Australia; c.menictas@unsw.edu.au
8. Applied Electrochemistry, Fraunhofer-Institute for Chemical Technology, 76327 Pfinztal, Germany; jens.noack@ict.fraunhofer.de
9. Centre for Advanced Energy Integration, University of Newcastle, Callaghan, NSW 2308, Australia; elham.doroodchi@newcastle.edu.au

* Correspondence: nicholas.gurieff@newcastle.edu.au; Tel.: +61-2-40339611

Received: 21 September 2020; Accepted: 13 October 2020; Published: 16 October 2020

Abstract: Human health is a key pillar of modern conceptions of sustainability. Humanity pays a considerable price for its dependence on fossil-fueled energy systems, which must be addressed for sustainable urban development. Public hospitals are focal points for communities and have an opportunity to lead the transition to renewable energy. We have reimagined the healthcare energy ecosystem with sustainable technologies to transform hospitals into networked clean energy hubs. In this concept design, hydrogen is used to couple energy with other on-site medical resource demands, and vanadium flow battery technology is used to engage the public with energy systems. This multi-generation system would reduce harmful emissions while providing reliable services, tackling the linked issues of human and environmental health.

Keywords: energy transitions; hydrogen; energy storage; vanadium; flow battery; industrial ecology; co-benefits; multi-generation; power-to-X; energy networks

---

1. Introduction: Health and Energy

The energy systems most modern economies rely on are outmoded and unhealthy, which has multiple significant negative impacts. In addition to anthropogenic climate change caused by greenhouse gas emissions, nitrogen and sulfur oxides and carbon particulates damage ecosystems and are harmful to human health. Excess mortality from outdoor air pollution due to fossil fuel use is estimated at four million people per year [1]. The economic costs of this air pollution were estimated at USD 2.9 trillion, or over three percent of global GDP in 2018 [2].

Discussion of the relationship between climate change and health is increasing, most recently as the COVID-19 pandemic has highlighted improvements in air quality from reductions in human activity. Addressing this issue more permanently is a challenge; however, it is not a wicked problem as solutions can be mutually beneficial. Tackling the climate emergency goes hand-in-hand with improving public health outcomes by reducing harmful air pollution and developing a circular economy based on renewable energy ecosystems. Hakovirta and Denuwara now suggest 'sustainability' be redefined as "the intersection of the economy, environment, society and human health" [3]. The link to health represents an opportunity to accelerate the transition to renewable energy, not least because the health system itself is in a unique position to lead.

Doctors have had considerable success framing this health-related sustainability challenge, raising awareness of the considerable price that humanity pays for an energy ecosystem that relies on fossil fuels [4]. The Sustainable Development Unit of the UK National Health Service (NHS) was established in 2008, and the NHS has since reduced its carbon footprint by 19 percent over the 10 year period to 2018 [5]. Analysis of these efforts was published in *The British Medical Journal* [6]. Other prestigious medical journals have been amplifying calls to action, with editorials published in *The Lancet* [7], *The New England Journal of Medicine* [8] and *The Medical Journal of Australia* [9].

These efforts are critical because the health sector has a significant environmental footprint. In Australia, health care represented seven percent of Australia's carbon dioxide (CO_2) emissions in 2014–2015, with over one third of that attributable to public hospitals [10]. Leadership here can have a more pervasive impact since public perception of technology and its safety is important for its acceptance and deployment [11,12]. Doctors and scientists are the two most trusted professions [13], and so have a vital role to play in driving the innovative and creative thinking that will move past just efficiency and recycling to deliver whole system sustainable health services [14].

Systems thinking research supports the concepts of industrial ecology and social ecology that acknowledge connections between organizations and society and the importance of this for driving long-term change [15]. Collective visions of promising techno-scientific futures can legitimize investments and transcend uncertainty [16]. Public energy installations and community energy services are already being used for community interaction with new energy technologies [17,18]. Communities have shown a willingness to invest directly in renewable energy installations and there is interest and receptivity of these installations specifically in the hospital context [19].

Hospitals, as critical and major piece of publicly funded infrastructure, are an excellent case study for energy ecosystems. A hospital is not simply an energy user, it is a community and industry hub. Hospitals are regarded as safe havens, resilient facilities for disaster and emergencies [20]. Large numbers of staff and public use them daily and on-site parking is necessary for patients, staff and for ambulances, as well as commercial delivery vehicles. The hospital facility itself requires extremely secure sources of heat and power, oxygen and water.

Using data from the NHS, heat and power accounted for only 17 percent of the carbon footprint of UK hospitals in 2017 [21]. Supply chain and services accounted for 54 percent, while travel and transport, including staff commuting, accounted for 16 percent. Manufactured fuels, chemicals and gases represented another four percent. This presents an opportunity to consider this power and resource demand holistically.

In Australia, backup power supply for hospitals has been identified as a government priority for projects to drive innovation and demonstrate capability [22]. Renewable technologies can do more than provide backup power, however—they can play a critical role in reimagining a sustainable energy ecosystem. Integrating production and storage solutions in distributed systems presents an opportunity to optimize hybrid systems including the use of hydrogen and batteries [23].

The following multi-generation design concept shows how we envision that sustainable energy technologies can transform a hospital from a resource sink to the centerpiece of a new reliable and healthy energy ecosystem. We assess relevant technologies and integrate them for a hypothetical hospital in New South Wales (NSW), Australia. This located approach provides some grounding for the design and discussion, though we note the same approach is widely applicable and we aim to inspire similar developments elsewhere.

2. Materials and Methods: Technology Assessment and Design Approach

There are a range of technologies available to fulfill the multitude of resource requirements of a hospital. Presented here is a selection of technologies available to supply reliable power and the other associated needs for this healthcare system. We frame the design research to revolve around three key atoms: hydrogen, oxygen and carbon. The strengths, weaknesses and potential co-benefits of the individual technologies are discussed and summarized. This assessment is then used to develop sustainable hospital power system concepts through multidisciplinary design that simultaneously responds to carbon emissions, health impacts and material sustainability.

2.1. Power Generation

2.1.1. Diesel Combustion

Coal has been the mainstay for the supply of power since the first Industrial Revolution, powering centralized electricity grids. Oil helped power the second Industrial Revolution, and a typical hospital relies on diesel internal combustion engine (ICE) generators for emergency power as shown in Figure 1. These stand-by engines are not designed to run for extended periods and so remain idle for most of their life. When the added burden of maintenance is considered, this is an expensive means of meeting mandated requirements.

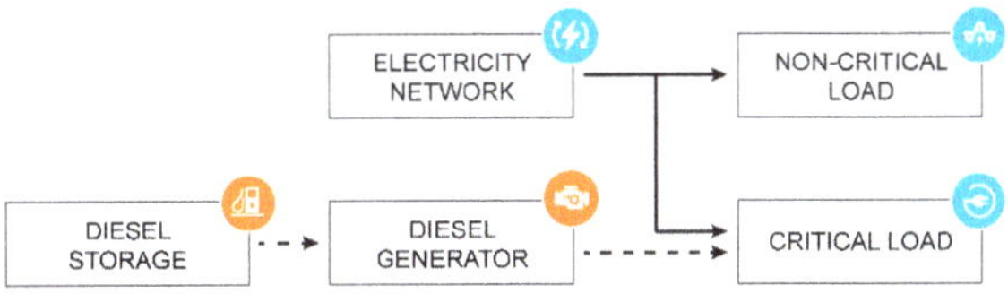

Figure 1. Conventional system with diesel backup during normal operation.

2.1.2. Gas Combustion

Co-generation or tri-generation captures thermal energy from combustion that would otherwise be wasted and deploys it for heating and cooling using absorption chillers. As hospitals have balanced power requirements and their heat requirements do not typically exceed the temperature for steam sterilization (below 160 °C), they make ideal candidates for using combined heat and power systems (CHP) [24]. This opportunity was identified more than a decade ago [25] and many hospitals have used this opportunity to improve energy efficiency, reduce costs and reduce emissions [26]. Hospital CHP installations use engines or gas turbines burning natural gas from existing networks as shown in Figure 2. This is an improvement on coal-fired power, though these systems still produce air pollution.

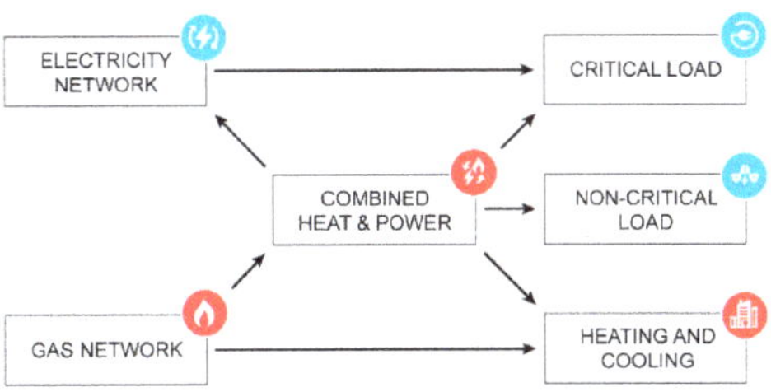

Figure 2. Combined heat and power system connected to both gas and electricity networks.

2.1.3. Fuel Cells

Commercial solid oxide fuel cell (SOFC) units in operation today generally include integrated steam methane reforming (SMR) equipment to use natural gas as a fuel source, which is broken apart to extract hydrogen. The high operating temperature and water 'exhaust' from the hydrogen fuel cell make this combination efficient and well suited for CHP systems. An advantage of this approach compared to simply burning the natural gas in an engine or turbine is the near complete elimination of harmful air pollutants (NO_X, SO_X and particulates).

Another fuel cell technology is the proton exchange membrane fuel cell (PEMFC), the twin of a proton exchange membrane (PEM) electrolyzer which applies the same principles but in reverse. The conversion process is shown in Figure 3. In a fuel cell, molecular hydrogen is recombined with oxygen from the air to recover stored potential energy. The only emission from this process is water. The round-trip efficiency of an electrolyzer and fuel cell system is low compared to a battery; however, hydrogen can be transported more readily so is more appropriate for extended duration emergency power.

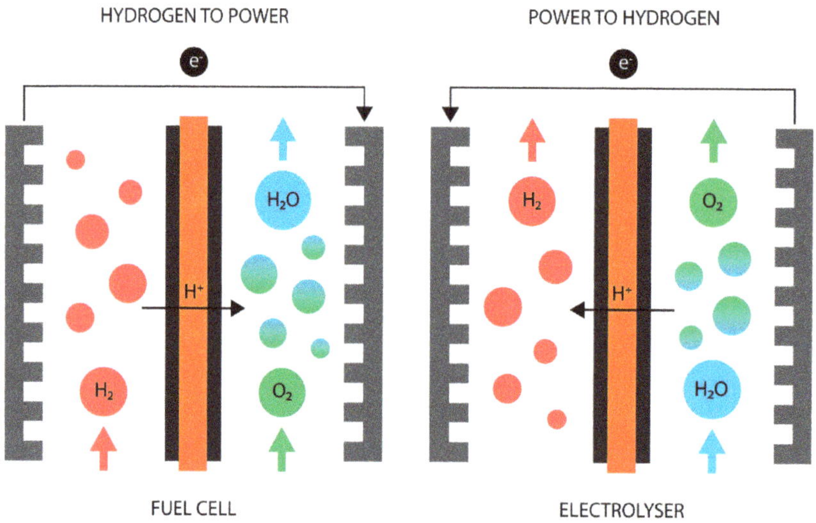

Figure 3. Simplified schematics of a proton exchange membrane (PEM) electrolyzer cell and a proton exchange membrane fuel cell (PEMFC) during operation, showing hydrogen and electron flows.

2.1.4. Renewable Energy Technologies

Renewable energy technologies integrated with digital grids are the new paradigm for electricity networks. Solar photovoltaic (PV) panels have become ubiquitous around the world and, accompanied by on- and off-shore wind turbines, are driving the transition to distributed non-fossil fuel based energy. Due to their solid-state nature, they require limited maintenance over their 25-year life, and economies of scale have resulted in spectacular cost reductions in recent years, which is continuing for large installations.

Despite the potential for large hospitals, only 13 of the 695 public and 497 private hospitals in Australia have been identified as having installed mid-scale solar PV systems [27]. An industrial 850-kW rooftop solar installation for a hospital in New South Wales (NSW) can be expected to achieve a capacity factor of 17 percent [27]. High-quality large-scale renewable resources in Australia supplied through the grid can increase capacity to 30 percent and 45 percent for solar and wind, respectively [28].

2.1.5. Comparison of Technologies

A comparison of renewables and the other power generation technologies discussed above is summarized in Table 1. Fuel cells and renewables present an opportunity to reduce emissions with flexible technology where energy can be stored and the value of co-benefits can be realized.

Table 1. Energy generation technologies for healthcare and their strengths, weaknesses and opportunities.

Technology	Strengths	Weakness	Opportunities
Diesel combustion	Well understood and easily refueled	Expensive and polluting	Existing system
Gas combustion	High energy efficiency	Polluting	Existing system
Hydrogen fuel cell	Clean and can be refueled	Expensive for conventional backup	Fuel flexibility
Renewables	Sustainable and scalable	Variable output requires storage	Low marginal cost power

2.2. *Energy Storage*

That you cannot turn on and off variable renewable energy (VRE) generators at will as you can with fossil fuel combustion generation is not always a weakness; the low marginal cost of renewable power is an opportunity to be creatively used in a symbiotic energy ecosystem.

2.2.1. Hydrogen

One of the greatest opportunities from low-cost renewable electricity technologies is to produce renewable hydrogen from water. PEM electrolyzers are now emerging as a preferred technology for this opportunity. The key advantage to PEM is their flexibility—PEM electrolyzers can accept partial, dynamic loads and are available from kW to MW scale. The technology functions by applying a current across a cell with two halves separated by a selective polymer that allows only hydrogen to move between the two, as shown in Figure 3. Water is fed into one side and the electrical energy splits the liquid into gaseous oxygen and hydrogen that can be collected and used in a range of applications.

Hydrogen is a good candidate for long-term energy storage to meet emergency requirements and the seasonal variation in energy demand which occurs in hospitals [29] and energy networks. There are many ways to store hydrogen, the most mature forms being as a compressed gas in high-pressure tanks and as a cryogenic liquid in insulated low-pressure vessels. Either of these storage systems can be situated on-site as stationary installations or mounted on truck trailers or rail cars. Solid-state hydrogen storage is an alternative with the potential for much greater energy density and it is now being demonstrated at scale [30]. Hydrogen can also be blended into the natural gas network, which will have a role to play in areas where it currently exists as complete fuel switching from gas to electricity is likely to cost more than rethinking existing infrastructure [31].

2.2.2. Flow Batteries

Carbon-free hydrogen gas production is extremely flexible; however, inherent energy losses mean it is not always the most efficient means of providing secure and reliable power from renewable energy technologies [32]. Batteries have a significant role to play, although there are serious sustainability concerns for the widespread adoption of lithium batteries [33,34]. Large-scale deployment of this incumbent technology will face battery materials constraints in a global-scale energy transition [35]. There are other battery technologies suitable for grid-scale energy storage, such as the vanadium redox flow battery (VRB/VRFB).

VRBs are a hybrid between fuel cell and conventional battery technology. Energy is stored in liquid electrolyte tanks for power conversion in cell stacks that operate in a similar manner to a fuel cell. The power is stored (or recovered) from the change in state of vanadium ions in an aqueous sulfuric acid solution which changes color based on its state of charge. Figure 4 provides a simplified visual explanation of this process. By separating power (kW/MW) from energy storage (kWh/MWh), a VRB system is highly scalable and can readily be configured to suit the needs of the application.

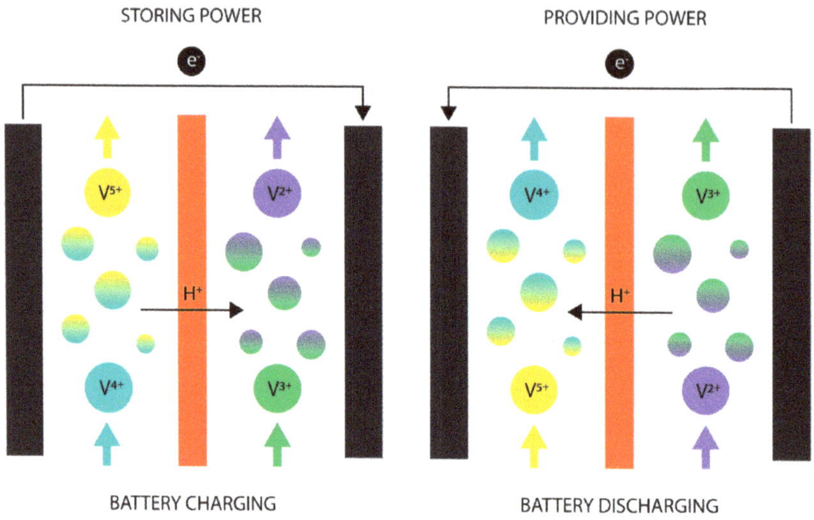

Figure 4. Simplified schematics of a vanadium redox flow cell during charge and discharge showing electrolyte color change with hydrogen and electron flows.

The aqueous electrolyte is non-flammable by nature, eliminating the potential fire safety hazard of lithium batteries. The reduced fire risk means these systems are more suitable for enclosed spaces, such as underground car parks. Electrolyte contamination is eliminated because the electrolyte is the same on both sides of the system, giving VRBs a long lifetime (>25 years) with low capacity fade. All-vanadium batteries' tolerance for practically unlimited charge–discharge cycles over their lifetime makes the technology ideal for high-use applications, such as supporting renewables and electric vehicles. The valuable vanadium in the electrolyte can be easily recovered for use in a new battery system or other applications. The balance of material is predominately carbon, metals and polymers that can also be recovered and recycled at end of life [36].

2.2.3. Hybrid Batteries

The ubiquitous lead-acid battery (LAB) has no moving parts and is the standard for uninterruptible power supplies (UPS). The conventional format of this technology, however, is not well suited to new power demands. Hybrid battery installations are being used to take advantage of the strengths of different technologies [37]. Pairing battery components with integrated supercapacitors creates new opportunities for mature LAB technology by improving its peak power capacity.

This hybrid technology was invented by the Australian Commonwealth Scientific and Industrial Research Organisation (CSIRO) and commercially acquired in 2010. It has since been deployed for a range of kW- and MW-scale storage applications. Similar to vanadium-sulfuric acid electrolyte flow batteries, hybrid lead-sulfuric acid batteries are almost completely recyclable. The Environmental Protection Agency in the United States found that lead-acid batteries are consistently one of the most recycled products [38]. This and the other energy storage technologies discussed above are summarized in Table 2 below.

Table 2. Storage technologies for healthcare and their key strengths, weaknesses, and opportunities.

Technology	Strengths	Weakness	Opportunities
Hydrogen electrolysis	Sustainable and flexible	Expensive when considered alone	Oxygen supply and demand management
Lithium-ion battery	High power density	Sustainability concerns and limited storage	Electricity network support
Lead-acid hybrid battery	Readily recyclable and high peak power	Limited storage capacity	Electricity network support
Vanadium flow battery	Recyclable, stable and non-flammable	Low power and energy density	Electricity network and electric vehicle support

2.3. Related Technologies and Resource Considerations

In addition to being a facility with heat and power demands, healthcare requires disinfectant and oxygen, and a hospital precinct is also a transport hub. Technologies related to energy generation and storage can help meet these additional sustainability challenges and realize the associated opportunities.

2.3.1. Transport

Society is transitioning away from fossil-fueled passenger cars and commercial vehicles to the use of electric vehicles (EVs). If managed poorly, charging battery electric vehicles (BEVs) could strain electricity grids [39]. Smart charging can reduce this impact by creating a large demand response capability. One step further is vehicle-to-grid (VtG/V2G) technology which allows BEVs or fuel cell electric vehicles (FCEVs) with hydrogen to act as networked energy storage devices [40]. In this way, an EV fleet can become a virtual power plant (VPP) that is profitable to owners and beneficial for power networks [41]. The potential to support system security [42] is driving trials with government fleets [43]. Fleet operators are one of the main stakeholders driving the electric vehicle market [44] which is an opportunity for health services looking to reduce air pollution with more environmentally friendly vehicles [45].

2.3.2. Oxygen

Non-energetic gas demands are an important component of energy transitions [46,47]. Oxygen is critical for health care and has been in short supply in some regions during the coronavirus pandemic in 2020 [48]. Hospitals commonly procure oxygen in bulk from suppliers as a compressed gas or as a cryogenic liquid, which is produced on a commercial scale through liquefaction and distillation. Oxygen is sold at a premium to the health sector, so generating it locally could save costs [49]. On-site production using compressed air with pressure swing adsorption (PSA) oxygen concentrators (OCs) has been found to reduce costs for hospitals [50]. High-quality medical grade 99.5% purity oxygen can be supplied from water as a by-product of producing hydrogen with electrolysis. Using this oxygen stream from renewable hydrogen in a multi-generation system can help support a new hydrogen economy.

2.3.3. Water

Water use is one of the more significant aspects of sustainability management considered by the health sector, so it is important that water use issues are not exacerbated when designing a sustainable energy ecosystem [51]. To close the hydrogen water cycle loop, it is necessary to consider the water supply for electrolysis. This is commonly expected to be obtained through the desalination of seawater, which will thus be an important component of renewable energy networks [47]. Alternatively, water can be drawn from the air using an atmospheric water generator (AWG). These adsorption-based devices can utilize thermal energy from solar or waste heat even in arid climates [52]. The other advantage is that the water they produce is of high purity, reducing the need for purification for electrolysis [53].

2.3.4. Disinfectant

Hydrogen peroxide is widely used in hospitals for disinfection and has been in particularly high demand during the COVID-19 pandemic [54,55]. The majority of hydrogen peroxide is produced using the industrial anthraquinone process, which is only viable at a large scale. Issues relating to highly concentrated solutions and concerns about this process from a green chemistry perspective have led to considerable interest in alternative means of supplying hydrogen peroxide [56]. Local production of dilute hydrogen peroxide reduces waste products and the risks associated with transporting and storing high concentrations in bulk [57].

Processes have been developed for direct production of hydrogen peroxide from hydrogen and oxygen [58] or from water and oxygen [59]. Convergent PEM electrochemical synthesis uses water and oxygen as inputs to a fuel cell and produces dilute hydrogen peroxide. This approach is particularly promising in a hospital context if oxygen is made readily available using on-site production for medical use. Other electrochemical synthesis processes are under development for on-site production of hydrogen peroxide along with other useful products, such as ozone [60], hydrogen [61,62] and oxygen [63].

2.4. Design Approach

A full engineering design study would be necessary for any specific site; however, it is informative to explore system-level constraints and capabilities, which is the approach we have taken in this conceptual research.

2.4.1. Regulation Constraints

Regulations relating to hospital power have been considered in our conceptual scenario of a multi-generation health precinct. Standards vary at the intra- and inter-national scale. For example, in the US, the updated code NFPA 99 allows for fuel cell systems as alternate sources of power since 2015 [64]. The code also now allows for oxygen concentrators as central supply sources for hospital medical oxygen systems, which could be integrated to support on-site production from electrolysis. Returning to the Australian context, the relevant standard for emergency power supplies in hospitals from 1998 (AS/NZS 3009:1998) states the power source may be provided by central battery systems, provided they are of a type specifically designed for continuous float charging conditions [65]. Lead-acid batteries meet this criterion but others do not, despite alternative battery technologies and contemporary digital battery management systems.

The N + 1 supply configuration is a commonly accepted practice that shapes the design of emergency power systems, including those for hospitals [66] in NSW [20], to ensure enough redundancy is built in to ensure a highly resilient system. This means building in one additional piece of key equipment than is strictly required, so that one sub-system can fail and operations can continue. Providing capacity modularity to meet an N + 1 guideline is straightforward with battery packs and fuel cells stacks; the most critical point of failure would relate to the inverter. The need for multiple inverters can be seen as a prohibitive cost or as an opportunity to install valuable assets able to provide ancillary services to electricity networks [67]. Demonstrations of the ability to replace the mechanical inertia of conventional power plants with battery-powered digital inertia are underway [68].

2.4.2. Specification Constraints

The ongoing trend towards modularity facilitated by batteries and other clean technology in the energy sector brings flexibility to concept design which can be readily adjusted for specific requirements and translated to different sites. To provide an accessible example, we present a design concept for a hypothetical 550-bed hospital for a small city such as Newcastle, NSW, Australia.

We consider average energy demand, assumed to be 41 MWh per day based on an annual average of 27 MWh per bed [69]. Total energy use consists of electrical and thermal demands which will vary based on climate and facilities. Using an average value from previous studies, it is assumed here that 49 percent of the total energy use is electrical demand and the remainder is thermal [24,29,69,70]. There are analytical data available for medical oxygen in hospitals [49] and the demand for this scale is assumed to be 708 kg per day.

2.4.3. Design Goals

The key constraints discussed above are:

- 41 MWh per day total average energy demand.
- 49 percent of total demand is electrical energy.
- 708 kg per day of medical oxygen.
- Uninterruptible power supply with N + 1 redundancy.
- Battery systems designed for continuous float charging.

Meeting hospital demands sustainably within these constraints is the goal of our design work presented below. Re-imagining a healthcare precinct as a renewable energy hub in this way uses public infrastructure to build resilience, improve public health and accelerate the energy transition.

3. Results and Discussion: Design Specification for Hospital Renewable Energy Ecosystem

We re-imagined a multi-generation energy system for a sustainable hospital precinct that integrates renewable hydrogen and battery energy technologies to reduce harmful emissions while supporting reliable operations. To present the integrated systems, we break down the concept design into two sections. The first replaces fossil fuel combustion with fuel cells and batteries for reliable power with redundancy. The second broadens the scope, presenting a networked multi-generation system to sustainably provide other resources in addition to heat and power for deep decarbonization.

3.1. Replacing Engines and Turbines

In this case, we consider only the requirement for electrical power. The conventional generator setup could be replaced by a hydrogen fuel cell with storage to meet the set number of hours for the given location. This would take the form of a hybrid energy storage system with a battery, as shown in Figure 5.

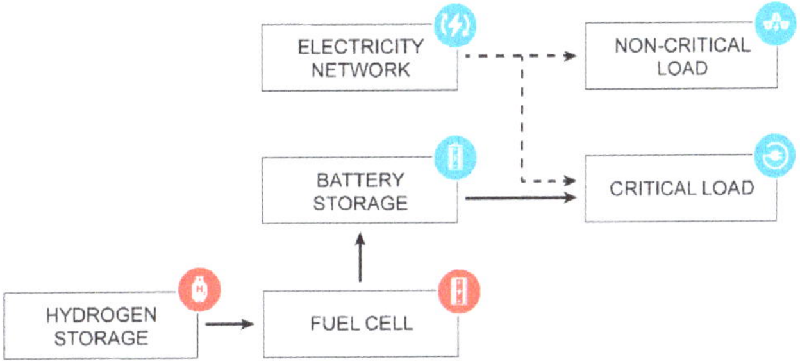

Figure 5. Electrical backup system with hydrogen fuel cells during emergency operation.

This energy storage system for emergency power could consist of 1.5 h in lead to comply with AS/NZS 3009 and 24 h in hydrogen. If the system is configured with a suitable amount of spare capacity, this secondary power supply need not be reserved solely for emergencies. The hybrid supercapacitor technology broadens the usefulness of the lead-acid battery cells, and additional flexible capacity can be added with VRBs.

Starting with the fuel cell backup, six containerized 200-kW units [71] would provide 1.2 MW of power to meet the N + 1 redundancy guideline for an additional unit. 24 h storage would require 850 kg of hydrogen, which could be replenished by a high capacity 300-bar truck trailer. This could be stored on site at 165 bar in approximately 40 tubes with mature technology, or two of the type of containerized solid state 17-MWh units planned for demonstration in Australia [30].

Three commercially available containerized hybrid lead-acid battery units would provide 2.5 MW of peak power with 1.5 MWh of storage [72] to meet a 1.5-h specification for tertiary power supply. An emergency capacity of 2.4 MWh exceeds this by almost 90 percent, meeting the AS/NZS required margin of 1.33 times minimum capacity at install. This also meets the N + 1 redundancy guideline, as two of three units could meet the demand if required.

A vanadium battery secondary supply could consist of four containerized 250-kW units, with a range of capacity options [73]. In a three-hour configuration, this would translate to 3 MWh of capacity and 1 MW of nominal power. This could provide a limited secondary power supply alone, whilst together it could contribute to electricity system security and reliability. For BEVs, this could support up to 100 vehicles with 10-kW vehicle-to-grid connections.

An alternative to this setup is a CHP installation which is typically matched to the heat demand. Two containerized SOFC units would deliver 880 kW of power and 900 kW of heat [74]. Heat pumps, boilers and chillers would support the tri-generation system to reliably meet the variable thermal demands of the facility. In this scenario, a gas-fueled CHP system can be the normal supply, and the electricity grid the backup.

Batteries and stored hydrogen are still desirable for a system like this, connected to both gas and electricity networks, providing greater redundancy whilst making a larger contribution to network security and reliability. Twenty four hours of storage is suitable for VRE penetration of 90 percent [75] and building this in to provide reserves will help deliver fast-responding power assets in the grid that are missing incentives [76]. To achieve the full potential of this approach, though, hydrogen should be enabled to act as a two-way resource as with the battery system.

3.2. Multi-Generation for Coupled Power and Resources

Smart energy networks (SNE) integrate electricity, gas and heat under common Information and Communications Technology (ICT) with power-to-gas technology (PtG) [77]. An 850-kW rooftop solar installation generating 3.5 MWh per day for an electrolyzer system of five 30-Nm^3/h units would be expected to produce 54 kg of hydrogen per day [78]. Two tonnes of adsorption material (sorbent) in an AWG system could be used to capture the approximately 500 L per day of water required for this electrolysis [52].

Theoretically, this size system could also produce 427 kg of oxygen per day, enough for approximately 45 medium-concentration oxygen therapy devices or 60 percent of the anticipated demand. Low-cost energy from grid connected solar or wind during periods of excess supply could power additional electrolysis to increase this to 80 or 120 percent of the demand. Efficient on-site oxygen concentrators could fill any supply gaps with cylinder backup [50]. Excess oxygen can feed hydrogen peroxide synthesis for disinfectant supply, as discussed in Section 2.

Excess hydrogen surplus to the hospital's energy storage requirements could be fed into the gas network. Existing methane infrastructure could accept up to 10 or 20 percent hydrogen [79,80], or more as synthetic methane after being combined with carbon dioxide extracted from the atmosphere [81]. Hydrogen could, alternatively, replace natural gas altogether. In addition to its use as fuel for electricity generation, hydrogen can be used in industry as a chemical feedstock and to supplement thermal energy primarily provided by heat pumps.

Hydrogen could also be used for vehicles, either in its pure molecular form for cars and commercial vehicles or as a feedstock for synthetic fuel [82] to support aeromedical services. A hospital campus of this size may have thousands of parking spaces [83]. Just 50 medium size EVs represent over 3 MWh of energy storage and at least 500 kW of two-way power with vehicle-to-grid connections. Covering the car park with solar PVs could provide a much larger on-site solar precinct.

Figure 6 shows how all these systems interact to meet hospital requirements. In addition to redundancy, all these systems would be backed up independently from an infrastructure network and/or by road transport.

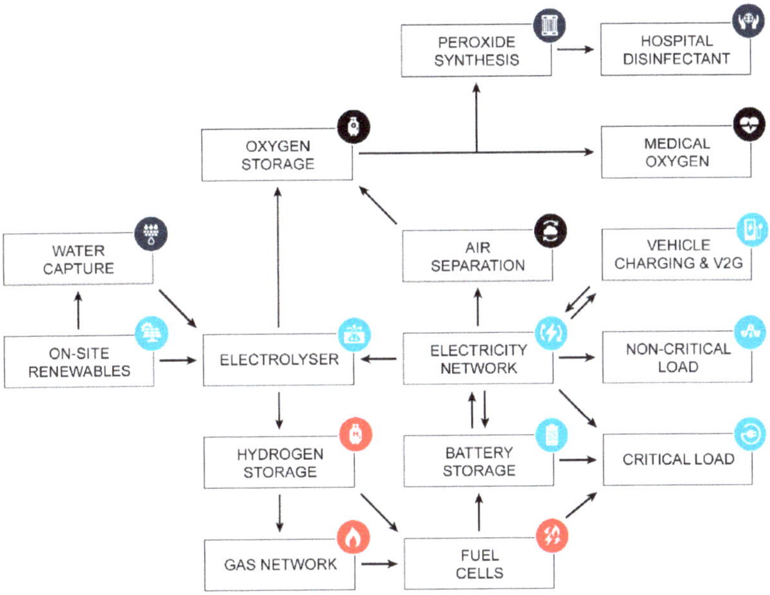

Figure 6. Multi-generation hospital precinct energy system concept diagram, with energy and resource flows for coupled energy and resource demands.

This high-level system design meets the goals outlined in Section 2.4 of this article, namely 'sustainably provisioning for reliable on-site power requirements to improve outcomes for the community'. Future work would need to examine specific site requirements in more detail. This would include techno-economic optimization of system components and could also explore other opportunities, such as wastewater treatment and mobile installations for field hospitals. Detailed engineering design would consider other applicable standards and regulation, including the location of gas storage. Hydrogen is not inherently more hazardous than conventional fuels but it must be managed appropriately [84–86]. The positioning of technology in this hybrid clean energy system presents an opportunity to engage the public with the vision presented here, particularly with less well known vanadium systems.

3.3. Illustration of Multi-Generation Hospital Precinct Energy System

Systems in plant equipment rooms and back-of-house containerized solutions provide large energy solutions; however, there is something to be gained from visible systems. A public installation could provide science education and potentially a useful distraction and on-site exhibition to the hospital patients and visitors. A dilute vanadium electrolyte solution or another water-based system with lighting would display the state of the energy system. Air or an inert gas, such as nitrogen, bubbled through the solution would simultaneously represent hydrogen and oxygen gas production from electrolysis. Battery modules, in a design suggestive of common household batteries [87], installed in locations such as car parks where solar PV panels are visible [88] would highlight the benefits of EV-to-grid technology for the hospital and the public. Figures 7 and 8 illustrate this design concept.

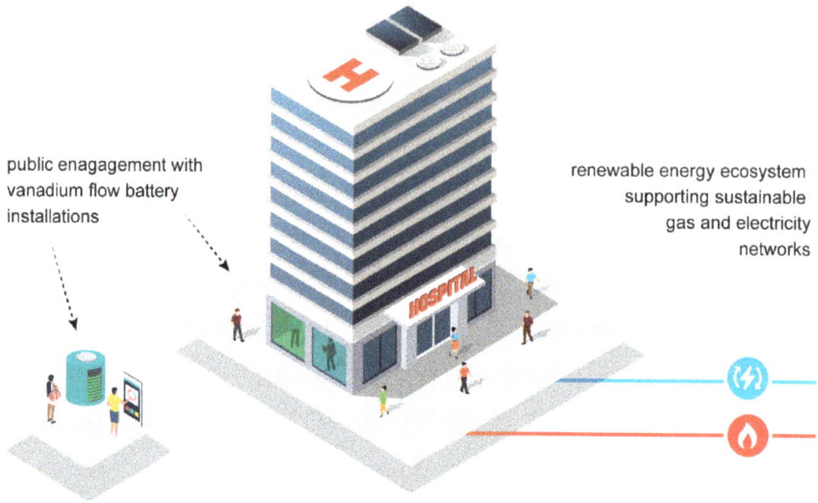

Figure 7. Illustration of hospital concept with on-site generation and storage connected to gas (red) and electricity (blue) networks. Color-changing vanadium flow battery installations are used to engage the public with the energy systems, shown here in the green and blue of V^{3+} and V^{4+}.

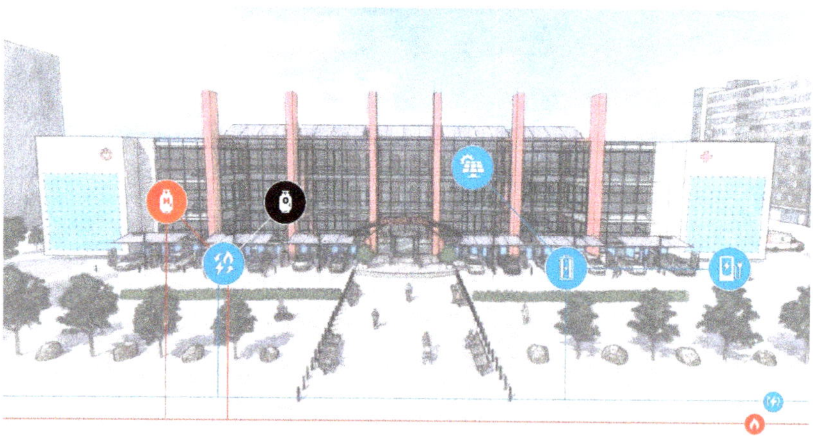

Figure 8. Model rendering of concept hospital powered by coupled on-site renewable multi-generation and storage connected to gas (red) and electricity (blue) networks. Electrolysis produces hydrogen for fuel cells and oxygen for medical demands while batteries support electric vehicles and provide uninterruptible power for critical loads.

4. Conclusions: A New Energy Ecosystem

We have re-imagined healthcare precincts and presented a design concept for a hospital as a flagship community energy hub where sustainable networks are coupled with medical requirements.

This design shows how versatile and scalable exchange membrane cell systems, including flow batteries and fuel cells, replace combustion to meet emergency power requirements and improve resource security. Battery inverters help manage grid power quality while solar powered electrolysis supports medical oxygen requirements and feeds hydrogen into decarbonized gas pipelines. Fleet, staff and public transport become an asset with electric vehicle-to-grid integration. Visible vanadium electrolyte and modular battery systems are also used to engage the community with this energy system to help drive the energy transition. Together, this provides a vision for healthy power to help redefine sustainability.

Author Contributions: Conceptualization, N.G., D.G., I.K., A.M., C.M., J.N., B.M. and E.D.; Funding acquisition, N.G., D.G., I.K. and B.M.; Investigation, N.G. and M.B.; Methodology, N.G., D.G., I.K. and M.L.; Project administration, N.G., D.G. and I.K.; Supervision, D.G., I.K., B.M. and E.D.; Visualization, N.G.; Writing—original draft, N.G., D.G. and M.B.; Writing—review and editing, D.G., I.K., M.L., M.B., A.M., C.M., J.N., B.M. and E.D. All authors have read and agreed to the published version of the manuscript.

Funding: This research was supported by funding from the University of New South Wales (UNSW) Digital Grid Futures Institute, UNSW, Sydney, under a cross-disciplinary fund scheme. The views expressed herein are those of the authors and are not necessarily those of the institute.

Conflicts of Interest: The authors declare no conflict of interest.

References

1. Lelieveld, J.; Klingmüller, K.; Pozzer, A.; Burnett, R.T.; Haines, A.; Ramanathan, V. Effects of fossil fuel and total anthropogenic emission removal on public health and climate. *Proc. Natl. Acad. Sci. USA* **2019**, *116*, 7192–7197. [CrossRef]
2. Myllyvirta, L. Quantifying the Economic Costs of Air Pollution from Fossil Fuels. 2020. Available online: https://energyandcleanair.org/publications/costs-of-air-pollution-from-fossil-fuels/ (accessed on 23 July 2020).
3. Hakovirta, M.; Denuwara, N. How COVID-19 redefines the concept of sustainability. *Sustainability* **2020**, *12*, 3727. [CrossRef]
4. Health Care without Harm Case Studies from GGHH Members. Available online: https://www.greenhospitals.net/case-studies/ (accessed on 17 September 2020).
5. Wise, J. NHS makes good progress on sustainability, report shows. *BMJ* **2018**, *362*, k4032. [CrossRef]
6. Pencheon, D.; Wight, J. Making healthcare and health systems net zero. *BMJ* **2020**, *368*, 1–3. [CrossRef]
7. The Lancet Climate and health: Joining up the pieces, scaling up the action. *Lancet* **2016**, *388*, 1956. [CrossRef]
8. Salas, R.N.; Malina, D.; Solomon, C.G. Prioritizing health in a changing climate. *N. Engl. J. Med.* **2019**, *381*, 773–774. [CrossRef]
9. Madden, D.L.; Capon, A.; Truskett, P.G. Environmentally sustainable health care: Now is the time for action. *Med. J. Aust.* **2020**, 361–362. [CrossRef]
10. Malik, A.; Lenzen, M.; McAlister, S.; McGain, F. The carbon footprint of Australian health care. *Lancet Planet. Health* **2018**, *2*, e2–e3. [CrossRef]
11. Boudet, H.S. Public perceptions of and responses to new energy technologies. *Nat. Energy* **2019**, *4*, 446–455. [CrossRef]
12. Ashworth, P.; Witt, K.; Ferguson, M.; Sehic, S. *Developing Community Trust in Hydrogen*; University of Queensland: Brisbane, Australia, 2019.
13. Skinner, G.; Clemence, M. *Global Trust in Professions*; Ipsos MORI Social Research Institute: London, UK, 2019.
14. Sainsbury, P.; Charlesworth, K.; Madden, L.; Capon, A.; Stewart, G.; Pencheon, D. Climate change is a health issue: What can doctors do? *Intern. Med. J.* **2019**, *49*, 1044–1048. [CrossRef]
15. Williams, A.; Kennedy, S.; Philipp, F.; Whiteman, G. Systems thinking: A review of sustainability management research. *J. Clean. Prod.* **2017**, *148*, 866–881. [CrossRef]
16. Ballo, I.F. Imagining energy futures: Sociotechnical imaginaries of the future Smart Grid in Norway. *Energy Res. Soc. Sci.* **2015**, *9*, 9–20. [CrossRef]
17. The University of Newcastle Public Debut for Printed Solar. Available online: https://www.newcastle.edu.au/newsroom/featured/public-debut-for-printed-solar (accessed on 23 July 2020).
18. Western Power, PowerBank Community Battery Storage. Available online: https://westernpower.com.au/our-energy-evolution/projects-and-trials/powerbank-community-battery-storage/ (accessed on 23 July 2020).

19. Southern Staffordshire Community Energy Limited. *Saving Lives with Solar*; Oxford, UK, 2016. Available online: https://www.ethex.org.uk/medialibrary/2016/07/21/3e7e443a/SSCEUHNMShareOfferDocumentFinal.pdf (accessed on 23 July 2020).
20. NSW Health Infrastructure. *Engineering Services Guidelines*; Ministry of Health: North Sydney, Australia, 2016.
21. Sustainable Development Unit. *Reducing the Use of Natural Resources in Health and Social Care*; Public Health England: Cambridge, UK, 2018.
22. Newcastle Institute for Energy and Resources. *Positioning the Hunter Region as a Key Enabler of the Future Hydrogen Economy*; Newcastle Institute for Energy and Resources: Newcastle, Australia, 2020.
23. Srinivasan, V.; Temminghoff, M.; Charnock, S.; Hartley, P. *Hydrogen Research, Development and Demonstration: Priorities and Opportunities for Australia*; CSIRO: Canberra, Australia, 2019.
24. Kolokotsa, D.; Tsoutsos, T.; Papantoniou, S. Energy conservation techniques for hospital buildings. *Adv. Build. Energy Res.* **2012**, *6*, 159–172. [CrossRef]
25. Midwest CHP Application Center. *Combined Heat & Power (CHP) Resource Guide for Hospital Applications*; 2007. Available online: https://archive.epa.gov/region1/healthcare/web/pdf/ushospitalguidebook_111907.pdf (accessed on 23 July 2020).
26. United States Environmental Protection Agency CHP for Hospitals: Superior Energy for Superior Patient Care. Available online: https://www.epa.gov/chp/chp-hospitals-superior-energy-superior-patient-care (accessed on 22 July 2020).
27. Starkey, S. *Mid-Scale PV Uptake Forecasts*; Jacobs Australia Pty Limited: Melbourne, Australia, 2019.
28. Longden, T.; Jotzo, F.; Prasad, M.; Andrews, R. *Green Hydrogen Production Costs in Australia: Implications of Renewable Energy and Electrolyser Costs*; Centre for Climate & Energy Policy: Canberra, Australia, 2020.
29. Alexis, G.K.; Liakos, P. A case study of a cogeneration system for a hospital in Greece. Economic and environmental impacts. *Appl. Therm. Eng.* **2013**, *54*, 488–496. [CrossRef]
30. Templeton, L. UNSW Hydrogen Storage Technology in World-First Application of Its Kind. Available online: https://newsroom.unsw.edu.au/news/science-tech/unsw-hydrogen-storage-technology-world-first-application-its-kind (accessed on 22 July 2020).
31. KPMG. *2050 Energy Scenarios: The UK Gas Networks Role in a 2050 Whole Energy System*. 2016. Available online: https://www.energynetworks.org/creating-tomorrows-networks (accessed on 23 July 2020).
32. Guandalini, G.; Campanari, S.; Romano, M.C. Power-to-gas plants and gas turbines for improved wind energy dispatchability: Energy and economic assessment. *Appl. Energy* **2015**, *147*, 117–130. [CrossRef]
33. Bruce, S.; Temminghoff, M.; Hayward, J.; Schmidt, E.; Munnings, C.; Palfreyman, D.; Hartley, P. *National Hydrogen Roadmap*; CSIRO: Canberra, Australia, 2018.
34. Gyuk, I. Energy Storage Technology: Evolving towards Long Duration. In Proceedings of the IFBF Showcase. 2020. Available online: https://www.energy-storage.news/news/energy-storage-digital-summit-long-duration-essential-but-needs-appropriate (accessed on 23 July 2020).
35. Brouwer, J. *In-Rack Direct DC Powering of Servers with Solid Oxide and Proton Exchange Membrane Fuel Cells*; University of California Irvine: Irvine, CA, USA, 2019.
36. Díaz-Ramírez, M.C.; Ferreira, V.J.; García-Armingol, T.; López-Sabirón, A.M.; Ferreira, G. Battery Manufacturing Resource Assessment to Minimise Component Production Environmental Impacts. *Sustainability* **2020**, *12*, 6840. [CrossRef]
37. Griffith, C. redT vanadium battery starts operating at Melbourne's Monash University. *Aust. Bus. Rev.* **2019**.
38. Environmental Protection Agency. *Advancing Sustainable Materials Management: Facts and Figures Report*; USA. 2019. Available online: https://www.epa.gov/facts-and-figures-about-materials-waste-and-recycling/advancing-sustainable-materials-management-0 (accessed on 23 July 2020).
39. Hu, J.; Morais, H.; Sousa, T.; Lind, M. Electric vehicle fleet management in smart grids: A review of services, optimization and control aspects. *Renew. Sustain. Energy Rev.* **2016**, *56*, 1207–1226. [CrossRef]
40. Mukherjee, U.; Maroufmashat, A.; Ranisau, J.; Barbouti, M.; Trainor, A.; Juthani, N.; El-Shayeb, H.; Fowler, M. Techno-economic, environmental, and safety assessment of hydrogen powered community microgrids; case study in Canada. *Int. J. Hydrog Energy* **2017**, *42*, 14333–14349. [CrossRef]
41. Kahlen, M.T.; Ketter, W.; van Dalen, J. Electric Vehicle Virtual Power Plant Dilemma: Grid Balancing Versus Customer Mobility. *Prod. Oper. Manag.* **2018**, *27*, 2054–2070. [CrossRef]

42. Andrey, C.; Barberi, P.; Lacombe, L.; van Nuffel, L.; Gérard, F.; Dedecca, J.G.; Rademaekers, K.; El Idrissi, Y.; Crenes, M. *Study on Energy Storage—Contribution to the Security of the Electricity Supply in Europe*; European Commission: Brussels, Belgium, 2020; ISBN 978-92-76-03377-6.
43. Australian Renewable Energy Agency. *World Leading Electric Vehicle to Grid Trial in ACT. ArenaWire*; 2020. Available online: https://arena.gov.au/news/world-leading-electric-vehicle-to-grid-trial-in-act/ (accessed on 23 July 2020).
44. McKerracher, C.; Izadi-Najafabadi, A.; O'Donovan, A.; Albanese, N.; Soulopolous, N.; Doherty, D.; Boers, M.; Fisher, R.; Cantor, C.; Frith, J.; et al. Electric Vehicle Outlook. 2020. Available online: https://about.bnef.com/electric-vehicle-outlook/ (accessed on 22 July 2020).
45. NHS England Simon Stevens Challenges Ambulance Makers to Help 'Blue Lights go Green'. Available online: https://www.england.nhs.uk/2019/02/simon-stevens-challenges-ambulance-makers-to-help-blue-lights-go-green/ (accessed on 10 August 2020).
46. Kilickaplan, A.; Bogdanov, D.; Peker, O.; Caldera, U.; Aghahosseini, A.; Breyer, C. An energy transition pathway for Turkey to achieve 100% renewable energy powered electricity, desalination and non-energetic industrial gas demand sectors by 2050. *Sol. Energy* **2017**, *158*, 218–235. [CrossRef]
47. Gulagi, A.; Bogdanov, D.; Breyer, C. The Demand for Storage Technologies in Energy Transition Pathways Towards 100% Renewable Energy for India. *Energy Procedia* **2017**, *135*, 37–50. [CrossRef]
48. Santucci, J.; Siegel, B.; Folmer, K.; Katersky, A. Some hospitals are facing an oxygen shortage amid coronavirus crisis. *ABC News* **2020**.
49. Gómez-Chaparro, M.; García-Sanz-Calcedo, J.; Márquez, L.A. Analytical determination of medical gases consumption and their impact on hospital sustainability. *Sustainability* **2018**, *10*, 2948. [CrossRef]
50. Friesen, R.M.; Raber, M.B.; Reimer, D.H. Oxygen concentrators: A primary oxygen supply source. *Can. J. Anaesth.* **1999**, *46*, 1185–1190. [CrossRef] [PubMed]
51. UNESCO. *The United Nations World Water Development Report 2015: Water for a Sustainable World*; UNESCO: Paris, France, 2015.
52. Kim, H.; Rao, S.R.; Kapustin, E.A.; Zhao, L.; Yang, S.; Yaghi, O.M.; Wang, E.N. Adsorption-based atmospheric water harvesting device for arid climates. *Nat. Commun.* **2018**, *9*, 1–8. [CrossRef]
53. The University of Newcastle. *New Green Hydrogen Made from Solar Power and Air. University News*. 2020. Available online: https://www.newcastle.edu.au/newsroom/featured/new-green-hydrogen-made-from-solar-power-and-air (accessed on 23 July 2020).
54. Research and Markets Hydrogen Peroxide: COVID-19. Available online: https://www.researchandmarkets.com/issues/hydrogen-peroxide-demand-rises (accessed on 21 July 2020).
55. United States Food and Drug Administration. Decontamination Systems for Personal Protective Equipment EUAs. Available online: https://www.fda.gov/medical-devices/coronavirus-disease-2019-covid-19-emergency-use-authorizations-medical-devices/decontamination-systems-personal-protective-equipment-euas (accessed on 23 July 2020).
56. Ranganathan, S.; Sieber, V. Recent advances in the direct synthesis of hydrogen peroxide using chemical catalysis—A review. *Catalysts* **2018**, *8*, 379. [CrossRef]
57. Xia, C.; Xia, Y.; Zhu, P.; Fan, L.; Wang, H. Direct electrosynthesis of pure aqueous H_2O_2 solutions up to 20% by weight using a solid electrolyte. *Science* **2019**, *366*, 226–231. [CrossRef]
58. Flaherty, D.W. Direct Synthesis of H_2O_2 from H_2 and O_2 on Pd Catalysts: Current Understanding, Outstanding Questions, and Research Needs. *ACS Catal.* **2018**, *8*, 1520–1527. [CrossRef]
59. Perry, S.C.; Pangotra, D.; Vieira, L.; Csepei, L.I.; Sieber, V.; Wang, L.; Ponce de León, C.; Walsh, F.C. Electrochemical synthesis of hydrogen peroxide from water and oxygen. *Nat. Rev. Chem.* **2019**, *3*, 442–458. [CrossRef]
60. Tatapudi, P. Simultaneous Synthesis of Ozone and Hydrogen Peroxide in a Proton-Exchange-Membrane Electrochemical Reactor. *J. Electrochem. Soc.* **1994**, *141*, 1174. [CrossRef]
61. Fuku, K.; Sayama, K. Efficient oxidative hydrogen peroxide production and accumulation in photoelectrochemical water splitting using a tungsten trioxide/bismuth vanadate photoanode. *Chem. Commun.* **2016**, *52*, 5406–5409. [CrossRef] [PubMed]
62. Fuku, K.; Miyase, Y.; Miseki, Y.; Funaki, T.; Gunji, T.; Sayama, K. Photoelectrochemical Hydrogen Peroxide Production from Water on a $WO_3/BiVO_4$ Photoanode and from O_2 on an Au Cathode Without External Bias. *Chem. Asian J.* **2017**, *12*, 1111–1119. [CrossRef] [PubMed]

63. Li, W.; Bonakdarpour, A.; Gyenge, E.; Wilkinson, D.P. Production of Hydrogen Peroxide for Drinking Water Treatment in a Proton Exchange Membrane Electrolyzer at Near-Neutral pH. *J. Electrochem. Soc.* **2020**, *167*, 044502. [CrossRef]
64. National Fire Protection Association. *Health Care Facilities Code Handbook*; Hart, J.R., Rubadou, C.B., Eds.; NFPA: Quincy, MA, USA, 2018; ISBN 9781455914876.
65. Standards Australia/New Zealand. *AS/NZS 3009:1998 Electrical Installations—Emergency Power Supplies in Hospitals 1998*; Standards Australia: Sydney, Australia, 1998.
66. UK Department of Health. *Health Technical Memorandum 06-01: Electrical Services Supply and Distribution*; UK. 2017. Available online: https://assets.publishing.service.gov.uk/government/uploads/system/uploads/attachment_data/file/608037/Health_tech_memo_0601.pdf (accessed on 23 July 2020).
67. European Association for Storage of Energy. *Energy Storage Applications Summary*; European Association for Storage of Energy: Brussels, Belgium, 2020.
68. Australian Renewable Energy Agency. *World's Largest Scale Battery Set to Get Even Bigger. ArenaWire*; 2019. Available online: https://arena.gov.au/news/worlds-largest-scale-battery-set-to-get-even-bigger/ (accessed on 23 July 2020).
69. Ji, R.; Qu, S. Investigation and evaluation of energy consumption performance for hospital buildings in China. *Sustainability* **2019**, *11*, 1724. [CrossRef]
70. García-Sanz-Calcedo, J.; Gómez-Chaparro, M.; Sanchez-Barroso, G. Electrical and thermal energy in private hospitals: Consumption indicators focused on healthcare activity. *Sustain. Cities Soc.* **2019**, *47*, 101482. [CrossRef]
71. Horizon Fuel Cell Technologies Horizon ships 200 kW fuel cell for Ulsan Technopark in S Korea. *Fuel Cells Bull.* **2018**, *2018*, 7. [CrossRef]
72. Ecoult Product Overview. Available online: https://www.ecoult.com/products/product-overview (accessed on 22 July 2020).
73. Sumitomo Electric Group. The Energy Solution. Available online: https://global-sei.com/products/redox/ (accessed on 16 March 2020).
74. Doosan Babcock PureCell Model 400 Fuel Cell System. Available online: http://www.doosanbabcock.com/en/greenpower/system/ (accessed on 23 July 2020).
75. Solomon, A.A.; Child, M.; Caldera, U.; Breyer, C. How much energy storage is needed to incorporate very large intermittent renewables? *Energy Procedia* **2017**, *135*, 283–293. [CrossRef]
76. Duan, J.; van Kooten, G.C.; Liu, X. Renewable electricity grids, battery storage and missing money. *Resour. Conserv. Recycl.* **2020**, *161*, 105001. [CrossRef]
77. Kouchachvili, L.; Entchev, E. Power to gas and H2/NG blend in SMART energy networks concept. *Renew. Energy* **2018**, *125*, 456–464. [CrossRef]
78. Nel C Series Specifications Sheet. Available online: https://nelhydrogen.com/wp-content/uploads/2020/03/C-Series-Spec-Sheet-Rev-G.pdf (accessed on 20 July 2020).
79. GPA Engineering. *Hydrogen in the Gas. Distribution Networks*; GPA Engineering: Adelaide, Australia, 2019.
80. Isaac, T. HyDeploy: The UK's First Hydrogen Blending Deployment Project. *Clean Energy* **2019**, *3*, 114–125. [CrossRef]
81. Australian Renewable Energy Agency. APA Renewable Methane Demonstration Project. Available online: https://arena.gov.au/news/trialling-renewable-methane-in-australias-gas-pipelines/ (accessed on 10 August 2020).
82. Bruce, S.; Temminghoff, M.; Hayward, J.; Palfreyman, D.; Munnings, C.; Burke, N.; Creasey, S. *Opportunities for Hydrogen in Commercial Aviation*; CSIRO: Canberra, Australia, 2020.
83. Hunter New England Health. New Car Parking at John Hunter Hospital Opens Ahead of Schedule. Available online: http://www.hnehealth.nsw.gov.au/News/Pages/MR15-08.aspx (accessed on 27 July 2020).
84. Bain, A.; Van Vorst, D.W. Hindenburg tragedy revisited: The fatal flaw found. *Int. J. Hydrog Energy* **1999**, *24*, 399–403. [CrossRef]
85. Hord, J. Is hydrogen a safe fuel? *Int. J. Hydrog Energy* **1978**, *3*, 157–176. [CrossRef]
86. De Santoli, L.; Paiolo, R.; Lo Basso, G. An overview on safety issues related to hydrogen and methane blend applications in domestic and industrial use. *Energy Procedia* **2017**, *126*, 297–304. [CrossRef]

87. Gurieff, N.; Keogh, D.F.; Baldry, M.; Timchenko, V.; Green, D.; Koskinen, I.; Menictas, C. Mass Transport Optimization for Redox Flow Battery Design. *Appl. Sci.* **2020**, *10*, 2801. [CrossRef]
88. Gundersen Health System. *Solar Panels a Visible Reminder of Environmental Stewardship*; Gundersen Health System: La Crosse, WI, USA, 2008.

sustainability

MDPI

Viewpoint

The Sustainability of Artificial Intelligence: An Urbanistic Viewpoint from the Lens of Smart and Sustainable Cities

Tan Yigitcanlar [1,*] and Federico Cugurullo [2]

1 School of Built Environment, Queensland University of Technology, 2 George Street, Brisbane, QLD 4000, Australia

2 Department of Geography, School of Natural Sciences, Trinity College Dublin, University of Dublin, D02 PN40 Dublin 2, Ireland; cugurulf@tcd.ie

* Correspondence: tan.yigitcanlar@qut.edu.au; Tel.: +61-731-382-418

Received: 18 September 2020; Accepted: 6 October 2020; Published: 15 October 2020

Abstract: The popularity and application of artificial intelligence (AI) are increasing rapidly all around the world—where, in simple terms, AI is a technology which mimics the behaviors commonly associated with human intelligence. Today, various AI applications are being used in areas ranging from marketing to banking and finance, from agriculture to healthcare and security, from space exploration to robotics and transport, and from chatbots to artificial creativity and manufacturing. More recently, AI applications have also started to become an integral part of many urban services. Urban artificial intelligences manage the transport systems of cities, run restaurants and shops where every day urbanity is expressed, repair urban infrastructure, and govern multiple urban domains such as traffic, air quality monitoring, garbage collection, and energy. In the age of uncertainty and complexity that is upon us, the increasing adoption of AI is expected to continue, and so its impact on the sustainability of our cities. This viewpoint explores and questions the sustainability of AI from the lens of smart and sustainable cities, and generates insights into emerging urban artificial intelligences and the potential symbiosis between AI and a smart and sustainable urbanism. In terms of methodology, this viewpoint deploys a thorough review of the current status of AI and smart and sustainable cities literature, research, developments, trends, and applications. In so doing, it contributes to existing academic debates in the fields of smart and sustainable cities and AI. In addition, by shedding light on the uptake of AI in cities, the viewpoint seeks to help urban policymakers, planners, and citizens make informed decisions about a sustainable adoption of AI.

Keywords: artificial intelligence (AI); artificially intelligent city; climate change; planetary challenges; smart and sustainable cities; smart city; technological disruption; urban policy; sustainable urbanism; urban artificial intelligences

---

1. Introduction

Artificial intelligence (AI) is one of the most disruptive technologies of our time [1]. In simple terms, AI can be defined as machines or computers that mimic cognitive functions that humans associate with the human mind, such as learning and problem solving [2]. The field of AI is vast and constantly expanding, and such characterization concerns AI beyond its current capabilities, namely artificial narrow intelligence, thereby comprehending two potential future types of AI: artificial general intelligence and artificial super intelligence [3–5].

AI is already here. AI applications are being used in areas ranging from marketing to banking and finance, from agriculture to healthcare and security, from space exploration to robotics and transport, and from chatbots to artificial creativity and manufacturing [6,7]. In recent years, AI applications have been also started to become an integral part of the city. AIs manage the transport systems of cities in the shape of autonomous cars [8–10]. Robots run restaurants and shops where core aspects of urban life are everyday played out, and repair urban infrastructure [11,12]. Invisible intelligent platforms govern multiple urban domains ranging from traffic to safety, and from garbage collection to air quality monitoring [13,14]. We refer to this strand of AI as *urban artificial intelligences*—where AIs are embodied in urban spaces, urban infrastructures, and urban technologies, which together are turning cities into autonomous entities operating in an unsupervised manner [15].

Focusing mostly on artificial narrow intelligence and present AI technology, this viewpoint elaborates the rise of AI in cities and discusses the *sustainability of urban artificial intelligence* from the lens of *smart and sustainable cities* [16–19]—where such cities utilize digital technologies to make infrastructure services more efficient and reactive to reduce resource consumption, increase environmental quality, and cut down on carbon emissions [20]. In other words, this viewpoint investigates how AI is being utilized in urban domains, unpacking the sustainability potential and risks that AI technology poses for our cities and their citizens.

In terms of methodology, this viewpoint deploys a thorough review of the current status of AI and smart and sustainable cities literature, research, developments, trends, and applications. Following this introduction, Section 2 highlights the key challenges that humankind faces to achieve sustainability at a planetary scale. Next, Section 3 advocates smart and sustainable cities as a potential urban model to realize sustainable futures. Section 4 puts emphasis on the increasing role of AI as an emerging technology fitting the smart and sustainable city paradigm. Afterwards, Section 5 explores the idea of a possible symbiosis between AI and smart and sustainable cities, and its likely offspring—i.e., the artificially intelligent city. Section 6 discusses how urban AIs can be improved to reach more sustainable urban futures. Lastly, Section 7 concludes the viewpoint with a set of insights meant to orientate urban research, policy and development towards a sustainable adoption of AI in cities.

2. Living in Interesting Times: Planetary Sustainability Challenges

We live in "interesting times", where such period refers to—as in the legendary Chinese curse—a time of danger, uncertainty, and complexity [21]. Unless the underlining drivers behind such dangers, uncertainties, and complexities are not eliminated or brought to a manageable level, these interesting times might coincide with the end of human civilization [22]. The primary underlining reasons—which are the key challenges of humanity today—include: (a) rapidly increasing global population; (b) rapidly depleting natural resources and climate change; (c) technological inequality and disruption; (d) misuse of data and information; (d) ruthless neoliberal economies; (e) global, regional, local conflicts; (f) corrupt or ineffective governance. These challenges are illustrated in Figure 1, and further elaborated below.

Rapidly increasing global population: With the appearance of *Homo sapiens*, the origin of humankind goes back to about 300,000 years ago. However, it is only during the last 10,000 years that we have managed to establish safer living conditions thanks to progress in the spheres of technology, knowledge, and wisdom. Subsequently, in the year 1800, the world's population reached the one billion mark. During the same year, London was the only city in the world hosting a million people. Today over 220 years later, our population is over 7.8 billion, and London is home to 9.3 million people. But, London is no longer the largest city in the world. The metropolitan region of Tokyo is approaching 40 million people, and there are over 30 other megacities around the world with over 10 million people. Population projections suggest that by the end of the century the global population will range between 9 and 12 billion. Along with megacity developments, all major metropolitan regions are also experiencing rapid peri-urban expansion [23]. This dual human–urban growth is causing alarming water, food, and energy insecurity [24–26].

Rapidly depleting natural resources and climate change: Ever increasing populations, coupled with unsustainable development practices, are pushing the limits of the world's carrying capacity [27–30]. Heavy fossil fuel dependency and limited clean-energy options—only about 25% of all the world's energy comes from renewable resources—together with various other contributing factors, are triggering biodiversity loss and anthropogenic climate change, and increasing the frequency and severity of natural disasters dramatically [31–33].

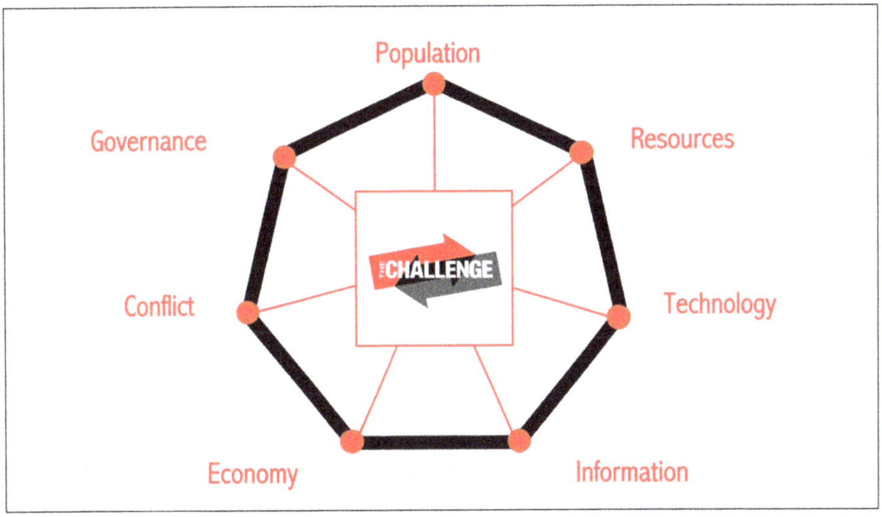

Figure 1. Key global sustainability challenges (Source: Authors).

Technological (or digital) inequality and disruption: Whilst there have been many positive technological inventions and developments, technology also creates disruption in our societies—particularly for those who cannot afford, access or adopt new technologies [34,35]. For instance, despite the fact that there are four billion smartphone users in the world, not everyone has access to the internet and mobile services at the same speed and bandwidth [36]. Particularly from an urban perspective, expensive urban technologies are often unevenly distributed across cities, thus contributing to the fracturing of urban societies and to the formation of high-tech *premium ecological enclaves* where only rich minorities can shield themselves from the burdens of climate change and environmental degradation [37–39].

Misuse of data and information: During the last two decades, with the raise of the second digital revolution and mass digitization, data and information have become more widely and easily accessible. Especially social media platforms and shared user-generated contents have provided large volumes of data. Nonetheless, this has also led to fake news and data integrity issues [40]. Furthermore, targeted Facebook and WhatsApp campaigns changed the results of the 2016 USA and 2018 Brazil presidential elections, and the 2016 Brexit referendum [41–43], thereby showing how data is being used not to inform, but rather to misinform and to protect the interests of certain political elites/groups.

Ruthless neoliberal economies: Today, the world is facing harsh economic challenges. Globally, we are moving towards another recession, if not already in. While some might blame the recent COVID-19 pandemic, the origin of the issue is neo-liberal capitalism and the consumeristic and materialistic practices that it reproduces [44,45] Only eight people, the richest in the world, have a net worth equivalent to that of the lower half of the world's population (about 3.8 billion people); this is the product of ruthless neoliberal economies [46]. Socioeconomic inequality is rapidly widening, and poverty and recession are making life harder for most people across the globe. Particularly with the existing COVID-19 pandemic, the situation is much more dramatic and unsustainable in developing countries, and for disadvantaged communities and individuals [47].

Global, regional and local conflicts: Human civilization has always experienced conflicts and wars over resources, land, or power. However, contemporary wars are not only taking place as trade, diplomatic and armed conflicts, but also as cyber warfare [48]. These multiple conflicts, together with climate change, are displacing many people, thus substantially increasing the number of refugees in the world [49,50].

Corrupt or ineffective governance: Governments should have supposedly addressed the aforementioned challenges. Instead, short termism in political circles, corporate influence, and various degrees of corruption make governments unable to be part of the solution [51]. An example is the Paris Agreement on climate change, which, although signed by 197 countries (and ratified by 189), has led to little or no tangible outcome due to government inaction [52].

3. Smart and Sustainable Cities: An Urban Focus to Achieve Sustainability

The aforementioned issues are extremely challenging to tackle, but they are not discouraging many scholars and thinkers from searching for solutions to realize more sustainable futures [53–55]. Today, approximately 55% of the global population lives in cities whose fabric is rapidly expanding across the planet [56]. The figure is over 85% in many countries—such as Australia, the UK, and the Netherlands [57]. This makes urban areas the prime focus of sustainability policy, not only because they house the majority of the world's population, but also because they contain the core of global socioeconomic activities [58,59]. The changing focus from *nation* to *city* has created new and alternative ideas for building sustainable futures by placing cities at the center of policy actions [60].

In recent years, one of the most prominent ideas in urban policy circles has been the imperative to employ information and communication technology (ICT), in order to address major urban and societal challenges [61]. This trend gave birth to the notion of 'smart city'. While the origin of the concept of smart city dates back to centuries ago, the practice of smart urbanism has been made popular only in the 2000s with urban projects led by private companies like IBM and Cisco [62–64]. Since then, many major technology, construction, and consultancy companies, together with policymakers and city planners, have jumped onto the smart city bandwagon [65,66]. This has resulted in a myriad of smart-city initiatives that are reshaping existing cities and building new ones all over the world [67,68]. In a nutshell, a smart city is, in theory, a locality that uses digital data and technology to improve efficiency in different interconnected urban domains (such as energy, transport and safety), eventually resulting in economic development, better quality of life and sustainability [69].

Nevertheless, in practice, this is not always the case. Numerous studies have shown that, actually, existing smart cities are often disproportionately driven by economic objectives and incapable of addressing social and environmental concerns [70–75]. This is why, in recent years, the focus of smart-city research has shifted towards the 'smart and sustainable city', in the attempt to rebalance the economic, social, and environmental dimensions of smart urbanism [76–78]. A conceptual framework is provided in Figure 2. A smart and sustainable city is defined as an urban locality functioning as a robust system of systems with sustainable practices, supported by community, technology, and policy, to generate desired outcomes and futures for all humans and non-humans [79].

This conceptualization utilizes the Input-Process-Output-Impact approach [80]. As the key 'input', we have the city and its indigenous assets. By using this asset base, three 'processes'—i.e., technology, policy, and community—generate strategies, actions, and initiatives. These result in 'outputs' in the economy, society, environment, and governance domains. When these outputs are aligned with knowledge-based and sustainable urban development goals, principles, and practices, they produce the desired 'impact' for a smart and sustainable city [79].

The framework underlines that, despite the prevalent technocentric perspective in the making of smart cities, in order to create cities that are smart *and* sustainable, we actually need a balanced view on the community, technology, and policy trio as the driver of transformation. It also highlights that cities should not be understood and treated as mere technological artefacts, but rather as social processes, and that sustainability should not be approached in a one-dimensional way, but rather holistically as the equilibrium among diverse social, environmental, and economic spheres [81–83]. In other words, technology will only lead to sustainability if its adequateness is thoroughly scrutinized via community engagement, and its implementation is carried out via a sound policy and government monitoring [79].

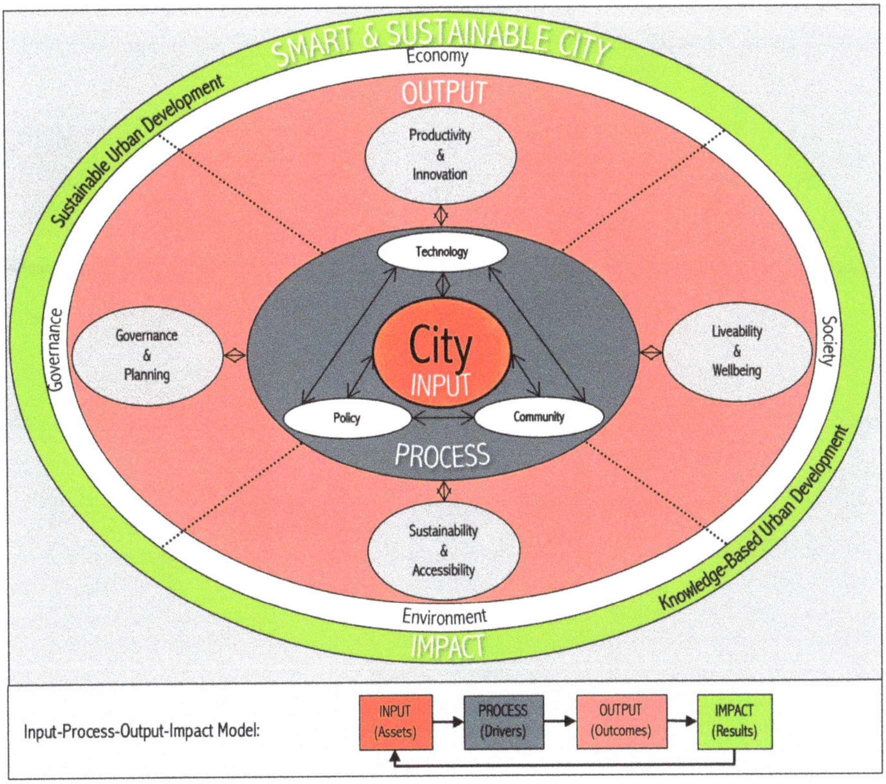

Figure 2. A conceptual framework of smart and sustainable cities, derived from [79].

4. Smart and Sustainable City Technologies: The Increasing Role of Artificial Intelligence

Digital technologies are increasingly offering new opportunities for cities in their journey to become smart and sustainable—especially in relation to issues of community engagement and participatory governance [84]. There is a large variety of smart and sustainable city technologies available today and their list is exhaustingly long [85,86]. For instance, in a recent study, Yigitcanlar et al. [87] have identified the most popular smart and sustainable city technologies in Australia by means of social media analytics. The study concentrated on determining what the key smart city concepts and technologies are, and how they are perceived and utilized in Australia. The results have shown that the concepts of innovation and sustainability, and Internet-of-things (IoT) and artificial intelligence (AI) technologies, are the dominant ones. Unsurprisingly, these top technologies are merging today to form artificial-intelligence-of-things (AIoT) [88] to achieve more efficient IoT operations, improve decision-making and human-machine interactions, and enhance data management and analytics [89].

There is neither a universal definition of AI, nor an established blueprint to build one [4,90]. In simple terms, an AI is a nonbiological intelligence that mimics the cognitive functions of the human mind, such as learning and problem solving [91,92]. More specifically, an artificially intelligent entity is supposed to possess the following capabilities: the ability to learn by acquiring information on the surrounding environment, the capacity to make sense of the data and extract concepts from it, the skill of handling uncertainty, and the power to make decisions and act without being supervised [15]. There are several types of machines and algorithms, which possess the above capabilities at different levels of development, meaning that there are various levels of AI [93]. These levels are illustrated in Figure 3 and described below.

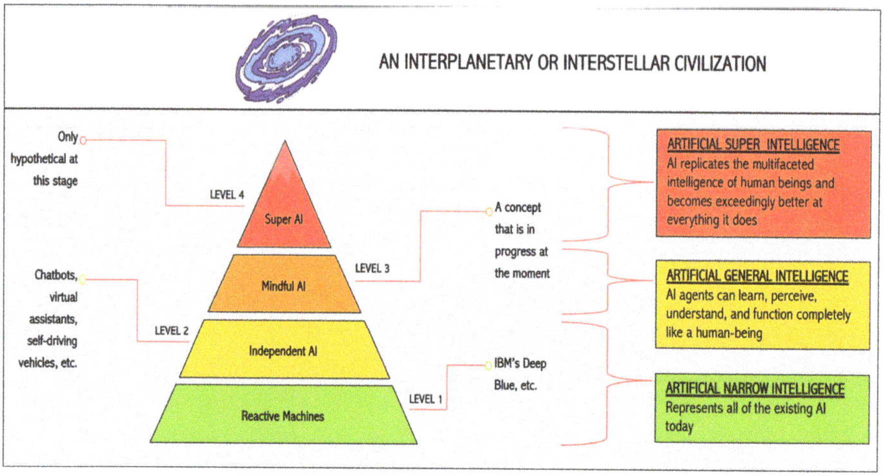

Figure 3. Levels of artificial intelligence (Source: Authors).

In 1997, IBM's Deep Blue defeated the then World Chess Champion Garry Kasparov—that was a remarkable twist in the story of AI and intelligent machines. However, it is more appropriate to classify Deep Blue as a 'reactive machine' (Level 1), since this AI is programmed to undertake one single task, and it does not have the capacity to learn and improve itself [94]. Above all, this type of AI does not take the initiative. It mostly *reacts* to human inputs, rather than planning and pursuing its own original agenda. Its actions and ideas are derivative and are triggered in response to external stimuli.

The next level (Level 2) is the 'Independent AI'. In 2016, Google's AlphaGo beat the international Go champion Lee Sedol. Go is arguably the most complex board game ever invented by mankind, and AlphaGo won thanks to its learning ability and capacity to take original actions that its human opponent could not foresee. This victory was an extraordinary outcome and boosted AI research world-wide. A similar, although less spectacular example, are now common AI chatbots which today many companies are using to interact with their customers on their websites. Other examples range from apps that regulate our phones and homes, to autonomous vehicles that are capable of determining and executing complex routes in chaotic urban environments [95–97]. What these AIs have in common is that they all operate independently. Human actions do not dictate their actions. Independent AIs proactively come up with their own agenda and implement it without humans leading the way.

The above categories constitute what is commonly referred to as 'artificial narrow intelligence'. This is the AI level that we have reached to date in practice, and that is becoming a common sight in contemporary cities and societies. However, R&D efforts are constantly leading to bolder and more innovative theories such as the 'theory of mind AI', which pictures an AI system that has beliefs, desires, and emotions [98]. A 'self-aware AI' is likely to be the next level of AI, thereby producing machines which actually function like us [99]. We call this level 'Mindful AI' (Level 3) to denote artificial intelligences which not only have a mind and are capable of thinking. They are also conscious of their own mind and thoughts which they apply to multiple domains of knowledge. This is the level of 'artificial general intelligence' at which machine behavior is almost indistinguishable from human behavior.

Mindful AIs, and artificial general intelligence more in general, are hypothetical stages of development, which could become the steppingstone to further technological progress in the field of AI. The ultimate level of AI that has so far been imagined is the 'artificial super intelligence'. Here at the 'Super AI' level (Level 4), the AI does everything and anything better than us humans [100]. The opinions of scholars on superintelligence are mixed. While some believe that this could be mankind's last invention leading to the end of human civilization, others posit that this technology could be the beginning of a new era as our only chance of leaving this planet and establishing an interplanetary or interstellar civilization [101–103].

As urbanists interested in the present and near future of urban development, we deal with those existing technologies that are already in the process of altering the sustainability of cities. The rest of the viewpoint will, therefore, focus on artificial narrow intelligence. This vast field of AI includes technologies with at least one of the following capabilities: (a) *perception* including audio/visual/textual/tactile (e.g., face recognition); (b) *decision-making* (e.g., medical diagnosis systems); (c) *prediction* (e.g., weather forecast); (d) *automatic knowledge extraction and pattern recognition* (e.g., discovery of fake news); (e) *interactive communication* (e.g., social robots or chat bots); (f) *logical reasoning and concept extraction* (e.g., theory development from premises) [104]. Mapping out the state of the art in AI is highly useful to better understand the capacities and impact of artificial narrow intelligence. Figure 4 illustrates the key AI problem domains and paradigms.

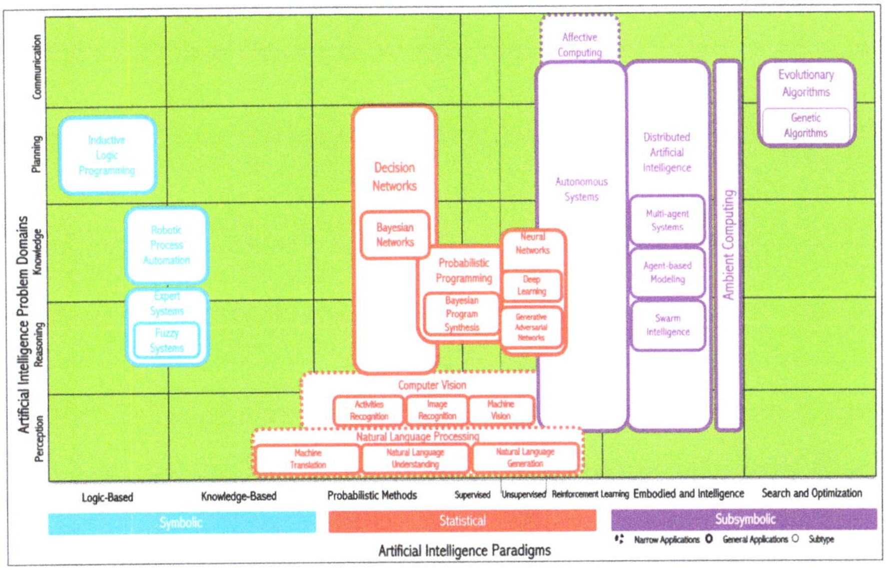

Figure 4. Artificial intelligence knowledge map, derived from [105].

Artificial narrow intelligence is increasingly becoming part of our lives, and an integral element of our cities. For instance, in many parts of the world, states are trialing AI-driven cars to prepare their cities and citizens for the disruptions that autonomous driving will generate [97,106–108]. Robotic dogs are employed in places like Singapore for monitoring social distancing in the era of COVID-19 [109]. A couple of years ago, Dubai has started robot police services meant to stop petty crime [110]. Hospitals in a number of countries, such as Japan, are employing robot doctors [111]. Many homes are getting safer and more energy efficient due to smart home technology and services, and home automation, or *domotics*, is becoming a big part of the construction industry [112]. Websites of both major corporations and ordinary companies have now chatbots to respond to clients' inquiries [113]. In China and Malaysia, large-scale urban artificial intelligences called *city brains* are managing the transport, energy and safety systems of several cities [15].

Additionally, AI is an integral part of environmental research in a number of countries such as Australia, where autonomous drones are detecting via machine learning environmental hazards and animals in danger of extinction [114,115]. Today, most smart phones offer an AI as a personal assistant [116]. Overall, these examples are only the tip of the AI iceberg, as the largest application of AI technology is in analytics. Many of the decisions impacting our life are being made as a result of descriptive, predictive, and prescriptive analyses of data collected and processed by AI [117,118]. In other words, AI-aided urban data science is being extensively used today in cities across the globe, to address the uncertainties and complexities of urbanity [119,120].

5. The Symbiosis: Towards an Artificially Intelligent City?

AI is one of the most powerful and disruptive technologies of our time, and its influence on urban settlements and activities is growing rapidly, ultimately affecting everyday life [121,122]. Given that cities are the main hubs and drivers of most socioeconomic activities, political actions, and environmental transformations, it is important to understand how the development of AI and the development of the city are intertwining [123]. This brings up the question of whether there is or could be a symbiotic relationship between them, and if this revolutionary technology could offer novel sustainability solutions feeding into new urban models. After all, AI has already entered our cities, and it is therefore essential to critically examine and question its urban sustainability potential [15].

A study by Yigitcanlar et. al. [124] investigated these questions through a thorough systematic literature review—99 peer-reviewed research articles concentrating on both smart cities and AI. The study arranged the findings under four smart city domains, as shown in Figure 2—i.e., economy, society, environment, governance.

In terms of the 'economy' domain of smart cities, the AI focus is predominately on technological innovation, and business productivity, profitability and management. Some of the most typical contributions of AI to this domain include [124]:

- Enhancing firm productivity and innovation by automating data management and analysis processes;
- Increasing the efficiency and effectiveness of existing resources, and reducing additional costs through pattern recognition;
- Supporting decision-making by analyzing large volumes of data—e.g., big data analytics—from multiple sources;
- Drawing conclusions to facilitate informed decisions based on logic, reason, and intuition via deep learning.

In terms of the 'society' domain of smart cities, the AI focus is predominately on the public health, wellbeing, and education areas. The COVID-19 pandemic is particularly accelerating the use of AI in these areas. The main contributions of AI to this domain include [124]:

- Improving community health monitoring via smart sensors and analytics tools embedded in homes and/or workplaces;
- Enhancing public health diagnoses through medical imaging analytics, particularly in radiology and healthcare services;
- Providing autonomous tutoring systems to teach algebra, grammar, and other subjects to pupils and adults;
- Offering personalized learning options to facilitate students' progress and expand their curriculum.

In terms of the 'environment' domain of smart cities, the AI focus is predominately on the transport, energy, land use, and climate areas. Some of the key contributions of AI to this domain include [124]:

- Operationalizing smart urban transport systems via mobility-as-a-service (MaaS)— integration of various transport services into a single on-demand mobility service;
- Optimizing energy production and consumption via domotics—home technologies with a focus on environmental issues, energy saving, and lifestyle improvement;
- Monitoring changes in the natural and the built environment via remote sensing with autonomous drones—used for multiple-object detection and tracking in aerial videos;

- Predicting the risks of climate change via machine learning algorithms combined with climate models—employed to foresee potential disastrous events in specific geographical areas and act in advance.

Moreover, beyond urban environmental issues, AI is also being used for addressing planetary environmental challenges. Overall, as Vinuesa et al. [104] have argued, AI applications can potentially contribute to achieving 17 Sustainable Development Goals (SDGs). Below, we provide a summary of the application areas touched by AI technologies, specifically in relation to environmental sustainability.

- AI application areas for *climate change/crisis mitigation* include: research, urban, and regional planning, land use, home, mobility, energy production and consumption [125–127];
- AI application areas for *ocean health* include: sustainable fishery, pollution monitoring, reduction and prevention, habitat and species protection, and acidification reduction [128–130];
- AI application areas for *clean air* include: pollutant filtering and capture, pollution monitoring, reduction and prevention, early pollution and hazard warning, clean energy, and real-time, integrated, adaptive urban management [131–133];
- AI application areas for *biodiversity and conservation* include: habitat protection and restoration, sustainable trade, pollution monitoring, reduction and prevention, invasive species and disease control, and natural capital enhancement and protection [134–136];
- AI application areas for *clean water security* include: water supply quantity, quality and efficiency management, water catchment control, sanitation, and drought planning [137–139];
- AI application areas for *weather and disaster resilience* include: prediction and forecasting, early warning systems, resilient infrastructure and planning, and financial instruments [140–142].

In terms of the 'governance' domain of smart cities, the AI focus is predominately on national and public security, urban governance and decision-making in government. Some of the principal contributions of AI to this domain include [124]:

- Deploying smart poles as digital sensors, and providing technological tools for citizen scientists to act like human sensors, for making informed decisions—smart poles and volunteer citizens equipped with smart tech, generate big data that is processed by AI;
- Aiding management, planning, and operations related to disasters, pandemics and other emergencies via predictive analytics—using AI to make predictions about future events;
- Enhancing the operability of surveillance systems via smart poles with AIoT (although due to cyber-attacks and privacy issues, benefits exist together with major concerns);
- Improving cybersecurity by analyzing data and records on cyber incidents, identifying potential threats, and providing patches and options to improve cyber security.

Nonetheless, the above list of benefits should not obscure that of the many problems that AI is bringing. AI is a double-edged sword. This sentient sword can be used to fight against global sustainability issues, but it can also cause much collateral damage as well as harm those who wield it. The drawbacks of AI are equal to its potentials [143]. Below, we provide a summary of prospects and constraints of AI according to different smart city domains [144]. As pointed out earlier, we need more than *technology* to achieve urban sustainability. Particularly *policy* and *community*, which are the other two drivers of smart and sustainable cities (see Figure 2), should be refined and operationalized to neutralize the technological shortcomings of AI.

- On the one hand, the *prospects* of AI in the *economy* domain include: enhancing productivity and innovation, reducing costs and increasing resources, supporting the decision-making process, automating decision-making [145–147]. On the other hand, the *constraints* of AI involve: making biased decisions, having an unstable job market, losing revenue streams and employment, and generating economic inequality [148–150].

- On the one hand, the *prospects* of AI in the *society* domain include: improving healthcare monitoring, enhancing medical diagnoses, increasing the adaptability of education systems, personalizing teaching and learning, and optimizing tasks [151–153]. On the other hand, the *constraints* of AI involve: making biased decisions, making misdiagnoses, having an unstable job market, losing employment, and undermining data privacy and security [154–156].
- On the one hand, the *prospects* of AI in the *environment* domain include: assisting environmental monitoring, optimizing energy consumption and production, optimizing transport systems, and assisting the development of more environmentally efficient transport and logistic systems [157–159]. On the other hand, the *constraints* of AI involve: making biased decisions, increasing urban sprawl, leading to more motor vehicle kilometers traveled, destabilizing property values, establishing heavy energy dependency due to intensive use of technology, and increasing carbon footprints [160–162].
- On the one hand, the *prospects* of AI in the *governance* domain include: enhancing surveillance system capacity, improving cyber safety, aiding disaster management planning and operations, and assisting citizen scientists with new technologies in producing crowdsourced data/information [163–165]. On the other hand, the *constraints* of AI involve: making biased decisions including racial bias and discrimination, suppressing public voice/protests/rights, violating civil liberties, causing privacy concerns, using technology unethically, risking the spread of misinformation, and creating cybersecurity concerns [166–168].

The above prospects and constraints should be evaluated in relation to the five different levels of autonomy that characterize the decision-making power of AI [15,169]. Level 0 corresponds to no autonomy—meaning full human control on every decision. Levels 1 and 2 correspond to assisted decision-making, where in Level 2 AI offers moderate assistance or recommendation. In Level 3, decisions require human approval, whilst in Level 4 only human monitoring or human oversight is needed, to step in in case of a problem. Level 5 is equal to complete autonomy, meaning that decisions are taken by an AI in an unsupervised manner. As we progress to Level 5, both the magnitude of disruption and opportunity will become greater. With this greater power, AI will have to assume greater responsibility, and it will be thus crucial to develop 'responsible and ethical AI' before we get to Level 5 [170–172]. From an urban point of view, AI technology is progressing fast, thereby gaining more and more autonomy in cities. Especially in experimental cities, where the pace of technological innovation is usually rapid, we can already see parts of the built environment that are not *automated* but rather *autonomous*.

The key difference between *automation* and *autonomy* is that an automated technology repetitively follows patterns previously established by a human intelligence, while an autonomous technology establishes its own patterns, seldom repeating the exact same action [15]. Simply put, this is the difference between an elevator always going up or down stopping at invariable floors, and an autonomous car which can traverse entire cities and never follow the same route twice. The difference is critical because autonomous AIs operate in real-life environments where the life of real people is at risk. Not in a confined elevator shaft but in, for example, an urban road shared by hundreds of individuals. Here unsupervised, AIs have to make important decisions and take actions that can actually kill. This is the case of the first pedestrian fatality caused by an autonomous car in Tempe (Arizona) in March 2018. An autonomous Uber was incapable of dealing with the uncertainty that is typical of unconfined urban spaces, and its incapacity killed a woman that was crossing a road outside the designated crossing lane [173]. The greater the autonomy of AI is, the greater its constraints are, given that, to date, we do now have urban artificial intelligences that can fully understand what is right or wrong (the issue of ethics) and then answer for their behavior (the issue of responsibility).

Furthermore, it is important to recognize that both the fields of smart and sustainable cities and AI are in constant evolution. As Sections 3 and 4 have illustrated, numerous smart-city projects have been implemented and an even larger number is under development, while the evolution of AI has reached only two levels out of four. This means that we have seen only a small part of what smart urbanism and AI can potentially offer. Whether the best or the worst is yet to come, is an open question. For sure, at the moment there is neither an ideal AI system, nor an ideal smart and sustainable city that can serve as a universal model of development and, given the many geographical differences that exist in the world, the very idea of having a global paradigm is questionable in the first place [68,174,175]. This is to say that we need to continue researching both conceptualizations and practical applications of AI and smart and sustainable cities, across geographical spaces and scales [176]. Only then will we be able to analyze and fully evaluate the symbiosis between AI and the city and understand whether this can give birth in particular places to 'artificially intelligent cities' [144].

Lastly, there is the critical issue of how we define and construct artificially intelligent cities. In its current conceptualization, an *artificially intelligent city* "is a city where algorithms are the dominant decision-makers and arbitrators of governance protocols—the rules and frameworks that enable humans and organizations to interact, from traffic lights to tax structures—and where humans might have limited say in the choices presented to them for any given interaction" [177]. For such type of cities to achieve a condition of sustainability, the issues of transparency, fairness, ethics, and the preservation of human values need to be carefully considered. These unresolved issues are intrinsic to AI and thus hinder its sustainability. In other words, in order to improve the chances that the city of artificial intelligence becomes a sustainable city, we need better AI, and this will be the topic of the next section.

6. Discussion: Better Artificial Intelligence for Better Cities

Makridakis [178] asks the question of whether the AI revolution creates a utopian or dystopian future, or somewhere in between. The answer to this question fully depends on how we are going to tackle the drawbacks of AI, and how we are going to utilize AI in our cities, businesses and, more in general, lives. As Batty [179] remarks, it is hard to predict the exact future of cities, while it is possible to build future cities, meaning that we can actively work in the present to improve contemporary cities and our results will ultimately be the cities of the future. Following this line of thought, if we focus on the pitfalls of AI, we can then search for ways to actually make AI better. *Better* in the sense of more useful to make our cities and societies more sustainable. The key areas of improvement to reach AIs that are conducive to sustainability, are illustrated in Figure 5, and further elaborated below.

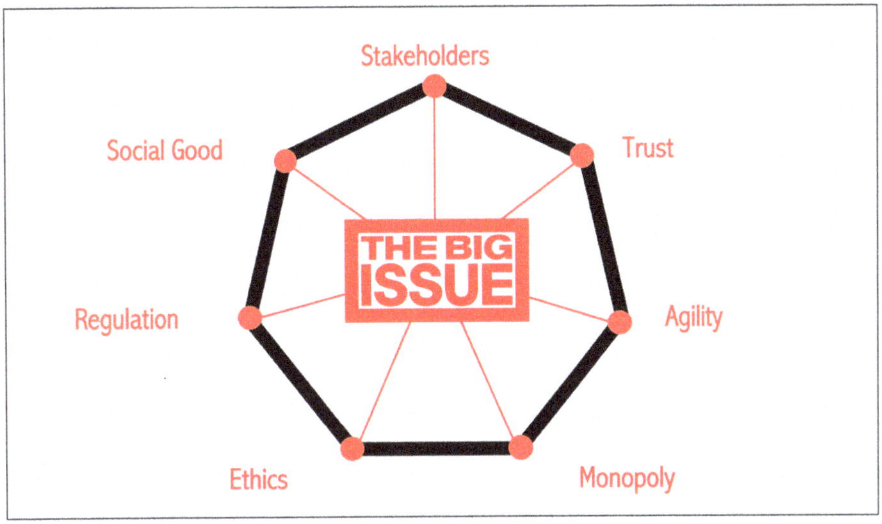

Figure 5. Areas of improvement for artificial intelligence (Source: Authors).

The first issue to consolidate a sustainability-oriented AI is *stakeholder engagement*. In general, AI technologies are created exclusively by technology companies without any or much consultation with wider interest groups or stakeholders. Active collaboration among a wide and inclusive range of stakeholders—ideally in the form of quadruple helix model participation of public, private, academia and community—in the development and deployment stages, in particular, will improve the caliber of the sustainability potential of AI [180,181]. This is, in essence, a matter of inclusion and democracy. Given that the ethos of sustainability is about achieving a *common future*, we argue that no common future can be envisioned and realized unless proper forms of democratic governance are in place. Specifically, in relation to AI, this means that each AI technology affecting cities should be discussed by all urban stakeholders, instead of being imposed in a top-down manner by influential tech companies.

The second issue is the *trust* problem. The blackbox nature of the decisions taken by AIs without much transparency (which, at times, are wrong), the possibility of AI failing in a life-or-death context, and cybersecurity vulnerabilities all limit public trust. AI technology needs to earn the trust not only in the public and the way people perceive it, but also in the minds of companies and government agencies that will be investing in AI [182–184]. This is a challenging problem because, as Greenfield [121] notes, AI is an arcane technology meaning that, although it is already part of the everyday of many people, its mechanics and actual functioning are understood by only a few.

The next area of improvement concerns the *agility* issue. AI systems should be competent enough to deal with complexity and uncertainty, which are extremely common features of contemporary cities [185]. Besides, AI systems should focus on the problem to be solved, rather than just on the data whose collection is arguably meaningless from a sustainability point of view, unless it serves the purpose of addressing a previously identified SDG. In addition, AI technology needs to be as frugal and affordable as possible. This is critical for a wider uptake of AI across cities through public sector funds [186,187]. Expensive AIs are ultimately elitist AIs, which only a rich minority can afford. Elitist AIs can only be unevenly distributed, thus creating a divide among richer and poorer cities, as well as internal fractures within individual cities where small premium enclaves coexist next to disadvantaged districts.

The fourth issue is the *monopoly*. A monopolistic structure behind technology development and deployment is problematic as a lack of competition limits technological variation. Avoiding AI monopolies can make AI technologies more affordable and support current efforts in 'open AI' development. This, in turn, would also promote the democratization of AI research and practice, as well as decrease the risk of the formation of a *singleton* [188,189]. According to Bostrom [4], a singleton is a world order in which one super intelligent agent is in charge. This is an unlikely situation when it comes to Level 1 and 2 AIs, but it might not be a remote possibility if only one tech company in the world has the capacity to build an artificial super intelligence.

Another critical issue is *ethics*. We need to develop AI in a way that it respects human rights, diversity, and the autonomy of individuals. The European Commission's recent ethical guidelines for AI development offer a good starting point [190]. However, as stated by Mittelstadt [191], principles alone cannot guarantee the development of an ethical AI. Hence, we need to develop globally an AI ethics—a multicultural system of moral principles that takes the risks of AI seriously—together with a mechanism to monitor ethics violations. Ethics should ensure the design of AI technologies for human flourishing around the world [192,193], but this is a very complex matter given that, as the work of Awad et al. [194,195] clearly demonstrates, universally valid and accepted ethical principles do not exist.

The sixth issue relates to *regulation* and regulatory challenges. AI cannot achieve sustainability and the common good if it is not regulated. In a situation in which different AI users (or potentially different mindful and super intelligent AIs) can do whatever they want, it is extremely unlikely that the common good will be achieved. Different actors will follow diverse trajectories and reach heterogenous (and not necessarily mutually beneficial) outcomes. This poses a big risk for society—particularly for disadvantaged groups, historically-marginalized groups, and low-income countries. Thus, we need well-regulated and responsible AIs with disruption mitigation mechanisms in place. Such regulation should also protect public values [196,197], and extent to the built environment. It is well documented in urban studies that, when urban development is unregulated, key sustainability themes (such as justice and environmental preservation) get neglected and overshadowed by economic interests [198,199]. Therefore, the regulation of AI and the regulation of the built environment should go hand in hand as a dual policy priority.

The last issue concerns the development of AI for *social good*, and for the benefit of every member of society [200]. AI and data need to be a shared resource employed for the good of society, rather than for serving the economic agenda of corporations and the interests of political elites. An *AI for all* would require establishing AI commons [201] and a similar attempt has been previously made to establish digital commons [202]. AI commons are supposed to allow anyone, anywhere, to enjoy the multiple benefits that AI can provide [203]. AI commons should be studied and pursued to enable AI adopters to connect with AI specialists and AI developers, with the overall aim of aligning every AI towards a shared common goal [204]. From an urbanistic perspective, this is arguably the biggest challenge, because opening up AI as a common good requires also opening up urban spaces, thinking about the city as a truly public resource rather than a territory balkanized by neoliberal ambitions.

7. Conclusions: The Next Big Sustainability Challenge

This viewpoint has explored the prospects and constraints of developing and deploying AI technology to make present and future cities more sustainable. The analysis has shown that, while AI technology is evolving and becoming an integral part of urban services, spaces, and operations, we still need to find ways to integrate AI in our cities in a sustainable manner, and also to minimize the negative social, environmental, economic, and political externalities that the increasingly global adoption of AI is triggering. In essence, the city of AI is not a sustainable city. Both the development of AI and the development of cities need to be refined and better aligned towards sustainability as the overarching goal. With this in mind, the viewpoint has generated the following insights, in the attempt to improve the sustainability of AI and that of those cities that are adopting it.

First of all, AI as part of urban informatics significantly advances our knowledge of computational urban science [205]. In the age of uncertainty and complexity, urban problems are being diagnosed and addressed by numerous AI technologies. However, from a sustainability perspective, the quality of our decisions about the future of cities heavily depends on this computational power (technology), *and* on the inclusivity of decision-making and policy processes. The greater computational power offered by AI, therefore, is not enough to achieve sustainability, unless it is coupled with systems of democratic governance and participatory planning.

Second, AI is being exponentially used to improve the efficiency of several urban domains such as business, data analytics, health, education, energy, environmental monitoring, land use, transport, governance, and security. This has a direct implication for our cities' planning, design, development, and management [206]. Yet, the different uses of AI tend to be fragmented, in the sense that heterogeneous AIs are targeting heterogeneous issues and goals without a holistic approach. Coordinating the many AIs present in our cities is thus necessary for a sustainable urbanism, given that sustainability is about thinking and acting in terms of *the whole* rather than single parts. On these terms, artificial narrow intelligences working on narrow tasks are missing the broad spectrum of social, environmental, and political issues, which is essential to achieve sustainability. We cannot and should not expect a hypothetical future artificial general intelligence to fill this lacuna [207]. Human initiative and coordination are needed now.

Third, the autonomous problem-solving capacity of AI can be useful in some urban decision-making processes. Still, the utmost care is needed to check and monitor the accuracy of any autonomous decisions made by an AI—human inputs and oversight are now critical in relation to artificial narrow intelligence, and they would be even more important should innovation reach the stage of artificial general intelligence [208]. AI can help us optimize various urban processes and can actually make cities smarter. We can move faster towards the goal of smart urbanism, but if we want to create smart *and* sustainable cities, then human intelligence must not be overshadowed by AI.

Fourth, AI can drive positive changes in cities and societies, and contribute to several SDGs [104,209]. Nonetheless, despite these positive prospects, we still need to be cautious about selecting the right AI technology for the right place and ensuring its affordability and alignment with sustainability policies, while also considering issues of community acceptance [210]. AI should not be imposed on society and cities, but rather discussed locally at the community level, taking into account geographical, cultural, demographic and economic differences. Sustainability can only be achieved with a healthy combination of technology, community and policy drivers, hence the urgent need to develop not only technologically, but also socially and politically.

Fifth, we need to be prepared for the upcoming and inevitable disruptions that AI will create in our cities and societies. The diffusion of AI will not be a black and white phenomenon. Many shades of grey will characterize the deployment of heterogeneous AIs in different parts of the world. Even in an optimistic scenario in which a 'benign AI' is promoting sustainability, somewhere someone/something will still be suffering. It is thus imperative to develop appropriate policies and regulations, and to allocate adequate funds, in order to mitigate the disruption that AI will cause to the most disadvantaged cities and social groups, and nature [211]. As we mentioned earlier, sustainability is not about single parts, but rather about the whole. Any form of development that fractures cities, societies, and the natural environment, producing winners and losers, is not sustainable. Like a hurricane, AI is likely to shake everything that we see, know, and care about. It should not be forgotten that we are only as strong as the weakest member of the society.

Sixth, a symbiotic relationship between AI and cities might become a concrete possibility in the future. Combined with progress in public policy and community engagement, progress in AI technology could mitigate the global sustainability challenges discussed in Section 2 [212]. In so doing, while the city might benefit from AI technologies and applications, AI might also benefit from the city to advance itself. This is a key aspect of the intersection between the development of AI and the development of the city. As we explained in Section 4, a key AI skill is learning. AIs learn by sensing the surrounding environment, thereby gaining and accumulating knowledge [15]. Learning is also how AIs improve themselves. AI is a technology that learns from the collected data, from its errors as well as from the mistakes made by other AIs and human intelligences. On these terms, the city represents the ideal learning environment for AI. Cities are the places where knowledge concentrates the most, where a wide-range of events occur, where numerous actors meet and where the biggest mistakes and greatest discoveries of mankind have been made. It this in this cauldron of ideas and experiences that we call *city* that contemporary artificial narrow intelligences can learn the most, potentially evolving into artificial general intelligences.

Seventh, we need to further decentralize political power and economic resources to make our local governments ready for the AI era that is upon us. While planning for a sustainable AI uptake in our cities is crucial, presently, almost all local governments in the globe are not ready—in terms of technical personnel, budget and gear—to thoroughly plan and implement AI projects city-wide [213,214]. Most AI technologies are expensive and it is therefore important to make them affordable, in order to avoid an uneven distribution and ultimately injustice. If AI is to become part of the city, then we need to think of AI not as an elitist technology, but rather as a common good on which everybody has a say. This is, in turn, a question of urban politics and a matter of politicizing AI so that its deployment in cities is discussed and agreed as inclusively and as democratically as possible, instead of being dictated by a handful of influential tech companies. Sustainability will not be achieved in a technocracy.

Eighth, some of the changes triggered by AI might be invisible and silent and, yet, their repercussions are likely to be tangible and loud from an urban perspective. For example, AI is clearly impacting on the economies of cities [215]. This impact will get deeper and wider as innovation keeps improving and expanding the capabilities of artificial narrow intelligences. *What is the role of humans in an economy in which artificial narrow intelligences, artificial general intelligences and artificial super intelligences can cheaply perform human tasks faster and better?* This is a recurring question in AI studies, to which we add a complementary urban question: *What is the role of cities as economic hubs in the era of AI?* A key reason why cities exist is that they provide the spaces that are necessary to perform and accommodate human labor and to train humans in many work-related fields. However, AI is undermining this *raison d'etre*. If human labor decreases or, worse, ceases to exist in cities, then cities are likely to decline and cease to exist too [1]. Now more than ever it is therefore vital to reimagine, replan and redesign cities in a way that their function and shape are not dictated by and dependent on human economies. This is both a matter of rethinking the economic dimension of cities and galvanizing the social, cultural, psychological, political, and environmental dimensions of urban spaces.

Lastly, in the context of smart and sustainable cities, AI is an emerging area of research. Further investigations, both theoretical and empirical, from various angles of the phenomenon and across disciplines, are required to build the knowledge base that is necessary for urban policymakers, managers, planners, and citizens to make informed decisions about the uptake of AI in cities and mitigate the inevitable disruptions that will follow. This will not be an easy task because AI is a technology while the city is not. Cities are primarily made of humans and are the product of human intelligence. The *merging of artificial and human intelligences in cities* is the world's *next big sustainability challenge*.

Author Contributions: T.Y. designed the study, conducted the analysis, and prepared the first draft of the manuscript. F.C. expanded the manuscript, and improved the rigor, relevance, critical perspective and reach of the study. All authors have read and agreed to the published version of the manuscript.

Funding: This research received no external funding.

Acknowledgments: This research did not receive any specific grant from funding agencies in the public, commercial or not-for-profit sectors. The authors thank the anonymous referees for their invaluable comments on an earlier version of the manuscript.

Conflicts of Interest: The authors declare no conflict of interest.

References

1. Kassens-Noor, E.; Hintze, A. Cities of the future? The potential impact of artificial intelligence. *Artif. Intell.* **2020**, *1*, 192–197. [CrossRef]
2. Schalkoff, R.J. *Artificial Intelligence: An Engineering Approach*; McGraw-Hill: New York, NY, USA, 1990.
3. Yampolskiy, R.V. *Artificial Superintelligence: A Futuristic Approach*; CRS Press: New York, NY, USA, 2015.
4. Bostrom, N. *Superintelligence*; Oxford University Press: Oxford, UK, 2017.
5. Kak, S.C. Can we define levels of artificial intelligence? *J. Intell. Syst.* **1996**, *6*, 133–144. [CrossRef]
6. Yun, J.; Lee, D.; Ahn, H.; Park, K.; Lee, S.; Yigitcanlar, T. Not deep learning but autonomous learning of open innovation for sustainable artificial intelligence. *Sustainability* **2016**, *8*, 797. [CrossRef]
7. Faisal, A.; Yigitcanlar, T.; Kamruzzaman, M.; Paz, A. Mapping two decades of autonomous vehicle research: A systematic scientometric analysis. *J. Urban. Technol.* **2020**. [CrossRef]
8. Acheampong, R.A.; Cugurullo, F. Capturing the behavioural determinants behind the adoption of autonomous vehicles: Conceptual frameworks and measurement models to predict public transport, sharing and ownership trends of self-driving cars. *Transp. Res. Part. F* **2019**, *62*, 349–375. [CrossRef]
9. Milakis, D.; Van Arem, B.; Van Wee, B. Policy and society related implications of automated driving: A review of literature and directions for future research. *J. Intell. Transp. Syst.* **2017**, *21*, 324–348. [CrossRef]
10. Nikitas, A.; Michalakopoulou, K.; Njoya, E.T.; Karampatzakis, D. Artificial intelligence, transport and the smart city: Definitions and dimensions of a new mobility era. *Sustainability* **2020**, *12*, 2789. [CrossRef]
11. Macrorie, R.; Marvin, S.; While, A. Robotics and automation in the city: A research agenda. *Urban. Geogr.* **2020**. [CrossRef]
12. Mende, M.; Scott, M.L.; van Doorn, J.; Grewal, D.; Shanks, I. Service robots rising: How humanoid robots influence service experiences and elicit compensatory consumer responses. *J. Mark. Res.* **2019**, *56*, 535–556. [CrossRef]
13. Caprotti, F.; Liu, D. Emerging platform urbanism in China: Reconfigurations of data, citizenship and materialities. *Technol. Forecast. Soc. Chang.* **2020**, *151*, 119690. [CrossRef]

14. Barns, S. *Platform Urbanism: Negotiating Platform Ecosystems in Connected Cities*; Palgrave Macmillan: Singapore, 2019.
15. Cugurullo, F. Urban artificial intelligence: From automation to autonomy in the smart city. *Front. Sustain. Cities* **2020**, *2*, 38. [CrossRef]
16. Yigitcanlar, T.; Kamruzzaman, M. Planning, development and management of sustainable cities: A commentary from the guest editors. *Sustainability* **2015**, *7*, 14677–14688. [CrossRef]
17. Voda, A.I.; Radu, L.D. Artificial intelligence and the future of smart cities. *Broad Res. Artif. Intell. Neurosci.* **2018**, *9*, 110–127.
18. Walshe, R.; Casey, K.; Kernan, J.; Fitzpatrick, D. AI and big data standardization: Contributing to United Nations sustainable development goals. *J. Ict Stand.* **2020**, *8*, 77–106. [CrossRef]
19. Yigitcanlar, T. *Sustainable Urban and Regional Infrastructure Development: Technologies, Applications and Management*; IGI Global: Hersey, PA, USA, 2010.
20. Evans, J.; Karvonen, A.; Luque-Ayala, A.; Martin, C.; McCormick, K.; Raven, R.; Palgan, Y.V. Smart and sustainable cities? Pipedreams, practicalities and possibilities. *Local Environ.* **2019**, *24*, 557–564. [CrossRef]
21. Coaffee, J.; Therrien, M.C.; Chelleri, L.; Henstra, D.; Aldrich, D.P.; Mitchell, C.L. Urban resilience implementation: A policy challenge and research agenda for the 21st century. *J. Contingencies Crisis Manag.* **2018**, *26*, 403–410. [CrossRef]
22. Yigitcanlar, T.; Foth, M.; Kamruzzaman, M. Towards post-anthropocentric cities: Reconceptualising smart cities to evade urban ecocide. *J. Urban. Technol.* **2019**, *26*, 147–152. [CrossRef]
23. Mortoja, M.G.; Yigitcanlar, T.; Mayere, S. What is the most suitable methodological approach to demarcate peri-urban areas? A systematic review of the literature. *Land Use Policy* **2020**, *95*, 104601. [CrossRef]
24. Tscharntke, T.; Clough, Y.; Wanger, T.C.; Jackson, L.; Motzke, I.; Perfecto, I.; Whitbread, A. Global food security, biodiversity conservation and the future of agricultural intensification. *Biol. Conserv.* **2012**, *151*, 53–59. [CrossRef]
25. Rasul, G. Food, water, and energy security in South Asia: A nexus perspective from the Hindu Kush Himalayan region. *Environ. Sci. Policy* **2014**, *39*, 35–48. [CrossRef]
26. Cohen, J.E. Human population: The next half century. *Science* **2003**, *302*, 1172–1175. [CrossRef] [PubMed]
27. Arbolino, R.; De Simone, L.; Carlucci, F.; Yigitcanlar, T.; Ioppolo, G. Towards a sustainable industrial ecology: Implementation of a novel approach in the performance evaluation of Italian regions. *J. Clean. Prod.* **2018**, *178*, 220–236. [CrossRef]
28. Berck, P.; Levy, A.; Chowdhury, K. An analysis of the world's environment and population dynamics with varying carrying capacity, concerns and skepticism. *Ecol. Econ.* **2012**, *73*, 103–112. [CrossRef]
29. Mortoja, M.; Yigitcanlar, T. Local drivers of anthropogenic climate change: Quantifying the impact through a remote sensing approach in Brisbane. *Remote Sens.* **2020**, *12*, 2270. [CrossRef]
30. Mahbub, P.; Goonetilleke, A.; Ayoko, G.A.; Egodawatta, P.; Yigitcanlar, T. Analysis of build-up of heavy metals and volatile organics on urban roads in Gold Coast, Australia. *Water Sci. Technol.* **2011**, *63*, 2077–2085. [CrossRef]
31. Konikow, L.F.; Kendy, E. Groundwater depletion: A global problem. *Hydrogeol. J.* **2005**, *13*, 317–320. [CrossRef]
32. Sotto, D.; Philippi, A.; Yigitcanlar, T.; Kamruzzaman, M. Aligning urban policy with climate action in the global south: Are Brazilian cities considering climate emergency in local planning practice? *Energies* **2019**, *12*, 3418. [CrossRef]
33. Prior, T.; Giurco, D.; Mudd, G.; Mason, L.; Behrisch, J. Resource depletion, peak minerals and the implications for sustainable resource management. *Glob. Environ. Chang.* **2012**, *22*, 577–587. [CrossRef]
34. Robinson, L.; Cotten, S.R.; Ono, H.; Quan-Haase, A.; Mesch, G.; Chen, W.; Stern, M.J. Digital inequalities and why they matter. *Inf. Commun. Soc.* **2015**, *18*, 569–582. [CrossRef]
35. Ragnedda, M. *The Third Digital Divide: A Weberian Approach to Digital Inequalities*; Taylor & Francis: New York, NY, USA, 2017.
36. Riddlesden, D.; Singleton, A.D. Broadband speed equity: A new digital divide? *Appl. Geogr.* **2014**, *52*, 25–33. [CrossRef]
37. Anguelovski, I.; Irazábal-Zurita, C.; Connolly, J.J. Grabbed urban landscapes: Socio-spatial tensions in green infrastructure planning in Medellín. *Int. J. Urban. Reg. Res.* **2019**, *43*, 133–156. [CrossRef]

38. Cugurullo, F. How to build a sandcastle: An analysis of the genesis and development of Masdar City. *J. Urban. Technol.* **2013**, *20*, 23–37. [CrossRef]
39. Hodson, M.; Marvin, S. Urbanism in the anthropocene: Ecological urbanism or premium ecological enclaves? *City* **2010**, *14*, 298–313. [CrossRef]
40. Guess, A.; Nagler, J.; Tucker, J. Less than you think: Prevalence and predictors of fake news dissemination on Facebook. *Sci. Adv.* **2019**, *5*, eaau4586. [CrossRef]
41. Bastos, M.; Mercea, D. The public accountability of social platforms: Lessons from a study on bots and trolls in the Brexit campaign. *Philos. Trans. R. Soc. A* **2018**, *376*, 20180003. [CrossRef]
42. Isaak, J.; Hanna, M.J. User data privacy: Facebook, Cambridge Analytica, and privacy protection. *Computer* **2018**, *51*, 56–59.
43. Evangelista, R.; Bruno, F. WhatsApp and political instability in Brazil: Targeted messages and political radicalisation. *Internet Policy Rev.* **2019**, *8*, 1–23. [CrossRef]
44. Rapley, J. *Globalization and Inequality: Neoliberalism's Downward Spiral*; Lynne Rienner Publishers: London, UK, 2004.
45. Regilme, S.S., Jr. The decline of American power and Donald Trump: Reflections on human rights, neoliberalism, and the world order. *Geoforum* **2019**, *102*, 157–166. [CrossRef]
46. Gould-Wartofsky, M.A. *The Occupiers: The Making of the 99 Percent Movement*; Oxford University Press: London, UK, 2015.
47. Grigoryev, L.M. Global social drama of pandemic and recession. *Popul. Econ.* **2020**, *4*, 18–25. [CrossRef]
48. Taplin, R. *Cyber Risk, Intellectual Property Theft and Cyberwarfare: Asia, Europe and the USA*; Routledge: London, UK, 2020.
49. Atapattu, S. Climate change and displacement: Protecting 'climate refugees' within a framework of justice and human rights. *J. Hum. Rights Environ.* **2020**, *11*, 86–113. [CrossRef]
50. Berchin, I.I.; Valduga, I.B.; Garcia, J.; de Andrade, J.B. Climate change and forced migrations: An effort towards recognizing climate refugees. *Geoforum* **2020**, *84*, 147–150. [CrossRef]
51. Rothstein, B. Corruption and social trust: Why the fish rots from the head down. *Soc. Res.* **2013**, *80*, 1009–1032. [CrossRef]
52. Accord, C. Trump decision on climate change 'major disappointment': United Nations. *Waste Water Manag. Aust.* **2017**, *44*, 35.
53. Jury, W.A.; Vaux, H. The role of science in solving the world's emerging water problems. *Proc. Natl. Acad. Sci. USA* **2005**, *102*, 15715–15720. [CrossRef]
54. Yigitcanlar, T. *Rethinking Sustainable Development: Urban Management, Engineering, and Design*; IGI Global: Hersey, PA, USA, 2010.
55. Wheeler, S.M. *Planning for Sustainability: Creating Livable, Equitable and Ecological Communities*; Routledge: New York, NY, USA, 2013.
56. Chen, G.; Li, X.; Liu, X.; Chen, Y.; Liang, X.; Leng, J.; Huang, K. Global projections of future urban land expansion under shared socioeconomic pathways. *Nat. Commun.* **2020**, *11*, 1–12. [CrossRef]
57. Metaxiotis, K.; Carrillo, J.; Yigitcanlar, T. *Knowledge-Based Development for Cities and Societies: Integrated Multi-Level Approaches*; IGI Global: Hersey, PA, USA, 2010.
58. Praharaj, S.; Han, J.H.; Hawken, S. Urban innovation through policy integration: Critical perspectives from 100 smart cities mission in India. *City Cult. Soc.* **2018**, *12*, 35–43. [CrossRef]
59. Yigitcanlar, T.; Dur, F. Making space and place for knowledge communities: Lessons for Australian practice. *Australas. J. Reg. Stud.* **2013**, *19*, 36–63.
60. Chu, E.K. The governance of climate change adaptation through urban policy experiments. *Environ. Policy Gov.* **2016**, *26*, 439–451. [CrossRef]
61. Trencher, G. Towards the smart city 2.0: Empirical evidence of using smartness as a tool for tackling social challenges. *Technol. Forecast. Soc. Chang.* **2019**, *142*, 117–128.
62. Angelidou, M. Smart cities: A conjuncture of four forces. *Cities* **2015**, *47*, 95–106. [CrossRef]
63. Cugurullo, F. The origin of the smart city imaginary: From the dawn of modernity to the eclipse of reason. In *The Routledge Companion to Urban Imaginaries*; Routledge: London, UK, 2018; pp. 113–124.
64. Desouza, K.; Hunter, M.; Jacop, B.; Yigitcanlar, T. Pathways to the making of prosperous smart cities: An exploratory study on the best practice. *J. Urban. Technol.* **2020**. [CrossRef]

65. Yigitcanlar, T. *Technology and the City: Systems, Applications and Implications*; Routledge: New York, NY, USA, 2016.
66. Yigitcanlar, T.; Inkinen, T. *Geographies of Disruption: Place Making for Innovation in the Age of Knowledge Economy*; Springer International Publishing: Cham, Switzerland, 2019.
67. Coletta, C.; Evans, L.; Heaphy, L.; Kitchin, R. *Creating Smart Cities*; Routledge: London, UK, 2019.
68. Karvonen, A.; Cugurullo, F.; Caprotti, F. *Inside Smart Cities: Place, Politics and Urban Innovation*; Routledge: London, UK, 2018.
69. Allam, Z.; Newman, P. Redefining the smart city: Culture, metabolism and governance. *Smart Cities* **2018**, *1*, 4–25. [CrossRef]
70. Cugurullo, F. Urban eco-modernisation and the policy context of new eco-city projects: Where Masdar City fails and why. *Urban. Stud.* **2016**, *53*, 2417–2433. [CrossRef]
71. Cugurullo, F. Exposing smart cities and eco-cities: Frankenstein urbanism and the sustainability challenges of the experimental city. *Environ. Plan. A* **2018**, *50*, 73–92. [CrossRef]
72. Kaika, M. Don't call me resilient again! The new urban agenda as immunology or what happens when communities refuse to be vaccinated with 'smart cities' and indicators. *Environ. Urban.* **2017**, *29*, 89–102. [CrossRef]
73. Perng, S.Y.; Kitchin, R.; Mac Donncha, D. Hackathons, entrepreneurial life and the making of smart cities. *Geoforum* **2018**, *97*, 189–197. [CrossRef]
74. Vanolo, A. Is there anybody out there? The place and role of citizens in tomorrow's smart cities. *Futures* **2016**, *82*, 26–36. [CrossRef]
75. Shelton, T.; Zook, M.; Wiig, A. The 'actually existing smart city'. *Camb. J. Reg. Econ. Soc.* **2015**, *8*, 13–25. [CrossRef]
76. Haarstad, H.; Wathne, M.W. Are smart city projects catalyzing urban energy sustainability? *Energy Policy* **2019**, *129*, 918–925. [CrossRef]
77. Machado, J.C.; Ribeiro, D.M.; da Silva, P.R.; Bazanini, R. Do Brazilian cities want to become smart or sustainable? *J. Clean. Prod.* **2018**, *199*, 214–221. [CrossRef]
78. Martin, C.J.; Evans, J.; Karvonen, A. Smart and sustainable? Five tensions in the visions and practices of the smart-sustainable city in Europe and North America. *Technol. Forecast. Soc. Chang.* **2018**, *133*, 269–278. [CrossRef]
79. Yigitcanlar, T.; Hoon, M.; Kamruzzaman, M.; Ioppolo, G.; Sabatini-Marques, J. The making of smart cities: Are Songdo, Masdar, Amsterdam, San Francisco and Brisbane the best we could build? *Land Use Policy* **2019**, *88*, 104187. [CrossRef]
80. Noori, N.; de Jong, M.; Janssen, M.; Schraven, D.; Hoppe, T. Input-output modeling for smart city development. *J. Urban. Technol.* **2020**. [CrossRef]
81. James, P. *Urban Sustainability in Theory and Practice: Circles of Sustainability*; Routledge: London, UK, 2014.
82. Elmqvist, T.; Andersson, E.; Frantzeskaki, N.; McPhearson, T.; Olsson, P.; Gaffney, O.; Takeuchi, K.; Folke, C. Sustainability and resilience for transformation in the urban century. *Nat. Sustain.* **2019**, *2*, 267–273. [CrossRef]
83. Robertson, M. *Sustainability Principles and Practice*; Routledge: London, UK, 2017.
84. Zhuravleva, N.A.; Nica, E.; Durana, P. Sustainable smart cities: Networked digital technologies, cognitive big data analytics, and information technology-driven economy. *Geopolit. Hist. Int. Relat.* **2019**, *11*, 41–47.
85. Chaurasia, V.K.; Yunus, A.; Singh, M. An overview of smart city: Observation, technologies, challenges and blockchain applications. In *Blockchain Technology for Smart Cities*; Springer: Singapore, 2020; pp. 133–154.
86. Ullah, Z.; Al-Turjman, F.; Mostarda, L.; Gagliardi, R. Applications of artificial intelligence and machine learning in smart cities. *Comput. Commun.* **2020**, *154*, 313–323. [CrossRef]
87. Yigitcanlar, T.; Kankanamge, N.; Vella, K. How are the smart city concepts and technologies perceived and utilized? A systematic geo-twitter analysis of smart cities in Australia. *J. Urban. Technol.* **2020**. [CrossRef]
88. Adly, A.S.; Adly, A.S.; Adly, M.S. Approaches based on artificial intelligence and the internet of intelligent things to prevent the spread of COVID-19: Scoping review. *J. Med. Internet Res.* **2020**, *22*, e19104. [CrossRef]
89. Mohamed, E. The relation of artificial intelligence with internet of things: A survey. *J. Cybersecur. Inf. Manag.* **2020**, *1*, 30–34.
90. Clifton, J.; Glasmeier, A.; Gray, M. When machines think for us: The consequences for work and place. *Camb. J. Reg. Econ. Soc.* **2020**, *13*, 3–23. [CrossRef]

91. Smith, T.R. Artificial intelligence and its applicability to geographical problem solving. *Prof. Geogr.* **1984**, *36*, 147–158. [CrossRef]
92. Russell, S.J.; Norvig, P. *Artificial Intelligence: A Modern Approach*; Pearson Education Limited: Harlow, UK, 2016.
93. Bach, J. When artificial intelligence becomes general enough to understand itself. Commentary on Pei Wang's paper "on defining artificial intelligence". *J. Artif. Gen. Intell.* **2020**, *11*, 15–18.
94. Girasa, R. AI as a disruptive technology. In *Artificial Intelligence as a Disruptive Technology*; Palgrave Macmillan: Cham, Switzerland, 2020; pp. 3–21.
95. Butler, L.; Yigitcanlar, T.; Paz, A. How can smart mobility innovations alleviate transportation disadvantage? Assembling a conceptual framework through a systematic review. *Appl. Sci.* **2020**, *10*, 6306. [CrossRef]
96. Hassani, H.; Silva, E.S.; Unger, S.; TajMazinani, M.; Mac Feely, S. Artificial intelligence (AI) or intelligence augmentation (IA): What is the future? *Artif. Intell.* **2020**, *1*, 143–155. [CrossRef]
97. Cugurullo, F.; Acheampong, R.A.; Gueriau, M.; Dusparic, I. The transition to autonomous cars, the redesign of cities and the future of urban sustainability. *Urban. Geogr.* **2020**. [CrossRef]
98. Cuzzolin, F.; Morelli, A.; Cîrstea, B.; Sahakian, B.J. Knowing me, knowing you: Theory of mind in AI. *Psychol. Med.* **2020**, *50*, 1057–1061. [CrossRef]
99. Gonzalez-Jimenez, H. Taking the fiction out of science fiction: (Self-aware) robots and what they mean for society, retailers and marketers. *Futures* **2018**, *98*, 49–56. [CrossRef]
100. Pueyo, S. Growth, degrowth, and the challenge of artificial superintelligence. *J. Clean. Prod.* **2018**, *197*, 1731–1736. [CrossRef]
101. Gurzadyan, G.A. *Theory of Interplanetary Flights*; CRC Press: New York, NY, USA, 1996.
102. Lovelock, J. *Novacene: The Coming Age of Hyperintelligence*; Allen Lane: London, UK, 2019.
103. Tegmark, M. *Life 3.0: Being Human in the Age of Artificial Intelligence*; Penguin: London, UK, 2017.
104. Vinuesa, R.; Azizpour, H.; Leite, I.; Balaam, M.; Dignum, V.; Domisch, S.; Nerini, F.F. The role of artificial intelligence in achieving the sustainable development goals. *Nat. Commun.* **2020**, *11*, 233. [CrossRef] [PubMed]
105. Corea, F. AI Knowledge Map: How to Classify AI Technologies. 2018. Available online: https://www.forbes.com/sites/cognitiveworld/2018/08/22/ai-knowledge-map-how-to-classify-aitechnologies/#5e99db627773 (accessed on 11 May 2020).
106. Faisal, A.; Yigitcanlar, T.; Kamruzzaman, M.; Currie, G. Understanding autonomous vehicles: A systematic literature review on capability, impact, planning and policy. *J. Transp. Land Use* **2019**, *12*, 45–72. [CrossRef]
107. Golbabaei, F.; Yigitcanlar, T.; Bunker, J. Shared autonomous vehicles in the context of smart urban mobility: A systematic review of the literature. *Int. J. Sustain. Transp.* **2020**. [CrossRef]
108. Narayanan, S.; Chaniotakis, E.; Antoniou, C. Shared autonomous vehicle services: A comprehensive review. *Transp. Res. Part. C* **2020**, *111*, 255–293. [CrossRef]
109. Schellin, H.; Oberley, T.; Patterson, K.; Kim, B.; Haring, K.S.; Tossell, C.C.; de Visser, E.J. Man's new best friend? Strengthening human-robot dog bonding by enhancing the doglikeness of Sony's Aibo. In Proceedings of the 2020 Systems and Information Engineering Design Symposium, Charlottesville, VA, USA, 24 April 2020; pp. 1–6.
110. Lakshmi, V.; Bahli, B. Understanding the robotization landscape transformation: A centering resonance analysis. *J. Innov. Knowl.* **2020**, *5*, 59–67. [CrossRef]
111. Suwa, S.; Tsujimura, M.; Kodate, N.; Donnelly, S.; Kitinoja, H.; Hallila, J.; Ishimaru, M. Exploring perceptions toward home-care robots for older people in Finland, Ireland, and Japan: A comparative questionnaire study. *Arch. Gerontol. Geriatr.* **2020**, *91*, 104178. [CrossRef] [PubMed]
112. Jaihar, J.; Lingayat, N.; Vijaybhai, P.S.; Venkatesh, G.; Upla, K.P. Smart home automation using machine learning algorithms. In Proceedings of the 2020 International Conference for Emerging Technology, Belgaum, India, 5–7 June 2020; pp. 1–4.
113. Brandtzaeg, P.B.; Følstad, A. Chatbots: Changing user needs and motivations. *Interactions* **2018**, *25*, 38–43. [CrossRef]
114. Aziz, K.; Haque, M.M.; Rahman, A.; Shamseldin, A.Y.; Shoaib, M. Flood estimation in ungauged catchments: Application of artificial intelligence-based methods for Eastern Australia. *Stoch. Environ. Res. Risk Assess.* **2017**, *31*, 1499–1514. [CrossRef]

115. Wearn, O.R.; Freeman, R.; Jacoby, D.M. Responsible AI for conservation. *Nat. Mach. Intell.* **2019**, *1*, 72–73. [CrossRef]
116. Kaplan, A.; Haenlein, M. Siri, Siri, in my hand: Who's the fairest in the land? On the interpretations, illustrations, and implications of artificial intelligence. *Bus. Horiz.* **2019**, *62*, 15–25. [CrossRef]
117. Wu, N.; Silva, E.A. Artificial intelligence solutions for urban land dynamics: A review. *J. Plan. Lit.* **2010**, *24*, 246–265.
118. El Morr, C.; Ali-Hassan, H. Descriptive, predictive, and prescriptive analytics. In *Analytics in Healthcare*; Springer: Cham, Switzerland, 2019; pp. 31–55.
119. Allam, Z.; Dhunny, Z.A. On big data, artificial intelligence and smart cities. *Cities* **2019**, *89*, 80–91. [CrossRef]
120. Engin, Z.; Treleaven, P. Algorithmic government: Automating public services and supporting civil servants in using data science technologies. *Comput. J.* **2019**, *62*, 448–460. [CrossRef]
121. Greenfield, A. *Radical Technologies: The Design of Everyday Life*; Verso Books: London, UK, 2018.
122. Lu, H.; Li, Y.; Chen, M.; Kim, H.; Serikawa, S. Brain intelligence: Go beyond artificial intelligence. *Mob. Netw. Appl.* **2018**, *23*, 368–375. [CrossRef]
123. Boenig-Liptsin, M. AI and robotics for the city: Imagining and transforming social infrastructure in San Francisco, Yokohama, and Lviv. *Field Actions Sci. Rep.* **2017**, *17*, 16–21.
124. Yigitcanlar, T.; Desouza, K.; Butler, L.; Roozkhosh, F. Contributions and risks of artificial intelligence (AI) in building smarter cities: Insights from a systematic review of the literature. *Energies* **2020**, *13*, 1473. [CrossRef]
125. Barnes, E.A.; Hurrell, J.W.; Ebert-Uphoff, I.; Anderson, C.; Anderson, D. Viewing forced climate patterns through an AI Lens. *Geophys. Res. Lett.* **2019**, *46*, 13389–13398. [CrossRef]
126. Huntingford, C.; Jeffers, E.S.; Bonsall, M.B.; Christensen, H.M.; Lees, T.; Yang, H. Machine learning and artificial intelligence to aid climate change research and preparedness. *Environ. Res. Lett.* **2019**, *14*, 124007. [CrossRef]
127. Jha, S.K.; Bilalovic, J.; Jha, A.; Patel, N.; Zhang, H. Renewable energy: Present research and future scope of Artificial Intelligence. *Renew. Sustain. Energy Rev.* **2017**, *77*, 297–317. [CrossRef]
128. Wang, P.; Yao, J.; Wang, G.; Hao, F.; Shrestha, S.; Xue, B.; Peng, Y. Exploring the application of artificial intelligence technology for identification of water pollution characteristics and tracing the source of water quality pollutants. *Sci. Total Environ.* **2019**, *693*, 133440. [CrossRef]
129. Lu, H.; Li, H.; Liu, T.; Fan, Y.; Yuan, Y.; Xie, M.; Qian, X. Simulating heavy metal concentrations in an aquatic environment using artificial intelligence models and physicochemical indexes. *Sci. Total Environ.* **2019**, *694*, 133591. [CrossRef]
130. Probst, W.N. How emerging data technologies can increase trust and transparency in fisheries. *J. Mar. Sci.* **2020**, *77*, 1286–1294. [CrossRef]
131. AlOmar, M.K.; Hameed, M.M.; AlSaadi, M.A. Multi hours ahead prediction of surface ozone gas concentration: Robust artificial intelligence approach. *Atmos. Pollut. Res.* **2020**, *11*, 1572–1587. [CrossRef]
132. Schürholz, D.; Kubler, S.; Zaslavsky, A. Artificial intelligence-enabled context-aware air quality prediction for smart cities. *J. Clean. Prod.* **2020**, *271*, 121941. [CrossRef]
133. Sun, W.; Bocchini, P.; Davison, B.D. Applications of artificial intelligence for disaster management. *Nat. Hazards* **2020**. [CrossRef]
134. Jahani, A.; Rayegani, B. Forest landscape visual quality evaluation using artificial intelligence techniques as a decision support system. *Stoch. Environ. Res. Risk Assess.* **2020**. [CrossRef]
135. Granata, F.; Gargano, R.; de Marinis, G. Artificial intelligence-based approaches to evaluate actual evapotranspiration in wetlands. *Sci. Total Environ.* **2020**, *703*, 135653. [CrossRef] [PubMed]
136. Santangeli, A.; Chen, Y.; Kluen, E.; Chirumamilla, R.; Tiainen, J.; Loehr, J. Integrating drone-borne thermal imaging with artificial intelligence to locate bird nests on agricultural land. *Sci. Rep.* **2020**, *10*, 1–8. [CrossRef]
137. Martínez-Santos, P.; Renard, P. Mapping groundwater potential through an ensemble of big data methods. *Groundwater* **2020**, *58*, 583–597. [CrossRef]
138. Singh, T.P.; Nandimath, P.; Kumbhar, V.; Das, S.; Barne, P. Drought risk assessment and prediction using artificial intelligence over the southern Maharashtra state of India. *Modeling Earth Syst. Environ.* **2020**. [CrossRef]
139. Tung, T.M.; Yaseen, Z.M. A survey on river water quality modelling using artificial intelligence models: 2000–2020. *J. Hydrol.* **2020**, *585*, 124670.

140. Pham, B.T.; Le, L.M.; Le, T.T.; Bui, K.T.; Le, V.M.; Ly, H.B.; Prakash, I. Development of advanced artificial intelligence models for daily rainfall prediction. *Atmos. Res.* **2020**, *237*, 104845. [CrossRef]
141. Ji, L.; Wang, Z.; Chen, M.; Fan, S.; Wang, Y.; Shen, Z. How much can AI techniques improve surface air temperature forecast? A report from AI Challenger 2018 Global Weather Forecast Contest. *J. Meteorol. Res.* **2019**, *33*, 989–992. [CrossRef]
142. Raza, M.; Awais, M.; Ali, K.; Aslam, N.; Paranthaman, V.V.; Imran, M.; Ali, F. Establishing effective communications in disaster affected areas and artificial intelligence-based detection using social media platform. *Future Gener. Comput. Syst.* **2020**, *112*, 1057–1069. [CrossRef]
143. Turchin, A.; Denkenberger, D. Classification of global catastrophic risks connected with artificial intelligence. *Ai Soc.* **2020**, *35*, 147–163. [CrossRef]
144. Yigitcanlar, T.; Butler, L.; Windle, E.; Desouza, K.; Mehmood, R.; Corchado, J. Can building 'artificially intelligent cities' protect humanity from natural disasters, pandemics and other catastrophes? An urban scholar's perspective. *Sensors* **2020**, *20*, 2988. [CrossRef] [PubMed]
145. Agrawal, A.; Gans, J.; Goldfarb, A. *Prediction Machines: The Simple Economics of Artificial Intelligence*; Harvard Business Press: Boston, MA, USA, 2018.
146. Li, B.H.; Hou, B.C.; Yu, W.T.; Lu, X.B.; Yang, C.W. Applications of artificial intelligence in intelligent manufacturing: A review. *Front. Inf. Technol. Electron. Eng.* **2017**, *18*, 86–96. [CrossRef]
147. Jarrahi, M.H. Artificial intelligence and the future of work: Human-AI symbiosis in organizational decision making. *Bus. Horiz.* **2018**, *61*, 577–586. [CrossRef]
148. Korinek, A.; Stiglitz, J.E. Artificial intelligence and its implications for income distribution and unemployment. *Natl. Bur. Econ. Res.* **2017**, *w24174*. [CrossRef]
149. Truby, J.; Brown, R.; Dahdal, A. Banking on AI: Mandating a proactive approach to AI regulation in the financial sector. *Law Financ. Mark. Rev.* **2020**, *14*, 110–120. [CrossRef]
150. Dauvergne, P. Is artificial intelligence greening global supply chains? Exposing the political economy of environmental costs. *Rev. Int. Political Econ.* **2020**. [CrossRef]
151. Chatterjee, S.; Bhattacharjee, K.K. Adoption of artificial intelligence in higher education: A quantitative analysis using structural equation modelling. *Educ. Inf. Technol.* **2020**. [CrossRef]
152. Kerasidou, A. Artificial intelligence and the ongoing need for empathy, compassion and trust in healthcare. *Bull. World Health Organ.* **2020**, *98*, 245. [CrossRef] [PubMed]
153. Yu, K.H.; Beam, A.L.; Kohane, I.S. Artificial intelligence in healthcare. *Nat. Biomed. Eng.* **2018**, *2*, 719–731. [CrossRef] [PubMed]
154. Hoffmann, A.L. Where fairness fails: Data, algorithms, and the limits of antidiscrimination discourse. *Inf. Commun. Soc.* **2019**, *22*, 900–915. [CrossRef]
155. Noble, S.U. *Algorithms of Oppression: How Search Engines Reinforce Racism*; New York University Press: New York, NY, USA, 2018.
156. O'Neil, C. *Weapons of Math Destruction: How Big Data Increases Inequality and Threatens Democracy*; Penguin: London, UK, 2016.
157. Bottarelli, L.; Bicego, M.; Blum, J.; Farinelli, A. Orienteering-based informative path planning for environmental monitoring. *Eng. Appl. Artif. Intell.* **2019**, *77*, 46–58. [CrossRef]
158. Guériau, M.; Cugurullo, F.; Acheampong, R.; Dusparic, I. Shared autonomous mobility-on-demand: Learning-based approach and its performance in the presence of traffic congestion. *IEEE Intell. Transp. Syst. Mag.* **2020**. [CrossRef]
159. Lu, J.; Feng, L.; Yang, J.; Hassan, M.M.; Alelaiwi, A.; Humar, I. Artificial agent: The fusion of artificial intelligence and a mobile agent for energy-efficient traffic control in wireless sensor networks. *Future Gener. Comput. Syst.* **2019**, *95*, 45–51. [CrossRef]
160. Brevini, B. Black boxes, not green: Mythologizing artificial intelligence and omitting the environment. *Big Data Soc.* **2020**, *7*, 2053951720935141. [CrossRef]
161. Hawkins, J.; Nurul Habib, K. Integrated models of land use and transportation for the autonomous vehicle revolution. *Transp. Rev.* **2019**, *39*, 66–83. [CrossRef]
162. Dauvergne, P. The globalization of artificial intelligence: Consequences for the politics of environmentalism. *Globalizations* **2020**. [CrossRef]
163. Zeadally, S.; Adi, E.; Baig, Z.; Khan, I.A. Harnessing artificial intelligence capabilities to improve cybersecurity. *Ieee Access* **2020**, *8*, 23817–23837. [CrossRef]

164. Zhang, J.; Hua, X.S.; Huang, J.; Shen, X.; Chen, J.; Zhou, Q. City brain: Practice of large-scale artificial intelligence in the real world. *Iet Smart Cities* **2019**, *1*, 28–37. [CrossRef]
165. Shneiderman, B. Human-centered artificial intelligence: Reliable, safe & trustworthy. *Int. J. Hum. Comput. Interact.* **2020**, *36*, 495–504.
166. Dignam, A. Artificial intelligence, tech corporate governance and the public interest regulatory response. *Camb. J. Reg. Econ. Soc.* **2020**, *13*, 37–54. [CrossRef]
167. Taddeo, M.; McCutcheon, T.; Floridi, L. Trusting artificial intelligence in cybersecurity is a double-edged sword. *Nat. Mach. Intell.* **2019**. [CrossRef]
168. Taeihagh, A.; Lim, H.S. Governing autonomous vehicles: Emerging responses for safety, liability, privacy, cybersecurity, and industry risks. *Transp. Rev.* **2019**, *39*, 103–128. [CrossRef]
169. Teoh, E.R. What's in a name? Drivers' perceptions of the use of five SAE Level 2 driving automation systems. *J. Saf. Res.* **2020**, *72*, 145–151. [CrossRef] [PubMed]
170. Arrieta, A.B.; Díaz-Rodríguez, N.; Del Ser, J.; Bennetot, A.; Tabik, S.; Barbado, A.; Chatila, R. Explainable artificial intelligence (XAI): Concepts, taxonomies, opportunities and challenges toward responsible AI. *Inf. Fusion* **2020**, *58*, 82–115. [CrossRef]
171. Burton, S.; Habli, I.; Lawton, T.; McDermid, J.; Morgan, P.; Porter, Z. Mind the gaps: Assuring the safety of autonomous systems from an engineering, ethical, and legal perspective. *Artif. Intell.* **2020**, *279*, 103201. [CrossRef]
172. Matthias, A. The responsibility gap: Ascribing responsibility for the actions of learning automata. *Ethics Inf. Technol.* **2004**, *6*, 175–183. [CrossRef]
173. Stilgoe, J. *Who's Driving Innovation? New Technologies and the Collaborative State*; Springer Nature: Berlin, Germany, 2019.
174. Yigitcanlar, T. Smart city policies revisited: Considerations for a truly smart and sustainable urbanism practice. *World Technopolis Rev.* **2018**, *7*, 97–112.
175. Yigitcanlar, T. Planning for smart urban ecosystems: Information technology applications for capacity building in environmental decision making. *Theor. Empir. Res. Urban. Manag.* **2009**, *4*, 5–21.
176. Leitheiser, S.; Follmann, A. The social innovation–(re) politicisation nexus: Unlocking the political in actually existing smart city campaigns? The case of SmartCity Cologne, Germany. *Urban. Stud.* **2020**, *57*, 894–915. [CrossRef]
177. Desouza, K. Governing in the Age of the Artificially Intelligent City. 2017. Available online: https://www.governing.com/commentary/col-governing-age-artificially-intelligent-city.html (accessed on 15 September 2020).
178. Makridakis, S. The forthcoming artificial intelligence (AI) revolution: Its impact on society and firms. *Futures* **2017**, *90*, 46–60. [CrossRef]
179. Batty, M. *Inventing Future Cities*; MIT Press: Cambridge, MA, USA, 2018.
180. Erskine, M. Artificial intelligence, the emerging needs for human factors engineering, risk management and stakeholder engagement. In Proceedings of the World Engineers Convention, Engineers Australia, Melbourne, Australia, 20–22 November 2019; pp. 9–10.
181. Loi, D.; Wolf, C.T.; Blomberg, J.L.; Arar, R.; Brereton, M. Co-designing AI futures: Integrating AI ethics, social computing, and design. In Proceedings of the 2019 on Designing Interactive Systems Conference, San Diego, CA, USA, 23–28 June 2019; pp. 381–384.
182. Ahmad, M.A.; Teredesai, A.; Eckert, C. Fairness, accountability, transparency in AI at scale: Lessons from national programs. In Proceedings of the 2020 Conference on Fairness, Accountability, and Transparency, Barcelona, Spain, 27–30 January 2020; pp. 690–699.
183. Chen, S.Y.; Kuo, H.Y.; Lee, C. Preparing society for automated vehicles: Perceptions of the importance and urgency of emerging issues of governance, regulations, and wider impacts. *Sustainability* **2020**, *12*, 7844. [CrossRef]
184. Larsson, S.; Heintz, F. Transparency in artificial intelligence. *Internet Policy Rev.* **2020**, *9*, 1–12. [CrossRef]
185. Kaker, S.A.; Evans, J.; Cugurullo, F.; Cook, M.; Petrova, S. Expanding cities: Living, planning and governing uncertainty. In *The Politics of Uncertainty*; Scoones, I., Stirling, A., Eds.; Routledge: London, UK, 2020; pp. 85–98.
186. Masanja, N.; Mkumbo, H. The application of open source artificial intelligence as an approach to frugal innovation in Tanzania. *Int. J. Res. Innov. Appl. Sci.* **2020**, *5*, 36–46.

187. Brock, J.K.; Von Wangenheim, F. Demystifying AI: What digital transformation leaders can teach you about realistic artificial intelligence. *Calif. Manag. Rev.* **2019**, *61*, 110–134. [CrossRef]
188. Allen, B.; Agarwal, S.; Kalpathy-Cramer, J.; Dreyer, K. Democratizing AI. *J. Am. Coll. Radiol.* **2019**, *16*, 961–963. [CrossRef]
189. Moreau, E.; Vogel, C.; Barry, M. A paradigm for democratizing artificial intelligence research. In *Innovations in Big Data Mining and Embedded Knowledge*; Springer: Cham, Switzerland, 2019; pp. 137–166.
190. Floridi, L. Establishing the rules for building trustworthy AI. *Nat. Mach. Intell.* **2019**, *1*, 261–262. [CrossRef]
191. Mittelstadt, B. Principles alone cannot guarantee ethical AI. *Nat. Mach. Intell.* **2019**, *1*, 501–507. [CrossRef]
192. Jobin, A.; Ienca, M.; Vayena, E. The global landscape of AI ethics guidelines. *Nat. Mach. Intell.* **2019**, *1*, 389–399. [CrossRef]
193. Hagendorff, T. The ethics of AI ethics: An evaluation of guidelines. *Minds Mach.* **2020**, *30*, 1–22. [CrossRef]
194. Awad, E.; Dsouza, S.; Kim, R.; Schulz, J.; Henrich, J.; Shariff, A.; Bonnefon, J.; Rahwan, I. The moral machine experiment. *Nature* **2018**, *563*, 59–64. [CrossRef] [PubMed]
195. Awad, E.; Dsouza, S.; Shariff, A.; Rahwan, I.; Bonnefon, J.F. Universals and variations in moral decisions made in 42 countries by 70,000 participants. *Proc. Natl. Acad. Sci. USA* **2020**, *117*, 2332–2337. [CrossRef] [PubMed]
196. Scherer, M.U. Regulating artificial intelligence systems: Risks, challenges, competencies, and strategies. *Harv. J. Law Technol.* **2015**, *29*, 353. [CrossRef]
197. Reed, C. How should we regulate artificial intelligence? *Philos. Trans. R. Soc. A* **2018**, *376*, 20170360. [CrossRef]
198. Cugurullo, F. Speed kills: Fast urbanism and endangered sustainability in the Masdar City project. In *Mega-Urbanization in the Global South: Fast Cities and New Urban Utopias of the Postcolonial State*; Datta, A., Shaban, A., Eds.; Routledge: London, UK, 2016; pp. 78–92.
199. Imrie, R.; Street, E. Regulating design: The practices of architecture, governance and control. *Urban. Stud.* **2009**, *46*, 2507–2518. [CrossRef]
200. Floridi, L.; Cowls, J.; King, T.C.; Taddeo, M. How to design AI for social good: Seven Essential factors. *Sci. Eng. Ethics* **2020**, *26*, 1771–1796. [CrossRef] [PubMed]
201. Tzimas, T. Artificial intelligence as global commons and the "international law supremacy" principle. In Proceedings of the 10th International RAIS Conference on Social Sciences and Humanities, Princeton, NJ, USA, 22–23 August 2018; pp. 83–88.
202. Rottz, M.; Sell, D.; Pacheco, R.; Yigitcanlar, T. Digital commons and citizen coproduction in smart cities: Assessment of Brazilian municipal e-government platforms. *Energies* **2019**, *12*, 2813. [CrossRef]
203. Cath, C.; Wachter, S.; Mittelstadt, B.; Taddeo, M.; Floridi, L. Artificial intelligence and the 'good society': The US, EU, and UK approach. *Sci. Eng. Ethics* **2018**, *24*, 505–528.
204. ITU News. Introducing 'AI Commons': A Framework for Collaboration to Achieve Global Impact. 2020. Available online: https://news.itu.int/introducing-ai-commons (accessed on 20 September 2020).
205. Kontokosta, C.E. Urban informatics in the science and practice of planning. *J. Plan. Educ. Res.* **2018**. [CrossRef]
206. Quan, S.J.; Park, J.; Economou, A.; Lee, S. Artificial intelligence-aided design: Smart design for sustainable city development. *Environ. Plan. B* **2019**, *46*, 1581–1599. [CrossRef]
207. Bundy, A. Preparing for the future of artificial intelligence. *Ai Soc.* **2017**, *32*, 285–287. [CrossRef]
208. Kirsch, D. Autopilot and algorithms: Accidents, errors, and the current need for human oversight. *J. Clin. Sleep Med.* **2020**. [CrossRef]
209. Dwivedi, Y.K.; Hughes, L.; Ismagilova, E.; Aarts, G.; Coombs, C.; Crick, T.; Galanos, V. Artificial intelligence (AI): Multidisciplinary perspectives on emerging challenges, opportunities, and agenda for research, practice and policy. *Int. J. Inf. Manag.* **2019**. [CrossRef]
210. Sohn, K.; Kwon, O. Technology acceptance theories and factors influencing artificial intelligence-based intelligent products. *Telemat. Inform.* **2020**, *47*, 101324. [CrossRef]
211. Donald, M. *Leading and Managing Change in the Age of Disruption and Artificial Intelligence*; Emerald Group Publishing: London, UK, 2019.
212. Musikanski, L.; Rakova, B.; Bradbury, J.; Phillips, R.; Manson, M. Artificial intelligence and community well-being: A proposal for an emerging area of research. *Int. J. Community Well-Being* **2020**, *3*, 39–55.

213. Mikhaylov, S.J.; Esteve, M.; Campion, A. Artificial intelligence for the public sector: Opportunities and challenges of cross-sector collaboration. *Philos. Trans. R. Soc. A* **2018**, *376*, 20170357. [CrossRef]
214. Sousa, W.G.; de Melo, E.R.; Bermejo, P.H.; Farias, R.A.; Gomes, A.O. How and where is artificial intelligence in the public sector going? A literature review and research agenda. *Gov. Inf. Q.* **2019**, *36*, 101392. [CrossRef]
215. Furman, J.; Seamans, R. AI and the economy. *Innov. Policy Econ.* **2019**, *19*, 161–191. [CrossRef]

Publisher's Note: MDPI stays neutral with regard to jurisdictional claims in published maps and institutional affiliations.

sustainability

MDPI

Article

Social Learning and Transdisciplinary Co-Production: A Social Practice Approach

Kimberley Slater [1,*] **and John Robinson** [2]

1 Department of Geography, University of Toronto, Toronto, ON M5S 3G3, Canada
2 Munk School of Global Affairs & Public Policy, University of Toronto, Toronto, ON M5S 3K7, Canada; johnb.robinson@utoronto.ca
* Correspondence: kim.slater@mail.utoronto.ca; Tel.: +1-604-698-7697

Received: 17 August 2020; Accepted: 3 September 2020; Published: 11 September 2020

Abstract: To address the challenge of achieving social learning in support of transformative change to sustainability, this paper develops an analytical framework that applies a social practice theory (SPT) lens to illuminate the constituent elements and dynamics of social learning in the context of transdisciplinary coproduction for sustainability transitions. Adopting an SPT approach affords a means of interpreting concrete practices at the local scale and exploring the potential for scaling them up. This framework is then applied to a real-world case at the urban neighbourhood scale in order to illustrate how social learning unfolded in a grassroots transdisciplinary coproduction process focused on climate action. We find that a social practice perspective illuminates the material and nonmaterial dimensions of the relationship between social learning and transdisciplinary coproduction. In decoupling these properties from individual human agency, the SPT perspective affords a means of tracing their emergence among social actors, generating a deeper understanding of how social learning arises and effects change, and sustainability can be reinforced.

Keywords: social learning; transdisciplinary coproduction; sustainability transitions

1. Introduction

The urgency and immensity of challenges like climate change and social inequality call for new ways of understanding the world and effecting change. Such "wicked problems" [1] are difficult to solve, as they are complex, contested and ambiguous with respect to their underlying values and causes [2] and display complex interdependencies with prevailing economic, technological and social systems. In confronting these societal challenges, transitions scholars advocate moving beyond incremental improvements, which have proven ineffectual, to find ways of achieving fundamental transitions or transformations in core systems in the direction of sustainability [3]. Such transitions entail "profound changes in dominant institutions, practices, technologies, policies, lifestyles and thinking" [4] (p. 6), at the heart of which are novel processes for knowledge production and social learning [5,6].

One such process is transdisciplinary (TD) coproduction, a knowledge production process in which individuals with different disciplinary, professional and experiential backgrounds combine academic and practice-based knowledges in the shared production, interpretation and ultimate use of scientific knowledge and its products [7–12]. These attributes of TD coproduction suggest optimal conditions for the social learning deemed important for sustainability transitions [2]. In the context of sustainability transitions, social learning generally refers to collective learning that generates collective responses to a shared dilemma or societal challenge. While such learning is clearly important, more work is needed to conceptualize social learning within TD coproduction and to better understand precisely how learning unfolds and knowledge is produced in such configurations. Indeed, the varied use of the term "social learning" across multiple disciplines and the consequent ambiguity surrounding its causes and effects has resulted in a notable lack of conceptual clarity surrounding the concept. This makes it difficult to assess whether social learning has occurred and, if so, what kind of learning has taken place, to what extent, between whom and how [13].

To address this challenge, this paper develops an analytical framework that applies a social practice theory (SPT) lens to illuminate the constituent elements and dynamics of social learning in the context of TD coproduction. Adopting an SPT approach affords a means of interpreting concrete practices at the local scale and exploring the potential for scaling them up. This framework is then applied to a real-world case in order to illustrate how social learning unfolded in a grassroots TD coproduction process. The process under study took place over 2018–2019 and brought together researchers from the University of Toronto, two funders (The Atmospheric Fund (TAF) and the City of Toronto) and 11 community practitioners who each represented a different climate intervention located in the Greater Toronto and Hamilton Area (GTHA). The aim of this effort was to codevelop an evaluation framework that would enable the assessment of their processes, their short- to medium-term outcomes and any longer or deeper sustainability impacts.

The aim of this paper is to bring coproduction processes for transition into conversation with social learning in order to clarify and yield a deeper understanding of both. Doing so sheds light on social learning's plural and dynamic nature in the context of TD coproduction efforts, in turn potentially strengthening the TD coproduction effort in the process. Ultimately, this forms the basis for a future research agenda that explores and demonstrates how social learning might be operationalized or leveraged in service of sustainability transitions. This paper is structured as follows: a review of the TD coproduction, social learning and social practice theory literatures, which underpin the social learning analytical framework, leads into a discussion of the analytical framework. This is followed by a description of the TD coproduction case at the centre of this study and how it is illustrative of the processes and outcomes of social learning referenced in the framework. The paper concludes with a discussion of the significance of conceptualizing social learning and TD coproduction in this way, particularly as a foundation for future research.

1.1. Conceptual Framework

1.1.1. TD Coproduction

TD coproduction of knowledge refers to the relocation of research activities out into communities, enabling researchers and societal actors to produce knowledge together by sharing in the joint framing of problems and goals as well as the management and ownership of research processes and related products [14,15]. In bringing together actors with diverse backgrounds, experiences and worldviews, TD coproduction can be extremely difficult to undertake in practice but rewarding if new ideas, understandings, trust and commitment emerge. Coproduction processes are particularly useful for addressing sustainability challenges, as they offer an avenue for engaging with sustainability's "essentially contested" nature [16,17], meaning that the specific meaning and interpretation of sustainability is far from universally agreed upon despite widespread acceptance of its importance. Thus, by its very nature, sustainability demands ongoing, place-based conversations informed by the unique beliefs, values, interests and many knowledges of the diverse collectives exploring its meaning and address. Such a view is articulated through the concept of "procedural sustainability", whereby sustainability is an emergent property of dialogue and negotiation that addresses the inherently normative and ethical question of how we should live and what choices we want to make given the best available scientific knowledge [18,19]. Such processes must be designed to make sense of the multiplicity of (potentially messy) perspectives while mediating the inevitable conflicts and engaging deeply with dimensions of power toward ultimately producing new understandings about sustainability issues and their possible solutions. Social learning provides one such mechanism for sense-making through dialogue, reflexivity and experimentation that may help to improve processes while opening up opportunities for collective action.

1.1.2. Social Learning

Social learning emerges when individuals and groups employ dialogue to collectively problem-solve; surface assumptions through reflexivity; and use experimentation, improvisation and adaptation in their initiation of novel approaches [2,20,21]. Such learning is believed to have substantive value, producing new shared knowledge and actions with the potential for adapting and responding to complex challenges [22,23]. It also holds normative value with learning, particularly that which is aimed at achieving an environmental or social goal, being an end in and of itself [24]. In addition to the presumed substantive and normative benefits of social learning, some scholars argue that social learning also provides instrumental value by enhancing trust, governance, social legitimacy, attitudinal and behavioural change, stakeholder empowerment and social networks [6,11] (p. 45), and by producing new identities, as well as institutions and individual capacities, that are more socially and ecologically robust [24].

Yet as Parson and Clark [25] (p. 429) argue, tremendous ambiguity surrounds the concept:

> The term social learning conceals great diversity. That many researchers describe the phenomena they are examining as "social learning" does not necessarily indicate a common theoretical perspective, disciplinary heritage, or even language. Rather, the contributions employ the language, concepts, and research methods of a half-dozen major disciplines; focus on individuals, groups, formal organizations, professional communities, or entire societies; and use divergent definitions of learning, of what it means for learning to be "social," and of theory.

For instance, scholars like Kilvington, Allen [26], and Fernandez-Gimenez et al. [27] conceptualize social learning as a deliberative process, one characterized by dialogue, negotiation and reflexivity between actors within social networks, typically in service of a pro-environmental goal (encapsulating both process and purpose). Others, like Reed et al. [28], place the emphasis on outcomes, arguing that three distinct criteria must be met for social learning to be obtained: (1) a change in understanding in the individuals involved must be demonstrated (i.e., learning outcomes), (2) this change must go beyond the individual and become situated within wider social units or communities of practice (i.e., network effects) and (3) this occurs through social interactions between actors within a social network (i.e., processes or conditions for social learning).

To further muddy the conceptual waters, the "social" in social learning, which refers to the social context that shapes and is shaped by learning [29–31], implies multiple settings for and influences on learning. It includes interpersonal settings though which individuals informally and collaboratively learn from one another as well as the culture in which they live and the groups with which they interact [31]. This challenges the determination of learning causes and effects.

1.1.3. Social Practice

Grounding social learning in the real-world contexts (e.g., TD coproduction efforts) in which it unfolds is one way of bringing clarity to the concept. While TD coproduction offers a potential, site and mechanism for social learning, such learning is not inevitable solely on the basis of the convening of some actors. Both material and nonmaterial elements play an important role in whether and how social learning may (or may not) transpire, and this is where a social practice lens is helpful for illuminating such components and their interactions. Through this lens, TD coproduction is seen as a practice comprised of enmeshed materials, skills and meanings [32]. Materials, for instance, objects, infrastructures, tools and the body itself, are tangible elements or entities utilized in the practice [32]. Skills, which are learned through doing and stored in the body and as mental routines, consist of know-how; competences; and ways of feeling, appreciating and doing as well as inherently shared notions of what is (and is not) good, normal, acceptable and appropriate [33]. Meanings (and images) are concepts, constructs or ideas that are shared socially and provide social and symbolic significance of participation in the practice at any one moment [32]. Meanings hinge on and inform norms, values and ideologies [34]. Collectively, all three practice elements shape one another as well as the contexts in which they are used [35,36].

Bathing, cooking and driving are examples of common practices that combine materials, skills and meanings and that have been given much attention in the social practices literature. Through this lens, the social is conceptualized as a dense and mutable fabric of entangled practices that perpetually transform as skills, materials and meanings change. Following this notion, society's spatial and temporal rhythms are tied to the emergence, diffusion, decline and disappearance of practices [37]. Conceptualizing TD coproduction efforts in this way, which surfaces their material and cultural dimensions, offers a system of identifying and interpreting social phenomena [38] that is particularly valuable in determining the manner in which social learning unfolds.

1.2. Analytical Framework

The analytical framework developed here (Figure 1) offers a coherent conceptualization of social learning in TD coproduction by applying a social practice lens. It is intended to act as a tool for illuminating the entangled dimensions of social learning and coproduction such that they may be enhanced and/or more fully leveraged in service of sustainability-oriented aims. The analytical framework envisions social learning as in a coevolutionary relationship with coproduction, which is theorized as a practice combining materials (e.g., stakeholder bodies, meeting spaces, documents and tools), meanings (e.g., values, norms and rules) and competences (e.g., skills and knowledges) [32]. These elements are acted upon—shaped and even reconstituted—by the dynamic force of social learning characterized by dialogue, reflexivity and innovation/experimentation. Following Sols, Wal and Wals [2], this produces emergent properties like trust, commitment and reframing, which in turn give rise to social learning outcomes of new knowledge, new actions and new relations [2]. Such learning outcomes may then be transferred by networks and taken up in wider social units [28], representing a potential contribution to sustainability transitions. As discussed in more detail below, this more nuanced and expansive view of social learning attends to its complex and plural nature while also integrating other theoretical resources to contribute to a deeper understanding of coproduction processes that shape and are shaped by it. The analytical framework depicts the elements of social learning acting through the "practice" of coproduction. A few (non-exhaustive) instances of the ways in which they interact and coproduce one another are provided in the ovals in Figure 1.

From left to right, Figure 1 shows social learning processes (e.g., reflexivity, dialogue and experimentation/innovation) mutually shaping and being shaped by a TD coproduction effort (comprised of materials, meanings and competences), resulting in emergent properties (e.g., commitment, trust and reframing) that both deepen the coproduction effort as well as lead to social learning outcomes (e.g., new knowledge, new actions and/or new relations). Networks help to spread and increase the uptake of learning outcomes in other social units (e.g., organizations, institutions and societies), ultimately resulting in sustainability impacts.

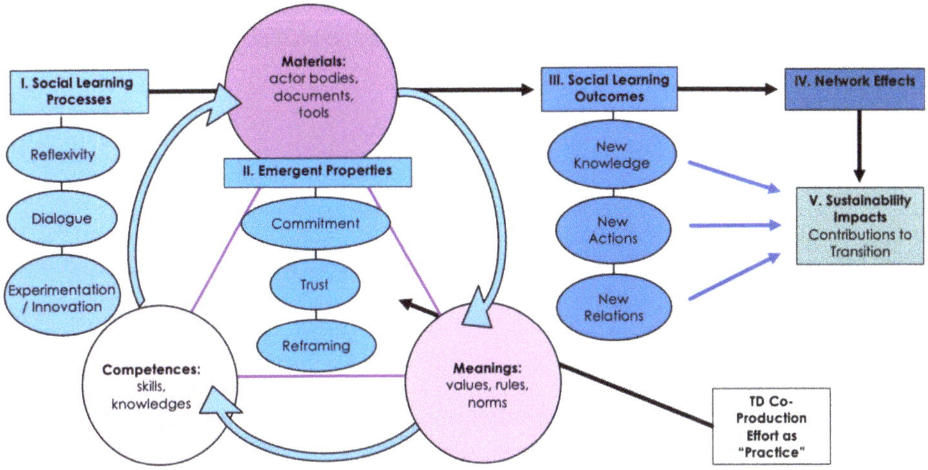

Figure 1. Social practice view of social learning in transdisciplinary (TD) coproduction for sustainability transitions.

1.2.1. Social Learning Processes in TD Coproduction

In the realm of sustainability transitions, social learning processes conceive of individual learning and interactive learning simultaneously as taking place "in a process of social change with effects on wider social-ecological systems" [28] (p. 2). In practice, this entails ongoing interaction, knowledge-sharing, deliberation, dialogue and problem-solving of diverse stakeholders in a trusting environment that is specifically directed at a resource management, governance or sustainability challenge in need of collective action [39,40]. Learning is prompted through iterative processes of dialogue, reflexivity and experimenting with solutions. In supplying the diversity of materials (actors and meeting spaces), meanings (beliefs and values) and competences (knowledges and skills), TD coproduction efforts go beyond offering a passive site at which social learning unfolds to actively supplying the essential ingredients for social learning to transpire. How such elements help to produce emergent properties that ultimately give way to social learning outcomes is described below.

1.2.2. Emergent Properties: Trust, Commitment and Reframing

One piece that often goes unaccounted for, and offers a rationale for the analytical framework presented here, is precisely how social learning processes beget social learning outcomes. In this regard, the Sol et al. [2] conceptualization is useful in that it sees them as being mediated by emergent properties, which they non-exhaustively identify as trust, commitment and reframing. Trust is a firm belief in the reliability, ability, strength or truth of someone or the expectation that others will act in an agreeable way without the need for intervening [41]. Commitment is evidenced by the degree to which participating actors (individuals and organizations) dedicate resources to achieving the goals of the project. Time, motivation and money are examples of resources. Reframing features the emergence of new, shared problem definitions regarding previously ill-defined issues faced by a relatively heterogeneous group [2]

Through an SPT stance, these dynamic components emerge from, and further shape, encounters between TD coproduction's practice dimensions of materials (stakeholders and participants, meeting spaces and tools), meanings (values, beliefs about the issue area and possible solutions) and competences (e.g., knowledges about the issue area, experience and skills for collaboration). In decoupling these properties from individual human agency, the SPT perspective offers a means of tracing their emergence between social units, generating a deeper understanding of how social learning arises and effects change. The outcomes of deepening trust and commitment and ultimately, reframing problems to produce innovative solutions are described below.

1.2.3. Social Learning Outcomes

Following Beers and van Mierlo [6], social learning outcomes are understood as resulting from the aforementioned learning processes, here described as created through TD coproduction efforts. These coalesce around (i) new knowledge (and knowledge products), (ii) new actions and/or behaviours and (iii) new relationships. Following Kuhn [42], Bloor [43], Latour and Woolgar [44] and Collins [45], knowledge here is understood as constituted by social practices, that is, socially constructed, and as such, is emergent, pluralistic, negotiated and coproduced through processes in which practice and knowledge are not separated [46]. New actions refer to collective approaches to shared challenges generated by the TD coproduction effort. Through an SPT lens, this entails changes in practices, which may be constituted as a pathway that unfolds over time across multiple settings and that is always situated within the evolution of broader social practices and institutions [47]. New relations are seen as new roles and identities between (new) actors within a TD coproduction effort as well as new relations that develop beyond such configurations. Through an SPT lens, relations also encompass the interactions between actions and events and resemble "a kind of chaotic network of habitual and non-habitual connections, always in flux, always reassembling in different ways" [48] (p. 19).

1.2.4. Network Effects

According to Reed et al. [28], the final criterion for social learning is that learning outcomes become dispersed via networks within wider social units, ostensibly beyond the original site of learning. As described in the organizational learning and communities of practice literatures, learning can take root in brains, bodies, routines, dialogue and symbols [49]. These literatures argue that it may be possible for social units to learn, whether they be institutions, organizations or communities of practice, as opposed to large numbers of individuals learning independently [13,50]. The aforementioned emergent properties of trust, commitment and reframing play important roles in forming and strengthening networks essential for transferring social learning outcomes to such social units. Over time, the learning that occurs across networks is thought to have the potential to transform complex situations [9,51].

1.2.5. Sustainability Impacts

Though discussed in much greater detail elsewhere [52], sustainability impacts in the context of sustainability interventions or climate actions are conceptualized as early markers of contributions to sustainability transitions. Such contributions include deep thinking or planning that connects the outcomes of the process/intervention to co-benefits in multiple sustainability domains' (e.g., health, justice, the economy, the environment, etc.) efforts to shift prevailing norms, values, rules or practices in the direction of sustainability and work on scaling the intervention out or deep.

1.3. Application of Analytical Framework

How this analytical framework might usefully reveal social learning's plural forms and even strengthen TD coproduction efforts is discussed in the context of a real-world case: a Toronto-based TD coproduction effort that took place over 2018–2019. This effort convened researchers from the University of Toronto, representatives of key funders (The Atmospheric Fund (TAF) and City of Toronto) and leaders of neighbourhood-scale interventions drawn from the region's climate action space, many of which carried out community engagements in their climate actions, campaigns and social change work. Their aim was to codevelop an evaluation framework that assesses the ongoing processes, short- to medium-term outcomes and longer and/or deeper sustainability impacts of small-scale sustainability-oriented interventions. This effort responded to the fact that much assessment of climate action and community engagement interventions looks primarily at process issues (e.g., numbers of participants), and to some extent, at direct and indirect outcomes (e.g., short- to medium-term results such as reports and changes in organizational or individual behaviour as well as new policies and programs), but rarely at sustainability impacts (e.g., deeper, long-term results that contribute to sustainability transitions). By looking at a full suite of results, framework users (e.g., funders like TAF as well as intervention leaders) get a better sense of the individual and collective impact such climate interventions are having as well as the progress being made on implementing TransformTO, the City of Toronto's climate action plan [53,54]. Such a framework also provides a critical tool for project managers, who are asked to perform evaluations of their projects but who typically have few resources and little expertise in doing so.

The result of this TD coproduction effort was a multipronged (assessing projects at different stages), light-touch (cost-effective and easy to use), utilization-focused [55] (the results are valuable to intervention leaders as well as just funders) evaluation framework that enables assessment of a broad suite of results, namely, the processes, outcomes and sustainability impacts, of neighbourhood- or smaller-scale climate change interventions (see Figure 2). The primary method of evaluation consisted of a self-evaluation questionnaire, which may have been supplemented with a document analysis and/or an interview with an intervention affiliate (someone who was aware of the intervention but was at arm's length to it). The results of these methods were probing questions, insights and recommendations customized to the evaluation needs of framework users. The framework also functions as a knowledge-sharing platform—a mechanism for generating and sharing lessons learned by projects and facilitating peer-to-peer coaching and learning—in support of a learning community comprised of people working throughout the Greater Toronto and Hamilton Area's climate action and community engagement spaces. Although evaluation was the explicit focus of this group, social learning was deeply embedded in both the process of developing the framework and the framework itself.

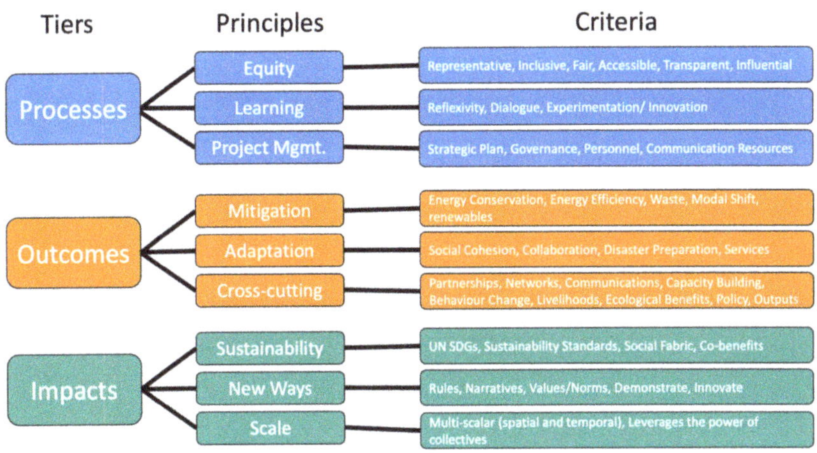

Figure 2. Evaluation framework for neighbourhood-scale climate interventions.

From left to right, Figure 2 displays three evaluative tiers (left column): the first (top row) assesses ongoing intervention processes (including external community engagement processes and internal project management processes), the second (middle row) assesses short- to medium-term outcomes of an intervention and the third (bottom row) assesses contributions to deep or long-term sustainability impacts. For each tier, principles have been identified (middle column)—drawn from the literature and from the input of coproducers—that climate action interventions might strive to realize along with assessment criteria for intervention leaders to self-evaluate the degree to which those principles are deemed important and apparent in their interventions, as evidenced by criteria (right column). Options for adding additional criteria and for stating that a criterion is not relevant are provided in the survey and interview instruments. Respondents' answers are checked and additional information is gleaned through a document analysis and interview with an arm's-length project affiliate.

2. Materials and Methods

Upon receipt of approval from the University of Toronto Research Ethics Board (Protocol 37210), the following methods were used to codevelop the evaluation framework and also form an illustrative case for applying the analytical framework described in this paper (all materials, data and protocols associated with the publication will be made available to readers upon request to the corresponding author):

(1) Review of defining documents (e.g., websites, funding proposals and interim/final reports) for 12 participating interventions (11 small-scale interventions and 1 funding program) whereby criteria in each of the evaluative tiers (processes, outcomes and impacts) helped to evaluate the design and implementation of interventions;
(2) In-depth semi-structured interviews (8 interviews conducted in March–May 2019) in which a combination of closed- and open-ended questions was posed to project leaders and their colleagues toward assessing their interventions; and
(3) Focus groups (2 Toronto-based workshops: 20 January 2019 and 6 June 2019, with approximately 15 individuals at each) in which facilitated dialogues helped to identify evaluation objectives, develop a corresponding evaluation framework and interpret the evaluation results. An unintended reimagining of the evaluation framework as a platform for a learning community also emerged.

Analysis

The process of data collection and analysis was iterative, meaning all three research instruments were administered, analyzed and refined at multiple stages through collective input from researchers and study participants. Data were coded manually using keywords from the analytical framework.

This case is illustrative in that it offers a basis for testing the analytical framework of social learning and transdisciplinary coproduction independent from the primary aims of the process (developing an evaluation framework). It reveals the entangled and emergent relationship between TD coproduction components (materials, meanings and competences) and social learning processes, which, as a result of their interactions (I), produce emergent properties (EP) and social learning outcomes (SLO).

3. Results

The results of applying the social learning analytical framework are categorized in the following three tables: materials, meanings and competences (See Tables 1–3). Within each, these codes are used: I—interactions between social learning processes and SPT elements in TD coproduction, EP—emergent property of coproduction and SLO—social learning outcome.

Table 1. Social learning analytical framework results—materials.

	Social Learning Processes	
Dialogue	**Reflexivity**	**Experimentation/Innovation**
Interactions (I): Dialogue shaped the creation of the evaluation tool. Tool affords opportunities for dialogue in evaluation of projects.	**I**: The meeting space that was used to convene the first meeting was not conducive to social learning (poor acoustics, distracting, cold, food order was missed—all presented barriers to focusing and, ultimately, learning). The facilitator also presented very different evaluation terminology, which created confusion. Attempts were made to modify the material conditions by the researcher, who recognized they were impeding learning (an alternative food order was placed, interceptions in the presentation were made to correct for misleading direction).	**I**: Open-ended question format in the evaluation tool re: sustainability was an experiment in determining respondents' preferred means of engaging with complex notions of sustainability. The results of this experimental approach drove multiple iterations of the evaluation tool and recommendations that multiple formats of the tool be offered to users and be connected (e.g., digital survey of individual forms basis for in-person conversation between colleagues).
Emergent Property (EP): Evaluation tool reframed by participants to operate as a networking and mentorship platform connecting a learning community.	**EP**: While the material conditions of this first meeting impeded the trust participants had in the coproduction effort and interfered with connections needed for social learning, it also (inadvertently) prompted reflexivity in the researcher leading the session to deeply examine what went wrong, including the (mis)direction given to the facilitator, which rendered a deeper understanding of the need for clarity in the proposed evaluation framework. The differences in evaluation perspectives prompted the questioning of assumptions about shared understandings and some slight reframing in the approach to ensure greater clarity moving forward (e.g., definitions and example indicators provided).	**EP**: Experimenting in this way opened up opportunities for offering multiple blended evaluative formats and interpretation components (in-person conversations, online surveys), a reframing of the original conception of the evaluation tool.
Social Learning Outcome (SLO): The learning community (if implemented) will generate a new set of relations characterized by co-mentorship. It will also enable the sharing of knowledge across a community of evaluation framework users to produce new individual and collective knowledges.	**SLO**: The disruptive space and confusing facilitation of the first meeting negated some learning outcomes (impinged on deep listening and dialogue) but ultimately generated some creative friction that resulted in new knowledge and actions to improve clarity of materials shared with participants.	**SLO**: Varied preferences for different formats led to the development of a workbook enabling users to self-select pieces of value and preferred depth (a set of new actions). Additionally, the notion of a learning community supporting mentorship between evaluation framework users reflects an innovative approach to evaluation and, if implemented, will entail new actions, relations and knowledges.

Table 2. Social learning analytical framework results—meanings.

Social Learning Processes		
Dialogue	**Reflexivity**	**Experimentation/Innovation**
I: Diverse meanings of different evaluation terms and evaluative indicators were uncovered through dialogue. Participants' learning expanded as a result and infused future dialogue with new terminology and richer set of meanings.	**I**: Feedback from participants revealed language in first iterations of evaluation framework was overly academic, which prompted exploration of positionality of researcher and corresponding adjustments to evaluation tool and engagement methods.	**I**: Experimentation with accessible terminology drove multiple iterations of the evaluation framework, which mutually shaped meanings in use in the group.
EP: Ongoing discursive dialogue built trust between participants and in the process itself. Also cultivated by researchers' commitment to continually modify terms to reflect participants' preferred terms and embed their values in evaluation framework.	**EP**: Incorporating feedback from participants built trust (between researcher and participants) and commitment to the participatory process.	**EP**: Experimentation of this kind made the process interesting and dynamic, increasing participant commitment.
SLO: New actions resulted from the dialogue and competences acting on one another and came in the form of new evaluative criteria and definitions.	**SLO**: As trust and commitment grew, opportunities for new relations (e.g., between participants and beyond with other potential evaluation framework users) and actions opened up (e.g., an enduring learning community or future collaborative work on a typology of actions).	**SLO**: The evaluation workbook that resulted from ongoing iteration of the tool supports a new evaluative approach (new actions).

Table 3. Social learning analytical framework results—competences.

Social Learning Processes		
Dialogue	**Reflexivity**	**Experimentation/Innovation**
I: Participants' knowledges (e.g., re: evaluation, strategies for achieving climate impact) were revealed and subsequently grew through dialogue. Researchers' competence with dialogue grew with practice over the course of the eight-month participatory process.	**I**: Some participants shared that they had limited capacity to evaluate their projects and strongly urged that an evaluation coach be brought on to support evaluative effort. This challenged assumptions that it was possible to create a light-touch evaluation tool.	**I**: Inclusion of participants' colleagues to build evaluation experience/capacity through practice with evaluation tool.
EP: In the course of dialogue, it became apparent that many of the participants had specific competences/knowledge sets in areas (e.g., behaviour change) that shaped the evaluation tool (showed up as explicit criteria). Grappling with the tension between individual agency triggered behaviour change, and social practice views of behaviour change contributed to the reframing of the evaluation tool to enable turning on or off sections that are not of relevance to the user.	**EP**: Engaging with the assumptions regarding users' evaluation capacity, commitment to improving user-friendliness of tool emerged (e.g., a decision tree was created to help users navigate the evaluation framework). Reframing the notion of "light-touch" also occurred whereby an evaluation coach was deemed to be the most user-friendly (though not most cost-effective) approach.	**EP**: Measuring and building capacity through practice with the evaluation tool highlighted the ways in which the evaluation approach could embody its own evaluative criteria, contributing to the reframing of the evaluation tool as one that promotes learning rather than measuring accountability. Providing this value back to framework cocreators also enhanced commitment to process.
SLO: Sharing knowledge via dialogue and embedding that knowledge into the evaluation framework built participant knowledge of evaluation and recommendations for new actions regarding format and functionality of the evaluation tool.	**SLO**: The result was new knowledge (re: capacity limitations) and recommendations to the funder to consider hiring an evaluation coach (a new action).	**SLO**: This contributed to new analytical approaches (actions and knowledges) exploring embodiment.

3.1. TD Coproduction Component—Materials

Examples of materials include participant bodies, tools and meeting implements.

3.2. TD Coproduction Component—Meanings

Examples of meanings include language, values and norms.

3.3. TD Coproduction Component—Competences

Examples of competences include skills, knowledges and capacities.

3.4. Interpretation

This analysis illustrates how various forms of social learning (processes and outcomes) unfolded within a TD coproduction effort, as framed by SPT, which is consistent with social learning's plural nature. Further, it highlights the coevolutionary relationship social learning has with TD coproduction, whereby they continually respond to and are changed by the other. It is too soon to glean network effects or fully realized contributions to sustainability transitions, but there are some early indications of both, which are summarized below.

3.4.1. Network Effects

The main output or knowledge product of this TD coproduction effort was an evaluation framework and supporting guidebook, which is now being utilized by a home retrofit project called BetterHomesTO, led by the City of Toronto. It may be further taken up by TAF and more broadly by other funders and projects in need of evaluation support (it is currently under consideration by several entities). This would represent a network effect whereby the embodied learning in these materials might shift organizations' practices, including adjustments to funding programs, project design and approaches for monitoring and evaluating to better reflect the assessment criteria. Correspondingly, the evaluation tool was designed to capture users' feedback on the tool itself and continually iterate and improve through use to embody network preferences.

Additionally, out of this coproduction effort emerged a desire for a learning community comprised of evaluation framework users who share their knowledge and mentor one another in support of their respective climate action and sustainability goals. If this community materializes, it will represent a strong network effect: a learning collective stimulating learning at and transmitting learning across multiple learning sites.

3.4.2. Contributions to Sustainability Transitions

The multipronged, light-touch, utilization-focused evaluation tool embeds the perspectives of participants regarding notions of sustainability transitions. A key part of this perspective entails a social practice orientation that prompts evaluators to look at the ways in which an intervention shifts practices (e.g., norms, values, rules and materials). Scalability (out/across/deep) and systems approaches that connect across sustainability domains and scales were other evaluative areas in the impacts tier. Surprisingly, despite the small scale of the interventions in the cohort who codeveloped the evaluation framework, the interviewees expressed deep interest in this most abstract and long-term level of evaluation, and this was, to some extent, reflected in their plans and actions. Indeed, the desire to form and join a learning community reflected a desire to magnify and deepen sustainability impacts. Following the analytical framework developed here, which calls attention to the need for social learning processes that feature dialogue, reflexivity and innovation as well as coproduction efforts reflecting as broad a diversity of views, knowledges and experience as possible, the effort to form a learning community might be enhanced by embedding these elements.

4. Discussion

This case study illustrates some of the ways that plural forms of social learning (processes and outcomes) can unfold through a TD coproduction effort, produced through interactions with social practice elements like diverse materials, meanings and competences. Such elements supply the raw resources for sense-making, challenging assumptions and producing new collective understandings and are simultaneously changed by the social learning processes of dialogue, reflexivity and experimentation; learning is taken up in human bodies (manifest in new practices), learning shifts meanings and learning foments new skills. Out of the mutual dynamism of this relationship comes the emergent properties of trust, commitment and reframing, which in turn have the potential to produce social learning outcomes like new relations, knowledges or actions, which may be taken up in broader social units via network effects. By creating discursive space for the exploration and deep engagement with contested notions of sustainability, entwined social learning and TD coproduction efforts operate as a powerful means of responding to sustainability challenges and conceiving of new pathways for transition.

Understanding that TD coproduction and social learning are in a tightly bound coevolutionary relationship offers two important insights. The first is that investing in social learning approaches, such as hiring skilled facilitators, allowing sufficient time for dialogue, thoughtfully prompting reflexivity and creating opportunities for experimenting with new ideas and learning by doing, not only supports social learning but also improves the participatory experience for TD coproduction participants. It does so by carving out space for them to feel heard, to more meaningfully relate to one another, to learn and grow and to see their ideas come to life. In the case study discussed here, the continual dialogue between coproduction participants regarding their respective evaluation capacities, needs and preferences fed an ongoing cycle of iteration (experimentation and innovation) in both the content (evaluation principles and criteria) and format (question type, online or in-person) of the evaluation framework. The group provided critical feedback on each framework iteration and prompted researchers to be reflexive, particularly in terms of using language that reflected practitioner, rather than academic, preferences. Over the course of the TD coproduction effort, reflexivity drove an exploration of how the evaluation framework might go beyond measuring a set of evaluation principles and criteria embodying them. For instance, efforts were made to ensure the framework not only measured accessibility in interventions being evaluated but was itself accessible by being cost-free and offered in multiple formats. While learning was occasionally frustrated during this TD coproduction effort, a result of deficiencies in facilitation and an initially inhospitable meeting space, these challenges were, for the most part, overcome by coproduction participants seeing their contributions reflected in the evaluation framework. This helped to grow trust between participants and commitment to the effort, which allowed for the reframing that saw this framework reimagined as a platform for a learning community of co-mentors. In short, social learning approaches deepened the degree of participation in the TD coproduction effort, and that participation in turn deepened the learning processes and outcomes that occurred.

The other key insight that emerges out of this analytical framework is the value, from a social learning perspective, of first of all, diversifying the materials, meanings and competences constituting the TD coproduction effort and secondly, mediating the inevitable conflicts and challenges that arise with such diversity. One method for ensuring efforts are rich with diverse individuals, ideas and values and are supported in engaging deeply with the differences such diversity brings is through the evaluation framework developed in the case described here. This framework offers a set of design considerations for new interventions and guideposts for existing interventions to steer in the direction in service of the greater climate or sustainability aims driving the effort. By embedding social learning approaches like dialogue and reflexivity into the assessment criteria of the evaluation framework and experimenting with innovative approaches for mentorship and sharing of learning across a community of framework users, this approach to evaluation helps to operationalize social learning. Indeed, the potential for a community of practice [56] to enable its members to share and learn through a series of interactions, "thus reflecting the social nature of human learning" [57], has generated much enthusiasm for this prospect amongst project participants, researchers and others in the region's climate action and social innovation space. More work is, however, needed to ensure fruitful germination of this idea and directs future research as discussed below.

In the case study detailed here, the group of coproduction participants held very different perspectives on evaluation and related terminology as well as had varying competences with evaluation (a result of participants' varying experiences in different sectors or settings, including municipal, nonprofit and academic). This generated some confusion and frustration amongst the group early on. Yet this tension represented a creative challenge rife with learning potential that ultimately spurred efforts to not only clarify language but to design the framework with enough flexibility to accommodate vastly different evaluation needs and competences. The end result is an evaluation tool that takes multiple material forms (as a relational learning network and in-person and online survey) and connects evaluation users with different competences, ideas and values in order to enrich learning, both in the context of their individual interventions and across an entire region's climate action space.

By embedding the SPT perspective and approaches articulated in the social learning analytical framework into the evaluation framework, both in terms of the criteria used to assess interventions and in the evaluation experience itself, a deeper understanding of social learning and social practice theory—gleaned in part by a learning by doing experience—might be imparted in evaluation users and help their respective interventions. Evaluation of this kind could become part of the design considerations for new small-scale sustainability interventions as well as offer important guideposts for existing interventions, potentially helping them learn and better interpret their practices such that they may eventually scale their efforts or better realize their climate or sustainability aims.

Future research should also carefully consider how this approach to evaluation might be implemented in the absence of researchers and within different funding parameters. The cocreation of the tool was resource-intensive, and its uptake by funders and organizations will require commitment and resources. However, mentorship across a learning community and open-source technologies might help to overcome some of the cost burden to ensure this kind of evaluation is accessible and continually improves.

5. Conclusions

This paper developed an analytical framework for clarifying the notion of social learning, which reveals its plural forms and teases out its coevolutionary relationship with TD coproduction efforts for sustainability transitions. A social practice perspective illuminates the material and nonmaterial dimensions of this relationship. In decoupling these properties from individual human agency, the SPT perspective affords a means of tracing their emergence in different groups and social contexts, generating a deeper understanding of how social learning arises and effects change.

This conceptualization of social learning was explored in the context of a real-world case—a Toronto-based TD coproduction effort that convened leaders of small-scale interventions working in the region's climate action space as well as key funders of their efforts and researchers. Applying the analytical framework to this process demonstrated that the diverse array of practice elements afforded by the TD coproduction effort helped to both spur and embody social learning while simultaneously being shaped and reconstituted by social learning.

Building on one of the core insights of this analytical framework—the need to operationalize social learning and the potential to do so through a novel evaluation approach—informs a future research agenda investigating how this might best be undertaken in the context of neighbourhood-scale interventions striving to realize bold sustainability aims.

Author Contributions: Conceptualization, K.S. and J.R.; data curation, K.S.; formal analysis, K.S. and J.R.; funding acquisition, J.R. and K.S.; investigation, K.S. and J.R.; methodology, J.R. and K.S.; project administration, K.S.; resources, J.R.; software, K.S.; supervision, J.R.; validation, J.R.; visualization, K.S. and J.R.; writing—original draft, K.S.; writing—review and editing, J.R. and K.S. Both authors have read and agreed to the published version of the manuscript. All authors have read and agreed to the published version of the manuscript.

Funding: This research was funded by The Atmospheric Fund (TAF), grant number 506108, and the University of Toronto by way of the corresponding author's PhD funding package.

Acknowledgments: The authors wish to gratefully acknowledge funding for this research project from The Atmospheric Fund and the University of Toronto. As a community-engaged research project employing a coproduced evaluation approach, we are grateful to our project partners who generously contributed their time, knowledge and insights to this project. Finally, we wish to thank members of our research group, namely, Grégoire Benzakin and Pani Pajouhesh at the University of Toronto, who were instrumental in developing the framework used in this paper.

Conflicts of Interest: In accordance with the principles of coproduction, all evaluation framework users, including the funder, contributed to the design of the study, the creation of the framework and the interpretation of results.

References

1. Rittel, H.W.J.; Webber, M.M. Dilemmas in a general theory of planning. *Policy Sci.* **1973**, *4*, 155–169. [CrossRef]
2. Sol, J.; Van Der Wal, M.M.; Beers, P.J.; Wals, A. Reframing the future: The role of reflexivity in governance networks in sustainability transitions. *Environ. Educ. Res.* **2018**, *24*, 1383–1405. [CrossRef]
3. Markard, J.; Raven, R.; Truffer, B. Sustainability transitions: An emerging field of research and its prospects. *Res. Policy* **2012**, *41*, 955–967. [CrossRef]
4. European Environmental Agency (EEA). Perspectives on Transitions to Sustainability. 2018. Available online: https://www.eea.europa.eu/publications/perspectives-on-transitions-to-sustainability/file (accessed on 6 November 2019).
5. Kemp, R.; Loorbach, D.; Rotmans, J. Transition management as a model for managing processes of co-evolution towards sustainable development. *Int. J. Sustain. Dev. World Ecol.* **2009**, *14*, 78–91. [CrossRef]
6. Beers, P.J.; Mierlo, B.; Hoes, A.C. Toward an integrative perspective on social learning in system innovation initiatives. *Ecol. Soc.* **2016**, *21*, 33. [CrossRef]
7. Caswell, C.; Shove, E. Introducing interactive social science. *Sci. Public Policy* **2000**, *27*, 154–157. [CrossRef]
8. Kates, W.; Thomas, R.; Robert, K.W.; Parris, T.M.; Leiserowitz, A. What is sustainable development? Goals, indicators, values, and practice. *Environ. Sci. Policy Sustain. Dev.* **2005**, *47*, 8–21. [CrossRef]
9. Wiek, A.; Talwar, S.; Robinson, J.; O'Shea, M. Toward a methodological scheme for capturing societal effects of participatory sustainability research. *Res. Eval.* **2014**, *23*, 117–132. [CrossRef]
10. Robinson, J.; Tansey, J. Co-production, emergent properties and strong interactive social research: The Georgia Basin futures project. *Sci. Public Policy* **2006**, *33*, 151–160. [CrossRef]
11. Popa, S.; Soto-Acosta, P.; Martinez-Conesa, I. Antecedents, moderators, and outcomes of innovation climate and open innovation: An empirical study in SMEs. *Technol. Forecast. Soc. Change* **2017**, *118*, 134–142. [CrossRef]
12. Hegger, D.; Lamers, M.; Van Zeijl-Rozema, A.; Dieperink, C. Conceptualising joint knowledge production in regional climate change adaptation projects: Success conditions and levers for action. *Environ. Sci. Policy* **2012**, *18*, 52–65. [CrossRef]
13. Armitage, D.; Marschke, M.; Plummer, R. Adaptive co-management and the paradox of learning. *Glob. Environ. Change* **2008**, *18*, 86–98. [CrossRef]
14. Polk, M. Transdisciplinary co-production: Designing and testing a transdisciplinary research framework for societal problem solving. *Futures* **2015**, *65*, 110–122. [CrossRef]
15. Schneider, F.; Giger, M.; Harari, N.; Moser, S.; Oberlack, C.; Providoli, I.; Schmid, L.; Tribaldos, T.; Zimmermann, A. Transdisciplinary co-production of knowledge and sustainability transformations: Three generic mechanisms of impact generation. *Environ. Sci. Policy* **2019**, *102*, 26–35. [CrossRef]
16. Robinson, J. Squaring the circle? Some thoughts on the idea of sustainable development. *Ecol. Econ.* **2004**, *48*, 369–384. [CrossRef]
17. Connelly, S. Mapping sustainable development as a contested concept. *Local Environ.* **2007**, *12*, 259–278. [CrossRef]
18. Robinson, J.; Cole, R.J. Theoretical underpinnings of regenerative sustainability. *Build. Res. Inf.* **2015**, *43*, 133–143. [CrossRef]
19. Maggs, D.; Robinson, J. International association for environmental philosophy recalibrating the anthropocene. *Environ. Philos.* **2016**, *13*, 175–194. [CrossRef]

20. Sol, J.; Beers, P.J.; Wals, A. Social learning in regional innovation networks: Trust, commitment and reframing as emergent properties of interaction. *J. Clean. Prod.* **2013**, *49*, 35–43. [CrossRef]
21. Wals, A.; Rodela, R. Social learning towards sustainability: Problematic, perspectives and promise. *NJAS Wagening. J. Life Sci.* **2014**, *69*, 1–3. [CrossRef]
22. Wells, J. *Complexity and Sustainability*; Routledge: Oxon, UK, 2012; pp. 1–354.
23. Folke, C.; Colding, J.; Berkes, F. Synthesis: Building resilience and adaptive capacity in social-ecological systems. In *Navigating Social-Ecological Systems: Building Resilience for Complexity and Change*; Berkes, F., Colding, J., Folke, C., Eds.; Cambridge University Press: Cambridge, UK, 2003; pp. 352–387. [CrossRef]
24. Pahl-Wostl, C. The importance of social learning in restoring the multifunctionality of rivers and floodplains. *Ecol. Soc.* **2006**, *11*, 1–10. [CrossRef]
25. Parson, E.A.; Clark, W.C. Sustainable development as social learning: Theoretical perspectives and practical challenges for the design of a research program. In *Barriers and Bridges to the Renewal of Ecosystems and Institutions*; Gunderson, L.H., Holling, C.S., Light, S.S., Eds.; Columbia University Press: New York, NY, USA, 1995; pp. 428–459.
26. Kilvington, M.J.; Allen, W. Social Learning: A basis for practice in environmental management. *Landcare Res. Manaaki Whenua* **2010**, 215–225. [CrossRef]
27. Fernandez-Gimenez, M.E.; Ballard, H.L.; Sturtevant, V.E. Adaptive management and social learning in collaborative and community-based monitoring: A study of five community-based forestry organizations in the western USA. *Ecol. Soc.* **2008**, *13*, 4. [CrossRef]
28. Reed, M.S.; Evely, A.C.; Cundill, G.; Fazey, I.; Glass, J.; Laing, A.; Newig, J.; Parrish, B.; Prell, C.; Raymond, C.; et al. What is social learning? *Ecol. Soc.* **2010**, *15*, r01. [CrossRef]
29. Bandura, A. *Social Learning Theory*; Prentice Hall: Englewood Cliffs, NJ, USA, 1977; p. 256.
30. Bandura, A. *Social Foundations of Thought and Action*; Prentice Hall: Englewood Cliffs, NJ, USA, 1986; p. 617.
31. Vygotsky, L. *Mind in Society: The Development of Higher Psychological Processes*; Michael, C., John-Steiner, V., Scribner, S., Souberman, E., Eds.; Harvard University Press: Boston, MA, USA, 1980; p. 176.
32. Shove, E.; Pantzar, M.; Watson, M. The dynamics of social practice: Everyday life and how it changes. *Nord. J. Sci. Technol. Stud.* **2013**, *1*, 41–43. [CrossRef]
33. Kuijer, L. Implications of Social Practice Theory for Sustainable Design. Ph.D. Thesis, Delft University of Technology, Delft, The Netherlands, 7 February 2014.
34. Shove, E.A.; Pantzar, M. Consumers, producers and practices: Understanding the invention and reinvention of Nordic Walking. *J. Consum. Cult.* **2005**, *5*, 43–64. [CrossRef]
35. Ingram, H.; Schneider, A.; Deleon, P. Social construction and policy design. In *Theories of the Policy Process*, 2nd ed.; Sabatier, P., Ed.; Routledge: New York, NY, USA, 2007; p. 34. [CrossRef]
36. Strengers, Y.; Moloney, S.; Maller, C.; Horne, R. Beyond behaviour change: Practical applications of social practice theory in behaviour change programmes. In *Social Practices, Interventions and Sustainability.*; Strengers, Y., Maller, C., Eds.; Routledge: London, UK, 2015; p. 208.
37. Caletrío, J. Article Review. The hypothesis of the mobility transition. *Mob. Lives Forum* **2015**, *61*, 219–249.
38. Reckwitz, A. Toward a theory of social practices: A development in culturalist theorizing. *Eur. J. Soc. Theory* **2002**, *5*, 243–263. [CrossRef]
39. Lee, E.; Krasny, M.E. The role of social learning for social-ecological systems in Korean village groves restoration. *Ecol. Soc.* **2015**, *20*, 1–42. [CrossRef]
40. Cundill, G.; Rodela, R. A review of assertions about the processes and outcomes of social learning in natural resource management. *J. Environ. Manag.* **2012**, *113*, 7–14. [CrossRef]
41. Peeman, T. *I Trust U. Managing with Trust*; Pearson Education Benelux: Amsterdam, The Netherlands, 2009.
42. Kuhn, T. *The Structure of Scientific Revolutions*; University of Chicago Press: Chicago, IL, USA, 1962; p. 210.
43. Bloor, D. *Knowledge and Social Imagery*; University of Chicago Press: Chicago, IL, USA, 1976; p. 211.
44. Latour, B.; Woolgar, S. *Laboratory Life: The Social Construction of Scientific Facts*; Sage Publication, Inc.: Beverly Hills, CA, USA, 1979; p. 291.
45. Collins, J. Some problems and purposes of narrative analysis in educational research. *J. Educ.* **1985**, *167*, 57–70. [CrossRef]
46. Salter, K.; Kothari, A.R. Knowledge 'Translation' as social learning: Negotiating the uptake of research-based knowledge in practice. *BMC Med. Educ.* **2016**, *16*, 76. [CrossRef] [PubMed]

47. Lave, J.; Wenger, E. *Situated Learning: Legitimate Peripheral Participation*; Cambridge University Press: Cambridge, UK, 1990; p. 138.
48. Potts, S.D.; Matuszewski, I.L. Ethics and corporate governance. *Corp. Gov. Int. Rev.* **2014**, *12*, 177–179. [CrossRef]
49. Blackler, F. Knowledge, knowledge work and organizations: An overview and interpretation. *Organ. Stud.* **1995**, *16*, 1021–1046. [CrossRef]
50. Wals, A.E.J. Learning in a changing world and changing in a learning world: Reflexively fumbling towards sustainability. *S. Afr. J. Environ. Educ.* **2007**, *24*, 35–45.
51. Ison, R.; Blackmore, C.; Iaquinto, B.L. Towards systemic and adaptive governance: Exploring the revealing and concealing aspects of contemporary social-learning metaphors. *Ecol. Econ.* **2013**, *87*, 34–42. [CrossRef]
52. Williams, S.; Robinson, J.B. Measuring sustainability: An evaluation framework for sustainability transition experiments. *Environ. Sci. Policy* **2020**, *103*, 58–66. [CrossRef]
53. City of Toronto. TransformTO: Climate Action for a Healthy Equitable, and Prosperous Toronto. 2016. Available online: https://www.toronto.ca/services-payments/water-environment/environmentally-friendly-city-initiatives/transformto/transformto-climate-action-strategy/ (accessed on 8 December 2016).
54. City of Toronto. Transform to: Climate Action for a Healthy Equitable, and Prosperous Toronto Community Conversation Guide. 2019. Available online: https://www.toronto.ca/wp-content/uploads/2019/10/974b-TTO-Conversation-Guide-V3.pdf (accessed on 4 November 2019).
55. Patton, M.Q. *Utilization-Focused Evaluation*, 4th ed.; Sage Publications Inc.: Thousand Oaks, CA, USA, 2008; p. 637.
56. Wenger, E. *Communities of Practice*; Cambridge University Press: Cambridge, UK, 1998; pp. 77–78.
57. Macfarlane, A. Community of Practice. Better Evaluation. Available online: https://www.betterevaluation.org/en/evaluation-option/community_of_practice (accessed on 5 May 2020).

sustainability MDPI

Article

In-Between 'Smart' Urban Growth and 'Sluggish' Rural Development? Reframing Population Dynamics in Greece, 1940–2019

Rosanna Salvia [1], Gianluca Egidi [2], Luca Salvati [3,*], Jesús Rodrigo-Comino [4,5] and Giovanni Quaranta [1]

1 Department of Mathematics, Computer Science and Economics Department, University of Basilicata, Viale dell'Ateneo Lucano, I-85100 Potenza, Italy; rosanna.salvia@unibas.it (R.S.); giovanni.quaranta@unibas.it (G.Q.)
2 Department of Agricultural and Forestry Sciences (DAFNE), Tuscia University, Via San Camillo de Lellis, I-01100 Viterbo, Italy; egidi.gianluca@unitus.it
3 Department of Economics and Law, University of Macerata, Via Armaroli 43, I-62100 Macerata, Italy
4 Soil Erosion and Degradation Research Group, Department of Geography, University of Valencia, 46010 Valencia, Spain; jesus.rodrigo@uv.es
5 Department of Physical Geography, Trier University, 54296 Trier, Germany
* Correspondence: luca.salvati@unimc.it; Tel.: +39-06-615-710; Fax: +39-06-61-57-10-36

Received: 8 July 2020; Accepted: 28 July 2020; Published: 31 July 2020

Abstract: Multifaceted demographic dynamics have shaped population growth in Mediterranean Europe, reflecting a metropolitan cycle from urbanization to re-urbanization. To assess the distinctive impact of economic downturns on population dynamics, the present study illustrates the results of an exploratory analysis that assesses urban expansion and rural decline at various temporal scales in Greece, a peripheral country in southeastern Europe. Statistical analysis based on multivariate exploratory techniques outlined the persistent increase of regional populations, evidencing the distinctive role of agglomeration/scale with urbanization and early suburbanization phases (1940–1980) and accessibility/amenities with late suburbanization and re-urbanization phases (1981–2019). Recession accompanied (and, in some way, consolidated) the decline of agglomeration economies, leading to counter-urbanization in some cases. As an indirect result of counter-urbanization, the population increased more rapidly in low-density coastal areas with moderate accessibility and tourism specialization. Consistently, settlement expansion has altered the persistent gap in central and peripheral locations. A polarized urban hierarchy centered on the capital city, Athens, was replaced with a more diffused growth of medium-sized cities and attractive rural locations, depicting a new development path for lagging countries in the European Union and other socioeconomic contexts worldwide.

Keywords: polycentrism; socioeconomic resilience; recession; Mediterranean Europe

1. Introduction

With economic downturns influencing population structures and consolidating a spatially polarized distribution of jobs and activities, the spatial outcomes of demographic transitions diverged largely in affluent societies [1–4]. Scholars have frequently demonstrated how demographic change reflects socioeconomic processes exalting urban–rural divides [5–9]. In this regard, the city life cycle theory has been traditionally proposed with the aim at delineating and characterizing long-term metropolitan trends in advanced countries [10–13]. This theory describes four stages of urban development: urbanization, suburbanization, de-urbanization and re-urbanization through the processes of concentration/de-concentration and growth/decline of entire functional urban regions [14–18]. However, a comprehensive investigation of metropolitan cycles linking, e.g., urbanization to suburbanization or counter-urbanization to re-urbanization [10–13] suggests that a linear interpretation of socioeconomic forces underlying population growth would be inappropriate to illustrate and understand regional demography patterns [19–21]. Apparent and latent mechanisms of population redistribution within countries and regions were investigated adopting multi-disciplinary approaches, distinctive indicators and refined statistical methodologies [22–25]. Economic downturns, internal and international migrations, social impulses, urban cycles, enhanced volatility in land and housing prices—together with the progressive gentrification of inner cities and latent social filtering in peri-urban areas—were recognized as factors responsible for complex (and less predictable) patterns of population redistribution over larger areas [26–28]. These forces have been investigated at different geographic levels, outlining (i) demographic dynamics that leverage heterogeneous impacts on population expansion under specific social contexts and (ii) economic processes influencing urban–rural demographic structures at wider spatial scales [29–32].

Although spatial inequalities persist at both continental and country scales, reducing territorial disparities and containing density divides were fundamental objectives of national strategies of regional development in European countries [33–35]. For instance, important divides have been observed among neighboring regions in Mediterranean Europe, providing a paradigmatic example of structural gaps that were (and still are) alimented by differential production structures [36–40], the unequal development of metropolitan hierarchies, asymmetric market-state interactions [41,42], demographic transitions and political instability [43–45]. Internal divides have been even more intense in regions with traditional economic structures and secularized sociocultural contexts, limited access to infrastructure and reduced accessibility, aging, fertility, unemployment, as well as low-quality human capital [21,46–51]. Analysis of territorial divides in population density has sometimes demised the role of external shocks shaping socioeconomic dynamics at local scales [52–54]. Assuming a variable impact of these shocks across regions [55–57], local systems were more (or less) able to resist short-term disturbances when confronted with long-term demographic shrinkage and economic stagnation. The intrinsic ability to overcome shocks was often demonstrated to depend on the socioeconomic diversification of local contexts. Earlier works have estimated the linkage between economic downturns, metropolitan cycles and population dynamics to identify the socioeconomic profile of demographically resilient regions [21] and desertification risk [58].

The combined effect of population dynamics at multiple spatial levels has been rarely investigated in light of demographic transitions, international migrations and metropolitan cycles, from urbanization to re-urbanization [59,60]. By reconnecting applied economics to regional demography, results of this analysis shed light on the latent mechanisms underlying territorial disparities and local systems' resilience. In this line of thinking, the present study identifies distinctive factors shaping population growth over different time windows. These findings may inform the design of fine-tuned development policies and spatial planning in Mediterranean Europe. Focusing on Greece, the present study specifically tests different models of population growth over both longer and shorter time scales, assuming a nonlinear evolution toward a complex metropolitan hierarchy with increasingly asymmetric spatial structures [61,62].

By this way, our work integrates a traditional investigation of metropolitan growth in Greece with a multitemporal analysis of population dynamics, reconnecting demographic processes over longer and shorter time scales. More specifically, our study adopted multiple statistical techniques prefiguring a comprehensive picture of population expansion (and shrinkage) in 51 Greek prefectures between 1940 and 2019. By deriving population dynamics from annual vital statistics, socioeconomic forces influencing demographic rebalance at the national scale were identified, contributing to a better knowledge of demographically resilient regions. The present study relates demographic growth with socioeconomic resilience of regional and local systems, classifying them on the base of long-term population expansion or decline. Socioeconomic resilience is an intrinsic property of complex systems and was traditionally estimated using different indicators and approaches. In the present study, socioeconomic resilience was estimated according to long-term population dynamics. Considering a sufficiently long and representative time window, assumptions on the level of resilience of local districts and communities were delineated based on three different demographic contexts: (i) continuously attracting population, (ii) maintaining a stable population stock and (iii) losing population. By explaining (apparent and latent) mechanisms that underlie population redistribution and demographic restructuring over larger and larger regions [63–65], these contexts (i to iii) were hypothesized to be associated with a decreasing level of regional socioeconomic resilience.

2. Methodology

2.1. Study Area

Being partitioned in 51 districts ('nomoi' or prefectures) corresponding to the Eurostat NUTS-3 level, Greece (131,982 km^2) coincided with the area investigated in the present study. Prefectures in Greece are a sufficiently detailed administrative spatial level to evaluate changes in population distribution as a function of socioeconomic transition (Figure 1). The regions of Athens and Thessaloniki concentrate nearly 50% of the Greek population [66,67]. Medium-sized cities (Iraklio, Patras) and prefectural head towns (e.g., Larisa, Volos, Kalamata, Chania, Kavala, Ioannina) grew substantially [27,32]. By displaying population dynamics contrasting with what was characteristic of tourism-specialized districts in the Aegean region, marginal districts experienced diffused land abandonment, depopulation and economic decline [68].

2.2. Data and Indicators

The resident population was derived from ten-year censuses of population and buildings whose results were disseminated by the Hellenic Statistical Authority (ELSTAT, the former ESYE, National Statistical Service of Greece). The analyzed period encompasses eight decades between 1940 and 2019 reflecting a cycle from urbanization to re-urbanization with sequential economic expansion and stagnation waves. For a few prefectures, the population at the beginning of the study period (1940) was reconstructed using published census reports of population count at municipal and village scale for the respective geographic area or administrative region.

Despite small changes over time, census sources were widely selected in the analysis of long-term population dynamics at the regional scale. In our case, the population census was the most authoritative source of demographic information for Greece, being under the continuous scientific supervision and technical realization of the national statistical service since a very long time. An intermediate spatial scale of the investigation was selected (prefectures) to assure a refined comparison of demographic trends over time instead of more detailed domains (e.g., municipalities), which revealed sometimes less stable in the first two decades of study. Annual population growth rates over each decade (1940–1951, 1951–1961, 1961–1971, 1971–1981, 1981–1991, 1991–2001, 2001–2011, 2011–2019) were calculated and normalized subtracting the column mean and then dividing by the column standard deviation before analysis. Population density (inhabitants/km^2) at prefectural level was computed for each year in the time series, deriving the total surface area of each analysis' spatial unit from a shapefile provided by ELSTAT. The same shapefile was used to regionalize demographic indicators and selected results of the analysis run in our work.

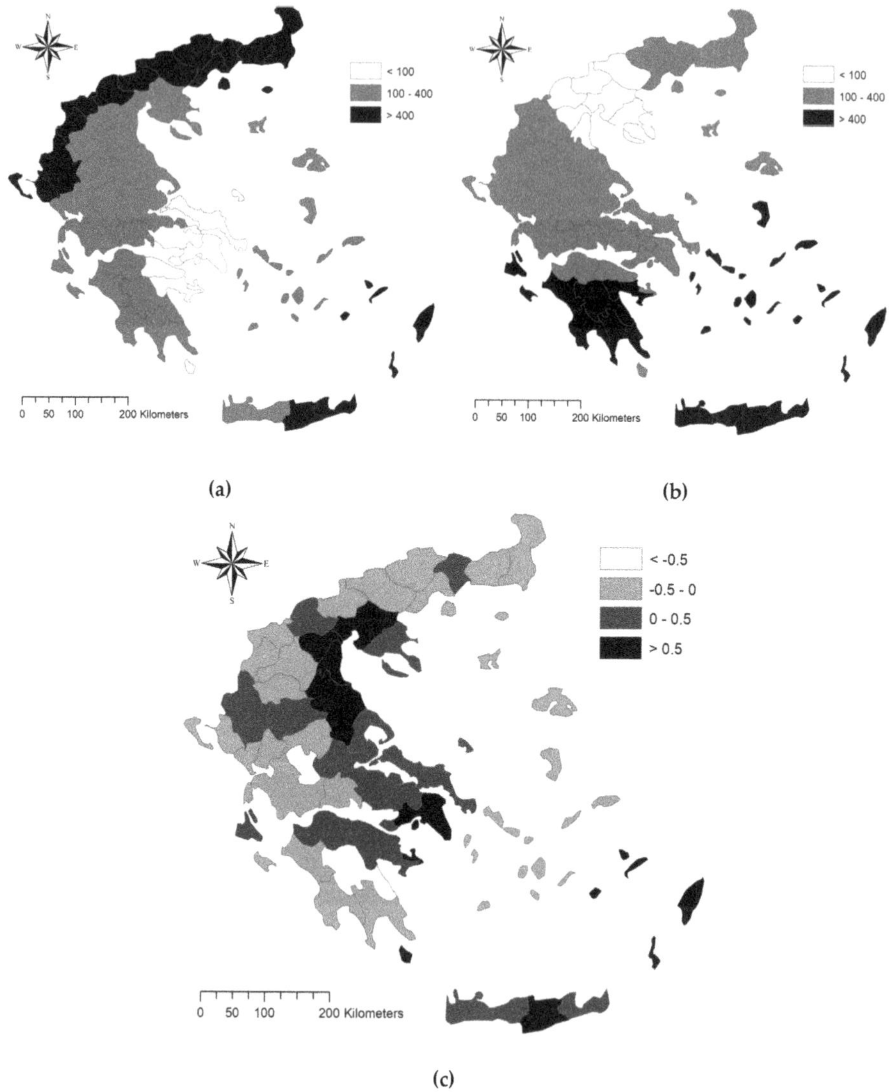

Figure 1. Maps of Greek prefectures illustrating (**a**) the linear distance (km) from downtown Athens; (**b**) the linear distance (km) from downtown Thessaloniki; (**c**) annual population growth rates (%) between 1940 and 2019.

2.3. Contextual Variables

The territorial context that characterizes each prefecture in Greece was delineated considering eight ancillary variables: (i) population density ('Den'), linear distance from (ii) downtown Athens ('DistAth') and (iii) downtown Thessaloniki ('DistSal'), (iv) a dummy illustrating the North–South gradient in Greece and classifying prefectures as belonging to Northern regions (1: Trace, Macedonia, Epirus) or Southern regions (0: the rest of the country), (v) a dummy illustrating the East–West (Aegean-Ionian gradient) and classifying prefectures bordering the Aegean Sea with 1 and the remaining prefectures with 0, (vi) a dummy identifying prefectures hosting the head town of 13 administrative regions of Greece, considered as the largest cities in the country (1) compared with the remaining prefectures

(0), (vii) a dummy identifying internal districts, i.e., mainland prefectures that do not have borders in common with the Ionian or the Aegean Sea (1) in respect with coastal prefectures (0) and, finally, (viii) a dummy identifying prefectures that include only islands (1) in respect with all the remaining prefectures (0). All these variables were made available from official statistics or derived from digital maps (www.geodata.gov.gr) using tools available in ArcGIS software (Redlands, USA). Considering the long time interval investigated in this study and the evident heterogeneity in the time series of many other relevant indicators, these variables—sometimes based on proxies of more complex demographic and socioeconomic processes—represent a satisfactory information ensemble contributing to clarify spatiotemporal population dynamics in Greek prefectures.

2.4. Data Analysis

Maps were prepared on the base of elementary data including annual rates of demographic increase by time window and prefecture, with the aim at identifying spatial similarities in population dynamics over shorter and longer time scales. A simplified framework was proposed with the aim at classifying prefectures on the base of long-term population trends. Criteria were oriented toward the identification of resilient territories under the assumption that prefectures attracting population over a sufficiently long time interval are considered demographically resilient and vice versa. Two criteria were adopted, considering 80 years (1940–2019) and two sub-periods of 40 years each (1940–1980; 1981–2019), representing, respectively compact-dense/radio-centric urbanization and decentralized suburbanization/counter-urbanization in Greece. Prefectures were classified as demographically resilient over the whole time interval (80 years) if the resident population increased for six, seven or eight decades; the reverse pattern characterized nonresilient districts with persistent population shrinkage. A similar framework was adopted over the two sub-periods, classifying prefectures as 'resilient' or 'nonresilient' if population, respectively grew (or declined) continuously for all the four decades in each sub-period. A pair-wise correlation analysis (based on Spearman nonparametric rank coefficients) was computed between each contextual variable (Section 2.3) and the annual rate of population variation (%) over (i) the whole study interval (1940–2019) or, separately, (ii) the two sub-periods (1940–1980 and 1981–2019). Significant correlations were delineated at $p < 0.05$ applying a Bonferroni's correction for multiple comparisons [69].

A hierarchical clustering (based on Euclidean similarity matrix under Ward's amalgamation approach) was run on a database constituted of population growth rates with the aim at classifying temporal units (years) and spatial units (prefectures). Assumed as a relevant dimension of demographic resilience, similar spatiotemporal structures characterize long-term population trends. Persistence (or change) in specific demographic trends was considered indicative of different background conditions. A principal component analysis (PCA) was carried out on a collection of variables including (i) population growth rates at eight decades, (ii) population density at the beginning of each decade and (iii) the remaining eight contextual variables (Section 2.3) at each prefecture. PCA was aimed at containing redundancy, evaluating changes over time in the multivariate relationship between variables at the same time. The PCA characterized research dimensions and distinctive demographic structures in Greece. components with eigenvalue > 1 were selected according to the results of the spectral decomposition of the correlation matrix [70]. The latent structure of variables and prefectures was finally illustrated adopting a biplot that depicts component loadings and scores. Multiple linear regression models were run for each decade and identified the predictors most associated with population growth in Greece. For each decade, predictors include the variables described in Section 2.3 and were standardized before analysis (e.g., [29]). Model's goodness-of-fit was estimated via adjusted R^2 tested for significance (against the null hypothesis of a statistically insignificant model) at $p < 0.01$ using a Fisher–Snedecor F statistic. Slope coefficient estimates and the related significance level at $p < 0.1$ were reported testing for the null-hypothesis of statistically insignificant coefficient based on a Student's t statistic.

3. Results

3.1. A Descriptive Analysis of Population Growth and Decline in Greece, 1940–2019

The resident population grew continuously in Greece between 1940 and 2001, declining slightly in the subsequent two decades (2001–2019). Figure 2 classified the investigated decades on the base of spatial diffusion (or concentration) of the population in Greek prefectures. A particularly rapid population increase in a few prefectures was observed during three decades (1951–1961, 1961–1971 and 1971–1981). Conversely, the population increase in the remaining five decades was slower and more dispersed over space. The largest diffusion of positive growth rates at the prefectural level was observed for the last two decades of investigation, corresponding with slightly negative growth rates at the national scale.

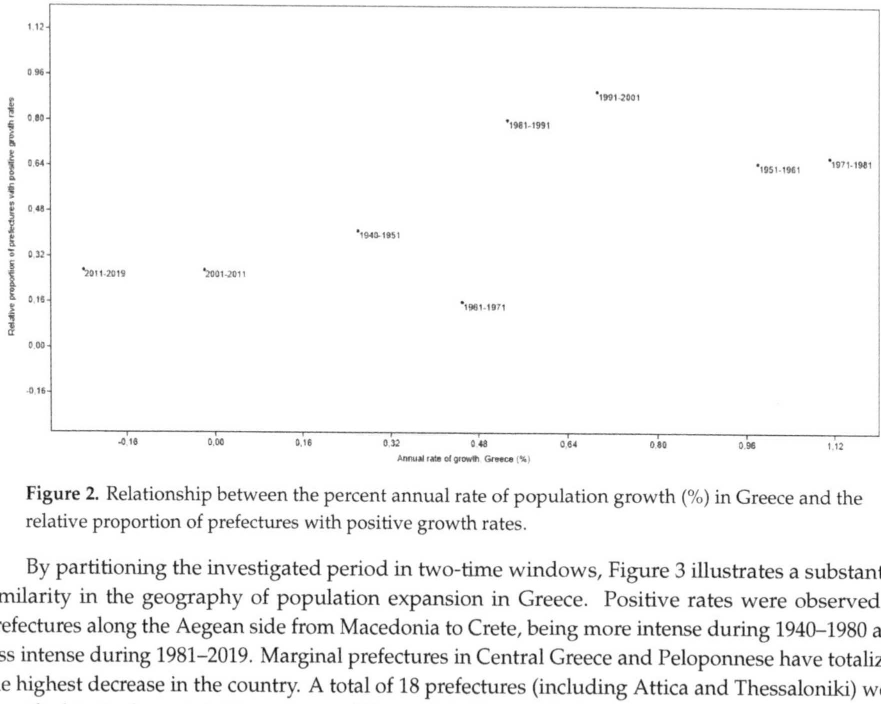

Figure 2. Relationship between the percent annual rate of population growth (%) in Greece and the relative proportion of prefectures with positive growth rates.

By partitioning the investigated period in two-time windows, Figure 3 illustrates a substantial similarity in the geography of population expansion in Greece. Positive rates were observed in prefectures along the Aegean side from Macedonia to Crete, being more intense during 1940–1980 and less intense during 1981–2019. Marginal prefectures in Central Greece and Peloponnese have totalized the highest decrease in the country. A total of 18 prefectures (including Attica and Thessaloniki) were classified in the lowest right quadrant of Figure 3, indicating higher growth rates during 1940–1980 than during 1981–2019.

These prefectures include urban, high-density areas (Athens, Thessaloniki, Iraklio, Patras, Larisa, Viotia, Argolida) expanding mostly with compact urbanization (1940–1960). A total of 33 prefectures were classified in the highest left quadrant of Figure 3, indicating higher growth rates during 1981–2019 than during 1940–1980. These prefectures include rural, coastal districts and islands with low and moderate density settlements, expanding mostly in the subsequent suburbanization phase (Figure 4).

Spatial persistence in growth rates was illustrated in Figure 5 considering together the whole study period and two separate time intervals. Prefectures with continuous population growth over six, seven or eight decades concentrated along the Aegean side from Macedonia to Crete. These prefectures (n = 13 out of 51) host cities constituting the highest rank of the Greek metropolitan hierarchy. Prefectures with continuous population decline (n = 4) were located in Central-Western Greece and Peloponnese. Between 1940 and 1980, continuously positive population growth rates at the decadal scale were observed in six prefectures that include urban areas around Athens and Thessaloniki, as well as in Crete (Iraklio), reflecting the tumultuous expansion of the three largest cities in Greece.

Continuously negative rates of population increase were observed in Central Greece, Peloponnese and Northern Aegean region ($n = 7$). Between 1981 and 2019, continuously positive rates of population expansion were recorded only in the Cyclades and Eastern Macedonia ($n = 2$ prefectures). Continuously negative population growth rates were recorded in one prefecture of Western Greece.

Figure 3. Relationship between the annual population growth rate (%) over two-time intervals in the recent history of Greece.

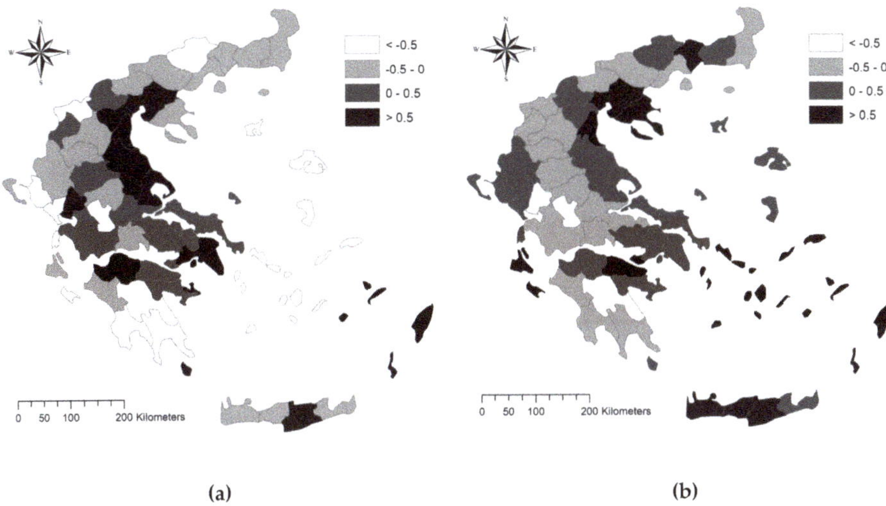

Figure 4. Spatial distribution of annual population growth rates (%) in Greek prefectures over 1940–1980 (**a**) and 1981–2019 (**b**).

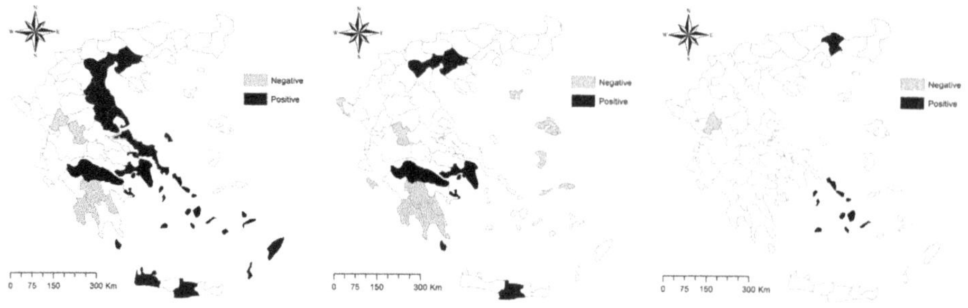

(a) (b) (c)

Figure 5. Temporal persistence of population growth rates in Greek prefectures (**a**) prefectures with at least six decades with positive (or negative) growth rates over 1940–2019; (**b**) prefectures with four decades with positive (or negative) growth rates over 1940–1980; (**c**) prefectures with four decades with positive (or negative) growth rates over 1981–2019).

3.2. Spatial Regime of Long-Term Population Growth (or Decline) in Greece

Hierarchical clustering illustrates similarities in long-term population dynamics across Greek prefectures. The analysis separated Attica and Thessaloniki, the largest urban areas in the country, from the remaining prefectures. Coastal areas and islands (from Kefallinia to Iraklio), expanding the most during the last decades, were also separated from the rest of the sample. However, island prefectures (Chios, Samos, Lesvos) in more peripheral locations of Northern Aegean were clustered together with coastal and inland prefectures in Peloponnese, depicting areas that experienced a more recent population expansion in response to suburbanization, decentralization of economic activities and development policies improving the attractiveness of peripheral coastal places (Figure 6). The remaining prefectures, representing the 'core' of the dendrogram, were mainly internal areas with variable population dynamics that are less correlated with the main waves of urbanization and suburbanization at the national scale.

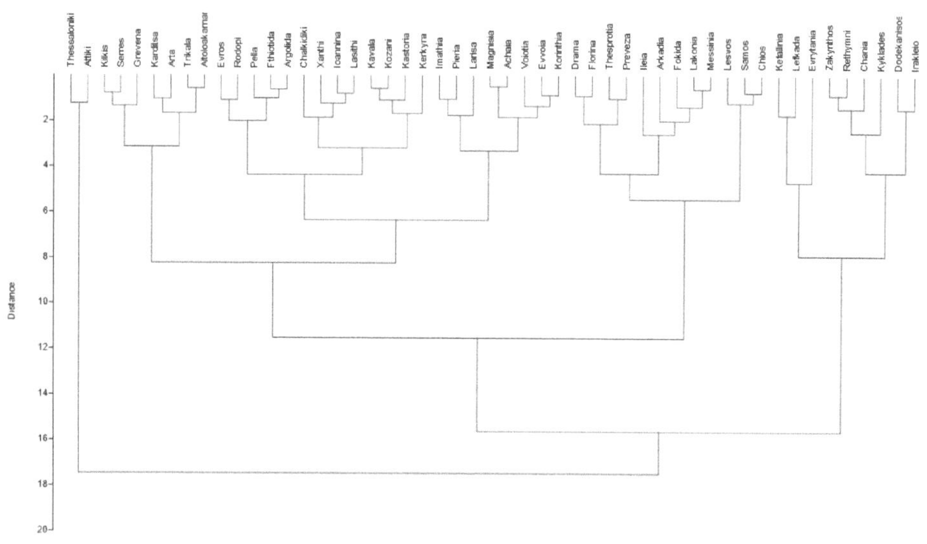

Figure 6. Hierarchical clustering (Ward's agglomeration rule, Euclidean distance) classifying Greek prefectures on the base of long-term population dynamics.

The relationship between population increase and background conditions in each prefecture was investigated considering nonparametric Spearman coefficients (Table 1). The West–East gradient and the presence of a regional capital city had the highest (positive) impact on population growth in Greece. Population density, the distance from Thessaloniki and being an island district had a positive impact only in the most recent period (1981–2019). Internal districts harmed population growth during 1981–2019.

Table 1. Spearman rank correlation analysis between the percent rate of population growth by time interval and selected predictors of the local context (* indicates significance at $p < 0.05$).

Variable	Percent Rate of Population Growth		
	1940–2019	1940–1980	1981–2019
Density	0.05	−0.07	0.36 *
North–South	0.04	0.19	−0.23
West–East	0.44 *	0.31 *	0.37 *
DistAthens	−0.04	−0.05	0.00
DistSalonika	−0.06	−0.25	0.32 *
RegCapital	0.42 *	0.39 *	0.31 *
Internal dis.	−0.22	0.01	−0.49 *
Island	0.04	−0.25	0.55 *

3.3. Population Growth and Territorial Background

A principal component analysis of population growth rates and territorial variables corroborates earlier results of this work (Figure 7). principal component analysis (Table 2) selected two axes accounting for 57.5% of the total variance. This analysis decomposed the most relevant processes of urban expansion (and the underlying drivers) in two independent gradients based on (i) population density and distance from Athens (component 1) and (ii) population dynamics and distance from Thessaloniki (component 2). component 2 discriminated between earlier phases of population growth (positive loadings) in turn associated with the North–South gradient, and later phases of growth (negative loadings).

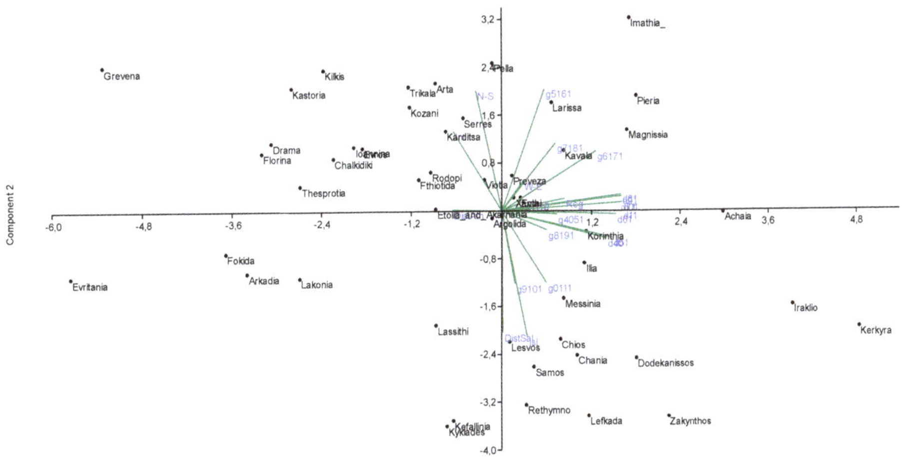

Figure 7. Biplot of a principal component analysis decomposing population dynamics (1940–2019) in two independent dimensions of metropolitan growth in Greece.

Table 2. Results of multiple regression models between the annual population growth rate (%) and predictors delineating the socioeconomic characteristics of prefectures in Greece (* indicates significant coefficient at $p < 0.1$).

Variable	Coeff.	Std. err.	t	p	Variable	Coeff.	Std. err.	t	p
1940–1951: Adj-R^2 = 0.19; F = 2.43; p = 0.03					1951–1961: Adj-R^2 = 0.47; F = 6.53; p < 0.01				
Constant	0.00	0.13	0.01	0.99	Constant	0.00	0.10	−0.01	0.99
Density	−0.22	0.17	−1.31	0.20	Density	0.26 *	0.15	1.74	0.09
North–South	0.19	0.28	0.69	0.49	North–South	0.43 *	0.22	1.91	0.06
West–East	−0.17	0.15	−1.13	0.26	West–East	0.22 *	0.12	1.83	0.07
DistAthens	−0.55 *	0.22	−2.48	0.02	DistAthens	−0.15	0.18	−0.81	0.42
DistSalonika	−0.43 *	0.21	−2.07	0.04	DistSalonika	0.01	0.17	0.04	0.97
RegCapital	0.36 *	0.13	2.70	0.01	RegCapital	0.20 *	0.11	1.79	0.08
Internal dist.	−0.03	0.16	−0.19	0.85	Internal dist.	0.13	0.13	0.94	0.35
Island	0.71 *	0.25	2.84	0.01	Island	−0.30	0.21	−1.42	0.16
1961–1971: Adj-R^2 = 0.60; F = 10.48; p < 0.001					1971–1981: Adj-R^2 = 0.33; F = 4.12; p < 0.001				
Constant	0.00	0.09	0.01	0.99	Constant	0.00	0.11	0.00	1.00
Density	0.37 *	0.12	2.96	0.01	Density	0.15	0.17	0.88	0.38
North–South	0.10	0.20	0.50	0.62	North–South	0.10	0.25	0.40	0.69
West–East	0.21 *	0.10	2.00	0.05	West–East	0.36 *	0.13	2.71	0.01
DistAthens	−0.34 *	0.16	−2.12	0.04	DistAthens	−0.02	0.21	−0.11	0.91
DistSalonika	-0.18	0.15	−1.25	0.22	DistSalonika	−0.01	0.19	−0.06	0.96
RegCapital	0.31 *	0.10	3.16	0.00	RegCapital	0.40 *	0.13	3.08	0.00
Internal dist.	−0.07	0.11	−0.66	0.51	Internal dist.	−0.01	0.15	−0.05	0.96
Island	−0.07	0.18	−0.38	0.70	Island	−0.21	0.23	−0.93	0.36
1981–1991: Adj-R^2 = 0.08; F = 1.57; p = 0.16					1991–2001: Adj-R^2 = 0.14; F = 1.98; p = 0.08				
Constant	0.00	0.13	0.00	1.00	Constant	0.00	0.13	0.01	0.99
Density	−0.05	0.20	−0.23	0.82	Density	−0.27	0.20	−1.36	0.18
North–South	−0.05	0.30	−0.17	0.87	North–South	−0.15	0.29	−0.52	0.61
West–East	0.12	0.15	0.79	0.43	West–East	−0.01	0.15	−0.07	0.95
DistAthens	−0.19	0.25	−0.76	0.45	DistAthens	−0.06	0.24	−0.23	0.82
DistSalonika	−0.19	0.22	−0.84	0.41	DistSalonika	−0.39 *	0.22	−1.78	0.08
RegCapital	0.23	0.16	1.49	0.14	RegCapital	0.24	0.15	1.53	0.13
Internal dist.	−0.26	0.17	−1.47	0.15	Internal dist.	−0.14	0.17	−0.83	0.41
Island	0.17	0.26	0.64	0.53	Island	0.56 *	0.25	2.22	0.03
2001–2011: Adj-R^2 = 0.56; F = 9.08; p < 0.001					2011–2019: Adj-R^2 = 0.22; F = 2.81; p < 0.01				
Constant	0.00	0.09	0.00	1.00	Constant	0.00	0.12	−0.01	0.99
Density	0.09	0.14	0.66	0.51	Density	−0.22	0.20	−1.12	0.27
North–South	0.36 *	0.21	1.78	0.08	North–South	0.57 *	0.28	2.07	0.05
West–East	0.12	0.11	1.15	0.26	West–East	0.25 *	0.14	1.76	0.09
DistAthens	0.07	0.17	0.39	0.70	DistAthens	−0.43 *	0.23	−1.88	0.07
DistSalonika	0.14	0.16	0.86	0.39	DistSalonika	−0.13	0.21	−0.63	0.53
RegCapital	0.29 *	0.11	2.60	0.01	RegCapital	0.01	0.15	0.09	0.93
Internal dist.	−0.14	0.12	−1.18	0.25	Internal dist.	−0.46 *	0.16	−2.85	0.01
Island	0.70 *	0.18	3.79	0.00	Island	0.40	0.25	1.58	0.12

Multiple linear regression was carried out to delineate the predictors of population growth over each investigated decade in Greece (Table 2). Regression models were particularly significant for specific decades (1951–1961, 1961–1971, 1971–1981, 2001–2011) and less significant for the remaining time intervals. Regression analysis identified distance from Athens and Thessaloniki (negative impact), the regional capital city and insular contexts (positive impact) as the main factors associated with population growth during 1940–1951. Geographic gradients including the North–South and the East–West gradient and the presence of a regional capital city were identified as the main drivers of population increase during 1951–1961. Density, distance from Athens, the regional capital city and the East–West gradient were found significant predictors of population growth for 1961–1971. These last two variables were found significant also for 1971–1981. No significant variables were detected for 1981–1991. Distance from Thessaloniki and insular contexts were found significant during 1991–2001. North–south gradient, the regional capital city and insular contexts were significant during 2001–2011.

Conversely, internal districts, together with the two geographic gradients and the distance from Athens, influenced population growth rates during the last decade (2011–2019).

4. Discussion

Advanced economies experienced intense demographic changes leveraging sequential urbanization and suburbanization waves [71–75]. In Europe, multifaceted migration trends, new household structures and a particularly heterogeneous natural balance over space shaped regional population trends [76–79]. In these regards, understanding how demographic processes, metropolitan cycles and economic downturns interact shaping regional competitiveness and location attractiveness, may contribute to a refined understanding of local development mechanisms [80–85]. When comparing population dynamics over sequential waves of urban expansion and rural shrinkage [86–88], Greece is exemplificative of traditional Mediterranean societies with a polarized economy along urban–rural gradients that reflect territorial divides between coastal and inland districts. Despite a persistent spatial configuration at the country scale, such divides had reduced during the last 80 years, fortifying—especially between 1940 and 1980—the role of few urban nodes (Athens-Piraeus and Thessaloniki) inserted in broader continental (or global) networks. This process consolidated in the subsequent four decades (1980–2019) reflecting the West–East gradient in Greece and the importance of (low-density) coastal districts attracting new population and economic activities (Figure 8).

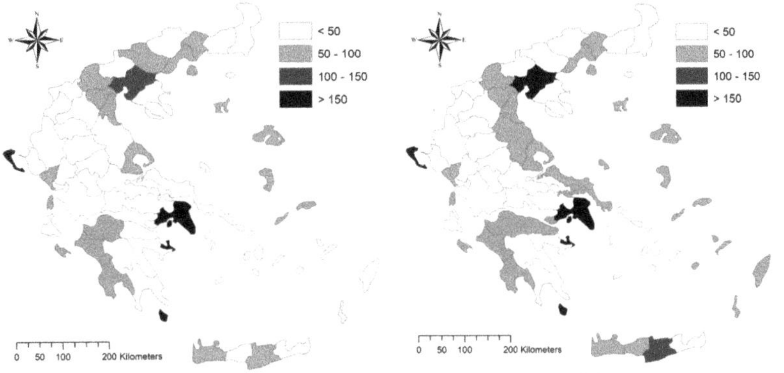

(a) (b)

Figure 8. Population growth rate (%) in Greek municipalities by time interval (**a**) 1940–1980; (**b**) 1981–2019).

A spatially heterogeneous population growth has been observed in Greece, being related with a comprehensive set of background predictors. Regional population growth in Greece has reflected mechanisms of local development leveraging spatial disparities (e.g., [12]). As a matter of fact, the empirical results of our study outlined a strong difference in population dynamics between urban and rural areas, especially from 1940 to 1980 [26]. Accessibility, as well as economic specialization and advanced productive functions were assumed to be significant factors shaping demographic dynamics in such decades, since they have alimented a compact and radio-centric settlement growth in metropolitan regions and rural-to-urban mobility determining rural shrinkage [89]. The following dynamics (1981–2019) were less associated with central cities, suggesting a less important role for agglomeration and scale factors. After a continuous increase of resident population, a less intense gap between urban and rural areas was observed since the early 1990s.

Population redistribution over wider regions was evident in Greece since the late 1990s, reflecting spatially heterogeneous socioeconomic transformations increasingly decoupled from traditional geographic gradients (e.g., urban–rural, coastal-inland). Since the 1990s, population increased in rural districts and low-density touristic coastal areas, especially large and medium-sized islands, as well as lowland, internal districts devoted to intensive cropland. More intense urban-to-rural internal movements were specifically observed in the last decade as a result of progressive counter-urbanization and shrinkage of central cities associated with worse socioeconomic conditions in metropolitan regions, mainly in Athens and, partly, Thessaloniki [13,32,90]. However, the overall impact of these processes on the Greek urban hierarchy was rather modest [21]. In such a context, the possible role of unregistered or partially registered (i) emigration abroad and (ii) illegal international immigration should be evaluated further.

As a result of economic processes and mixed socio-demographic contexts, spatial heterogeneity in population growth rates was persistently observed in both urban and rural districts. However, this process was clearly asymmetric over time, being more intense in the second time interval (1980–2019) corresponding with late suburbanization, counter-urbanization and early re-urbanization and less intense in the first time interval (1940–1980) corresponding with urbanization and early suburbanization in Greece. Tapia et al. [91] investigated population trends in Spain at different time windows between 1860 and 1991 with the aim at testing if population at the beginning of each time window affects population growth during the same time window. According to the authors' results (p. 81) "while such a relationship between these two variables hardly existed during the second half of the nineteenth century, this link increased significantly between 1910 and 1970, although this trend was abruptly interrupted by the Civil War and the autarkic period that followed [92,93]. The intensity of this relationship decreased in the 1970s, a process that continued during the 1980s (...); agglomeration economies were stronger in medium-sized districts, especially from 1960 onwards".

In Greece, a slightly different spatial model was observed, based on the specific socioeconomic context and the distinctive development path. Population decline in urban, industrial and intensive agricultural districts was more intense than in rural districts experiencing long depopulation, suggesting that central locations display a lower demographic attractiveness than peripheral locations with dynamic economic sectors, e.g., tourism [94–96]. In these regards, a mix of factors that includes industrial decline, informal economies and dependence on external funds, especially in agriculture, determined population shrinkage in Southern Europe [97–102], suggesting a lower demographic resilience of the locations (mostly rural) experiencing this kind of development path.

A better knowledge of new demographic scenarios opens up a key reflection on the intimate mechanisms of long-term change in population redistribution over wider regions [7,103–106] and may delineate opportunities for regional development policy. In these regards, newly emergent issues such as resurging internal migration, declining immigration flows from abroad and re-approaching the lowest-low fertility levels, suggest a thorough rethinking of sustainable urbanization [107–111]. Planning strategies stimulating a polycentric expansion of human settlements should incorporate policy measures improving local development in regions exposed to demographic and economic stagnation [29,112,113]. A comparative, long-term analysis of population distribution over space proved to be a necessary tool identifying the emergent socioeconomic dynamics and delineating the most appropriate policies to face with [114,115]. A refined investigation of the negative impact of recession on local socioeconomic structures is therefore meaningful to shed light on future population dynamics in a post-crisis Mediterranean Europe.

Taken together, results of this study, at least for the first time interval (1940–1980), suggest how "being closely intertwined with the policy debate, the concept of agglomeration economies and its relation with spatial economic performance has maintained a central role" in regional science [116]. However, the empirical results for the most recent time horizon (1981–2019) reflect more complex demographic and settlement processes over space. According with Berliant and Wang [117], "while basic questions such as 'does urbanization cause growth or does growth cause urbanization?' or 'does supporting urbanization imply neglecting rural areas?' are still valid and need a more comprehensive research and immediate policy response, our study definitely contributes to a better understanding of the role of urbanization in population distribution and economic growth, informing policies that tackle the formidable challenges it poses".

5. Conclusions

Results of our study illustrate long-term demographic dynamics in 51 Greek prefectures (1940–2019) providing evidence to settlement (re)distribution along basic geographic gradients. Prefectures are comprehensive and homogeneous spatial units when investigating patterns of population (re)distribution in Greece under different economic conditions, having empirical relevance to demographic issues. Exploratory data analysis indicates a particularly complex regional framework with diversified metropolitan dimensions. Going beyond traditional theories linking urbanization with scale/agglomeration economies, these findings suggest the importance of a comparative analysis of demographic processes aimed at confirming (or confuting) such trends. This analysis contributes to interpret the other phases of the urban cycle (from suburbanization to re-urbanization) as primarily associated with 'soft' economic and alternative non-economic aspects of local development, including amenities, specialization in advanced services, the increased impact of economic downturns, as well as gentrification, class segregation and social filtering.

By reverting the paradigm of growth in high-density areas, the specificity of the 'Greek' model delineates new developmental paths going beyond the traditional dichotomy in compact and dispersed settlements. Highlighting the role of 'intermediate' districts, in-between large cities and economically depressed rural areas, our findings outline how a long process of spatial redistribution of settlements led—directly or indirectly—to a sort of (functional) polycentric development. Non-urban prefectures in coastal regions and in accessible, rural areas were the engine of such development path. These districts were also regarded as resilient, since they attract population in a context of demographic stagnation at the country scale, with intense shrinkage of the main urban centers. Our study definitely confirms the informative power of a multiscale investigation of the differential impact of metropolitan cycles and economic downturns in past, present and future population growth. In such perspectives, future studies are increasingly required to develop a refined analysis of the intrinsic linkage between demographic growth and long-term economic development at an enough detailed scale of investigation (prefectures, local districts, municipalities) and based on appropriate socioeconomic indicators.

Author Contributions: Conceptualization, L.S. and R.S.; methodology, G.Q. and G.E.; software, G.E.; validation, R.S.; formal analysis, L.S., J.R.C. and G.E.; investigation, G.Q. and R.S.; resources, G.E. and L.S.; data curation, G.E. and R.S.; writing—original draft preparation, L.S., J.R.C. and G.E.; writing—review and editing, R.S., J.R.C. and G.Q.; visualization, G.E.; supervision, G.Q.; project administration, G.Q.; funding acquisition, G.E. All authors have read and agreed to the published version of the manuscript.

Funding: This work received no funding.

Conflicts of Interest: The authors declare no conflicts of interest.

References

1. Tapinos, G.; Mason, A.; Bravo, J. *Demographic Responses to Economic Adjustment in Latin America*; Oxford University Press: New York, NY, USA, 1997; ISBN 978-0-19-829210-4.
2. Chen, Y. Urban chaos and perplexing dynamics of urbanization. *Lett. Spat. Resour. Sci.* **2009**, *2*, 85. [CrossRef]
3. Cechella, C.; Dentinho, T.; Silva, J.R. World centralities for political, demographic and market purposes. *Lett. Spat. Resour. Sci.* **2012**, *5*, 73–83. [CrossRef]
4. Arapoglou, V.P.; Sayas, J. New Facets of Urban Segregation in Southern Europe: Gender, Migration and Social Class Change in Athens. *Eur. Urban Reg. Stud.* **2009**. [CrossRef]
5. Dyson, T. On Demographic and Democratic Transitions. *Popul. Dev. Rev.* **2013**, *38*, 83–102. [CrossRef]
6. Bono, E.D.; Weber, A.; Winter-Ebmer, R. Fertility and economic instability: The role of unemployment and job displacement. *J. Popul. Econ.* **2015**, *28*, 463–478. [CrossRef]
7. Walford, N.; Kurek, S. Outworking of the Second Demographic Transition: National Trends and Regional Patterns of Fertility Change in Poland, and England and Wales, 2002–2012. *Popul. Space Place* **2016**, *22*, 508–525. [CrossRef]

8. Li, Z.; Cheng, J.; Wu, Q. Analyzing regional economic development patterns in a fast developing province of China through geographically weighted principal component analysis. *Lett. Spat. Resour. Sci.* **2016**, *9*, 233–245. [CrossRef]
9. Benyaklef, M. Socio-economic disparities in the Mediterranean. *Mediterr. Politics* **1997**, *2*, 93–112. [CrossRef]
10. Kulu, H.; Boyle, P.; Andersson, G. High Suburban Fertility: Evidence from Four Northern European Countries. *Demogr. Res.* **2009**, *21*, 915–944. [CrossRef]
11. Kabisch, N.; Haase, D. Diversifying European agglomerations: Evidence of urban population trends for the 21st century. *Popul. Space Place* **2011**, *17*, 236–253. [CrossRef]
12. Salvati, L.; Sateriano, A.; Grigoriadis, E. Crisis and the city: Profiling urban growth under economic expansion and stagnation. *Lett. Spat. Resour. Sci.* **2016**, *9*, 329–342. [CrossRef]
13. Gkartzios, M.; Scott, K. A Cultural Panic in the Province? Counterurban Mobilities, Creativity, and Crisis in Greece. *Popul. Space Place* **2015**, *21*, 843–855. [CrossRef]
14. Berry, B.J.L. Urbanization and Counterurbanization in the United States. *ANNALS Am. Acad. Political Soc. Sci.* **1980**, *451*, 13–20. [CrossRef]
15. Roberts, S. A Critical Evaluation of the City Life Cycle Idea. *Urban Geogr.* **1991**, *12*, 431–449. [CrossRef]
16. Pumain, D. Settlement systems in the evolution. *Geogr. Ann. Ser. B Hum. Geogr.* **2000**, *82*, 73–87. [CrossRef]
17. Berg, M. van den City Children and Genderfied Neighbourhoods: The New Generation as Urban Regeneration Strategy. *Int. J. Urban Reg. Res.* **2013**, *37*, 523–536. [CrossRef]
18. Bonavero, P.; Sforzi, F.; Dematteis, G. *The Italian Urban System: Towards European Integration*; Routledge: London, UK, 2009.
19. Işik, O.; Pinarcioğlu, M.M. Geographies of a silent transition: A geographically weighted regression approach to regional fertility differences in Turkey. *Eur. J. Popul.* **2006**, *22*, 399–421. [CrossRef]
20. Falagas, M.E.; Vouloumanou, E.K.; Mavros, M.N.; Karageorgopoulos, D.E. Economic crises and mortality: A review of the literature. *Int. J. Clin. Pract.* **2009**, *63*, 1128–1135. [CrossRef]
21. Salvati, L. The Dark Side of the Crisis: Disparities in per Capita income (2000–12) and the Urban-Rural Gradient in Greece. *Tijdschr. Voor Econ. en Soc. Geogr.* **2016**, *107*, 628–641. [CrossRef]
22. Haase, A.; Kabisch, S.; Steinführer, A.; Bouzarovski, S.; Hall, R.; Ogden, P. Emergent spaces of reurbanisation: Exploring the demographic dimension of inner-city residential change in a European setting. *Popul. Space Place* **2010**, *16*, 443–463. [CrossRef]
23. Kroll, F.; Kabisch, N. The Relation of Diverging Urban Growth Processes and Demographic Change along an Urban–Rural Gradient. *Popul. Space Place* **2012**, *18*, 260–276. [CrossRef]
24. Bosker, M. Growth, Agglomeration and Convergence: A Space-time Analysis for European Regions. *Spat. Econ. Anal.* **2007**, *2*, 91–100. [CrossRef]
25. Carson, D.B.; Carson, D.A.; Porter, R.; Ahlin, C.Y.; Sköld, P. Decline, Adaptation or Transformation: New Perspectives on Demographic Change in Resource Peripheries in Australia and Sweden. *Comp. Popul. Stud.* **2016**, *41*.
26. Serra, P.; Vera, A.; Tulla, A.F.; Salvati, L. Beyond urban–rural dichotomy: Exploring socioeconomic and land-use processes of change in Spain (1991–2011). *Appl. Geogr.* **2014**, *55*, 71–81. [CrossRef]
27. Carlucci, M.; Grigoriadis, E.; Rontos, K.; Salvati, L. Revisiting a Hegemonic Concept: Long-term 'Mediterranean Urbanization' in Between City Re-polarization and Metropolitan Decline. *Appl. Spat. Anal.* **2017**, *10*, 347–362. [CrossRef]
28. Cuadrado-Ciuraneta, S.; Durà-Guimerà, A.; Salvati, L. Not only tourism: Unravelling suburbanization, second-home expansion and "rural" sprawl in Catalonia, Spain. *Urban Geogr.* **2017**, *38*, 66–89. [CrossRef]
29. Stockdale, A. Contemporary and 'Messy' Rural In-migration Processes: Comparing Counterurban and Lateral Rural Migration. *Popul. Space Place* **2016**, *22*, 599–616. [CrossRef]
30. Biasi, R.; Colantoni, A.; Ferrara, C.; Ranalli, F.; Salvati, L. In-between sprawl and fires: Long-term forest expansion and settlement dynamics at the wildland–urban interface in Rome, Italy. *Int. J. Sustain. Dev. World Ecol.* **2015**, *22*, 467–475. [CrossRef]
31. Mavrakis, A.; Papavasileiou, C.; Salvati, L. Towards (Un) sustainable urban growth? Industrial development, land-use, soil depletion and climate aridity in a Greek agro-forest area. *J. Arid. Environ.* **2015**, *121*, 1–6. [CrossRef]

32. Zambon, I.; Serra, P.; Sauri, D.; Carlucci, M.; Salvati, L. Beyond the 'Mediterranean city': Socioeconomic disparities and urban sprawl in three Southern European cities. *Geogr. Ann. Ser. B Hum. Geogr.* **2017**, *99*, 319–337. [CrossRef]
33. Davies, S. Regional resilience in the 2008–2010 downturn: Comparative evidence from European countries. *Camb. J. Reg. Econ. Soc.* **2011**, *4*, 369–382. [CrossRef]
34. Connaughton, J.E. Local Economic Impact of the Great Recession of 2008/2009. *RRS* **2010**, *40*, 1–4.
35. Di Feliciantonio, C.; Salvati, L.; Sarantakou, E.; Rontos, K. Class diversification, economic growth and urban sprawl: Evidences from a pre-crisis European city. *Qual. Quant.* **2018**, *52*, 1501–1522. [CrossRef]
36. Ferrara, C.; Carlucci, M.; Grigoriadis, E.; Corona, P.; Salvati, L. A comprehensive insight into the geography of forest cover in Italy: Exploring the importance of socioeconomic local contexts. *For. Policy Econ.* **2017**, *75*, 12–22. [CrossRef]
37. Petrakos, G.C. Urban Concentration and Agglomeration Economies: Re-examining the Relationship. *Urban Stud.* **1992**. [CrossRef]
38. Dijkstra, L.; Garcilazo, E.; McCann, P. The effects of the global financial crisis on European regions and cities. *J. Econ. Geogr.* **2015**, *15*, 935–949. [CrossRef]
39. Chorianopoulos, I.; Tsilimigkas, G.; Koukoulas, S.; Balatsos, T. The shift to competitiveness and a new phase of sprawl in the Mediterranean city: Enterprises guiding growth in Messoghia—Athens. *Cities* **2014**, *39*, 133–143. [CrossRef]
40. Bonifazi, C.; Heins, F. Testing the differential urbanisation model for Italy. *Tijdschr. Voor Econ. en Soc. Geogr.* **2003**, *94*, 23–37. [CrossRef]
41. Frick, S.A.; Rodríguez-Pose, A. Change in urban concentration and economic growth. *World Dev.* **2018**, *105*, 156–170. [CrossRef]
42. Rodríguez-Pose, A.; Fratesi, U. Between Development and Social Policies: The Impact of European Structural Funds in Objective 1 Regions. *Reg. Stud.* **2004**, *38*, 97–113. [CrossRef]
43. Salvati, L.; Carlucci, M. In-Between Stability and Subtle Changes: Urban Growth, Population Structure, and the City Life Cycle in Rome. *Popul. Space Place* **2016**, *22*, 216–227. [CrossRef]
44. Schneider, D. The Great Recession, Fertility, and Uncertainty: Evidence from the United States. *J. Marriage Fam.* **2015**, *77*, 1144–1156. [CrossRef]
45. Modena, F.; Rondinelli, C.; Sabatini, F. Economic Insecurity and Fertility Intentions: The Case of Italy. *Rev. Income Wealth* **2014**, *60*, S233–S255. [CrossRef]
46. Souliotis, N. Cultural economy, sovereign debt crisis and the importance of local contexts: The case of Athens. *Cities* **2013**, *33*, 61–68. [CrossRef]
47. Wheaton, W.C.; Shishido, H. Urban Concentration, Agglomeration Economies, and the Level of Economic Development. *Econ. Dev. Cult. Chang.* **1981**, *30*, 17–30. [CrossRef]
48. Caltabiano, M. A turning point in Italian fertility. *J. Pop. Res.* **2016**, *33*, 379–397. [CrossRef]
49. Caltabiano, M.; Castiglioni, M.; Rosina, A. Lowest-Low Fertility: Signs of a recovery in Italy? *Demogr. Res.* **2009**, *21*, 681–718. [CrossRef]
50. Cazzola, A.; Pasquini, L.; Angeli, A. The relationship between unemployment and fertility in Italy: A time-series analysis. *Demogr. Res.* **2016**, *34*, 1–38. [CrossRef]
51. Rosti, L.; Yamaguchi, C.; Castagnetti, C. Educational Performance as Signalling Device: Evidence from Italy. *Econ. Bull.* **2005**, *9*, 1–7.
52. Storper, M. Why do regions develop and change? The challenge for geography and economics. *J. Econ. Geogr.* **2011**, *11*, 333–346. [CrossRef]
53. Rosen, A.M. *Civilizing Climate: Social Responses to Climate Change in the Ancient Near East*; Rowman Altamira: Lanham, MD, USA, 2007; ISBN 0-7591-0494-8.
54. Lamonica, G.R.; Chelli, F.M. The performance of non-survey techniques for constructing sub-territorial input-output tables. *Pap. Reg. Sci.* **2018**, *97*, 1169–1202. [CrossRef]
55. Sobotka, T.; Skirbekk, V.; Philipov, D. Economic Recession and Fertility in the Developed World. *Popul. Dev. Rev.* **2011**, *37*, 267–306. [CrossRef] [PubMed]
56. De Rosa, S.; Salvati, L. Beyond a 'side street story'? Naples from spontaneous centrality to entropic polycentricism, towards a 'crisis city'. *Cities* **2016**, *51*, 74–83. [CrossRef]
57. Vignoli, D.; Drefahl, S.; De Santis, G. Whose job instability affects the likelihood of becoming a parent in Italy? A tale of two partners. *Demogr. Res.* **2012**, *26*, 41–62. [CrossRef]

58. Salvati, L.; Zitti, M.; Ceccarelli, T. Integrating economic and environmental indicators in the assessment of desertification risk: A case study. *Appl. Ecol. Env. Res.* **2007**, *6*, 129–138. [CrossRef]
59. Lerch, M. Internal and International Migration Across the Urban Hierarchy in Albania. *Popul. Res. Policy Rev.* **2016**, *35*, 851–876. [CrossRef]
60. Maloutas, T. Segregation, Social Polarization and Immigration in Athens during the 1990s: Theoretical Expectations and Contextual Difference. *Int. J. Urban Reg. Res.* **2007**, *31*, 733–758. [CrossRef]
61. Di Feliciantonio, C.; Salvati, L. 'Southern' Alternatives of Urban Diffusion: Investigating Settlement Characteristics and Socio-Economic Patterns in Three Mediterranean Regions. *Tijdschr. Voor Econ. en Soc. Geogr.* **2015**, *106*, 453–470. [CrossRef]
62. Tragaki, A.; Bagavos, C. Male fertility in Greece: Trends and differentials by education level and employment status. *Demogr. Res.* **2014**, *31*, 137–160. [CrossRef]
63. Copus, A.; Mantino, F.; Noguera, J. Inner Peripheries: An oxymoron or a real challenge for territorial cohesion? *Ital. J. Plan. Pract.* **2017**, *7*, 24–49.
64. Combes, P.-P.; Duranton, G.; Gobillon, L. The identification of agglomeration economies. *J. Econ. Geogr.* **2011**, *11*, 253–266. [CrossRef]
65. Lopez-Gay, A. Population growth and re-urbanization in Spanish inner cities: The role of internal migration and residential mobility. *Quetelet J.* **2014**, *2*, 67–92. [CrossRef]
66. Pili, S.; Grigoriadis, E.; Carlucci, M.; Clemente, M.; Salvati, L. Towards sustainable growth? A multi-criteria assessment of (changing) urban forms. *Ecol. Indic.* **2017**, *76*, 71–80. [CrossRef]
67. Rontos, K. Demographic Trends, Young People's Attitudes towards Marriage and Socio-economic Changes Related to Family Formation in Greece and in the European Union: A Comparative Analysis Based on Official and Survey Research Data. *Int. J. Econ. Bus. Adm. (IJEBA)* **2015**, *III*, 11–33. [CrossRef]
68. Couch, C.; Petschel-Held, G.; Leontidou, L. *Urban Sprawl in Europe: Landscape, Land-Use Change and Policy Wiley*; Wiley-Blackwel: London, UK, 2008.
69. Colantoni, A.; Grigoriadis, E.; Sateriano, A.; Venanzoni, G.; Salvati, L. Cities as selective land predators? A lesson on urban growth, deregulated planning and sprawl containment. *Sci. Total. Environ.* **2016**, *545–546*, 329–339. [CrossRef] [PubMed]
70. Salvati, L.; Zitti, M. Regional convergence of environmental variables: Empirical evidences from land degradation. *Ecol. Econ.* **2008**, *68*, 162–168. [CrossRef]
71. Zambon, I.; Benedetti, A.; Ferrara, C.; Salvati, L. Soil Matters? A Multivariate Analysis of Socioeconomic Constraints to Urban Expansion in Mediterranean Europe. *Ecol. Econ.* **2018**, *146*, 173–183. [CrossRef]
72. Billari, F.C.; Kohler, H.-P. Patterns of Low and Lowest-Low Fertility in Europe. *Popul. Stud.* **2004**, *58*, 161–176. [CrossRef]
73. Surkyn, J.; Lesthaeghe, R. Value Orientations and the Second Demographic Transition (SDT) in Northern, Western and Southern Europe: An Update. *Demogr. Res.* **2004**, *3*, 45–86. [CrossRef]
74. Lesthaeghe, R.J.; Neidert, L. The Second Demographic Transition in the United States: Exception or Textbook Example? *Popul. Dev. Rev.* **2006**, *32*, 669–698. [CrossRef]
75. Peck, J. Austerity urbanism. *City* **2012**, *16*, 626–655. [CrossRef]
76. Billari, F.C.; Kohler, H.-P.; Andersson, G.; Lundström, H. Approaching the Limit: Long-Term Trends in Late and Very Late Fertility. *Popul. Dev. Rev.* **2007**, *33*, 149–170. [CrossRef]
77. Pison, G. Two children per woman in France in 2010: Is French fertility immune to economic crisis? *Popul. Soc.* **2011**, *476*, 4.
78. Reher, D.S. Economic and Social Implications of the Demographic Transition. *Popul. Dev. Rev.* **2011**, *37*, 11–33. [CrossRef]
79. Champion, T.; Hugo, G. *New Forms of Urbanization: Beyond the Urban-Rural Dichotomy*; Routledge: London, UK, 2017; ISBN 978-1-315-24807-3.
80. Cherlin, A.; Cumberworth, E.; Morgan, S.P.; Wimer, C. The Effects of the Great Recession on Family Structure and Fertility. *ANNALS Am. Acad. Political Soc. Sci.* **2013**, *650*, 214–231. [CrossRef]
81. Goldstein, J.; Kreyenfeld, M.; Jasilioniene, A.; Karaman Örsal, D.D. Fertility Reactions to the "Great Recession" in Europe: Recent Evidence from Order-Specific Data. *Demogr. Res.* **2013**, *29*, 85–104. [CrossRef]
82. Lee, K.O.; Painter, G. What happens to household formation in a recession? *J. Urban Econ.* **2013**, *76*, 93–109. [CrossRef]

83. Palloni, A.; Tienda, M. Demographic Responses to Economic Recessions in Latin America Since 1900. *Sociol. Inq.* **1992**, *62*, 246–270. [CrossRef]
84. García, M. The Breakdown of the Spanish Urban Growth Model: Social and Territorial Effects of the Global Crisis. *Int. J. Urban Reg. Res.* **2010**, *34*, 967–980. [CrossRef]
85. Pérez, J.M.G. The Real Estate and Economic Crisis: An Opportunity for Urban Return and Rehabilitation Policies in Spain. *Sustainability* **2010**, *2*, 1571–1601. [CrossRef]
86. Rontos, K.; Grigoriadis, E.; Sateriano, A.; Syrmali, M.; Vavouras, I.; Salvati, L. Lost in protest, found in segregation: Divided cities in the light of the 2015 "Oχι" referendum in Greece. *City Cult. Soc.* **2016**, *7*, 139–148. [CrossRef]
87. Gavalas, V.S.; Rontos, K.; Salvati, L. Who Becomes an Unwed Mother in Greece? Sociodemographic and Geographical Aspects of an Emerging Phenomenon. *Popul. Space Place* **2014**, *20*, 250–263. [CrossRef]
88. Remoundou, K.; Gkartzios, M.; Garrod, G. Conceptualizing Mobility in Times of Crisis: Towards Crisis-Led Counterurbanization? *Reg. Stud.* **2016**, *50*, 1663–1674. [CrossRef]
89. Copus, A.; Melo, P.C.; Kaup, S.; Tagai, G.; Artelaris, P. Regional poverty mapping in Europe—Challenges, advances, benefits and limitations. *Local Econ.* **2015**, *30*, 742–764. [CrossRef]
90. Gkartzios, M. 'Leaving Athens': Narratives of counterurbanisation in times of crisis. *J. Rural. Stud.* **2013**, *32*, 158–167. [CrossRef]
91. Tapia, F.J.B.; Díez-Minguela, A.; Martinez-Galarraga, J. Tracing the Evolution of Agglomeration Economies: Spain, 1860–1991. *J. Econ. Hist.* **2018**, *78*, 81–117. [CrossRef]
92. Gil-Alonso, F.; Bayona-i-Carrasco, J.; Pujadas-i-Rúbies, I. From boom to crash: Spanish urban areas in a decade of change (2001–2011). *Eur. Urban Reg. Stud.* **2016**, *23*, 198–216. [CrossRef]
93. Goerlich, F.J.; Mas, M. La concentration spatiale de la population espagnole de 1900 à 2001: Confirmations empiriques. *Population* **2008**, *63*, 731–745. [CrossRef]
94. Karamesouti, M.; Detsis, V.; Kounalaki, A.; Vasiliou, P.; Salvati, L.; Kosmas, C. Land-use and land degradation processes affecting soil resources: {Evidence} from a traditional {Mediterranean} cropland ({Greece}). *CATENA* **2015**, *132*, 45–55. [CrossRef]
95. Stuckler, D.; Reeves, A.; Karanikolos, M.; McKee, M. The health effects of the global financial crisis: Can we reconcile the differing views? A network analysis of literature across disciplines. *Health Econ. Policy Law* **2015**, *10*, 83–99. [CrossRef]
96. Fuchs, R.J.; Demko, G.J. Rethinking population distribution policies. *Popul. Res. Policy Rev.* **1983**, *2*, 161–187. [CrossRef]
97. Chelli, F.; Rosti, L. Age and gender differences in Italian workers' mobility. *Int. J. Manpow.* **2002**, *23*, 313–325. [CrossRef]
98. Monastiriotis, V. Examining the consistency of spatial association patterns across socio-economic indicators: An application to the Greek regions. *Empir. Econ.* **2009**, *37*, 25–49. [CrossRef]
99. Monastiriotis, V. Making geographical sense of the Greek austerity measures: Compositional effects and long-run implications. *Camb. J. Reg. Econ. Soc.* **2011**, *4*, 323–337. [CrossRef]
100. Koutsampelas, C.; Tsakloglou, P. The distribution of full income in Greece. *Int. J. Soc. Econ.* **2013**, *40*, 311–330. [CrossRef]
101. Kosmas, C.; Karamesouti, M.; Kounalaki, K.; Detsis, V.; Vassiliou, P.; Salvati, L. Land degradation and long-term changes in agro-pastoral systems: An empirical analysis of ecological resilience in Asteroussia—Crete (Greece). *CATENA* **2016**, *147*, 196–204. [CrossRef]
102. Lusardi, A.; Schneider, D.; Tufano, P. The Economic Crisis and Medical Care Use: Comparative Evidence from Five High-Income Countries. *Soc. Sci. Q.* **2015**, *96*, 202–213. [CrossRef]
103. Carlucci, M.; Chelli, F.M.; Salvati, L. Toward a New Cycle: Short-Term Population Dynamics, Gentrification, and Re-Urbanization of Milan (Italy). *Sustainability* **2018**, *10*, 3014. [CrossRef]
104. Ciommi, M.; Chelli, F.M.; Carlucci, M.; Salvati, L. Urban Growth and Demographic Dynamics in Southern Europe: Toward a New Statistical Approach to Regional Science. *Sustainability* **2018**, *10*, 2765. [CrossRef]
105. Ciommi, M.; Chelli, F.M.; Salvati, L. Integrating parametric and non-parametric multivariate analysis of urban growth and commuting patterns in a European metropolitan area. *Qual. Quant.* **2019**, *53*, 957–979. [CrossRef]

106. Salvati, L.; Ciommi, M.T.; Serra, P.; Chelli, F.M. Exploring the spatial structure of housing prices under economic expansion and stagnation: The role of socio-demographic factors in metropolitan Rome, Italy. *Land Use Policy* **2019**, *81*, 143–152. [CrossRef]
107. Kreyenfeld, M.; Andersson, G.; Pailhé, A. Economic Uncertainty and Family Dynamics in Europe: Introduction. *Demogr. Res.* **2012**, *S12*, 835–852. [CrossRef]
108. Hofmann, B.; Hohmeyer, K. Perceived Economic Uncertainty and Fertility: Evidence From a Labor Market Reform. *J. Marriage Fam.* **2013**, *75*, 503–521. [CrossRef]
109. Simou, E.; Koutsogeorgou, E. Effects of the economic crisis on health and healthcare in Greece in the literature from 2009 to 2013: A systematic review. *Health Policy* **2014**, *115*, 111–119. [CrossRef]
110. Lamonica, G.R.; Recchioni, M.C.; Chelli, F.M.; Salvati, L. The efficiency of the cross-entropy method when estimating the technical coefficients of input–output tables. *Spat. Econ. Anal.* **2020**, *15*, 62–91. [CrossRef]
111. Klasen, S.; Nestmann, T. Population, population density and technological change. *J. Popul. Econ.* **2006**, *19*, 611–626. [CrossRef]
112. Dyson, T. The Role of the Demographic Transition in the Process of Urbanization. *Popul. Dev. Rev.* **2011**, *37*, 34–54. [CrossRef]
113. Lerch, M. The Role of Migration in the Urban Transition: A Demonstration from Albania. *Demography* **2014**, *51*, 1527–1550. [CrossRef]
114. Caldwell, J.C. *Demographic Transition Theory*; Springer: Amsterdam, The Netherlands, 2006; ISBN 978-1-4020-4373-4.
115. Zitti, M.; Ferrara, C.; Perini, L.; Carlucci, M.; Salvati, L. Long-Term Urban Growth and Land Use Efficiency in Southern Europe: Implications for Sustainable Land Management. *Sustainability* **2015**, *7*, 3359–3385. [CrossRef]
116. Camagni, R.; Capello, R.; Caragliu, A. Static vs. dynamic agglomeration economies. Spatial context and structural evolution behind urban growth. *Pap. Reg. Sci.* **2016**, *95*, 133–158. [CrossRef]
117. Berliant, M.; Wang, P. Dynamic Urban Models: Agglomeration and Growth. In *Contributions to Economic Analysis*; Elsevier: London, UK, 2004; Volume 266, pp. 531–581. ISBN 978-0-444-51481-3.

sustainability

MDPI

Article

A Social Network Analysis of the Spanish Network of Smart Cities

Ivan Serrano [1,2,*], Laura Calvet-Mir [2], Ramon Ribera-Fumaz [2], Isabel Díaz [2] and Hug March [2,3,*]

1 Estudis de Dret i Ciència Política, Universitat Oberta de Catalunya, Castelldefels, E-08860 Barcelona, Spain
2 Internet Interdisciplinary Institute (IN3), Castelldefels, E-08860 Barcelona, Spain; lcalvetmir@gmail.com (L.C.-M.); rriberaf@uoc.edu (R.R.-F.); twoixa@gmail.com (I.D.)
3 Estudis d'Economia i Empresa, Universitat Oberta de Catalunya, E-08035 Barcelona, Spain
* Correspondence: iserranoba@uoc.edu (I.S.); hmarch@uoc.edu (H.M.); Tel.: +34-93-450-5200 (I.S.)

Received: 29 May 2020; Accepted: 23 June 2020; Published: 26 June 2020

Abstract: This paper explores the relations of centrality and hierarchy between cities and firms implementing Smart City strategies in the context of the Spanish Network of Smart Cities (RECI). While the literature has usually focused on the global dimension of cities and firms networks, exploring a national case offers interesting insights about the presence of multinational firms in these contexts and the role played by medium-sized cities in their market expansion. The analysis is based on a two-mode network of cities and firms participating in Smart City projects with the usual measures of betweenness, in-degree and closeness, as well as computing the Gini index for each of them to assess the levels of inequality. We then explore whether the structural advantages of participating in these networks have a leveling effect or rather reinforce existing hierarchies of cities. Second, we explore how firms are intertwined in Smart City projects and whether medium-sized local firms have a relevant presence. Our findings suggest these networks become a regional gateway for multinational firms to expand their presence in Smart City national markets, rather than empowering medium-sized cities and small national firms.

Keywords: smart cities; Spain; networks; firms

---

1. Introduction

The Smart City has become a key urban paradigm for cities across the world to transform themselves into successful competitive enclaves within the global digital capitalism geographies, through technological transformations. Embedded within the rhetoric of urban competitiveness in a globalized world, the Smart City has become an institutional policy promoted not only by local governments but also by national governments and supranational actors such as the European Union (EU). Moreover, Smart City policies are also presented as a factor of success not only for cities with a prominent role at the world level but also for middle-sized cities often competing in regional rather than global networks [1–3]. Central to this argument is the fact that the development of Smart City projects will enhance the local innovation ecosystems and the creation and/or consolidation of a local smart industry with locally based companies [4]. Yet very often these projects tend to focus on technology rather than the citizens [5]. Critical studies point out that the development of Smart City strategies is part of the neoliberalization of urban governance and instrumental in the colonization of the urban technology markets by transnational corporations [6–9]. In a nutshell, the deployment of Smart City strategies across the globe would reinforce current patterns of uneven development between global and unconnected cities, and between transnational companies and small and medium-sized companies. In contrast, recent critical enquiries on the consolidation of the Smart City have shown that the picture is more nuanced [10–12]. A growing literature, based on paradigmatic case studies, has incorporated alternative smart strategies taking into account more inclusive and sustainable possibilities [13,14]. Yet, in both cases, Smart City studies need to pay attention to the networked insertion of Smart Cities in the global flows of the digital economy, avoiding "over-territorialization" [15,16]. In other words, move from place-based studies to the relational geographics of the Smart City to explore the limits and potentialities of Smart City models beyond local processes and factors.

In this regard, Wall and Stavrapoulos [15] propose the introduction of network analysis methodologies to understand, among other things, the articulation between local Smart City developments and the global urban and firm dynamics (see, for example, [17]). Following this call for incorporating urban networks analysis, this paper focuses on the articulation between cities, smart strategies and firms. However, rather than focusing on the global level, we will focus on the interrelation between Smart City strategies and national networks, an area that has received little attention so far. Unlike global cities, which usually have the capacity to implement their own agendas, for many regional and provincial cities the capacity to develop Smart City strategies depends on the mediation and support from nation-states or national networks, such as China's or India's Smart City programs. Additionally, the connection of these cities with global flows is often mediated at the national scale by firms that, at the most, establish local offices subordinated to national headquarters (see, for instance, [18] for a good description of the Indian case). Thus, this national mediation can open the door for different Smart City economies not controlled by usual global players.

To contribute to the scholarship documenting the network inequalities of cities and firms in the development of Smart City projects, in this paper we take as case study the Spanish Network of Smart Cities (*Red Española de Ciudades Inteligentes*), known as RECI, and focus on the relations of centrality and hierarchy among cities and firms in Spain participating of an institutionalized network of Smart Cities. First, we explore whether the structural advantages of participating in these networks have a leveling effect or rather reinforce existing hierarchies of cities. Second, we explore how firms are intertwined in Smart City projects and whether medium-sized local firms have a relevant presence, or whether these networks become a regional gateway for multinational firms to expand their presence, reinforcing their dominance.

Our results suggest that medium-sized cities involved in the network of Smart City projects can have a similar presence to that of bigger cities, but connectedness may come at the price of multinational firms being the main gateway that connect cities to the network. Indeed, our analysis suggests that the network of firms involved in Smart City projects in Spain is dominated by multinational companies. These findings support the claim that Smart City projects at national levels can be regarded as instrumental to the globalization process, by means of which big firms penetrate and maximize their presence in national markets. Moreover, we find limited evidence for a counter-argument often found in the public policy arena, according to which the participation of big firms in national networks, including medium-sized projects and cities, may foster the participation of small national firms that could improve their viability and scale.

2. Literature Review

The Smart City paradigm is often optimistically presented as a strategic driver for success in the global network of city competition [19]. With the dominance of information and communication, and technologies as the key strategic vector of economic growth, cities must become "smart", integrating innovative hardware, software and the intensive use of big data and social networks [20]. It can be argued that the rhetoric of the Smart City is the evolution of those positions claiming the need for cities to become competitive agents attracting talent and fostering a creative environment, with economic innovation as the only path towards minimally inclusive societies [21,22]. The "smart" strategy would allow cities not only to compete in a globalized market but also to become more efficient and sustainable, offering economic progress and well-being not only to the more skilled segments of the workforce.

Indeed, this vision follows the dominant approach in urban studies and globalization. It is argued that cities can be better understood as nodes of a network [23,24]. Instead of classical hierarchical models, Castells [25] defined cities participating in global networks as "spaces of flows". Within contemporary capitalism, the network approach can reinterpret Friedman's definition of cities [26] as basing points of "spatial organization and articulation of production and markets". In the public policy sphere, the metaphor of cities as nodes in a network has resulted in an optimistic expectation about the emergence of Information and Communication Technologies (ICTs) and contemporary globalization taking over the determinant role that "place" had traditionally played in shaping the potential success of cities [21,27]. To do so, cities must cooperate with public and private organizations providing infrastructure and an adequate environment to promote start-ups and attract innovative firms.

In contrast, critical studies highlight that, as a consequence of this dominant Smart City rhetoric, the goal of urban sustainability and economic development becomes the commodification of the city and urban policy [7,28]. The existing evidence suggests that being "smart" and integrated into global flows is associated with higher levels of wealth and competitiveness, and to stronger positions in the global network of cities. For instance, Caragliu and Nijkamp [1] find that the intensity of ICT use is correlated with urban wealth, while, in a similar vein, Wall and Stavrapoulos [15] show that GDP levels and the attraction of foreign direct investments are associated to the "smartness" of a city. The intense pursuit of public policies aimed at increasing the competitive capacities of cities might then show that embracing the standardized practices of Smart Cities would not alter existing hierarchies. On the contrary, the consolidation of this paradigm can reinforce the power of big firms instead of balancing the potential of cities, an even deepen the existing inequalities. All in all, big firms are well equipped to succeed in this competitive, innovative and technological paradigm. Cities' and firms' networks reflect the structure of the current process of globalization, but they are fundamental agents of globalization that play a key role in defining which cities become global [17,29]. Indeed, the study of city and corporate relations as a dual network has suggested that inequality is greater among cities than among firms, with the former more conditioned by their geographical situation and other classical factors than an optimistic view of ICTs would allow, while the latter more flexible in seeking competitive advantages.

Thus, whether the Smart City epitomizes a deregulated global capitalism, increasing inequalities, commodification, privatization and deregulated and depoliticized public institutions is a matter of controversy. There are critical voices that refer to the Smart City paradigm as an empty concept, reflecting the worldview of the neoliberal project [6,19,30]. Under such a dominant market-oriented conception of global networks, cities can become an instrumental device for the intra-national expansion of corporations. From this perspective, cities would have a subordinated role in the configuration of current geographies of globalization, with the Smart City rhetoric permeating urban strategies not only in large but also medium-sized cities.

However, there is limited evidence on how this emerging paradigm affects regional or national networks (see, for instance, [31] for a longitudinal perspective on Smart City networks in China). Research has often focused on global city networks and big firms from advanced business services, and much less on medium-sized and technology-oriented enterprises. Other contributions focus on qualitative case studies or small comparative analyses [32,33], but little has been done on regional networks taking into consideration medium-sized cities and firms. In this sense, the research presented here provides evidence on how cities and firms operate in an institutionalized national network. In doing so, it adds some insights that can be useful to explore the limits and possibilities of the deployment of Smart City strategies to develop local/national smart technologies and economies.

3. Materials and Methods

Our case study is the Spanish Network of Smart Cities (RECI), established in June 2012. As mentioned in their mission statement, the network aimed to become "an association of local territories whose entities are representative of the territory and lead innovation systems in their own field by promoting their own local network of agents related to research and innovation" [34]. Even though the network was formally promoted by different city council members and currently includes more than 70 cities, it was initially funded by a private foundation (FUNDETEC) sponsored by big national and international firms with a strong presence in Spain, such as INDRA, Hewlett-Packard or Telefónica. Later, in 2015, this foundation was dissolved and the RECI network assumed part of its infrastructure and goals. According to the RECI manifesto, Smart cities are those with "innovation and networking" systems aimed at improving "economic and political efficiency allowing social, cultural and urban development" [34]. From this perspective, the RECI aims to foster a dynamic of cooperation between cities and firms leading to the consolidation of a "Spanish network of smart cities". This policy should promote the automatic and efficient management of urban infrastructure and services, as well as the reduction of public expenditure and the improvement of the quality of services, thus attracting economic activity and generating progress [34]. A key instrument in achieving these goals is "economic growth" by fostering public-private cooperation among cities and creative and technological industries [34].

Many cities have embraced this paradigm as a key strategy that affects different policy areas at the urban level, and the European Union has incorporated the concept in a variety of publicly funded programs for public administrations, research institutions and private companies. Spain's Smart City policy is a good example of a national policy that follows EU's guidelines and other quality standards in the industry (such as the Spanish Agency for Normalization, AENOR), aimed at promoting sustainable growth by fostering the participation of local councils.

3.1. Methods

To achieve our research objective we used social network analysis (SNA), a standard tool to study social structures, relations and flows [35]. A growing body of work has applied SNA to understand the relational patterns between firms and cities, mostly at the global/world level [17,22,36,37], but also at the regional [31] and multi-scalar levels [8]. Despite some critiques and limitations of networks as a metaphor to understand relational patterns and the resulting hierarchies in world cities [38–40], SNA has been providing relevant insights for assessing and modeling urban networks through corporate networks [41]. We draw on this body of literature, opting for a two-mode (also known as affiliation) network of cities and firms. This option allows us to explore how both groups of actors are associated to each other by means of Smart City projects in Spain. In this sense, the analysis conducted here expands the available evidence about Smart City networks using SNA to compare cities and firms at a national level, a dimension that has not received attention in the existing literature.

The data on the cities and the corporations involved in urban development through the RECI were compiled through Internet searches of the websites of the RECI, of the city councils, and of news and press releases between July and November 2016. The cities were identified by consulting RECI's official website; at the time of data gathering, the RECI was made up of 65 cities. The firms involved in Smart City projects in these 65 cities were identified through Internet searches of each city's official website (some cities, like Barcelona or Santander, have specific websites for the smart city development projects) and by using key search words such as the cities' names followed by "smart city", "smart city projects" or "urban innovation". Overall, we followed a broad inclusion criterion, searching for corporations that were taking part in Smart City related projects or actions in each city. This strategy resulted in 553 corporations involved in Smart City projects of RECI members, including firms taking part in Smart City actions at a local, regional and international level, for example as part of European consortiums. Of these 553 initially identified firms, 66 were participating in Smart City initiatives in at least two cities of the RECI and were included in the analysis. By doing so, we removed pendant nodes that would be connected to the graph by only one tie. Moreover, these pendant nodes corresponded mostly to cities participating in EU intercity projects, so this decision allowed us to focus on cases actually connected by the RECI association [42], which may help to understand the processes within the RECI at a local level. The number of initial firms was also reduced, as we consider different branches of the same firm as a single firm. For example, Acciona Agua, Acciona Energía, Acciona Infraestructuras and Acciona Instalaciones were all considered as Acciona, for clarity purposes. As a result, the final network consisted of 47 cities and 66 firms.

3.2. The Two-Mode or Affiliation Network of Cities and Corporations

To assess the firms' articulation in the Smart City development through corporate networks, we constructed a two-mode network or affiliation network. In social network analysis, the term "affiliation" is used to refer to membership or participation data, such as actors participating in events, which consist of a set of binary relationships between members of two sets of items [35]. Examples of affiliation data include corporate board memberships [43,44], attendances of events [45] or patterns of authorships and institutional pathways of scientists [46]. Two-mode networks allowed to capture the interaction among cities and firms and to test an urban network formation hypothesis to deepen the understanding of how individual cities and firms interact locally and form observed corporate networks [41].

We built an affiliation matrix where the rows and the columns correspond to the corporations (a total of 66) and the cities (the 65 cities members of the RECI). The matrix allowed us to determine which corporations participate in which cities' smart projects, and which cities are related to which corporations through Smart City projects. The matrix computes z_{ij} as 1 when a corporation i is present in city j, and otherwise as 0. Moreover, we used Gephi to visualize the bipartite graph resulting from the affiliation network. Links connect firms with cities in which they are involved through Smart City initiatives, and there is no direct linkage within the same set of nodes (i.e., between cities or between firms), which enables a clear visualization of linkages between the two sets. In the visualization of the bipartite graph, the nodes' size represents the betweenness centrality in the network [35]. Two-mode betweenness centrality refers to the number of geodesic paths that pass through a given node, weighted inversely by the total number of equivalent paths between the same two nodes, including those that do not pass through the given node [47]. Accordingly, links between firms always include cities, and links between cities always include firms, which means that cities are on geodesic paths between firms and firms on geodesic paths between cities. A city's betweenness centrality increases when the pairs of firms are linked only through that city, and if a firm is involved in that city, all geodesic paths from that firm must include that particular city. In the same vein, a firm gains betweenness centrality when two cities are only connected between them because of the firm's participation in Smart city projects in those cities. Thus, betweenness centrality is an indicator of a city's (or firm's) strategic power by acting as bridge between firm pairs (or city pairs) that would otherwise be unconnected.

3.3. Co-Affiliation Networks: Inter-Corporation and Inter-City Linkages

In order to explore the connections between firms and between cities separately, we transformed the affiliation matrix of corporations and cities (66 × 65), in which there were no direct linkages within the same set of nodes, into two one-mode squared matrices of overlaps, through the cross-product method for corporations and cities separately [42]. Applied to our binary matrix, each product is 1 only if two corporations were "present" in a city, and the sum across cities yields the number of cities in which the two corporations participated. Then, the first one-mode matrix obtained for corporations (66 × 66) accounts for the number of cities in which each pair of corporations is present; and the cities (65 × 65) matrix indicates how many corporations were present in each pair of cities. The co-participation (or co-affiliation) of corporations in Smart City projects of the same city can be considered as an indicator of the extent to which firms get involved in these cities. Then, in this case, co-affiliation would be present if the same firm was involved in Smart City projects in two cities. In other words, the more cities in which two firms co-participate, the greater the possibility that these two firms are somehow "inter-linked" [36]. Because most network methods are designed for the analysis of binary data, we followed the procedures by Neal [36] and Field et al. [48] to dichotomize the resulting valued matrices of overlaps. Inter-firm and inter-city relationships were recoded as present if they were two or more. That is, if two corporations have two or more cities in common, the matrix score will be 1, and 0 if two corporations have less than 2 cities in common. Following the same logic, the inter-city matrix scores were recoded as 1 (present) if two cities had two or more firms in common and 0 otherwise. Setting the threshold of connections at 2 is rather a conservative decision, according to the literature. However, we must be aware of its limitations, particularly when making comparative claims [29]. In order to prevent this overreaching, we tried to emphasize the characterization of the network as a whole while trying to be cautious about its capacity to equalize the structural advantages of cities.

We proceeded by representing the inter-firm and inter-city valued networks, to visualize the most important corporations and cities involved in Smart City projects within the RECI. Subsequently, we computed three normalized measures of centrality degree, closeness and betweenness for each city and firm, the conventional indicators to assess the relative structural position and importance of a node within a network, which in turn creates hierarchies [17,36,49]. Following the usual practice in this field, we calculated normalized measures to allow for the comparability between the different measures of centrality. Degree centrality accounts for the number of direct links within a network. In this case, degree centrality refers to the direct links (both incoming and outgoing) between cities or between firms, which indicates the direct involvement in the network. Closeness centrality captures the extent to which a city is directly connected to other cities in the network or separated from them by only short indirect linkages, reflecting the notion that both direct and indirect linkages contribute to a city's centrality and to opportunities for capital accumulation or innovation diffusion [36]. The closer two firms are in the network, the less dependent one is on intermediary channels. Thus, firms with high closeness centrality can provide their producer clients with a more independent and rapid access to information from a wider ranges of sources [36]. Betweenness centrality focuses on the extent to which a city or a firm serves as an intermediary that facilitates the flow of resources for other cities (or other firms) in the network. A strong position in terms of betweenness would reflect a brokering or gatekeeping position, which would provide an actor with a unique ability to monitor and control resource flows, acting as an interlocking agent. Finally, we also calculated the Gini coefficient as a measure of inequality within each of these centrality measures of hierarchies, with values closer to 0 indicating more equality. Under the logic of network analysis suggested by Neal [36], this inequality index complements the rank provided by centrality measures by reflecting structural inequalities among cities and firms.

4. Results

The bipartite graph from the RECI's two-mode network (Figure 1) shows to what extent cities and corporations are in structural advantage positions when acting as interlocking agents. Bigger nodes have greater betweenness centrality and function as "brokers" in the network. Spanish firms such as Telefónica, Indra, Acciona or Tecnalia have a strong presence in the network, but foreign capital firms such as Philips or Endesa (which was acquired by the Italian group Enel in 2009) with a relevant position can also be found. Among cities, Madrid, Barcelona, Málaga or the association between Palencia and Valladolid have a strong position within the network. A bipartite graph such as the one represented in Figure 1 offers a general approach to a two-mode network, indicating that some companies seem to play a prominent role and that certain cities—mostly big Spanish agglomerations—seem to act as connecting nodes among firms and the rest of the cities. However, it does not fully reflect the internal features of the network, that is, how their presence is structured.

To further explore the multiple dimensions of a two-mode network, the data informing the visualization offers relevant insights about the characteristics of cities' and firms' participation. For instance, according to the number of cities where each corporation is present (shown in Table S1, in the Supplementary Data section) Telefónica is involved in 17 cities, followed by corporations such as Acciona, Indra or Phillips, which are involved in eight cities. This defines a network structure with a strong presence of major companies of Spanish origin. Conversely, over 65% of the corporations identified were only present in two cities, which indicates that a small number of corporations articulate the core of the Smart City development within the RECI. However, this is not to say multinational companies are absent. It is worth mentioning that multinational companies not based in Spain represent four out of the first 11 companies present in at least five cities. Moreover, when considering the total amount of companies which are at least present in two cities, more than half of them (54.5%) are multinational companies with non-Spanish capital, which suggests that the RECI can be seen as a network that facilitates the entrance into the Spanish market, where these companies benefit from a structural advantage by being associated to Spanish firms. This possibility is consistent with the data displayed in Table S4 (Supplementary Material section), where we find that foreign multinationals are present in nine out of the 14 top pairs of companies with a shared presence in the same city.

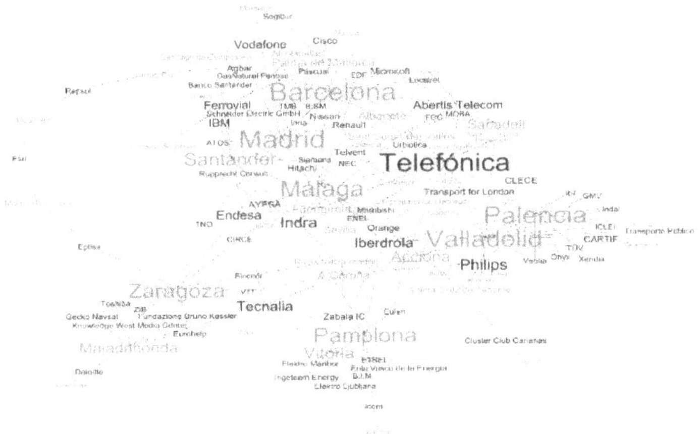

Figure 1. Bipartite graph of the two-mode network city-by-firm of the RECI. Font size represents betweenness centrality. Darker color is used to identify firms, while lighter correspond to cities.

Turning to the cities' presence in the network, we find cities such as Barcelona, Madrid or Valladolid/Palencia, with at least 16 corporations present in two cities of the RECI, taking part in Smart City projects (see Table S2 in the Supplementary Materials). Other important cities in terms of the number of corporations involved are Málaga, with 15, or Pamplona and Santander, with 12 corporations each. A key finding is that a remarkable percentage (29%) of cities belonging to the network—a total of 19—do not participate in Smart City projects despite being institutional members of the RECI, and 13 other cities have a minimal participation, with only one corporation involved in a Smart City project. In most cases, cities in the network that do not participate correspond to less populated cities, even though there are some cases of relatively populated ones with just one participation or not participating at all, such as Las Palmas or Cordoba, both with a population above 300,000. However, the highest concentration of firms in Smart City projects is found among the most populated cities. There are some relative exceptions, such as Palencia, but this can be considered as a single case, since it is associated with Valladolid in a joint partnership for Smart City projects.

4.1. The Cities Network

Cities are mapped in a network graph (Figure 2) with label sizes proportional to their betweenness. Barcelona, Málaga or the association between Palencia and Valladolid appear as the main nodes of the network in terms of their brokering capacity. The graph shows that bigger cities not only have a higher presence of corporations and more connections but also tend to be connected among them. Furthermore, a closer look also suggests that in some cases these cities may have a neighboring effect, playing a leading role for medium-sized cities in their metropolitan areas.

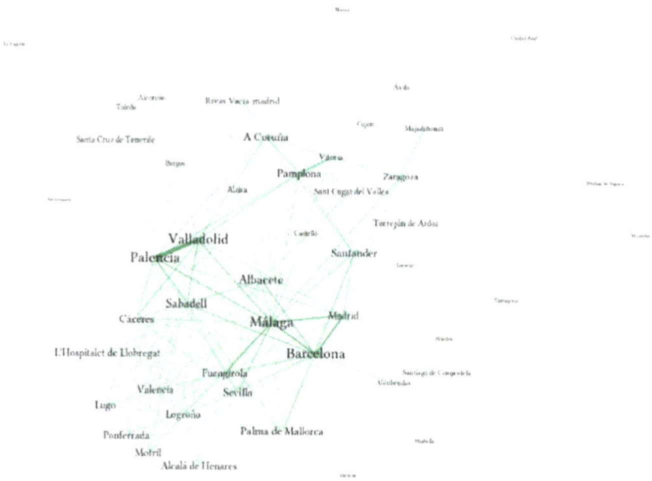

Figure 2. Inter-city networks. Node size is based on betweenness centrality in the network. A tie between nodes means they have at least two corporations in common.

Table S3 displays pairs of cities that share at least three corporations. This threshold allowed us to focus on those cases where participation in the network would provide a structural advantage for cities offering the possibility of a high level of interaction with companies and other cities. The pairs of cities with the highest number of corporations shared are Valladolid and Palencia, which are the capital of neighboring provinces. However, the reason for their close connection is that they established a platform known as Smart City Valladolid and Palencia to foster public-private collaborations for both joint and individual innovations in relation to Smart City development. As for the rest of cities with high numbers of shared corporations, we find different yet complementary situations. On the one hand, big cities such as Madrid, Barcelona or Malaga are usually well connected among them; on the other hand, they could benefit from a neighboring effect, that is, cities from the same region have more options to share the presence of the same firms. This is the case of cities such as Sabadell (near Barcelona) or Fuengirola (near Malaga), but there are opposite situations such as Madrid, which is in general poorly connected to its metropolitan cities. Table S3 in the Supplementary Materials illustrates these possible relations for cities sharing at least three firms. A total of 11 out of 15 pairs include at least one of the main Spanish cities, while eight cases feature some kind of connection, either top-to-top city, regional or metropolitan—variations that would deserve further research combining a quantitative and qualitative analysis and exploring causal factors such as the socioeconomic structure or the political parties in government. However, while factors such as regional and metropolitan links could help explain the cities' connections, it seems that the cities' links seem to follow a hierarchical logic, with bigger cities more often connected among them.

To capture the whole complexity of the data informing the visualization, Table S6 includes individual measures of degree, betweenness and closeness. Moreover, a Gini coefficient is computed for each measure to assess the inequality of the three indicators. These measures reveal two further features of the network as a whole. First, betweenness indicates that even the most successful brokering cities acting as the shortest path between two other cities do not play this role very often: Malaga scores 3.3, followed by Barcelona or Santander, with scores below 3. Accordingly, these cities tend to have low scores of degree and closeness, indicating they are poorly connected to the other cities. Second, although the network is characterized by low centrality levels, there are high disparities among cities, as reflected in the high Gini's coefficient on betweenness (0.88) and degree (0.79), given that most cities never act as brokers between another pair of cities and tend to have few connections to other cities.

A possible explanation for this situation is that firms may feature higher levels of connectivity and brokering capacity, aspects that we explore in the next subsection.

4.2. The Firms Network

In the case of cities, following the same criteria, we have identified pairs of corporations that share their presence in at least three cities (Table S4). Telefónica is the company with more presence in this list, participating in five out of 14 pairs, followed by Endesa, with four pairs. Other companies such as Acciona, Iberdrola or Philips are present in three pairs. Accordingly, these companies are also those with more shared participations in cities, and they are particularly associated with each other. For instance, the pairs Acciona-Iberdrola, Telefónica-Acciona and Telefónica-Endesa all share a presence in four cities. This evidence suggests interesting conclusions about the corporate network of Smart City projects in Spain. On the one hand, most presences in the same city correspond mostly to multinational Spanish capital firms, while foreign capital corporations tend to share their presence in RECI's member cities with a Spanish multinational. It must be noted here that the case of Endesa, which was acquired by the Italian Enel in 2009, could be seen in this sense as a strategy pursued by multinational firms to gain access to new markets by acquiring local companies. Thus, in all cases, most companies are well-established multinational firms, either with Spanish or foreign capital. This would call into question the expectation that Smart City networks would foster the participation of local and medium-to-small companies. On the contrary, big firms seem to enjoy greater structural advantages from the network. On the other hand, this pattern could also indicate that participating in this network offers foreign firms a mediated access to the growing Spanish Smart City market.

The firm network (Figure 3) shows the prominent role of Telefónica and Iberdrola in terms of betweenness, followed by Acciona, Tecnalia, Iberdrola and Endesa. With the exception of Endesa (probably having to do with its acquisition, mentioned above), all correspond to Spanish multinational firms, which is consistent with the idea that they play a key role as gateways for international companies to enter the Spanish market.

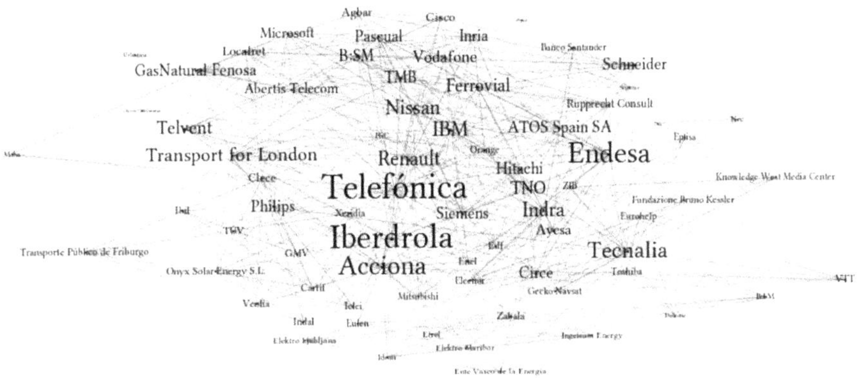

Figure 3. Inter-firm networks. Node size is based on betweenness centrality in the network. A tie between nodes means they have at least two cities in common.

Their capacity for mediation is confirmed when analyzing the specific scores for the main firms in the list, as displayed in Table S5. Telefónica and Iberdrola display higher measures not only in betweenness but also in the other two measures of centrality (in-degree and closeness), confirming their prominent role as fundamental nodes in the network. The other corporations appear in different positions for each centrality measure. Acciona and Endesa have similar in-degree scores, but Endesa appears before Acciona with regard to closeness. Tecnalia, for example, has more betweenness than Endesa. These variations suggest that, in terms of hierarchies, there are remarkable levels of inequality, and the Gini score of betweenness confirms that a small set of companies concentrate most of the intermediary role.

Thus, the firm network offers two interesting conclusions. Some companies combine a relatively high level of connections with a low brokering capacity. We find this situation in multinational companies such as Philips, which could suggest that participating in the network is a way of accessing the Spanish Smart City market. As the participation of multinational companies is mediated by local multinational firms, this presence does not provide them with a brokering capacity. Conversely, some companies have relatively low levels of connections but a relevant betweenness capacity, suggesting that they act as strategic hubs in the network because of their specialization (like Tecnalia, a technological corporation mostly focused on applied research) and even regardless of their origins, if they have been established in Spain for a long period of time (the case of IBM).

5. Discussion

The network analysis carried out for the Spanish Network of Smart-Cities (RECI) indicates that the latter is not able to involve a significant amount of its members. A high percentage of member cities only participate in a single Smart City project or in no project at all. The centrality measures computed in the analysis show a network characterized by a combination of low centrality scores and a high level of inequality. Only a handful of cities are able to have a brokering role, but they very rarely find themselves in this situation. As for the degree measures, complemented with the overall Gini index, we obtained a picture where inequality is indeed a key feature of the city network. Therefore, the data revealed that a small number of cities concentrate most of the connections, well above the other member cities. Moreover, they mostly correspond to Spain's larger cities, such as Madrid, Barcelona or Malaga, suggesting that cities that are already more powerful obtain better structural advantages from participating in the network. Thus, the network actually reinforces the existing hierarchies of cities, instead of leveling the field for medium-sized cities.

Concerning the role of companies, the core of Smart City development within the RECI network consists of a small number of corporations, with Spanish-capital multinational firms having the strongest presence. They represent key economic sectors with a strong technological component, from telecommunications (Telefónica and Abertis) to energy (Iberdrola and Endesa) and urban services and infrastructure (Acciona and Ferrovial). In this sense, the Spanish ecosystem of Smart City projects seems an opportunity to reinforce national companies, but only those with a strong international dimension. These elements are reflected by the Gini coefficient, with a small number of companies benefiting from the structural advantages of the network. This indicates that structural advantages are more available to the biggest companies, while medium-sized and national companies seem to have a poorer access to the network's possibilities. Moreover, the brokering role of bigger national companies seems to act as a hub for non-Spanish multinationals to gain access to the national market. Rather than having links with other non-Spanish firms, they tend to be related mostly with Spanish multinationals. Thus, the high proportion of international firms that are present in at least two cities reinforces the claim that participating in RECI's member cities is a way of accessing Spain's growing Smart City market, which, according to specialized reports, can reach billions of euros in the next years.

Thus, firms seem to benefit more from the network than cities, even though the original goal of the RECI is to strengthen cities' capabilities. In fact, this kind of institutionalized networks seems to strengthen a small group of cities and companies enjoying greater structural advantages, reinforcing the concentration dynamic and increasing the inequalities in the development of Smart Cities in Spain. As for non-central actors, neither medium-sized cities nor Spanish capital firms seem to obtain the same benefits from participating in the network, the exception being foreign-capital firms that may see an opportunity to access an emerging regional market. These contradictory outcomes show the necessity to move beyond discursive analysis based on the acceptance or rejection of the rhetoric underpinning Smart City discourse, based on market notions such as competitiveness, survival and efficiency. In order to understand how more progressive, inclusive and sustainable smart strategies can be deployed, it is necessary to explore the geographical relationalities in which the actually existing Smart City is produced [50] and, more particularly, the necessity to link place-based dynamics with supra-local network developments [15]. To sum up, this evidence has implications for the design and evaluation of actions aimed at developing Smart City initiatives. Our findings point to the necessity of establishing clear goals and indicators of evaluation when adopting Smart City initiatives. Otherwise, unexpected outcomes may arise which are actually detrimental to the formal objectives of the institutional policy.

6. Conclusions

Institutionalized Smart City initiatives are aimed at delivering better structural advantages to cities [1]. It is argued that, in the context of an increasingly competitive globalized world, the Smart City model in its mainstream formulation can deliver a better city governance, taking advantage of the ICT possibilities developed by means of public-private partnerships [51,52]. In national contexts such as our case study, in Spain, institutionalized networks are expected to be particularly beneficial to medium-sized cities and firms. However, from a more critical perspective, it could be argued that the expansion of the Smart City paradigm in national arenas can become a gateway for bigger firms to expand their market possibilities, without changing the hierarchy of cities, if not reinforcing existing inequalities [53–55].

This paper has explored the relations of centrality and hierarchy between cities and firms implementing Smart City strategies in the context of the Spanish Network of Smart Cities (RECI) as of 2018. Since the literature has usually focused on the global dimension of cities and firms networks [15], exploring a national case offers interesting insights about the presence of multinational firms in these contexts and the role played by medium-sized cities in their market expansion. Building on a two-mode network of cities and firms participating in Smart City projects, we have explored whether the structural advantages from participating in these networks have a leveling effect or rather reinforce existing hierarchies of cities. We have also analyzed how national and multinational corporations are interrelated in Smart City projects and whether medium-sized local companies have a relevant presence or not. Our findings suggest that these networks become a regional gateway for multinational firms to expand their presence in Smart City national markets, rather than empowering medium-sized cities and small national firms. Thus, our results suggest that an institutionalized network in a national context can indeed reflect the existing hierarchy of cities, while acting as a gateway for foreign multinational companies to have an emerging access to a growing market, rather than providing better structural advantages to medium-sized cities and local companies.

These elements should be taken into consideration when designing and implementing institutionalized policies on Smart cities, taking into account that national Smart City policies (with specific objectives such as promoting Smart City industry clusters) may not be fully aligned with local Smart City strategies (centered on implementing the best solutions available). While it might be difficult to reconcile the multiple rationalities behind the promotion of Smart City initiatives, instruments such as networks of cities may help to reconcile both perspectives, if their functioning is carefully analyzed and there is the clear objective of preserving public interests over private ones.

Supplementary Materials: The following are available online at http://www.mdpi.com/2071-1050/12/12/5219/s1, Table S1, Corporations identified as participating in smart city projects in cities member of the RECI by the number of cities in which they are involved; Table S2, RECI city members by the number of corporations identified as participating in smart cities initiatives that were present in at least 2 cities of the network; Table S3, Pairs of cities that shared at least three corporations involved in smart city projects; Table S4, Pairs of corporations that shared participation in three or more cities; Table S5, Centrality and hierarchy in the RECI corporate networks; Table S6, Centrality and hierarchy in the RECI city network.

Author Contributions: Conceptualization, I.S., R.R.-F., H.M. and L.C.-M.; methodology; I.S. and L.C.-M.; formal analysis; I.S. and L.C.-M.; investigation, I.S.; data curation, I.D., I.S. and L.C.-M.; writing-original draft preparation, I.S.; writing-review and editing; I.S., R.R.-F., H.M. and L.C.-M.; visualization, I.S. All authors have read and agreed to the published version of the manuscript.

Funding: The authors wants to express their gratitude for the internal funding of the Internet Interdisciplinary Institute (IN3)-Universitat Oberta de Catalunya (UOC) that contributed to develop this project.

Conflicts of Interest: The authors declare no conflict of interest.

References

1. Caragliu, C.; Nijkamp, P. Smart Cities in Europe. *J. Urban Technol.* **2011**, *18*, 65–82. [CrossRef]
2. Lopes, I.M.; Oliveira, P. Can a small city be considered a smart city? *Procedia Comput. Sci.* **2017**, *121*, 617–624. [CrossRef]
3. Kollar, M.; Bubbico, R.L.; Arsalide, N. *Smart Cities, Smart Investment in Central, Eastern and South-Eastern Europe. Economics—Thematic Studies*; European Invesment Bank: Kirchberg, Luxemburg, 2018.
4. Almirall, E.; Wareham, J.; Ratti, C.; Conesa, P.; Bria, F.; Gaviria, A.; Edmondson, A. Smart Cities at the Crossroads. *Calif. Manag. Rev.* **2016**, *59*, 141–152. [CrossRef]
5. Neirotti, P.; De Marco, A.; Cagliano, A.C.; Mangano, G.; Scorrano, F. Current trends in Smart City initiatives: Some stylised facts. *Cities* **2014**, *38*, 25–36. [CrossRef]
6. Hollands, R.G. Will the real smart city please stand up? Intelligent, progressive or entrepreneurial? *City* **2008**, *12*, 303–320. [CrossRef]
7. Hollands, R.G. Critical interventions into the corporate smart city. *Camb. J. Reg. Econ. Soc.* **2014**, *8*, 61–77. [CrossRef]
8. Martinus, K.; Sigler, T.; Searle, G.; Tonts, M. Strategic globalizing centers and sub-network geometries: A social network analysis of multi-scalar energy networks. *Geoforum* **2015**, *64*, 78–89. [CrossRef]
9. Datta, A. New urban utopias of postcolonial India. *Dialog. Hum. Geogr.* **2015**, *5*, 3–22. [CrossRef]
10. Grossi, G.; Pianezzi, D. Smart cities: Utopia or neoliberal ideology? *Cities* **2017**, *69*, 79–85. [CrossRef]
11. Mora, L.; Bolici, R.; Deakin, M. The First Two Decades of Smart-City Research: A Bibliometric Analysis. *J. Urban Technol.* **2017**, *24*, 3–27. [CrossRef]
12. Mora, L.; Deakin, M.; Reid, A.; Angelidou, M. How to Overcome the Dichotomous Nature of Smart City Research: Proposed Methodology and Results of a Pilot Study. *J. Urban Technol.* **2018**, *26*, 89–128. [CrossRef]
13. Cardullo, P.; Cesare, F.; Kitchin, R. *The Rigth to the Smart City*; Emerald Group Publishing: Bingley, UK, 2019.
14. March, H.; Ribera-Fumaz, R. Barcelona: From corporate smart city to technological sovereignty. In *Inside Smart Cities: Place, Politics and Urban Innovation*; Karvonen, A., Caprotti, F., Cugurullo, F., Eds.; Routledge: Abingdon, UK, 2018; pp. 227–242.
15. Wall, R.; Stavropoulos, S. Smart cities within world city networks. *Appl. Econ. Lett.* **2016**, *23*, 1–5. [CrossRef]
16. Charnock, G.; March, H.; Ribera-Fumaz, R. From smart to rebel city? Worlding, provincialising and the Barcelona Model. *Urban Stud.* **2019**, 0042098019872119. [CrossRef]
17. Alderson, A.S.; Beckfield, J. Power and Position in the World City System. *Am. J. Sociol.* **2004**, *109*, 811–851. [CrossRef]
18. Praharaj, S.; Han, J.H.; Hawken, S. Urban innovation through policy integration: Critical perspectives from 100 smart cities mission in India. *City Cult. Soc.* **2018**, *12*, 35–43. [CrossRef]
19. Söderström, O.; Paasche, T.; Klauser, F. Smart cities as corporate storytelling. *City* **2014**, *18*, 307–320. [CrossRef]
20. Kitchin, R. Big Data, new epistemologies and paradigm shifts. *Big Data Soc.* **2014**, *1*, 2053951714528481. [CrossRef]
21. Florida, R. *The Rise of the Creative Class—Revisited: Revised and Expanded*; Basic Books (AZ): New York, NY, USA, 2014.
22. Krätke, S. Global Pharmaceutical and Biotechnology Firms' Linkages in the World City Network. *Urban Stud.* **2013**, *51*, 1196–1213. [CrossRef]
23. Sassen, S. Locating cities on global circuits. *Environ. Urban.* **2002**, *14*, 13–30. [CrossRef]
24. Sassen, S. *Global Networks, Linked Cities*; Routledge: London, UK, 2002.
25. Castells, M. *The Rise of the Network Society*; Blackwell: Malden, MA, USA, 1996.
26. Friedmann, J. The world city hypothesis. *Dev. Chang.* **1986**, *17*, 69–83. [CrossRef]
27. Mitchell, W.J. *City of Bits: Space, Place, and the Infobahn*; MIT Press: Cambridge, MA, USA, 1996.
28. March, H.; Ribera-Fumaz, R. Smart contradictions: The politics of making Barcelona a Self-sufficient city. *Eur. Urban Reg. Stud.* **2016**, *23*, 816–830. [CrossRef]
29. Neal, Z.P. Well connected compared to what? Rethinking frames of reference in world city network research. *Environ. Plan. A Econ. Space* **2016**, *49*, 2859–2877. [CrossRef]

30. Coppola, A.; Vanolo, A. Normalising autonomous spaces: Ongoing transformations in Christiania, Copenhagen. *Urban Stud.* **2014**, *52*, 1152–1168. [CrossRef]
31. Lu, D.; Tian, Y.; Liu, Y.; Zhang, Y. The Performance of the Smart Cities in China—A Comparative Study by Means of Self-Organizing Maps and Social Networks Analysis. *Sustainability* **2015**, *7*, 7604–7621. [CrossRef]
32. Bakıcı, T.; Almirall, E.; Wareham, J. A Smart City Initiative: The Case of Barcelona. *J. Knowl. Econ.* **2012**, *4*, 135–148. [CrossRef]
33. Anthopoulos, L. Smart utopia VS smart reality: Learning by experience from 10 smart city cases. *Cities* **2017**, *63*, 128–148. [CrossRef]
34. Red Española de Ciudades Inteligentes. Available online: http://reddeciudadesinteligentes.es/ (accessed on 19 June 2020).
35. Borgatti, S.P.; Halgin, D.S. Analyzing affiliation networks. *Sage Handb. Soc. Netw. Anal.* **2011**, *1*, 417–433.
36. Neal, Z.P. The duality of world cities and firms: Comparing networks, hierarchies, and inequalities in the global economy. *Glob. Netw.* **2007**, *8*, 94–115. [CrossRef]
37. Hennemann, S.; Derudder, B. An Alternative Approach to the Calculation and Analysis of Connectivity in the World City Network. *Environ. Plan. B Plan. Des.* **2014**, *41*, 392–412. [CrossRef]
38. Robinson, J. Cities in a World of Cities: The Comparative Gesture. *Int. J. Urban Reg. Res.* **2010**, *35*, 1–23. [CrossRef]
39. Robinson, J. Global and world cities: A view from off the map. *Int. J. Urban Reg. Res.* **2002**, *26*, 531–554. [CrossRef]
40. Smith, R.; Doel, M.A. Questioning the Theoretical Basis of Current Global-City Research: Structures, Networks and Actor-Networks. *Int. J. Urban Reg. Res.* **2010**, *35*, 24–39. [CrossRef]
41. Derudder, B.; Taylor, P.J.; Hoyler, M.; Ni, P.; Liu, X.; Zhao, M.; Shen, W.; Witlox, F. Measurement and interpretation of connectivity of Chinese cities in world city network, 2010. *Chin. Geogr. Sci.* **2013**, *23*, 261–273. [CrossRef]
42. Hanneman, R.A.; Riddle, M. *Introduction to Social Network Methods*; University of California Riverside: Riverside, CA, USA, 2005.
43. Lester, R.H.; Cannella, A.A. Interorganizational Familiness: How Family Firms Use Interlocking Directorates to Build Community-Level Social Capital. *Entrep. Theory Pract.* **2006**, *30*, 755–775. [CrossRef]
44. Robins, G.; Alexander, M. Small Worlds among Interlocking Directors: Network Structure and Distance in Bipartite Graphs. *Comput. Math. Organ. Theory* **2004**, *10*, 69–94. [CrossRef]
45. Jones, L.W.; Davis, A.; Gardner, B.B.; Gardner, M.R. Deep South: A Social Anthropological Study of Caste and Class. *South. Econ. J.* **1942**, *9*, 159. [CrossRef]
46. Corbera, E.; Calvet-Mir, L.; Hughes, H.; Paterson, M. Patterns of authorship in the IPCC Working Group III report. *Nat. Clim. Chang.* **2015**, *6*, 94–99. [CrossRef]
47. Borgatti, S.P.; Everett, M.G.; Johnson, J.C. *Analyzing Social Networks*; Sage: London, UK, 2018; ISBN 978-1-5264-1848-7.
48. Field, S.; Frank, K.A.; Schiller, K.; Riegle-Crumb, C.; Muller, C. Identifying positions from affiliation networks: Preserving the duality of people and events. *Soc. Netw.* **2006**, *28*, 97–123. [CrossRef]
49. Wall, R.; Van Der Knaap, G.A. Sectoral Differentiation and Network Structure Within Contemporary Worldwide Corporate Networks. *Econ. Geogr.* **2011**, *87*, 267–308. [CrossRef]
50. Shelton, T.; Zook, M.; Wiig, A. The 'actually existing smart city'. *Camb. J. Reg. Econ. Soc.* **2014**, *8*, 13–25. [CrossRef]
51. Bank, T.W.; Group, W.B. *World Development Report 2016: Digital Dividends*; World Bank Publications: Washington, DC, USA, 2016.
52. Habitat, U.N. *World Cities Report 2016: Urbanization and Development—Emerging Futures*; UN-Habitat: Nairobi, Kenya, 2016.
53. Mosco, V. *The Smart City in a Digital World*; Emerald Publishing: Bingley, UK, 2019.
54. Clark, J. *Uneven Innovation: The Work of Smart Cities*; Columbia University Press: New York, NY, USA, 2020.
55. Walker, R.A. *Pictures of a Gone City: Tech. and the Dark Side of Prosperity in the San Francisco Bay Area*; PM Press: Oakland, CA, USA, 2018.

sustainability

MDPI

Article

New Town Development and Sustainable Transition under Urban Entrepreneurialism in China

Yun Song [1,*], Dominic Stead [2] and Martin de Jong [3,4]

1 Faculty of Technology, Policy and Management, Delft University of Technology, 2628 BX Delft, The Netherlands
2 Faculty of Architecture and the Built Environment, Delft University of Technology, 2628 BL Delft, The Netherlands; D.Stead@tudelft.nl
3 Rotterdam School of Management and Erasmus School of Law, Erasmus University Rotterdam, 3062 PA Rotterdam, The Netherlands; w.m.jong@law.eur.nl
4 Institute for Global Public Policy, Fudan University, Shanghai 200433, China
* Correspondence: y.song-3@tudelft.nl

Received: 8 June 2020; Accepted: 23 June 2020; Published: 25 June 2020

Abstract: New towns are a major form of urban growth in China. In recent years, increasing numbers of large new town projects have been planned and built in and around existing cities. These new town projects have frequently been employed by city governments as central elements of pro-growth strategies, based on ideas of urban entrepreneurialism, which seek to promote economic growth, project a dynamic city image, and increase urban competitiveness. This article studies how the pro-growth, urban entrepreneurial approach affects the planning and development of Chinese megacities. A conceptual framework focusing on land-leasing revenue and new town development strategies is employed to explore the linkages between urban growth mechanisms and urban outcomes. Empirical material from four cities in the Pearl River Delta—Guangzhou, Shenzhen, Foshan, and Zhuhai—is presented. The analysis indicates that new town developments in these cities have different levels of dependency on spatial expansion and land revenue, and emphasize different issues of sustainable development in their plans. Cities with a lower dependency on physical and economic growth are be more likely to emphasize the quality of the built environment and address issues of sustainable urban development more closely when planning and implementing new town projects.

Keywords: new town development; urban entrepreneurialism; land-driven economy; Pearl River Delta

1. Introduction

China has undergone high levels of urbanization and economic growth for several decades since major economic reform in 1978, during which its urban population grew from 172 million to 831 million by the end of 2018 [1]. In the early part of the 21st century, the Chinese government announced its plan to build 20 new towns every year before 2020 [2,3] and new town development has gradually become a key form of urban growth [4]. Various types of new towns have been planned and developed in recent decades including university towns, administrative new towns, high-speed new towns, financial city, smart city, eco-city, and low-carbon city [5]. New town development with progressive and sustainable urban concepts not only creates new economic development poles in the city, but is also used by city authorities to symbolise local urban and economic achievement, better quality of life, and innovative urban transitions.

The increasing popularity of sustainable urban concepts in China is due to intensive competition among cities [6]. Capitalist cities tend to compete with each other to increase their attractiveness to capital, talent, and visitors [7,8]. Local governments have adopted pro-growth approaches through 'place making or promotion' [9] 'civic boosterism' [10], prestige projects [11] to stimulate local economic growth and increase their competitiveness. Similar entrepreneurial strategies have also been adopted in Chinese cities [12–14]. Recently, various theories used to explain pro-growth urban governance in capitalist cities, such as entrepreneurialism [7], growth machine [15], and urban regime [16], have been applied to the Chinese context [4,17,18].

In general, entrepreneurial governance in China is often led by local government and strongly influenced by the legacy of state socialism [12]. There are several defining characteristics of China's 'entrepreneurial local state', such as the 1994 tax-sharing system reform [19], land property and market reform [4], and cadre appointment system [20]. Each of these characteristics encourage local governments to adopt pro-growth urban development strategies to increase extra budgetary revenue, and enhance local economic and political performance. Furthermore, local governments are also protected by 'soft budget constraints' [21], which means that they are cushioned from financial losses if urban investments do not provide a return. This has generated a tendency for local governments to overspend and overdevelop [22]. For instance, Long (2019) identifies 180 cities in China with shrinking populations, which are nevertheless making pro-growth urban master plans based on ambitions for population and urban growth in the near future [23]. In extreme cases, some new town projects have resulted in 'ghost towns' where urban expansion has far outpaced population growth [24].

New town or 'new city' [4] development is China's main strategy of 'city making' [25], as new towns not only house residents and businesses, but also provide new centres of regional spatial reconfiguration [26], contributing to 'a globalizing central area that formed a unified global city region' [27]. The increasing popularity of sustainable and innovative urban concepts like eco-city, low-carbon city, sponge city, smart city, and knowledge city in Chinese new town development illustrate the attempts of local governments to create global cities and boost their attractiveness through city branding and marketing [6,28]. Using sustainability as a 'city branding' tactic highlights that local government is not only a market regulator or a unique player, but it can use market instruments to achieve its hidden political agendas. De Jong (2019) argues that this 'eco-civilisation' and 'new-type urbanisation' agenda is not likely to be genuinely implemented under its structural institutional mechanism encouraging the accumulation of land and power of local elites [5,29]. Through the transition from rural industrialism to new urbanism, new town development generally embraces a regime of accumulation and legitimation from land value [4], turning suburbs into spaces of capital accumulation [30]. These observations suggest that new town developments are often a feature of local governments' pro-growth strategies, which can lead to overspending and unnecessary investments, and potentially hamper the implementation of its sustainable and innovative urban objectives in the long run.

There is a large amount of literature on China's new town development and its urban entrepreneurial mechanism. At the same time, there is a lack of empirical evidence on how local pro-growth strategies affect the planning and development of city-level new towns and the implementation of sustainable and innovative objectives. This article aims to fill the gap by examining local government rationales and strategies of new town planning, development, and the potential urban outcomes, and to contribute to a deeper understanding of China's local politics and urban growth of new town development in both theoretical and empirical terms. The article is organized in six parts. Section 2 reviews urban theories that help to explain the cause and effects of land-driven economy and pro-growth mechanism in China's urbanisation and new town development. A conceptual framework is proposed to explore the linkages between urban growth mechanisms and urban outcomes. Section 3 outlines the methods used to select cases and gather data. Sections 4 and 5 examine and analyse the selected cases with the conceptual framework. Section 6 discusses the results and analysis from the examination of the case studies and Section 7 concludes on the main findings from the study.

2. Urban Entrepreneurialism in New Town Development

2.1. Urban Entrepreneurialism in China

Understanding China's urban governance and its urban transformation, including new town development, requires an understanding of its post-socialist state-market relation in facilitating urban growth [31]. As already illustrated above, two main paradigms of urban growth machine and urban entrepreneurialism are often used to explain China's growth mechanism [7,15]. For example, drivers of urban development in Shanghai have been explained in terms of post-socialist pro-growth coalitions [18,32]. From the perspective of neo-liberalism and entrepreneurialism, post-socialist reforms have led to the creation of new local governments in China that are more 'entrepreneurial' in nature [17,33,34], but lacking in financial discipline and public accountability, which promote urban growth for political and economic objectives [21]. To understand how China's entrepreneurialism differs from that in other parts of the world, and the relations of its growth mechanism with new town development, it is first necessary to provide some more detail about the institutional context.

China's 1994 tax-sharing system reform is a key foundation of the land-driven economy and its concomitant governance. The goal of the reform was to increase tax revenue of the central government from lower tier governments. After the reform, local municipal governments had to transfer a larger share of their tax revenue to the central government, but in return they had more decision-making power in local urban development and were allowed to keep all revenue by leasing land to developers [19]. However, there is a gap between local tax income and public expenditure. In 2018, for instance, local governments collected 53% of the 'general public budget revenue' (tax revenue), but were responsible for 85% of total tax expenditure. Although the 'transfer payment' system has enabled provincial and central government to allocate tax revenue to facilitate those local governments in severe deficit, local governments in general have to seek extra budgetary revenue to support local development, and the revenue by leasing land has become a major source (which is also known as 'government-managed funds revenue'). The tax-sharing reform resulted in a decentralization of decision-making power in land politics, which made the behaviour of local government more firm-like [35]. The fiscal decentralisation not only laid the foundation for its formation of pro-growth urban development mechanisms, but also provided a new motivation for local governments to lead and promote local urban development projects.

Another significant institutional change is the shift in China's cadre appointment and evaluation system, which generates local leaders. Under China's cadre appointment system, local leaders are appointed by upper-tier governments. Since 1978, the economic performance of cities became a key criterion for evaluating the suitability of local leaders for positions in upper-tier governments [20]. Urban growth, especially new town development, became a preferred way for local governments to demonstrate local economic growth and modernization achievements after an extensive wave of industrialization (*kaifaqu*) in 1990s [4]. Thus, local leaders are often zealous in promoting conspicuous local economic growth in order to secure career promotion in China's administrative hierarchy [36]. For key local officials, like those in megacities, urban achievement is often more about prestige and less about functionality or sustainability [37].

Land and housing reform is another key factor in shaping local government's pro-growth approach. The establishment of a land leasehold market and the enacting of the Land Administration Law and Planning Act in 1998 made the previously strictly state-owned land and properties tradeable [38]. Local governments therefore have the legitimate right in land requisition, leasing out to developers, and retaining most income. Soon thereafter, local governments, as *de facto* owners of land within their jurisdictions, were not solely land suppliers but also major market players [21]. Local governments became more willing to initiate and lead urban projects to inflate local economic and political performance through local state corporatism [39], local state entrepreneurialism [40], and private–public partnership with developers [41].

After the above institutional reforms, local governments acquired significant autonomy in the disposal of land and finance, two vital resources in urban development [21]. From an urban governance perspective, local governments rely on non-public investors and form coalitions as it is not financially feasible to undertake large projects alone [42]. A pro-growth approach is therefore adopted, which has heavy reliance on the property sector to promote economic and urban growth [36,43]. According to such pro-growth approaches, infrastructure projects and property-led development have become an essential mechanism to restructure urban areas, build a good city image, and attract investment. However, it has also been criticised for resulting in a diversion of public resources from social needs [21], lacking public accountability and social goals [44], and causing overheated property booms. The approach also creates a reliance on land-driven revenue (*tudi caizheng*) of local governments, especially of the relatively developed cities. Due to the increasingly strict regulation and macroeconomic control in first-tier cities like Beijing, Shanghai, and Guangzhou, pro-growth strategies based around maximising land revenue have become more prevalent in second-, third-, and even fourth-tier cities in recent years [45]. In other words, pro-growth approaches have become a default mechanism for many local governments in China when planning urban development. The nature of this pro-growth mechanism is the focus of analysis in this paper.

2.2. New Town Development

New town development in China is the main strategy of 'city making', and is the main form of urban development, which is closely connected to the formation of China's pro-growth mechanism. According to Wakeman (2016) "new towns are not a novelty, but have an established history and well-known experiences" [46], and this is also the case in China. New towns or 'new cities'(*xin cheng*) began to gain importance and popularity in the late 1990s, marking the major transition from industrialism to urbanism [4]. The history of new town development in China reflects the evolution of national land politics. The new towns of early 2000s reflected the efforts of local governments to consolidate their territorial authority over urban fringe and rural hinterland where rural governments used to enjoy a high degree of decentralised land control of *kaifaqu* (development zone, usually refers to industrial parks) during the 1980s and early 1990s [4]. Around this time, new towns tended to act as multifunctional satellite or commuter towns constructed to accommodate the rapid growth of the urban population (or university towns for the expansion of universities from late 1990s). However, as intercity competition intensified, the nature of new towns gradually shifted and increasingly became sites for spatial reconfiguration of the city and its wider region. Innovative and sustainable urban concepts such as financial city, eco-city, low-carbon city, smart-city, and knowledge city were used to justify (or simply to label) new town projects. These justifications or labels were designed to tap into the international urban discourses to increase global urban competitiveness and attractiveness.

In China, the term new town covers a wide range of urban development. It can be used to describe new communities, new towns, new districts, and new cities [47], from urban centres, urban fringes, to rural hinterlands. The ambiguity of new towns is not merely a linguistic problem; new towns in China are difficult to explain in terms of a single discourse, urban tradition, or historical era [25]. In practice, various actors, from municipalities to private developers, use the term 'new town' on their projects, from city-level urban projects up to tens of square kilometres, to neighbourhood-level gated communities. In this article, we only look at new town projects initiated by municipalities to see how local governments shape urban outcomes from a political economy perspective.

The study employs a conceptual framework with a two-step approach to explore the linkages between urban growth mechanisms and urban outcomes (Figure 1). In the first step, the dependency on land-leasing revenue in city's fiscal structure is examined, since this is the key indicator of urban growth process as capital accumulation. Chinese local governments have two major sources of revenue: Tax revenue and land-leasing revenues. To measure the dependency of land-leasing revenue, the ratio of land-leasing revenue to total revenue is measured (R). Where a city's land-leasing revenue is equivalent or even surpasses its tax revenue ($R \geq 0.5$), then this city relies on land-leasing revenue and can be regarded as high dependency. As a reference, in 2018, the total tax revenue of all local governments in China is 7595 billion yuan, and the total land-leasing revenue is 6291 billion yuan [48]. The R of all local governments is therefore 0.45. Using this as an average figure and considering the fact that most Chinese cities adopt pro-growth strategies, it is assumed that cities with R from 0.3 to 0.5 have medium dependency on land-leasing revenue and cities with R less than 0.3 have low dependency on land-leasing revenue. The assumption of distinguishing low, medium, and high dependency on land-leasing revenue is not generated through strict quantitative analysis (which is not the goal of the paper). Where a city has far more revenue from tax than from land-leasing, then this city no longer bound by the mechanism of urban growth to boost local fiscal revenue. Thus, it is likely that the main purpose of urban development of the city gradually changes from quantity to quality. This is then examined in more detail by referring to the overall strategies and urban outcomes of the new town projects of these cities. Cities with a high dependency on land-leasing revenue are likely to develop new towns for quantity growth both fiscally and physically. Cities with land revenue as priority have higher risk of overdevelopment and urban sprawl, and those new towns branded with sustainable and innovative concepts, if any, are unlikely to be implemented as they are subject to city branding for capital accumulation. Cities with medium dependency on land-leasing revenue may evolve a mixed growth pattern. For these cities, land-leasing revenue still matters, but they may not rely on massive spatial growth to sustain its land-leasing revenue growth. They may try to focus more on the quality of urban environment as it enhances city's attractiveness in the long run, but the implementation process is likely to be constantly challenged by the need to boost land revenue, especially in new town projects. Finally, cities with low dependency on land-leasing revenue should also have low dependency on spatial growth. New town projects in these cities are likely to aim for urban quality growth such as sustainable urban transition.

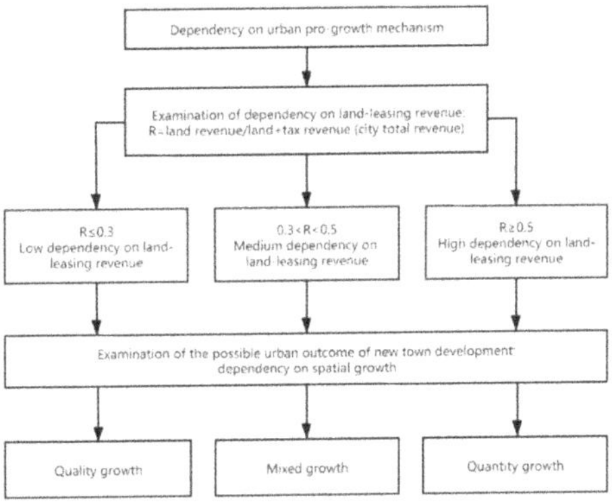

Figure 1. Conceptual framework with a two-step examination of dependency on pro-growth mechanism to possible urban outcomes.

3. Research Methods and Data Collection

The paper draws on empirical evidence from four cities in the Pearl River Delta (PRD) of Guangdong province: Guangzhou, Shenzhen, Foshan, and Zhuhai. They represent the four wealthiest cities in the PRD in terms of GDP per capita, but are very different in terms of city size and population (Table 1). Guangzhou and Shenzhen are first-tier cities and have similar levels of population and GDP, but Guangzhou's land area is much bigger than Shenzhen (i.e., Shenzhen has much higher population density). Zhuhai and Shenzhen have similar levels of GDP per capita but Zhuhai is a much smaller city than Shenzhen with a much lower population density. Foshan has the lowest GDP per capita of the four cities, and around the same level of population density as Guangzhou. They are also different in terms of administrative arrangements: Guangzhou is the provincial capital city of Guangdong province; Shenzhen and Zhuhai are special economic zones (SEZ) while Shenzhen is directly under the central government; Foshan is an ordinary city.

A major reason for comparing these four cities is they are all located within one highly competitive regional urban system: The Pearl River Delta. It is widely recognised that these cities are often in competition with each other [49–51]. According to Porter's competitive city concept [52], cities compete with each other and their competition does not fundamentally differ from national level competition. Thus, the competitiveness of a city is determined rather by indigenous factors than external ones, among which its local socio-economic environment works as an indispensable source of growth dynamics [21]. Guangzhou and Shenzhen are in competition for the leading role in the PRD region in terms of urban economic development, while Zhuhai as a much smaller city is competing with other megacities for urban environment, liveability, and sustainability. Foshan, on the other hand, as the neighbouring city of Guangzhou, is competing with its lower level of living costs and looser controls on industry. If intercity competition motivates cities to pursue more innovative and sustainable urban development to attract investment, residents, and visitors, then examining cities that interrelate within a highly competitive urban network can provide a richer understanding of the rationale of new town development strategies, how their growth mechanism works, and what this implies for urban sustainability.

Table 1. Basic information of Guangzhou, Shenzhen, Foshan, and Zhuhai and their rankings in the Guangdong province.

	Guangzhou	Shenzhen	Foshan	Zhuhai
Administrative Land Area (sq.km)	7434	1997	3798	1736
Population (2018)	15,305,900 (1st)	13,026,600 (2nd)	7,905,700 (4th)	1,891,100 (21st)
Density of Population (Person/sq.km)	2059	6522	2082	1089
GDP (billion yuan; 2019)	2363 (2nd)	2693 (1st)	877 (3rd)	344 (6th)
GDP/capita (yuan; 2019)	156,427 (3rd)	203,489 (1st)	117,985 (4th)	177,550 (2nd)

Source: Statistic Year Books of Guangzhou [53], Shenzhen [54], Foshan [55] and Zhuhai [56].

New-town projects in each of these cities were mapped (see Figure 2). The following rules for selecting new town projects were applied: (1) All new town projects mentioned in the urban master plans; (2) new town projects planned by district governments, but not included in urban master plan; (3) national districts emphasized in the urban master plans. Although some gated communities are also labelled as 'new towns', they are not included in this study. In very rare cases, some other public actors like state-owned enterprises also develop new towns. For example, the Guangzhou Iron and Steel Group developed Guanggang, a new town on its abandoned industrial site, but as it is closer to a gated community project, it is not included either.

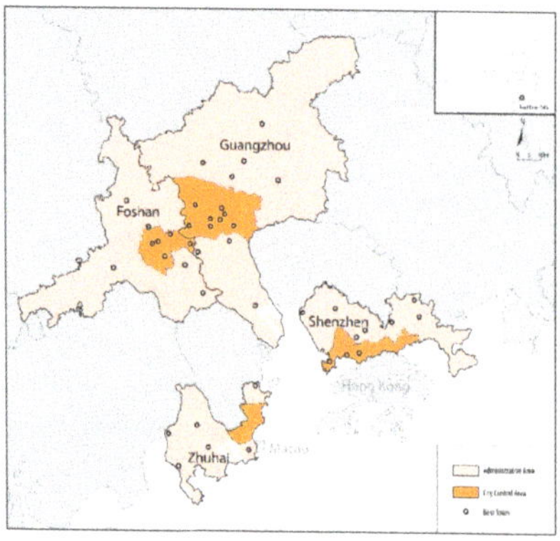

Figure 2. Guangzhou, Shenzhen, Foshan, and Zhuhai in Pearl River Delta (PRD) region and their new town projects.

Based on these selected cities and new towns, the following data to support the case studies were collected: (1) City-level fiscal data including tax revenue and land-leasing revenue from city-level statistic year book and budget performance reports; (2) the latest version of urban master plans; (3) basic information of each new town project including their size, concepts, locations, and initiating actors. In addition, interviews were conducted with local university researchers and urban planners to gather supplementary background information regarding new town development in these four cities. Lastly, field work was conducted on some new town projects to examine how they were implemented in practice.

Clearly, this contribution (and its approach) is not without its limitations, which are briefly outlined below. First is the limitation of the underlying conceptual framework. The assumed healthy fiscal revenue structure with a low dependency on land-leasing revenue only applies to relatively developed and prosperous cities. Underdeveloped cities (not represented in the selection) may experience very low shares of land-leasing revenue simply because they are in a recession or debt crisis. This situation is growing more common since in recent years the central government has begun to control the scale of local debts and curbed the growth mechanism by limiting the application of certain policy instruments. These cities relied heavily on land revenue before, and their sudden fall in land revenue certainly does not stimulate high-quality growth. Second, this study focuses more on the overall strategies of cities in new town development. In-depth empirical study on urban outcomes of these strategies is needed to further illustrate the physical impact of pro-growth mechanism.

4. Examination of Dependency on Land-Leasing Revenue

In this section, the dependency on land-leasing revenue of four selected cities is examined. According to data from 2008 to 2018, land revenue has clearly played an increasingly important role in each city's fiscal revenue structure (Figures 3 and 4). For instance, the ratio of land revenue to total revenue in Guangzhou rose from 0.26 in 2008 to 0.49 in 2018, meaning that the growth rate of Guangzhou's land-leasing revenue significantly surpassed the growth rate of tax revenue. Thus, Guangzhou has become increasingly dependent on land-leasing revenue. With almost half of its fiscal revenue from land leasing, Guangzhou has become a city with medium to high dependency on land-leasing revenues. Urban development in Guangzhou is not only a means to enhance its competitiveness but also a way of accumulating capital because of the economic revenues generated during the development process.

Shenzhen's fiscal revenue structure is quite different than that of Guangzhou. As illustrated in Figure 3, the ratio of Shenzhen's land-leasing revenue to its total fiscal revenue never exceeded 0.3 between 2008 and 2018, which represents a low level of dependency on land-leasing revenue of Shenzhen's local government. In general, Shenzhen has far less reliance on the economic contribution of land development. This can be partially explained in terms of land scarcity, which is an important concern in Shenzhen: Its population density is three times higher than Guangzhou. Because Guangzhou has medium dependency on land-leasing revenue, its overall approach to urban development might be assumed to be a combination of both quality and quantity growth criteria (according to the analytical framework presented above). Meanwhile, Shenzhen's relatively low dependency on land-leasing revenue is likely to result in an emphasis on the quality of growth (more than quantity) in its urban development strategies.

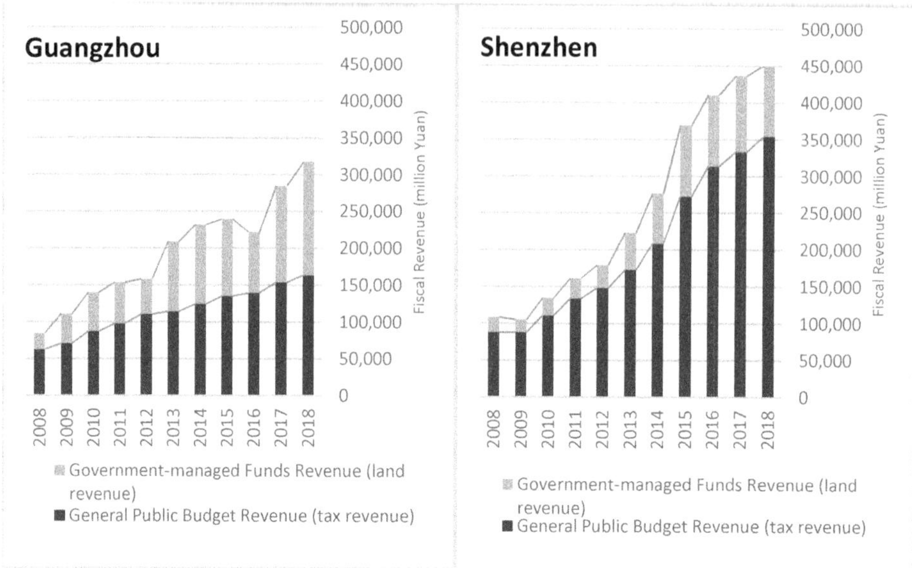

Figure 3. Land revenue versus tax revenue in Guangzhou (left) and Shenzhen (right) from 2008 to 2018. Sources: Statistic Year Books and Budgetary Reports of Guangzhou [53] and Shenzhen [54].

Foshan and Zhuhai, as lower-tier cities than Guangzhou and Shenzhen, have a higher dependency on land-leasing revenue in general. Foshan's land-leasing revenue was 42–52% of its total revenue between 2010 to 2016. In 2017, 60% of Foshan's fiscal revenue was from land-leasing (Figure 4). As such, Foshan is very dependent on the revenue through land development. The city's reliance on urban growth can be further observed in its strategies to develop new town projects. Zhuhai also witnessed significant increases of land-leasing revenue despite its smaller size. Apart from some missing data (the annual budgetary report of Zhuhai in 2013 is missing on its government website; Zhuhai began to release its budgetary report from 2011 so that land revenue data before that are missing), it can be seen that Zhuhai's land-leasing revenue surpassed its tax revenue in recent years. In 2014, 64% of total fiscal revenue was derived from land-leasing (Figure 4). The latest data for Foshan and Zhuhai show that both cities have more than 60% of their fiscal revenue was from land-leasing, which can be described as high dependency on land revenue according to the conceptual framework (Figure 1). According to the analytical framework presented above, both cities are likely to emphasise the quantity (rather than the quality) of growth in their urban development plans and practices. In the following section, the new town development plans and practices in each of the cities are examined to see whether their urban development strategies match the above propositions.

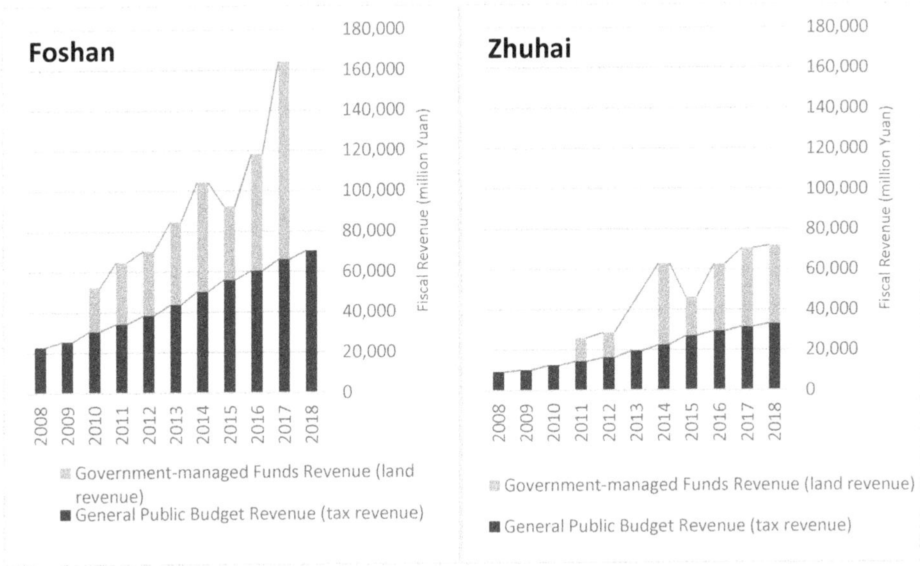

Figure 4. Land revenue versus tax revenue in Foshan (left) and Zhuhai (right) from 2008 to 2018. Sources: Statistic Year Books and Budgetary Reports of Foshan [55] and Zhuhai [56]. Note: The government-managed funds data of Foshan in 2008, 2009, and 2018 are missing, and the Foshan government change statistical calibre in 2016; the government-managed funds data of Zhuhai in 2008, 2009, 2010, and 2013 are missing.

5. Examination of New Town Development Strategies

In this section, new town development in each city is compared. The history of the urban development process is analysed to understand the role and significance of new town in the urban development process. The town projects are placed in the context of the urban development narrative to illustrate how local entrepreneurial governance affects new town development in practice.

5.1. *Guangzhou*

Guangzhou, the capital city of Guangdong province, is the political and economic centre of the PRD region. It played a key role in China's development as its historical southern gateway [57] but now faces similar challenges as many other hub cities, and is becoming overshadowed by the rapid growth of nearby Shenzhen. Nevertheless, Guangzhou has strong decision-making powers and economic resources for urban development, and the idea of developing large urban projects and deploying pro-growth strategies for enhancing its competitiveness can be easily justified.

In the process of urban development, new towns have played an increasingly important role in Guangzhou. The concept of new town appeared in the early 2000s when the city government initiated its university town and Zhujiang new town project (Figure 5 and Table 2). At that time, the concept of university town had just became popular, and the university town in Guangzhou was essentially an experimental project on the urban fringe [58]. Zhujiang new town was regarded as the core of the new city central axis of Guangzhou. The municipal government was quite cautious about this new town project and had several rounds of plans and revisions for almost a decade. The 2010 Asian Olympic Games was the event that pushed the development process of Zhujiang new town and finally implemented the project. Zhujiang new town soon became the most prestigious and successful urban project in Guangzhou. With several landmarks constructed, including the Guangzhou Opera House designed by the famous architect Zaha Hadid, Zhujiang new town demonstrates Guangzhou's latest urban achievement as the so-called 'city living room'. On the other hand, hosting the Asian Games also brought huge amount of debt to the local government due to lavish spending on landmark developments (e.g., Guangzhou Tower and Haixinsha Island where the Asian Games opening ceremony took place), sport stadiums, public transport facilities, and urban beautification projects [59].

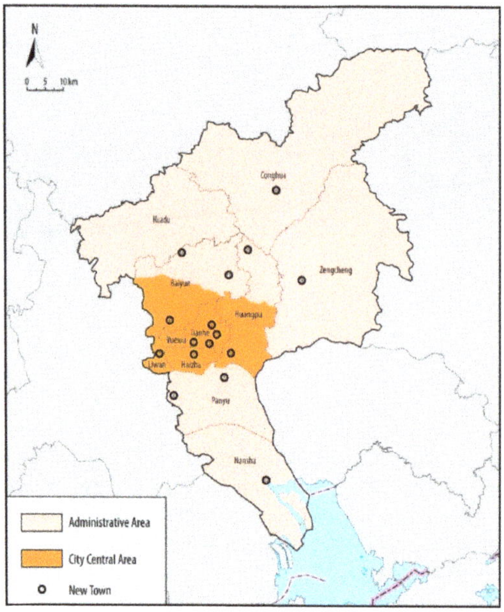

Figure 5. New town projects in Guangzhou.

Table 2. Information of new towns in Guangzhou.

	Name	Initiative	Branding	Overall/Starting Area	First Planned
1	Zhujiang New Town	Municipal	CBD	6.44 km^2	1993
2	Haizhu Ecological City	Municipal	Eco-city	92 km^2/8.9 km^2	2013
3	Guangzhou International Financial City	Municipal	Financial city	17.2 km^2/1.32 km^2	2011
4	Tianhe Smart Vally (Olympic New Town)	Municipal	Smart city	15.2 km^2	2018
5	Tianhe Smart City	Municipal	Smart city	63 km^2/20.69 km^2	2012
6	Baiyun New Town	Municipal	Second-CBD	9.3 km^2	2004
7	Huadi Ecological City	Municipal	Eco-city	3 km^2	2013
8	Southern High-speed Railway Station Business District	Municipal	High-speed rail new town	36 km^2/4 km^2	2013
9	Guangzhou International Innovative City (University Town)	Municipal	Knowledge city	5.67 km^2	2013
10	Huangpu Harbour Business District	Municipal	Second-CBD	25.04 km^2	2013
11	Guangzhou International Healthcare City	Municipal	Healthcare	33.1 km^2	2013
12	Huadu Airport Economic District	Municipal	Airport city	-	2011
13	Nansha New District	Central government	-	-	2011
14	Guangzhou Education City	District government	Knowledge city	10.79 km^2	2014
15	Sino-Singapore Knowledge City	Municipal	Knowledge city	5.86 km^2	2011
16	Conghua New City	District government	-	39 km^2	2013

However, the increased debts did not slow down the development of new towns, instead accelerating the process. Guangzhou's local government initiated and announced nine new town projects in 2013 alone (No. 2, 3, 5, 7, 8, 9, 10, 11, 12 in Table 2). Every new town in this set of projects had a 'world-class' vision and multifarious urban progressive concepts such as eco-city, financial-city, smart-city, knowledge city, healthcare city, and airport city. All these new towns carried the ambition of transforming Guangzhou's urban and industrial environment over the long term. However, the fiscal data in the same year (Figure 3) indicate a drastic increase of land-leasing revenue, more than double compared to one year before. It is highly likely that the sudden increase of land-leasing revenue was due to the release of nine new town projects and their subsequent land transaction from land market. Taking the example of Guangzhou international financial city (No. 3), after its detailed plan completed by the end of 2012, Guangzhou local government leased out four parcels of land of the 'boosting area' of the financial city in February 2013, gaining 13 billion CNY revenue, which was equivalent to one-third of the total land-leasing revenue of the year before [60]. In Chinese media, these land transactions were called the 'king of land deals' (*diwang*). Soon after, criticisms emerged accusing the Guangzhou local government for using new town projects as a means to 'sell land' (i.e., adopting a 'land-driven economy'). These new town projects continued to act as local government financing vehicles (LGFVs), attracting property investments and generating capital through the land market [32].

Many new town projects were hastily planned under entrepreneurial principles, resulting in problematic implementation. First, there were duplications of urban concepts. Tianhe smart valley (No. 4) and Tianhe smart city (No. 5) not only followed very similar development concepts, they were also in close proximity to each other. There are three other knowledge cities (No. 9, 14, 15), two second-CBD projects (No. 6, 10), and two eco-cities (No. 2, 7). Some of them adapted these concepts based on local conditions, like the former university town turning into international innovative city (No. 9), but some of them had nothing to do with their labels, such as Huadi ecological city (No. 7). This also relates to a second problem: Their planning lacked empirical justification. For example, Tianhe smart valley used to be branded as Olympic new town for a city-level sport stadium located there. But this place turned into 'Guangzhou eastern ecological and liveable district' later, and then it transformed again into a 'smart valley' for technological and innovative industries. It is hard not to question the scientific basis underlying these several plan versions. This reflects the negative effects of seeing entrepreneurial considerations outweigh urban transition targets in new town projects.

Two main observations can be made in light of the above analysis of new town practices in Guangzhou. First, the size of new town projects varies greatly depending on the function and location of the new town. Most new towns are located within or near the urban central area (Figure 5), and their starting areas are planned in details. Suburban expansion is not the major objective of new town projects in Guangzhou; they are more focused on small-area urban renovation and upgrading. However, Guangzhou is still highly dependent on land-leasing revenue, which is closely linked with new town projects. The sudden release of many new town projects and the subsequent increase of land-leasing revenue immediately follow each other. However, because of high property and land prices, Guangzhou can generate a large amount of land revenue by small-area development in the urban centre, rather than large area suburban growth. Second, the urban pro-growth approach in Guangzhou is driven more by land-driven income than spatial growth. Pro-growth approaches have resulted in several planning problems in new town practices, including redundant functions, duplication of activities, and weak linkage to urban master plans.

5.2. Shenzhen

Shenzhen is the youngest and the fastest growing megacity in the PRD. Before China's economic reform in 1980s, Shenzhen was a small fishing village. Because of its proximity to Hong Kong, it was chosen as a special economic zone (SEZ) to learn from and experiment with Hong Kong's capitalist market economy. Shenzhen has a unique urban structure defined by its borders with both Hong Kong and the mainland, known as the two-line borders (*erxianguan*). Shenzhen's SEZ was strictly confined to protect the socialist system in mainland China, which deeply influenced Shenzhen's urban structure. The central four districts Nanshan, Futian, Luohu, and Yantian, formed the original SEZ area (called '*guannei*') and later became the city's central area, while Baoan, Guangming, and Longgang were the buffer areas (called '*guanwai*') and now form Shenzhen's suburban area (Figure 6). The hard border between '*guannei*' and '*guanwai*' was removed in 2010 and Shenzhen's administrative area increased fivefold from 395 km^2 to 1948 km^2. In 2012, the redevelopment of land surpassed the new construction land, which marked the end of expansion era, and urban redevelopment and renovation became the main theme in Shenzhen's urban planning.

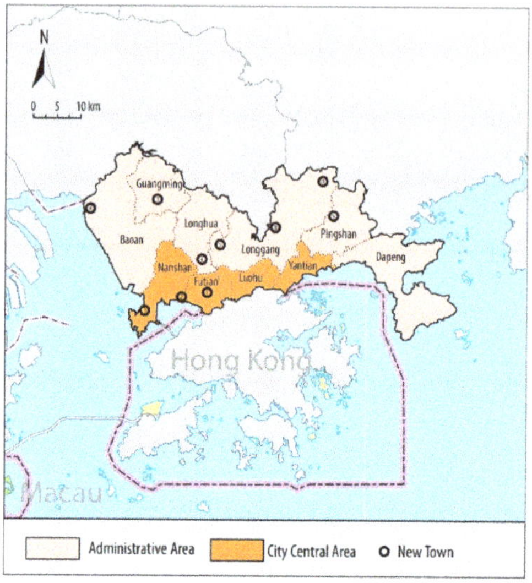

Figure 6. New town projects in Shenzhen.

Shenzhen's low dependency on land revenue and urban growth mechanism is also reflected in its new town development strategies. Unlike Guangzhou, which launched a number of new town projects that were not contained in its urban master plan, new town projects in Shenzhen are generally in alignment with its master plan, as most of them were listed and mapped on the 'key development area and projects in the near future'. In the urban central area, there are only three main new town projects (No. 1–3 in Table 3). Xiangmihu new financial centre (No. 1) and super headquarter base (No. 2) are the latest urban upgrade projects in the dense and developed urban centre. They have a small amount of land to redevelop, and aim to use it efficiently using very high floor area ratios. Qianhai (No. 3) is a long-developing free trade zone and harbour city in Shenzhen.

Table 3. Information of new towns in Shenzhen.

	Name	Initiative	Branding	Overall/Starting Area	First Planned
1	Xiangmihu New Financial Centre	Municipal	Financial city	4.9 km^2/1.9 km^2	2018
2	Shenzhen Bay Super HQ Base	Municipal	CBD	1.2 km^2	2013
3	Qianhai Central Zone	Municipal	Free Trade Zone	14.92 km^2	2012
4	Shenzhen North Station Business District	Municipal	High-speed Railway city	6.1 km^2	2014
5	Banxue Science and Technology City	Municipal	Smart city	21.9 km^2 + 10.88 km^2	2016
6	Shenzhen Airport City	Municipal	Airport city	95 km^2	2014
7	Guangming Phoenix City	District government	District centre	14.89 km^2	2006
8	Dayun New Town	District government	District centre	15.93 km^2	2006
9	Pingshan New Town	District government	District centre	-	2006
10	Shenzhen International Low Carbon City	Municipal	Eco-city	53 km^2/1 km^2	2012

In suburban areas, there are seven new town projects (No. 4–10). Three of them (No. 7–9) are new district centres initiated by district level governments. Normally, the main goal of these new towns is to accommodate the increasing population. Shenzhen north station business district (No. 4) is a typical high-speed rail station area development, which can also be found in Guangzhou and Foshan. The smart city of Shenzhen called Banxue science and technology city (No. 5) is also located near north station to fully make use of transport benefits, thus forming a transport and technology urban cluster. Lastly, the Shenzhen airport city (No. 6) and the international low carbon city (No. 10) are the two remaining special ones. They are located on the border area of Shenzhen where there is plenty of land to use and they are the biggest new town projects. Although sustainability is not the core value in Shenzhen's urban development, it remains as an important experimental base for ecological technologies.

Several characteristics of new town development in Shenzhen are apparent. First, most of the new town projects in Shenzhen are located in the suburban area (in Guangzhou most of them are located in central urban areas). However, it cannot be simply interpreted as Shenzhen having more suburbanisation and urban growth than Guangzhou. These two megacities have very different urban structures, as Guangzhou should be regarded as basically a network of cities. The central area of Shenzhen is denser and more urbanised than of Guangzhou. The land price of central areas in these megacities is much higher than it in suburban areas. Cities like Guangzhou and Shenzhen no longer need to develop large amounts of suburban land to generate capital, as small-scale projects in central areas may be more profitable. Thus, Shenzhen has small-scale new town projects in its central area for urban renovation and upgrading, mid-scale new towns in district cores as new development poles, and large-scale new towns in suburbs. Second, although similar urban concepts and functions are used in new town development in Shenzhen and Guangzhou, a much more organized and planned pattern of new towns can be observed in Shenzhen. The local government tends to have a more cautious attitude towards property development. Rather than boosting land-leasing revenue, the main goal of new town projects in Shenzhen is to build a leading model of innovative and sustainable urban development in the PRD region. The increase of competitiveness and attractiveness by prestigious projects outweighs the sheer economic benefits from land development.

5.3. Foshan

Foshan is the closest neighbour of Guangzhou. The two cities not only share common borders, but their urban areas are closely linked. Foshan's urban development history is similar to that of Guangzhou as Foshan also incorporated four independent cities (Shunde, Nanhai, Sanshui, and Gaoming) in 2003 (Figure 7). Like Guangzhou, its urban structure can be also regarded as a network of cities. However, the local government of Foshan is much weaker than Guangzhou. Foshan municipal government has less control over the former-city districts compared to Guangzhou, resulting in the fact that Foshan's district governments have higher levels of autonomy when it comes to the deployment of land and capital. As a result, Foshan district governments have more decision-making powers regarding urban development, including new towns. This is apparent in the case of Shunde, which became a special district level government directly under the administration of Guangdong provincial government. Thus, Foshan and Shunde, even after their integration, are still independent from each other and have separate fiscal and budgetary systems. As such, Foshan's municipal government has little administrative power over Shunde.

The local government of Foshan has two main goals in the area of urban development: To strengthen the urban development of central area; and to build closer relationship with Guangzhou for more cross-border business and communications with the concept of 'Guangzhou and Foshan as one city' (*Guang Fo tongcheng*). New towns play a key role in Foshan's urban development and are the main mechanism to achieve the two goals outlined above. By mapping new town development in Foshan, one of the noticeable differences between Foshan, Guangzhou, and Shenzhen is that Foshan's new towns are generally much bigger in size. Many of them are as big as the largest ones in Shenzhen and Guangzhou, and the average size is several times bigger. This suggests that Foshan is more ambitious in land development and urban expansion.

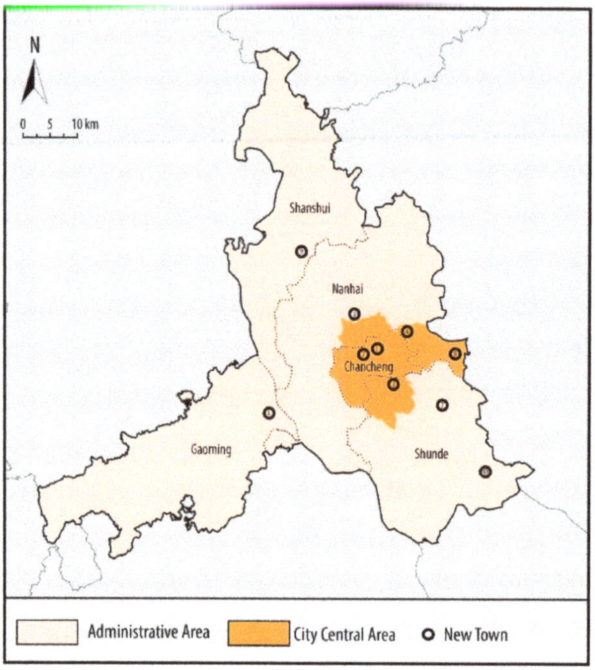

Figure 7. New town projects in Foshan.

There are six new towns (No. 1–5 and 8 in Table 4) located in and near the urban central area, and other four new towns (No. 6, 7, 9, 10) located in suburban district centres (Figure 7). Qiandenghu (No. 4) was the pilot project of Guangzhou and Foshan integration strategy as the two cities signed an official integration cooperating agreement in 2009. Qiandenghu is regarded as a successful bridge linking two cities and it also became Foshan's 'urban living room'. This has become a prestigious brand of new town development in Foshan. As a result, some new town projects have branded themselves as 'the second Qiandenghu'. Sanshan low-carbon city (No. 8) is one of these projects. It is considered as an integration project due to its proximity to Guangzhou's high-speed rail station area, which provides fast connections to the high-speed rail network. Although it was branded as the first low-carbon city in Guangdong province, it is not much different to other residential new towns close to railway stations. Foshan new town (No. 3) was one of the first new towns planned by Foshan local government but is located inside Shunde administrative area, resulting in conflicts about property rights and land-leasing revenues between the governments of Foshan and Shunde. In 2013, Foshan new town was transferred to the government of Shunde, together with land and fiscal rights.

Table 4. Information of new towns in Foshan.

	Name	Initiative	Branding	Overall Area/Booting Area	First Planned
1	Chanxi New Town	Municipal	CBD	36.8 km^2	2013
2	Zhangcha/Foshan Smart City	Municipal	Smart city	26.5 km^2	
3	Foshan/Dongping New Town	Municipal	CBD	88.6 km^2	2003
4	Qiandenghu/Guangdong Financial and High-tech Zone	Municipal	CBD axis	6.5 km^2/18 km^2	1999/2007
5	Foshan West Station New Town	Municipal	High-speed Railway new town	92 km^2/8.58 km^2	2015
6	Gaoming Xijiang New Town	District gov.	District centre	20 km^2	2009
7	Beijiao New Town	District gov.	-	-	2008
8	Sanshan low-carbon city	Municipal	Eco-city	23.8 km^2	2010
9	Sanshui New Town	District gov.	RBD	13.95 km^2/56.9 km^2/ 128.22 km^2	
10	Desheng/Shunde New Town	District gov.	District centre	70 km^2/6.5 km^2	2001

In general, new town development in Foshan is different from that in Guangzhou and Shenzhen in several respects. First, Foshan is still in a period of rapid urban expansion, in which new towns are the major mechanism for implementing urban growth. Most new towns in Foshan are large-scale projects. Unlike in Guangzhou and Shenzhen, new towns in Foshan are outlined in a district plan (*fenqu guihua*), a detailed version of an urban master plan for urban districts [61]. Second, Foshan is highly dependent on land-leasing revenue. Its pro-growth model of urban development resembles the model found in typical medium sized cities in China, which carry the risks of overdevelopment and local debt.

5.4. *Zhuhai*

Zhuhai is the smallest city among the four selected case studies and has only 1/10 of the population of Guangzhou. Within the PRD, however, Zhuhai has a unique and important position. Like Shenzhen, Zhuhai is also a special economic zone (SEZ) because of its proximity to Macau. However, just as the city of Macau can be hardly compared to world-financial centre of Hong Kong, Zhuhai has never attempted to become a second Shenzhen. Instead, it has developed its own reputation of liveability and lifestyle. In terms of its urban structure, Xiangzhou forms its central area and Hengqin joined as a new district in 2009. Doumen county has weaker connection to Zhuhai central area, although it has been a part of Zhuhai since the 1980 s. Doumen county split into Doumen district and Jinwan district in 2001, and Jinwan became a national industrial park with a harbour. However, the key area of urban development in Zhuhai remained in Xiangzhou and Hengqin.

The number of new town projects in Zhuhai is lower than in other cases, and all of them are contained in the master plan. However, the situation is more complicated as will be outlined below. In 2013, the Singaporean urban planner Liu Thai Ker was commissioned to develop a new spatial plan for Zhuhai up to 2060, setting Zhuhai's new CBD in Hezhou island, an uninhabited wetland located between the Jinwan and Hengqin districts. The idea was to provide a connection between the central area with the western Jinwan and Doumen districts. A year later, however, the idea of a new CBD in Hezhou was dropped from the revised Zhuhai master plan in 2014. Instead, the master plan identified six new town projects as shown in Figure 8. Some of the new town projects were poorly justified. For example, Fushan city (No. 6 in Table 5) has very little information about its development goals and roles in Zhuhai urban structure. Airport city (No. 3) and Binjiang city (No. 4) were together branded as 'western ecological new town', without explanation of how two new towns could be connected and related to ecology. Many of them were named with ambiguous concepts. Additionally, these new town projects are quite large compared with the small central area of Zhuhai. Strangely, the new CBD in Hezhou and Liu Thai Ker's plan are still referenced by Zhuhai official media, despite the fact that almost none of it was ever implemented and they are still excluded from the most recent urban master plan. This appears to be a branding technique designed to sell land.

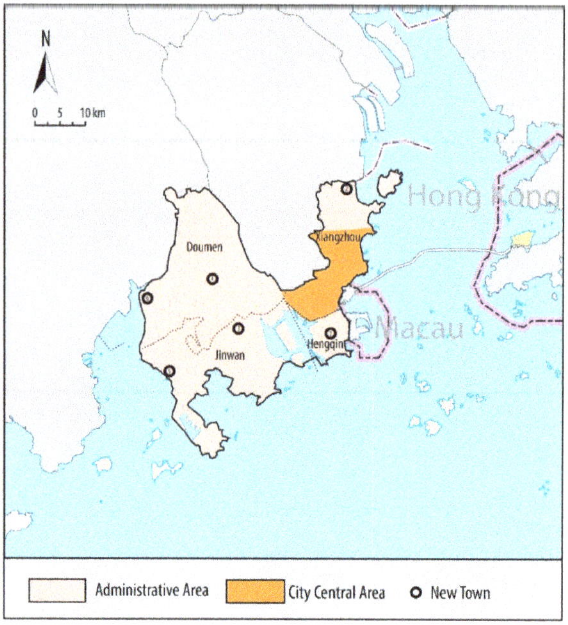

Figure 8. New town projects of Zhuhai.

Table 5. Information of new towns in Zhuhai.

	Name	Initiative	Branding	Overall Area/Booting Area	First Planned
1	Hengqin New District	Central gov.	Free Trade Zone	28 km^2	2009
2	Science and Education City/Tangjiawan New Town	Municipal	Technology city	17.07 km^2	2008
3	Airport City	Municipal	Airport city/eco-city	193 km^2	2004
4	Binjiang City	Municipal	Eco-city		2004
5	Harbour City/Pingsha New Town	Municipal	Harbour city	25 km^2	
6	Fushan Industrial City	Municipal	Industrial city	47.94 km^2	2016

In general, as the smallest city in our selection, Zhuhai is quite ambitious in terms of new town development and is similar to Foshan in several ways, such as its large-scale district-level planning for new towns and its high dependency on land-leasing revenues. Analysis indicates that urban development in Zhuhai aims to achieve spatial growth and generate land revenue. New town projects are used as the main vehicle to implement such growth. Although its new town projects are well connected with its urban master plan, it is likely that these new town plans serve entrepreneurial purposes than urban transition despite heavy use of sustainability-related concepts such as transit-oriented development and eco-city.

6. Discussion

The four case studies illustrate that cities adopt an 'entrepreneurial' stance to promote urban growth and compete with each other to attract capital. In the highly competitive PRD region, a city's competitiveness is not only defined by its infrastructure, institutional set-up, and physical attractiveness, but also about the 'differences in image' [62]. This is the reason that these cities are keen to embrace innovative and sustainable urban concepts, since being more innovative and sustainable can reconstruct city's image to enhance its importance of rankings in regional development. Since the 1990s, land and financial reforms have given local governments more scope for deploying two critical resources:

land and capital. Local governments have adopted pro-growth approaches in urban development, focusing on large projects like new towns to boost economic development, to project new city images and to accumulate capital. This has also resulted in a general dependency on urban growth mechanisms and consequently the risk of urban overdevelopment and local debt. A two-step examination has been used to illustrate the dependency on land-leasing revenue of four selected cities and their respective strategies of new town development.

Guangzhou and Shenzhen are in competition for the leading role in the PRD region. Although they share many similarities in terms of social and economic performance and sustainable urban development goals, Guangzhou has a much higher dependency on land-leasing revenue than Shenzhen. Their reliance on the urban pro-growth mechanism has impacts on their new town development strategies. Although Guangzhou does not need large-scale land development to generate capital, it has more small-scale new town projects than other cities and some of them are similarly branded. The hasty planning of new towns to boost land-lease revenue and cover the debts generated by hosting the Asian Games reveals a hidden agenda behind the sustainability-related concepts and labels. On the other hand, Shenzhen has tried to align its new town projects, partly due to its compact urban development tradition for land scarcity and high population density, and partly due to its low dependency on land-leasing revenue. New town projects in Shenzhen indicate more attention to fulfilling its innovation-driven ambitions and for generating more quality-based prestige in the region other than boosting short-term land revenue. Foshan and Zhuhai both have higher dependency on land-leasing revenue than Guangzhou. Their new town projects are much larger than those in Guangzhou and Shenzhen since they need more land development to achieve their land-leasing revenue targets. Prestige projects still matter in Foshan and Zhuhai, but since they are not competing for the best of the best in PRD, land revenue can easily eclipse other goals. Even in Zhuhai, with more attention to liveability, the city has adopted very ambitious pro-growth approaches in urban expansion. Its urban development strategies and plans are the results of a gap between its urban development objectives on paper and the hidden agenda of promoting urban growth. The innovative and sustainable brands it proposed are therefore unlikely to be realised. The study illustrates that local governments' land-leasing revenue can be used as an indicator for how a city depends on urban pro-growth mechanisms (Table 6). This has policy implications to decision makers, urban planners, and researchers that the reliance on urban pro-growth strategy can bring potential risk to any sustainable urban transition.

Table 6. Comparison of land-leasing revenue dependency and new town development strategies of four selected cities.

City	Dependency on Land-Leasing Revenue & Growth Pattern	Typical Location of New Town Projects	Average Size of New Town Projects	New Town Development Strategy
Guangzhou	Medium (Mixed)	Central & fringe	25 km^2	Compactness & revenue growth
Shenzhen	Low (Quality)	Suburban	25 km^2	Compactness
Foshan	High (Quantity)	Central & fringe	46 km^2	Spatial growth
Zhuhai	High (Quantity)	Suburban	51 km^2	Spatial growth

7. Conclusions

The results of the case study support the assumption that there is a linkage between urban growth mechanisms and urban outcomes in Chinese cities, as proposed in the analytical framework (Figure 1). The higher the dependency on land-leasing revenue in a city, the higher possibility it adopts a growth-at-all-costs approach, which increases the likelihood of overdevelopment. On the contrary, the lower the dependency on land-leasing revenue, the higher the possibility that the city pays more attention to promoting urban quality, where innovative and sustainable urban transition is more likely to be realised. Because the formation of urban growth approaches in China is deeply embedded in its institutional systems (e.g., tax-sharing system, land and housing reform, and cadre appointment system), institutional reforms are needed to achieve more sustainable forms of urban development.

Currently, central government tends to use macroeconomic controls and administrative orders to stop the overgrowth of land-driven economy and local debts, but these tools cannot solve the fundamental problem of fiscal deficit of local governments under the current tax-sharing system. The central government can easily fall in the dilemma of "control then it dies, leave it then it becomes chaotic" (*yiguanjiusi, yifangjiuluan*) [63]. Furthermore, the strategies and behaviours of local governments in urban development are only supervised by the upper-tier governments. The lack of local supervision contributes to the formation of systematic risks in urban development of Chinese cities.

This study also calls urban entrepreneurialism into question in China. Chinese local governments act as both market regulators and players, holding two critical resources of capital and land, using market instruments for the recreation and reconfiguration of space. China is not a rule-of-law based society, and local leaders are not elected but officials appointed by high level governments. Local governments can go beyond budgetary constraints and apply a market logic to recklessly mobilise resources for urban development and political objectives. This often creates a trend of overspending and excessive investment in infrastructure and urban development. In this sense, their 'entrepreneurial nature' has uncertain and debateable long-tern consequences. Even though local leaders are often proud of the 'Chinese speed' in urban development and use it for political performance, the risks of overdevelopment, misallocation of social resources, and potential debt crisis may eventually do more harm than good from the perspective of long-term sustainable urban development.

Author Contributions: Y.S. is first author. He developed the concepts and methods, collected the data, and wrote the first draft of the paper. D.S. and M.d.J. advised on the structure of the article, the analytical framework, and the overall content of the manuscript. D.S. edited the final version of the article and M.d.J. was responsible for project management. All authors have read and agreed to the published version of the manuscript.

Funding: The first author is funded by the China Scholarship Council (CSC).

Conflicts of Interest: The authors declare no conflict of interest.

References

1. Xinhua News China's Urbanization Rate Reached 58.52%. Available online: http://www.xinhuanet.com/2018-02/04/c_1122366246.htm (accessed on 4 February 2020).
2. Shepard, W. *Ghost Cities of China: The Story of Cities without People in the World's Most Populated Country*; Zed Books: London, UK, 2015.
3. Fang, C.; Yu, D. *China's New Urbanization*; Springer: Berlin, Germany, 2016.
4. Hsing, Y. *The Great Urban Transformation*; Oxford University Press: Oxford, UK, 2010; ISBN 9780199568048.
5. De Jong, M.; Joss, S.; Schraven, D.; Zhan, C.; Weijnen, M. Sustainable-smart-resilient-low carbon-eco-knowledge cities; Making sense of a multitude of concepts promoting sustainable urbanization. *J. Clean. Prod.* **2015**, *109*, 25–38. [CrossRef]
6. Han, M.; de Jong, M.; Cui, Z.; Xu, L.; Lu, H.; Sun, B. City Branding in China's Northeastern Region: How Do Cities Reposition Themselves When Facing Industrial Decline and Ecological Modernization? *Sustainability* **2018**, *10*, 102. [CrossRef]
7. Harvey, D. From Managerialism to Entrepreneurialism: The Transformation in Urban Governance in Late Capitalism. *Geogr. Ann. Ser. B* **1989**, *71*, 3. [CrossRef]
8. Jessop, B. The narrative of enterprise and the enterprise of narrative: Place marketing and the entrepreneurial city. In *The Entrepreneurial City: Geographies of Politics, Regime and Representation*; John Wiley & Sons Ltd.: Chichester, UK, 1998.
9. Kotler, P.; Haider, D.H.; Rein, I.J. *Marketing Places: Attracting Investment, Industry, and Tourism to Cities, States, and Nations*; Free Press: New York, NY, USA, 1990.
10. Ashworth, G.J.; Voogd, H. *Selling the City: Marketing Approaches in Public Sector Urban Planning*; Belhaven: London, UK, 1990.
11. Loftman, P.; Nevin, B. Going for Growth: Prestige Projects in Three British Cities. *Urban Stud.* **1996**, *33*, 991–1019. [CrossRef]
12. Wu, F. Place promotion in Shanghai, PRC. *Cities* **2000**, *17*, 349–361. [CrossRef]

13. Li, Y.; Wu, F. The transformation of regional governance in China: The rescaling of statehood. *Prog. Plann.* **2012**, *78*, 55–99. [CrossRef]
14. Young, C.; Kaczmarek, S. Changing the Perception of the Post-Socialist City: Place Promotion and Imagery in Łódź, Poland. *Geogr. J.* **1999**, *165*, 183. [CrossRef]
15. Molotch, H.L.; Logan, J.R. Urban Fortunes: The Political Economy of Place. In *Urban FortunesThe Political Economy of Place*; University of California Press: Berkeley, CA, USA, 1987; pp. 51–94. ISBN 0520254287.
16. Stoker, G. Regime theory and urban politics. In *Theories of Urban Politics*; Judge, D., Stoker, G., Wolman, H., Eds.; Sage: London, UK, 1995.
17. Wu, F. Planning Under Urban Entrepreneurialism. In *Planning for Growth: Urban and Regional Planning in China*; Taylor & Francis: Didcot, UK, 2015; pp. 79–117.
18. Zhang, T. Urban Development and a Socialist Pro-Growth Coalition in Shanghai. *Urban Aff. Rev.* **2002**, *37*, 475–499. [CrossRef]
19. Peng, J. 20 years after tax-sharing reform (Chinese literature). *Res. Financ. Econ. Issues* **2014**, *366*, 71–78.
20. Li, H.; Zhou, L.A. Political turnover and economic performance: The incentive role of personnel control in China. *J. Public Econ.* **2005**, *89*, 1743–1762. [CrossRef]
21. Xu, J.; Yeh, A.G.O. City repositioning and competitiveness building in regional development: New development strategies in Guangzhou, China. *Int. J. Urban Reg. Res.* **2005**, *29*, 283–308. [CrossRef]
22. Jing, J.; Zou, H. Soft-budget constraint on local governments in China. In *Fiscal Decentralization and the Challenge of Hard Budget Constraints*; Rodden, J., Eskeland, G.S., Litvak, J., Eds.; MIT Press: Cambridge, MA, USA, 2003.
23. Long, Y.; Gao, S. Shrinking Cities in China. In *Shrinking Cities in China: The Other Facet of Urbanization*; Long, Y., Gao, S., Eds.; Springer: Singapore, 2019; pp. 3–21. ISBN 9789811326462.
24. Sorace, C.; Hurst, W. China's Phantom Urbanisation and the Pathology of Ghost Cities. *J. Contemp. Asia* **2016**, *46*, 304–322. [CrossRef]
25. Governa, F.; Sampieri, A. Urbanisation processes and new towns in contemporary China: A critical understanding from a decentred view. *Urban Stud.* **2020**, *57*, 366–382. [CrossRef]
26. Xu, J.; Yeh, A.G.O. Governing global city regions in China and the West. *Prog. Plann.* **2010**, *73*, 17–22.
27. Wu, F. China's Emergent City-Region Governance: A New Form of State Spatial Selectivity through State-orchestrated Rescaling. *Int. J. Urban Reg. Res.* **2016**, *40*, 1134–1151. [CrossRef]
28. Lu, H.; de Jong, M.; Song, Y.; Zhao, M. The multi-level governance of formulating regional brand identities: Evidence from three Mega City Regions in China. *Cities* **2020**, *100*, 102668. [CrossRef]
29. De Jong, M. From Eco-Civilization to City Branding: A Neo-Marxist Perspective of Sustainable Urbanization in China. *Sustainability* **2019**, *11*, 5608. [CrossRef]
30. Shen, J.; Wu, F. The Suburb as a Space of Capital Accumulation: The Development of New Towns in Shanghai, China. *Antipode* **2017**, *49*, 761–780. [CrossRef]
31. Yeh, A.G.O.; Yang, F.F.; Wang, J. Economic transition and urban transformation of China: The interplay of the state and the market. *Urban Stud.* **2015**, *52*, 2822–2848. [CrossRef]
32. He, S.; Wu, F. Property-led redevelopment in post-reform China: A case study of Xintiandi redevelopment project in Shanghai. *J. Urban Aff.* **2005**, *27*, 1–23. [CrossRef]
33. Chien, S.-S. New local state power through administrative restructuring—A case study of post-Mao China county-level urban entrepreneurialism in Kunshan. *Geoforum* **2013**, *46*, 103–112. [CrossRef]
34. He, S.; Wu, F. China's emerging neoliberal urbanism: Perspectives from urban redevelopment. *Antipode* **2009**, *41*, 282–304. [CrossRef]
35. Walder, A.G. Local governments as industrial firms: An organizational analysis of China's transitional economy. *Am. J. Sociol.* **1995**, *101*, 263–301. [CrossRef]
36. Zhu, J. Local developmental state and order in China's urban development during transition. *Int. J. Urban Reg. Res.* **2004**, *28*, 424–447. [CrossRef]
37. Cartier, C. Transnational urbanism in the reform-era Chinese City: Landscapes from Shenzhen. *Urban Stud.* **2002**, *39*, 1513–1532. [CrossRef]
38. Yeh, A.G.O.; Wu, F. The new land development process and urban development in Chinese cities. *Int. J. Urban Reg. Res.* **1996**, *20*, 330–353. [CrossRef]
39. Oi, J.C. The Role of the Local State in China's Transitional Economy. *China Q.* **1995**, 1132–1149. [CrossRef]
40. Duckett, J. *The Entrepreneurial State in China*; Routledge: London, UK, 1998.

41. Zhao, Y. 'China's leading historical and cultural city'. Branding Dali City through public private partnerships in Bai architecture revitalization. *Cities* **2015**, *49*, 106–112. [CrossRef]
42. Wu, F. The "game" of landed-property production and capital circulation in China's transitional economy, with reference to Shanghai. *Environ. Plan. A* **1999**, *31*, 1757–1771. [CrossRef]
43. Wu, F. China's Changing Urban Governance. *Urban Stud.* **2002**, *39*, 1071–1093. [CrossRef]
44. Wilson, P.A. Embracing Locality in Local Economic Development. *Urban Stud.* **1995**, *32*, 645–658. [CrossRef]
45. Li, Y. Quanguo Maidi Shouru Mengzeng 43%, Sheizai Mengyan Kuangben [National revenue of leasing land increased 43%, who is runing with eyes closed]. *China Econ. Wkly.* **2018**. Available online: http://finance.people.com.cn/n1/2018/0724/c1004-30165563.html (accessed on 24 July 2018).
46. Wakeman, R. *Practicing Utopia: An Intellectual History of the New Town Movement*; The University of Chicago Press: Chicago, IL, USA, 2016.
47. Chan, K.W. Fundamentals of China's urbanization and policy. *China Rev.* **2010**, *10*, 63–94.
48. National Bureau of Statistics of China. *Statistic Year Book of China*; National Bureau of Statistics of China: Beijing, China, 2019.
49. Begg, I. Cities and Competitiveness. *Urban Stud.* **1999**, *36*, 795–809. [CrossRef]
50. Cheshire, P. Cities in Competition: Articulating the Gains from Integration. *Urban Stud.* **1999**, *36*, 843–864. [CrossRef]
51. Kresl, P.K.; Singh, B. Competitiveness and the urban economy: Twenty-four large US Metropolitan areas. *Urban Stud.* **1999**, *36*, 1017–1027. [CrossRef]
52. Porter, M. *The Competitive Advantage of Nations*; Macmillan: London, UK, 1990.
53. Guangzhou Bureau of Statistics. *Guangzhou Statistical Yearbook*; Guangzhou Bureau of Statistics: Guangzhou, China, 2019.
54. Shenzhen Bureau of Statistics. *Shenzhen Statistical Yearbook*; Shenzhen Bureau of Statistics: Shenzhen, China, 2019.
55. Foshan Bureau of Statistics. *Foshan Statistical Yearbook*; Foshan Bureau of Statistics: Foshan, China, 2019.
56. Zhuhai Bureau of Statistics. *Zhuhai Statistical Yearbook*; Zhuhai Bureau of Statistics: Zhuhai, China, 2019.
57. Xu, X. Guangzhou: China's Southern Gateway. In *Chinese Cities: The Growth of the Metropolis Since 1949*; Oxford University Press: Hong Kong, 1985.
58. Li, Z.; Li, X.; Wang, L. Speculative urbanism and the making of university towns in China: A case of Guangzhou University Town. *Habitat Int.* **2014**, *44*, 422–431. [CrossRef]
59. Wu, F. Planning centrality, market instruments: Governing Chinese urban transformation under state entrepreneurialism. *Urban Stud.* **2018**, *55*, 1383–1399. [CrossRef]
60. Liu, L. Nine new town plans released in Guangzhou to pay debt. *China Times*, 2013.
61. Chen, F. The design dimension of China's planning system: Urban design for development control. *Int. Plan. Stud.* **2016**, *21*, 81–100. [CrossRef]
62. Zhai, B.; Jia, Y.; Xu, Q. A Long Way to Go: The Coordinative Development in the Capital Region of China? In Proceedings of the 40th ISoCaRP Congress, Geneva, Switzerland, 18–22 September 2004; pp. 1–11.
63. XinhuaNet. How to Avoid "Yiguanjiusi, Yifangjiuluan". Available online: http://www.xinhuanet.com/politics/2019-12/25/c_1125385181.htm (accessed on 25 December 2019).

MDPI
St. Alban-Anlage 66
4052 Basel
Switzerland
Tel. +41 61 683 77 34
Fax +41 61 302 89 18
www.mdpi.com

Sustainability Editorial Office
E-mail: sustainability@mdpi.com
www.mdpi.com/journal/sustainability